Advances in Representation Theory of Algebras

CONTEMPORARY MATHEMATICS

761

Centre de Recherches Mathématiques Proceedings

Advances in Representation Theory of Algebras

Conference in Honor of
José Antonio de la Peña's 60th birthday
Advances in
Representation Theory of Algebras (ARTA VII)
September 24–28, 2018
Instituto de Matemáticas,
Universidad Nacional Autónoma de México,
Mexico

CRM

Ibrahim Assem
Christof Geiß
Sonia Trepode
Editors

AMS AMERICAN MATHEMATICAL SOCIETY

2020 *Mathematics Subject Classification.* Primary 16G20, 16G60, 16G70, 16G99, 18E30, 13F60.

Library of Congress Cataloging-in-Publication Data

Names: Assem, Ibrahim, editor. | Geiss, Christof, 1964– editor. | Trepode, Sonia, 1962– editor.
Title: Advances in representation theory of algebras / Ibrahim Assem, Christof Geiß, Sonia Trepode, editors.
Description: Providence, Rhode Island : American Mathematical Society, [2021] | Series: Contemporary mathematics, 0271-4132 ; volume 761 | "Conference in Honor of José Antonio de la Peña's 60th birthday on Advances in Representation Theory of Algebras (ARTA VII), September 24–28, 2018, Instituto de Matemáticas, Universidad Nacional Autónoma de México, Mexico City, Mexico." | Includes bibliographical references.
Identifiers: LCCN 2020042980 | ISBN 9781470451592 (paperback) | ISBN 9781470464172 (ebook)
Subjects: LCSH: Representations of algebras. | AMS: Associative rings and algebras – Representation theory of rings and algebras – Representations of quivers and partially ordered sets. | Associative rings and algebras – Representation theory of rings and algebras – Representation type (finite, tame, wild, etc.). | Associative rings and algebras – Representation theory of rings and algebras – Auslander-Reiten sequences (almost split sequences) and Auslander-Reiten quivers. | Category theory; homological algebra. | Commutative algebra – Arithmetic rings and other special rings – Cluster algebras.
Classification: LCC QA150 .A26 2021 | DDC 512/.46–dc23
LC record available at https://lccn.loc.gov/2020042980
DOI: https://doi.org/10.1090/conm/761

∞ The paper used in this book is acid-free and falls within the guidelines
established to ensure permanence and durability.
Visit the AMS home page at `https://www.ams.org/`

10 9 8 7 6 5 4 3 2 1   26 25 24 23 22 21

This volume consists of papers dedicated to José Antonio de la Peña
on the occasion of his 60th birthday.

Contents

Preface ix

List of Talks xi

Lectures on split-by-nilpotent extensions
I. Assem 1

A note on sequential walks
I. Assem, M. J. Redondo, and R. Schiffler 37

A characterization of representation infinite quiver settings
G. Bobiński 51

Dualities for modules of finite projective dimension
B. Huisgen-Zimmermann and M. Saorín 61

Frieze varieties are invariant under Coxeter mutation
K. Igusa and R. Schiffler 103

On tame strongly simply connected algebras
S. Kasjan and A. Skowroński 117

A remark on Hochschild cohomology and Koszul duality
B. Keller 131

Thick subcategories of the stable category of modules over the exterior algebra
O. Kerner and D. Zacharia 137

Finite direct sums of cyclic embeddings
J. Kosakowska and M. Schmidmeier 159

The algebraic theory of fuchsian singularities
H. Lenzing 171

Cycle-finite modules over artin algebras
P. Malicki and A. Skowroński 191

Piecewise hereditary incidence algebras
E. N. Marcos and M. Moreira 213

Derived categories of gentle algebras and surface dissections
P.-G. Plamondon 227

Socle deformations of selfinjective orbit algebras of tilted type
A. Skowroński and K. Yamagata 239

Preface

The Seventh ARTA ("Advances in Representation Theory of Algebras") was held at the Instituto de Matemáticas de la UNAM (Universidad Nacional Autónoma de México) in Mexico City from the 24th to the 28th of September 2018. It was dedicated to our colleague and friend José Antonio de la Peña on the occasion of his 60th birthday.

For over thirty-five years José Antonio has been one of the prominent figures in the representation theory of algebras. His work on topics as diverse as covering theory, tame algebras, the use of quadratic forms in representation theory, the spectral theory of graphs and others have inspired, and continue to inspire, several generations of mathematicians. He was also active in the chemical applications of the theory and in the dissemination of mathematics. José Antonio has won an impressive series of national and international prizes and medals. His work has had an important impact on mathematics in Mexico, where he held, and still holds, prestigious administrative or honourary positions. José Antonio has always been willing to help young researchers from all countries and, as a consequence, had a strong influence on them. As a person, he is very generous, always ready to help those of his students or colleagues who experience difficulties.

This volume contains fourteen papers dedicated to José Antonio. We would like to express our deepest gratitude to the authors who submitted their contributions and to the referees for their kind assistance. We also wish to extend our gratitude to Véronique Hussin of the CRM and to Christine Thivierge of the AMS for their help in including this volume in the AMS Contemporary Mathematics series.

While this volume was being prepared, we have learned with deep sorrow that our colleague and close friend Andrzej Skowroński has passed away on the 22nd of October, 2020 in Torun, after an infection with Covid-19. This marks the loss of an excellent friend, and a brillant mathematician in our field. He was one of the founders of the ARTA series of conferences and is coauthor of three papers in this volume, as well as a friend and close collaborator of José Antonio de la Peña, to whom this volume is dedicated .

Ibrahim Assem

Christof Geiß

Sonia Trepode

List of Talks

Advances in Representation Theory of Algebras VII

Helmut Lenzing, Universität Paderborn, Germany
Working with José Antonio

Andrea Solotar, Universidad de Buenos Aires, Argentina
The Gerstenhaber structure of Hochschild cohomology

Ralf Schiffler, University of Connecticut, USA
Frieze varieties: A characterization of the finite-tame-wild trichotomy for acyclic quivers

Gordana Todorov, Northeastern University, USA
Cyclic posets and triangulation clusters

Eduardo do Nascimento Marcos, Universidade de São Paulo, Brazil
Hochschild-Mitchell (co)homology of skew categories and of Galois coverings

Birge Huisgen-Zimmermann, University of California, Santa Barbara, USA
Truncated path algebras, a geometric and homological stepping stone. Part I

Ken Goodearl, University of California, Santa Barbara, USA
Truncated path algebras, a geometric and homological stepping stone. Part II

Bernhard Keller, Université Paris Diderot - Paris 7
Koszul duality and Hochschild cohomology

Lidia Angeleri-Hügel, Università degli Studi di Verona, Italy
Comparing localizations of rings

Thomas Brüstle, Bishop's University and Université de Sherbrooke, Québec, Canada.
Matrix reduction and exact structures

Raymundo Bautista, CCM-UNAM, Mexico
Representations of bocses and $A_\infty-$modules

Pierre-Guy Plamondon, Université de Paris-Sud XI, France
Derived categories of gentle algebras and surface dissections

Sonia Trepode, Universidad Nacional de Mar del Plata, Argentina
The representation dimension of selfinjective algebras of tilted type

Changchang Xi, Capital Normal University, China
Derived decompositions of abelian categories

Henning Krause, Universität Bielefeld, Germany
The Morita theory for derived categories revisited

Grzegorz Bobiński, Nicolaus Copernicus University, Poland
Birkhoff varieties

Kunio Yamagata, Tokyo University of Agriculture and Technology, Japan
Selfinjective orbit algebras induced from repetitive algebras

Piotr Malicki, Nicolaus Copernicus University, Poland
Cycle-finite module categories

Gustavo Jasso, Universität Bonn, Germany
Higher-dimensional Auslander algebras of type A and the higher-dimensional Waldhausen S-constructions

Stanislaw Kasjan, Nicolaus Copernicus University, Poland
Tame strongly simply connected algebras

Contemporary Mathematics
Volume **761**, 2021
https://doi.org/10.1090/conm/761/15306

Lectures on split-by-nilpotent extensions

Ibrahim Assem

Dedicated to José Antonio de la Peña for his 60th birthday

ABSTRACT. We survey known properties of split-by-nilpotent extensions of algebras, concentrating on their bound quivers, the change of rings functors and tilting theory.

Introduction

These notes are an updated version of a course given years ago at the Universidad Nacional del Sur in Bahía Blanca (Argentina). Split extensions are fascinating mathematical objects defined as follows. Let A be a finite dimensional algebra over a field $\mathbf{k}$, and E an A-A-bimodule, finite dimensional over $\mathbf{k}$, equipped with an associative product $E \otimes_A E \longrightarrow E$, then the split extension R of A by E is the $\mathbf{k}$-vector space $R = A \oplus E$ with the multiplication

$$(a\,,\,e)(a'\,,\,e') = (aa'\,,\,ea' + ae' + ee')$$

for $a, a' \in A$ and $e, e' \in E$, where ee' stands for the product in E. If E is nilpotent for its product, then R is called a split-by-nilpotent extension. Examples abound in the mathematical literature, the best known being trivial extension algebras. Thus, the study of split-by-nilpotent extensions connects with those of selfinjective algebras and, more recently, cluster tilted algebras.

The general problem of split-by-nilpotent extensions is to predict properties of R knowing properties of A and E, and conversely. In an abstract setting, this is a difficult problem and more information is needed to obtain concrete results.

The objective of these notes is to survey known results about split-by-nilpotent extensions. We tried to keep the notes as selfcontained as possible, providing proofs and examples whenever possible. The first section is devoted to the definition and basic properties of this class. The second section relates the bound quivers of R and A. In the third, we start comparing the module categories of E and A, using the classical change of rings functors of [**21**]. Finally, the fourth section is devoted to the comparison of the tilting theories of R and A, with a particular attention to the induced torsion pairs.

2010 *Mathematics Subject Classification.* Primary 16G10, 16G20, 16E10, 16S70.

Key words and phrases. Split-by-nilpotent extensions, bound quivers of algebras, change of rings functors, tilting modules, tilting torsion pairs.

The author gratefully acknowledges partial support from the NSERC of Canada.

1. Split-by-nilpotent extensions

1.1. Notation. Throughout, $\mathbf{k}$ denotes an algebraically closed field. By algebra is meant a basic finite dimensional associative $\mathbf{k}$-algebra with an identity.

A ***quiver*** Q is a quadruple $Q = (Q_0, Q_1, s, t)$, where Q_0, Q_1 are sets whose elements are respectively called ***points*** and ***arrows***, and $s, t\colon Q_1 \longrightarrow Q_0$ are maps which associate to an arrow α its ***source*** $s(\alpha)$ and its ***target*** $t(\alpha)$. Given a connected algebra A, there exists a (unique) connected quiver Q_A and (at least) a surjective algebra morphism $\eta\colon \mathbf{k}Q_A \longrightarrow A$, where $\mathbf{k}Q_A$ is the path algebra of Q_A. The ideal $I = \operatorname{Ker}\eta$ is then ***admissible***, that is, there exists $m \geq 2$ such that $\mathbf{k}Q_A^{+m} \subseteq I \subseteq \mathbf{k}Q_A^{+2}$, where $\mathbf{k}Q_A^{+i}$ is the two-sided ideal of $\mathbf{k}Q_A$ generated by the paths of length at least i. The isomorphism $A \cong \mathbf{k}Q_A/I$ (or the morphism η) is called a ***presentation*** of A, and A is said to be given by the ***bound quiver*** (Q_A, I), see [**12**]. The ideal I is generated by a finite set of elements called relations: given $x, y \in (Q_A)_0$, a ***relation*** from x to y is a linear combination $\rho = \sum_{i=1}^{m} c_i w_i$ where the c_i are nonzero scalars, and the w_i are paths of length at least two from x to y. The relation ρ is called ***monomial*** if $m = 1$, and ***minimal*** if $m \geq 2$ and, for every nonempty subset $J \subsetneq \{1, 2, \dots, m\}$, we have $\sum_{j \in J} c_j w_j \notin I$.

Following [**19**], we sometimes consider an algebra $A = \mathbf{k}Q_A/I$ as a $\mathbf{k}$-category, of which the object class A_0 is the set $(Q_A)_0$, while the set of morphisms from x to y is the quotient of the vector space $\mathbf{k}Q_A(x, y)$ of all $\mathbf{k}$-linear combinations of paths from x to y by the subspace $I(x, y) = I \cap \mathbf{k}Q_A(x, y)$. An algebra A is called ***triangular*** if Q_A is acyclic.

We denote by $\operatorname{mod} A$ the category of finitely generated right A-modules and by $\operatorname{ind} A$ a full subcategory containing exactly one representative from each isoclass (= isomorphism class) of indecomposable modules. When we speak about a module, or an indecomposable module, we mean implicitly that it belongs to $\operatorname{mod} A$, or $\operatorname{ind} A$, respectively. If $\mathscr{C}$ is a full subcategory of $\operatorname{mod} A$, we write $M \in \mathscr{C}$ to express that M is an object in $\mathscr{C}$. We denote by $\operatorname{add}\mathscr{C}$ the full subcategory of $\operatorname{mod} A$ having as objects the direct sums of direct summands of objects in $\mathscr{C}$. If there exists a module M such that $\mathscr{C} = \{M\}$, then we write $\operatorname{add} M$ instead of $\operatorname{add}\mathscr{C}$. Given a module M, we denote by $\operatorname{pd} M$ and $\operatorname{id} M$ its projective and injective dimensions, respectively. The global dimension of A is denoted by $\operatorname{gl.dim.} A$.

For $x \in (Q_A)_0$, we let e_x denote the corresponding primitive idempotent of A, and let S_x, P_x, I_x denote respectively the corresponding simple, indecomposable projective and indecomposable injective modules. The standard duality between right and left modules is denoted by $\mathrm{D} = \operatorname{Hom}_{\mathbf{k}}(-, \mathbf{k})$ and the Auslander-Reiten translations by $\tau_A = \mathrm{D}\mathrm{Tr}$ and $\tau_A^{-1} = \mathrm{Tr}\,\mathrm{D}$ (or simply τ, τ^{-1} if there is no ambiguity). For more notions and results about $\operatorname{mod} A$, we refer the reader to [**12, 15**].

1.2. Definition and examples. Let A be an algebra, and E an A-A-bimodule, which is finite dimensional as a $\mathbf{k}$-vector space. We say that E is equipped with an ***associative product*** if there exists a morphism $E \otimes_A E \longrightarrow E$ of A-A-bimodules, denoted as $e \otimes e' \longmapsto ee'$ for $e, e' \in E$ such that

$$e(e'e'') = (ee')e''$$

for all $e, e', e'' \in E$.

DEFINITION 1.2.1. Let A be an algebra and E an A-A-bimodule equipped with an associative product. The $\mathbf{k}$-vector space

$$R = A \oplus E = \{(a\,,\,e) \mid a \in A, e \in E\}$$

together with the multiplication

$$(a\,,\,e)(a'\,,\,e') = (aa'\,,\,ae' + ea' + ee')$$

for $a, a' \in A$ and $e, e' \in E$, is an algebra, called the ***split extension*** of A by E. If moreover E is nilpotent as a two-sided ideal of R, then R is called a ***split-by-nilpotent extension***.

Clearly, $\dim_{\mathbf{k}} R = \dim_{\mathbf{k}} A + \dim_{\mathbf{k}} E$ and there exists an exact sequence of vector spaces

$$0 \longrightarrow E \xrightarrow{\ \iota\ } R \xrightarrow{\ \pi\ } A \longrightarrow 0$$

where $\iota\colon e \mapsto (0\,,\,e)$ for $e \in E$, while $\pi\colon (a\,,\,e) \mapsto a$ for $(a\,,\,e) \in R$. Then π is an algebra morphism and admits as section the algebra morphism $\sigma\colon A \longrightarrow R$, $a \mapsto (a\,,\,0)$ for $a \in A$. Because ι, π, σ are also A-A-bimodule morphisms, the previous exact sequence may also be considered as a split exact sequence of A-A-bimodules and so, in particular, as a split exact sequence of right, or left, A-modules. Of course, it is also an exact sequence of R-R-bimodules, or of right, or left, R-modules. But then, it is generally not split.

Saying that E is nilpotent amounts to saying that $E \subseteq \operatorname{rad} R$. In the sequel, we always assume that E is nilpotent.

There may be several decompositions of ${}_AR_A$ as a direct sum isomorphic to $A \oplus E$. Therefore the data of an exact sequence as above does not suffice to determine a split extension: one must also fix a direct sum decomposition $R = A \oplus E$, or, equivalently, fix a section σ to π.

EXAMPLES 1.2.2.

(a) Because $\mathbf{k}$ is algebraically closed, any algebra can be written as a direct sum $R = (R/\operatorname{rad} R) \oplus \operatorname{rad} R$, so it is a split extension of the semisimple algebra $R/\operatorname{rad} R$ by the nilpotent bimodule $\operatorname{rad} R$.

(b) If $E^2 = 0$, then a split extension of A by E is called a ***trivial extension*** and denoted as $A \ltimes E$. This class plays a very important rôle in the classification results for selfinjective algebras. In this case, one takes E to be the minimal injective cogenerator bimodule $E = \mathrm{D}({}_AA_A)$ with its canonical bimodule structure, see [**31, 32**]. Another type of trivial extensions appeared in the theory of cluster algebras: it is indeed proved in [**5**] that an algebra is cluster tilted if and only if it is the trivial extension of a tilted algebra A by the so-called relation bimodule $E = \operatorname{Ext}^2_A(\mathrm{D}A\,,\,A)$ with its canonical bimodule structure.

Perhaps the smallest nontrivial example is the following: let $A = \mathbf{k}$, $E = \mathbf{k}$ with its canonical $\mathbf{k}$-$\mathbf{k}$-bimodule structure. The trivial extension $A \ltimes E$ is the vector space

$$\mathbf{k}^2 = \{(a\,,\,b) \mid a, b \in \mathbf{k}\}$$

with the multiplication

$$(a\,,\,b)(a'\,,\,b') = (aa'\,,\,ab' + ba')$$

for $a, a', b, b' \in \mathbf{k}$. Clearly, we have an algebra isomorphism $A \ltimes E \cong \mathbf{k}[t]/(t^2)$.

(c) We now give an example of a split extension which is not a trivial extension. Let $A = \mathbf{k}$, $E = \mathbf{k}^2$ with its canonical bimodule structure and equipped with the (obviously associative) product

$$(b\,,\,c)(b'\,,\,c') = (0\,,\,bb')$$

for $b, b', c, c' \in \mathbf{k}$. The split extension is the three-dimensional vector space

$$R = A \oplus E = \{(a\,,\,(b\,,\,c)) \mid a, b, c \in \mathbf{k}\}$$

with multiplication

$$(a\,,\,(b\,,\,c))(a'\,,\,(b'\,,\,c')) = (aa'\,,\,(ab' + ba'\,,\,ac' + ca' + bb'))$$

for $a, b, c, a', b', c' \in \mathbf{k}$. It is easy to see that actually $R \cong \mathbf{k}[t]/(t^3)$. One can realise in this way any truncated polynomial algebra $\mathbf{k}[t]/(t^n)$ as split extension of $A = \mathbf{k}$ by $E = \mathbf{k}^{n-1}$.

(d) Let A be given by the quiver

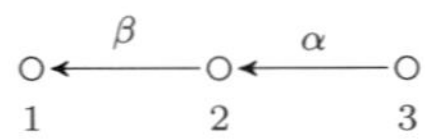

bound by $\alpha\beta = 0$, and R be given by the quiver

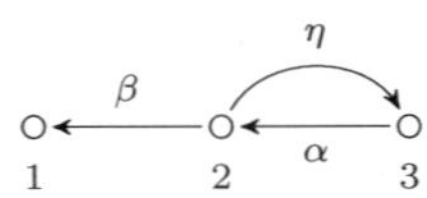

bound by $\alpha\beta = 0$, $\eta\alpha\eta\alpha\eta = 0$. Then R is the split extension of A by the bimodule E generated by the arrow η. To find a $\mathbf{k}$-basis of E, we construct those paths (more precisely, classes of paths modulo the binding ideal, but we identify the two) which contain η. This gives the following basis

$$\{\eta\,,\,\eta\alpha\,,\,\alpha\eta\,,\,\eta\alpha\eta\,,\,\alpha\eta\alpha\,,\,\eta\alpha\eta\alpha\,,\,\alpha\eta\alpha\eta\,,\,\alpha\eta\alpha\eta\alpha\}.$$

The right and left A-module structures of E are computed as follows. We have $A_A = 1 \oplus {\begin{smallmatrix}2\\1\end{smallmatrix}} \oplus {\begin{smallmatrix}3\\2\end{smallmatrix}}$ where indecomposable modules are represented by their Loewy series. Similarly, $R_R = 1 \oplus {\begin{smallmatrix} & 2 & \\ 1 & & 3\\ & & 2\\ & & 3\\ & & 2\end{smallmatrix}} \oplus {\begin{smallmatrix}3\\2\\3\\3\\2\\3\\2\end{smallmatrix}}$. We next compute R_A: deleting η from the indecomposable R-modules gives their A-module structure. We get $R_A = 1 \oplus {\begin{smallmatrix}2\\1\end{smallmatrix}} \oplus \left({\begin{smallmatrix}3\\2\end{smallmatrix}}\right)^2 \oplus {\begin{smallmatrix}3\\2\end{smallmatrix}} \oplus \left({\begin{smallmatrix}3\\2\end{smallmatrix}}\right)^2$ from where we deduce $E_A = \left({\begin{smallmatrix}3\\2\end{smallmatrix}}\right)^4$. Similarly $(\mathrm{D}A)_A = {\begin{smallmatrix}2\\1\end{smallmatrix}} \oplus {\begin{smallmatrix}3\\2\end{smallmatrix}} \oplus 3$ and $(\mathrm{D}R)_R = {\begin{smallmatrix}2\\1\end{smallmatrix}} \oplus {\begin{smallmatrix}3\\2\\3\\2\\3\\2\end{smallmatrix}} \oplus {\begin{smallmatrix}3\\2\\3\\2\\3\end{smallmatrix}}$ yields $(\mathrm{D}E)_A = \left({\begin{smallmatrix}3\\2\end{smallmatrix}}\right)^4$.

1.3. Properties. Our next objective is to describe the quiver Q_R of a split extension R of an algebra A by a nilpotent bimodule E, in terms of the quiver Q_A of A.

LEMMA 1.3.1. *Let R be a split extension of A by a nilpotent bimodule E, then* $\operatorname{rad} A = (\operatorname{rad} R)/E$.

PROOF. We have $E \subseteq \operatorname{rad} R$ and $(\operatorname{rad} R)/E$ nilpotent as an ideal in $R/E \cong A$. Moreover, $\frac{R/E}{(\operatorname{rad} R)/E} \cong \frac{R}{\operatorname{rad} R}$ is semisimple. Therefore $(\operatorname{rad} R)/E \cong \operatorname{rad}(R/E) \cong \operatorname{rad} A$. $\square$

THEOREM 1.3.2 ([**10**]1.2). *Let R be a split extension of A by a nilpotent bimodule E. The quiver Q_R of R is constructed as follows:*

(a) $(Q_R)_0 = (Q_A)_0$;

(b) *for $x, y \in (Q_R)_0$, the set of arrows in Q_R from x to y equals the set of arrows in Q_A from x to y plus*

$$\dim_{\mathbf{k}} e_x \left(\frac{E}{E \cdot \operatorname{rad} A + \operatorname{rad} A \cdot E + E^2} \right) e_y$$

additional arrows.

PROOF. Because of Lemma 1.3.1, Q_R and Q_A have the same points and moreover, $\operatorname{rad} R = \operatorname{rad} A \oplus E$ as a vector space. Hence

$$\operatorname{rad}^2 R = \operatorname{rad}^2 A \oplus \big(E \cdot \operatorname{rad} A + \operatorname{rad} A \cdot E + E^2 \big).$$

The arrows in Q_R from x to y are in bijection with vectors in a basis of the vector space $e_x \big(\frac{\operatorname{rad} R}{\operatorname{rad}^2 R} \big) e_y$. Because $\operatorname{rad}^2 A \subseteq \operatorname{rad} A$ and $E \cdot \operatorname{rad} A + \operatorname{rad} A \cdot E + E^2 \subseteq E$, the statement follows. □

Thus, if A is a connected algebra, then so is R. We now see how split extensions behave upon taking full subcategories.

LEMMA 1.3.3 ([**14**]1.4). *Let R be a split extension of A by a nilpotent bimodule E and e an idempotent in A. Then eRe is a split extension of eAe by eEe.*

PROOF. Clearly, eEe is a two-sided ideal of eRe. Its nilpotency follows from the fact that $eEe \subseteq E$. The map $\pi_e \colon eRe \longrightarrow eAe$, $e(a\,,\,x)e \longmapsto eae$ for $(a\,,\,x) \in R$ is a surjective algebra morphism having as section $\sigma_e \colon eAe \longrightarrow eRe$, $eae \longmapsto e(a\,,\,0)e$ for $a \in A$. Moreover, σ_e is an algebra morphism and $eEe \subseteq \operatorname{Ker} \pi_e$. Because $eRe = eAe \oplus eEe$ as vector spaces, we get the statement by comparing dimensions. □

As we now see, taking split extensions is a transitive procedure.

LEMMA 1.3.4 ([**10**]1.7). *Let R be a split extension of A by a nilpotent bimodule E and S a split extension of R by a nilpotent bimodule F. Then S is a split extension of A.*

PROOF. We have exact sequences of vector spaces

$$0 \longrightarrow E \xrightarrow{\ \iota\ } R \underset{\sigma}{\overset{\pi}{\rightleftarrows}} A \longrightarrow 0 \ , \ 0 \longrightarrow F \xrightarrow{\ \iota'\ } S \underset{\sigma'}{\overset{\pi'}{\rightleftarrows}} R \longrightarrow 0$$

where $\pi, \sigma, \pi', \sigma'$ are algebra morphisms and $\pi\sigma = \mathrm{id}_A$, $\pi'\sigma' = \mathrm{id}_R$. Also, there exist $m, n > 0$ such that $E^m = 0$, $F^n = 0$. We get an exact sequence

$$0 \longrightarrow {\pi'}^{-1}(E) \longrightarrow S \xrightarrow{\ \pi\pi'\ } A \longrightarrow 0.$$

Both $\pi\pi', \sigma'\sigma$ are algebra morphisms and $\pi\pi'\sigma'\sigma = \mathrm{id}_A$, so it suffices to prove that ${\pi'}^{-1}(E)$ is nilpotent. We claim that ${\pi'}^{-1}(E)^{mn} = 0$. Let $x_{ij} \in {\pi'}^{-1}(E)$ with $1 \leq i \leq n$, $1 \leq j \leq m$. Then $\pi'(x_{ij}) \in E$ for all i, j. Therefore, for each i, we have $\pi'\Big(\prod_{j=1}^m x_{ij} \Big) = \prod_{j=1}^m \pi'(x_{ij}) \in E^m = 0$. Thus, for each i, the product $\prod_{j=1}^m x_{ij}$ lies in $\operatorname{Ker} \pi' = F$. But then $\prod_{i=1}^n \prod_{j=1}^m x_{ij} \in F^n = 0$. □

2. The bound quiver of a split extension

2.1. Presentations. Let R be a split extension of A by a nilpotent bimodule E. Because of 1.3.2, the quiver Q_R of R is obtained from the quiver Q_A of A by adding arrows. It is therefore reasonable to think that E, as an ideal, is generated precisely by the added arrows. Let $\eta_R\colon \mathbf{k}Q_R \twoheadrightarrow R \cong \mathbf{k}Q_R/I_R$ and $\eta_A\colon \mathbf{k}Q_A \twoheadrightarrow A \cong \mathbf{k}Q_A/I_A$ be respectively bound quiver presentations of R and A. For $x, y \in (Q_A)_0$, it follows from the proof of 1.3.2 that there is an inclusion of vector spaces

$$e_x\left(\frac{\operatorname{rad} A}{\operatorname{rad}^2 A}\right)e_y \longrightarrow e_x\left(\frac{\operatorname{rad} R}{\operatorname{rad}^2 R}\right)e_y.$$

Therefore, there exists a basis of $e_x\left(\frac{\operatorname{rad} R}{\operatorname{rad}^2 R}\right)e_y$ which contains as a subset a basis of $e_x\left(\frac{\operatorname{rad} A}{\operatorname{rad}^2 A}\right)e_y$. When the arrows in Q_R are taken in bijection with vectors in such a basis, for any x, y, then we say that the presentation η_R (or (Q_R, I_R)) of R ***respects*** η_A (or (Q_A, I_A), or simply A). The previous comments show that there always exists a presentation of R respecting A.

LEMMA 2.1.1 ([**10**]1.5). *Let R be a split extension of A by a nilpotent bimodule E. Then there exists a presentation of R respecting A such that E is generated by the classes of arrows in Q_R which are not in Q_A.*

PROOF. Let (Q_R, I_R) be a bound quiver presentation of R which respects A and $\{\rho_1, \ldots, \rho_s\}$ be the preimage modulo I_R of any linearly independent set of generators for E. We may assume that each ρ_i is a linear combination of paths having the same source and the same target: for, if this is not the case, then we multiply each ρ_i on the left and on the right by stationary paths and we obtain such a set. Because Q_A, Q_R have the same points, all paths involved in the ρ_i have length at least one. Moreover, as seen in 1.3.2, the top of E is contained in $\operatorname{rad} R/\operatorname{rad}^2 R$, that is $\rho_i + \operatorname{rad}^2 R \in \operatorname{rad} R/\operatorname{rad}^2 R$ for all i with $1 \le i \le s$. So we have

$$\rho_i = \alpha_i + \sum_j \lambda_j w_j$$

where α_i is an arrow in Q_R and $\sum_j \lambda_j w_j$ a linear combination of paths of length at least one. Because the ρ_i are linearly independent modulo I_R, we define a new presentation by replacing α_i by

$$\alpha_i' = \alpha_i + \sum_j \lambda_j w_j.$$

In this presentation, E is indeed generated by $\alpha_1', \ldots, \alpha_s'$. □

COROLLARY 2.1.2 ([**10**]2.1). *Let R be a split extension of A by a nilpotent bimodule E. Given a presentation $\eta_A\colon \mathbf{k}Q_A \twoheadrightarrow A \cong \mathbf{k}Q_A/I_A$, there exists a presentation $\eta_R\colon \mathbf{k}Q_R \twoheadrightarrow R \cong \mathbf{k}Q_R/I_R$ respecting A such that:*

(a) *E is an ideal of R generated by classes of arrows,*
(b) *there exist algebra morphisms $\tilde{\pi}\colon \mathbf{k}Q_R \twoheadrightarrow \mathbf{k}Q_A$, $\tilde{\sigma}\colon \mathbf{k}Q_A \twoheadrightarrow \mathbf{k}Q_R$ such that $\tilde{\pi}\tilde{\sigma} = \mathrm{id}_{\mathbf{k}Q_A}$, $\eta_R\tilde{\sigma} = \sigma\eta_A$ and $\tilde{\sigma}(I_A) \subseteq I_R$,*

(c) *there exists a commutative diagram with exact rows and columns*

$$
\begin{array}{ccccccccc}
 & & 0 & & 0 & & 0 & & \\
 & & \downarrow & & \downarrow & & \downarrow & & \\
0 & \longrightarrow & \tilde{E}\cap I_R & \longrightarrow & I_R & \longrightarrow & I_A & \longrightarrow & 0 \\
 & & \downarrow & & \downarrow & & \downarrow & & \\
0 & \longrightarrow & \tilde{E} & \longrightarrow & \mathbf{k}Q_R & \underset{\tilde{\sigma}}{\overset{\tilde{\pi}}{\rightleftarrows}} & \mathbf{k}Q_A & \longrightarrow & 0 \\
 & & \downarrow & & \downarrow{\scriptstyle \eta_R} & & \downarrow{\scriptstyle \eta_A} & & \\
0 & \longrightarrow & E & \longrightarrow & R & \underset{\sigma}{\overset{\pi}{\rightleftarrows}} & A & \longrightarrow & 0. \\
 & & \downarrow & & \downarrow & & \downarrow & & \\
 & & 0 & & 0 & & 0 & &
\end{array}
$$

PROOF. Because of 1.3.2, we may (and shall) identify Q_A to a nonfull subquiver of Q_R. Applying [**12**](II.1.8), the inclusion $Q_A \hookrightarrow Q_R$ extends to an algebra morphism $\tilde{\sigma}\colon \mathbf{k}Q_A \longrightarrow \mathbf{k}Q_R$ preserving stationary paths and arrows. Letting $\eta_R\colon \mathbf{k}Q_R \longrightarrow \mathbf{k}Q_R/I_R \cong R$ be a presentation constructed as in 2.1.1, we then have $\eta_R\tilde{\sigma} = \sigma\eta_A$. Moreover, there exists a set S of arrows in Q_R such that E is the ideal generated by the classes of the elements of S. Let $\tilde{E}$ be the lifted ideal in $\mathbf{k}Q_R$, that is, the one generated by the elements of S. Applying again [**12**](II.1.8), there exists a surjective algebra morphism $\tilde{\pi}\colon \mathbf{k}Q_R \longrightarrow \mathbf{k}Q_A$ preserving stationary paths and such that, for an arrow β,

$$\tilde{\pi}(\beta) = \begin{cases} \beta & \text{if } \beta \in (Q_R)_1 \setminus S \\ 0 & \text{if } \beta \in S. \end{cases}$$

We deduce an exact sequence of vector spaces

$$0 \longrightarrow \tilde{E} \longrightarrow \mathbf{k}Q_R \longrightarrow \mathbf{k}Q_A \longrightarrow 0$$

and also $\eta_A\tilde{\pi} = \pi\eta_R$. A direct calculation shows that $\tilde{\pi}\tilde{\sigma} = \mathrm{id}_{\mathbf{k}Q_A}$ and (c) follows by passing to kernels. □

2.2. The relations. We have seen that, if R is a split extension of A by E, then E may be assumed to be generated by arrows in Q_R. But what is not clear is whether, if we choose an arbitrary set of arrows in Q_R, and call E the ideal they generate, then R is a split extension of R/E by E or not. Actually, this is not always the case, as the following example shows.

EXAMPLE 2.2.1. Let R be given by the quiver

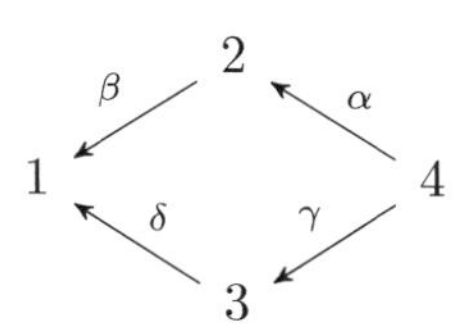

bound by $\alpha\beta = \gamma\delta$. Let E be the ideal generated by α. Then $A = R/E$ is not a subalgebra of R: indeed, the product of (the classes of) γ and δ is zero in A, but not in R. Thus, R is not a split extension of A by E.

If, on the other hand, we let E' be generated by α and γ, then it is easily seen that R is a split extension of R/E' by E'.

This example shows that, when passing from R to A, any deletion of arrows must take into account the relations.

LEMMA 2.2.2 ([**10**]2.1)(2.3). *Let $\eta_R\colon \mathbf{k}Q_R \twoheadrightarrow \mathbf{k}Q_R/I_R \cong R$ be a presentation, S a set of arrows in Q_R and E the ideal in R generated by S.*

(a) *Setting $A = R/E$, there exists a presentation $\eta_A\colon \mathbf{k}Q_A \twoheadrightarrow \mathbf{k}Q_A/I_A \cong A$ such that we have a commutative diagram with exact rows and columns*

$$\begin{array}{ccccccccc}
 & & 0 & & 0 & & 0 & & \\
 & & \downarrow & & \downarrow & & \downarrow & & \\
0 & \longrightarrow & \tilde{E} \cap I_R & \longrightarrow & I_R & \longrightarrow & I_A & \longrightarrow & 0 \\
 & & \downarrow & & \downarrow & & \downarrow & & \\
0 & \longrightarrow & \tilde{E} & \longrightarrow & \mathbf{k}Q_R & \overset{\tilde{\pi}}{\underset{\tilde{\sigma}}{\rightleftarrows}} & \mathbf{k}Q_A & \longrightarrow & 0 \\
 & & \downarrow & & \downarrow{\scriptstyle \eta_R} & & \downarrow{\scriptstyle \eta_A} & & \\
0 & \longrightarrow & E & \longrightarrow & R & \xrightarrow{\pi} & A & \longrightarrow & 0 \\
 & & \downarrow & & \downarrow & & \downarrow & & \\
 & & 0 & & 0 & & 0 & &
\end{array}$$

where $\tilde{\pi}$, $\tilde{\sigma}$ are algebra morphisms such that $\tilde{\pi}\tilde{\sigma} = \mathrm{id}_{\mathbf{k}Q_A}$.

(b) *If moreover $\tilde{\sigma}(I_A) \subseteq I_R$, then the lower sequence realises R as a split extension of A by E.*

PROOF. (a) Let Q be the quiver having the same points as Q_R and arrows all arrows of Q_R except those in S. Using [**12**](II.1.8) there exists a surjective algebra morphism $\tilde{\pi}\colon \mathbf{k}Q_R \twoheadrightarrow \mathbf{k}Q$, preserving stationary paths and such that

$$\tilde{\pi}(\beta) = \begin{cases} \beta & \text{if } \beta \in (Q_R)_1 \setminus S \\ 0 & \text{if } \beta \in S. \end{cases}$$

Let $\pi\colon R \twoheadrightarrow A$ be the projection and $\tilde{E} = \operatorname{Ker} \tilde{\pi}$. Then $\tilde{E}$ is the ideal of $\mathbf{k}Q_R$ generated by the arrows in S. Clearly, $\pi\eta_R(\tilde{E}) = 0$ hence there exists a unique algebra morphism $\eta_A\colon \mathbf{k}Q \twoheadrightarrow A$ such that $\eta_A\tilde{\pi} = \pi\eta_R$. Moreover, η_A is surjective, because so are π and η_R.

We claim that $I_A = \operatorname{Ker} \eta_A$ is an admissible ideal of $\mathbf{k}Q$. We first prove that $I_A \subseteq \mathbf{k}Q^{+2}$. If this is not the case, let $\gamma \in I_A \setminus \mathbf{k}Q^{+2}$. There exist $\alpha_1, \ldots, \alpha_t \in Q_1$, nonzero scalars $c_1, \ldots, c_t$ and $\gamma' \in \mathbf{k}Q^{+2}$ such that

$$\gamma = \sum_{i=1}^{t} c_i\alpha_i + \gamma'.$$

Considering γ as an element of $\mathbf{k}Q_R$, we have

$$\pi\eta_R(\gamma) = \eta_A\tilde{\pi}(\gamma) = \eta_A(\gamma) = 0.$$

Hence $\eta_R(\gamma) \in \operatorname{Ker} \pi = E$. Therefore there exist nonzero scalars $d_1, \dots, d_s$ and arrows $\beta_1, \dots, \beta_s \in S$ such that

$$\sum_{i=1}^{t} c_i \alpha_i + \gamma' + I_R = \gamma + I_R = \sum_{j=1}^{s} d_j \beta_j + I_R.$$

Because I_R is admissible and $\gamma' \in \mathbf{k}Q^{+2}$, this equality yields, because of the grading,

$$\sum_{i=1}^{t} c_i \alpha_i = \sum_{j=1}^{s} d_j \beta_j.$$

Now the β_j lie in S, while the α_i do not. This absurdity yields $I_A \subseteq \mathbf{k}Q^{+2}$. On the other hand, there exists $m \geq 2$ such that $\mathbf{k}Q_R^{+m} \subseteq I_R$. Because Q is a subquiver of Q_R, we have $\mathbf{k}Q^{+m} \subseteq \mathbf{k}Q_R^{+m}$ so that $\mathbf{k}Q^{+m} \subseteq I_R$. Because of the definition of η_R, the last inclusion reads as $\mathbf{k}Q^{+m} \subseteq I_A$. This establishes our claim.

Therefore $\eta_A \colon \mathbf{k}Q \longrightarrow \mathbf{k}Q/I_A \cong A$ is a presentation of A. Because the quiver of an algebra is uniquely determined, we have $Q = Q_A$. Moreover that $\eta_A \tilde{\pi} = \pi \eta_R$ and E, $\tilde{E}$ are the respective kernels of π, $\tilde{\pi}$ imply that the shown diagram is commutative with exact rows and columns.

Finally, the (non full) quiver inclusion $Q_A \hookrightarrow Q_R$ yields, because of [**12**](II.1.8), an algebra morphism $\tilde{\sigma} \colon \mathbf{k}Q_A \longrightarrow \mathbf{k}Q_R$ such that $\tilde{\pi}\tilde{\sigma} = \mathrm{id}_{\mathbf{k}Q_A}$.

(b) The hypothesis yields a morphism of abelian groups $\sigma \colon A \longrightarrow R$ such that $\sigma \eta_A = \eta_R \tilde{\sigma}$. Because $E \subseteq \operatorname{rad} R$, it suffices to prove that σ is an algebra morphism and a section to π. Let w, w' be paths in Q_A, then

$$\begin{aligned}\sigma\big((w + I_A)(w' + I_A)\big) &= \sigma(ww' + I_A) = \sigma\eta_A(ww') = \eta_R\tilde{\sigma}(ww') \\ &= \eta_R\tilde{\sigma}(w)\eta_R\tilde{\sigma}(w') = \sigma\eta_A(w)\sigma\eta_A(w') \\ &= \sigma(w + I_A)\sigma(w' + I_A).\end{aligned}$$

Thus, σ is an algebra morphism. Also, $\tilde{\pi}\tilde{\sigma} = \mathrm{id}_{\mathbf{k}Q_A}$ implies that $\pi\sigma\eta_A = \pi\eta_R\tilde{\sigma} = \eta_A\tilde{\pi}\tilde{\sigma} = \eta_A$. The surjectivity of η_A yields $\pi\sigma = \mathrm{id}_A$, as required. □

Let w be a path in a quiver and α an arrow on w, that is, such that there exist subpaths w_1, w_2 of w satisfying $w = w_1 \alpha w_2$, then we write $\alpha \mid w$. Let now S be a set of arrows and $\rho = \sum_{i=1}^{t} c_i w_i$ a relation, with the c_i nonzero scalars and the w_i paths. We say that ρ is ***consistently cut*** if, for any i, if there exists an arrow $\alpha_i \in S$ such that $\alpha_i \mid w_i$ then for every $j \neq i$, there exists $\alpha_j \in S$ such that $\alpha_j \mid w_j$. That is, if S cuts one branch of ρ, then it cuts all its branches.

In 2.2.1, the relation $\alpha\beta = \gamma\delta$ is not consistently cut by the set $\{\alpha\}$, but it is consistently cut by $\{\alpha, \gamma\}$.

Because relations in a bound quiver may be assumed monomial or minimal, and because monomial relations are trivially consistently cut, the definition above applies only to minimal relations.

THEOREM 2.2.3 ([**10**]2.5). *Let $\eta_R \colon \mathbf{k}Q_R \longrightarrow R \cong \mathbf{k}Q_R/I_R$ be a presentation, S a set of arrows in Q_R, E the ideal they generate and $\pi \colon R \longrightarrow R/E = A$ the projection. Then:*

(a) *If every minimal relation in I_R is consistently cut, then the exact sequence*

$$0 \longrightarrow E \longrightarrow R \xrightarrow{\pi} A \longrightarrow 0$$

realises R as a split extension of A by E.

(b) *Conversely, if the sequence in (a) is a split extension and η_R respects A, then every minimal relation is consistently cut.*

PROOF. (a) Assume that every minimal relation in I_R is consistently cut. As seen in 2.2.2(a), the projection π lifts to an algebra morphism $\tilde{\pi}\colon \mathbf{k}Q_R \longrightarrow \mathbf{k}Q_A$. Let $\rho \in I_R$ be a relation, then $\rho = \sum c_i w_i$ where the c_i are nonzero scalars and the w_i paths. Because ρ is consistently cut, if there exists i such that $\tilde{\pi}(w_i) = 0$ then, for each $j \neq i$, we have $\tilde{\pi}(w_j) = 0$. This proves that, for any relation ρ, we have either $\tilde{\pi}(\rho) = \rho$ or $\tilde{\pi}(\rho) = 0$.

In order to prove our statement, it suffices, because of 2.2.2(b), to prove that the algebra morphism $\tilde{\sigma}\colon \mathbf{k}Q_A \longrightarrow \mathbf{k}Q_R$ induced by the inclusion $Q_A \hookrightarrow Q_R$ satisfies $\tilde{\sigma}(I_A) \subseteq I_R$.

Let $\rho \in I_A$ be nonzero. We may assume, without loss of generality, that ρ is a relation. Because the restriction $\tilde{\pi}|_{I_R} : I_R \longrightarrow I_A$ is surjective, there exists $\rho' \in I_R$ such that $\tilde{\pi}(\rho') = \rho$. Then ρ' can be written as $\rho' = \sigma + \eta$, where $\sigma = \sum_i \sigma_i$ is a sum of monomial relations and $\eta = \sum_j \eta_j$ is a sum of minimal relations. We distinguish two cases:

1) Assume ρ is monomial. Because each of the $\tilde{\pi}(\sigma_i)$, $\tilde{\pi}(\eta_j)$ is a summand of $\rho = \tilde{\pi}(\rho')$, then $\tilde{\pi}(\eta_j) = 0$ for all j and there exists a unique i such that $\rho = \tilde{\pi}(\sigma_i) = \sigma_i$. We thus have $\tilde{\pi}(\rho) = \rho$ and so $\tilde{\sigma}(\rho) = \rho \in I_R$.
2) If ρ is minimal, then, for each i, we have $\tilde{\pi}(\sigma_i) = 0$ because otherwise, σ_i would be a summand of the minimal relation ρ, a contradiction. Similarly, if j, k are distinct indices such that $\tilde{\pi}(\eta_j) = \eta_j$ and $\tilde{\pi}(\eta_k) = \eta_k$, then $\eta_j + \eta_k$ would be a summand of ρ, another contradiction to minimality. Hence there exists a unique j such that $\rho = \tilde{\pi}(\eta_j) = \eta_j$. Again, we have $\rho = \tilde{\pi}(\rho)$ and $\tilde{\sigma}(\rho) = \rho \in I_R$.

This completes the proof of (a).

(b) Conversely, assume that the given sequence is a split extension and that η_R respects A. Let $\rho = \sum_i c_i w_i$ be a minimal relation in I_R with the c_i nonzero scalars and the w_i paths. Assume there exist i and $\alpha_i \in S$ such that $\alpha_i \mid w_i$. Let J be the proper subset of $\{1, \ldots, t\}$ consisting of those j such that there is no arrow $\alpha_j \in S$ such that $\alpha_j \mid w_j$. We must prove that $J = \varnothing$. If not, then we can write

$$\rho = \sum_{i \notin J} c_i w_i + \sum_{j \in J} c_j w_j.$$

The commutative diagram of 2.2.2 yields $\tilde{\pi}(\rho) = \sum_{j\in J} c_j w_j$ in $\mathbf{k}Q_A$. Because $\rho \in I_R$, we have $\eta_A \tilde{\pi}(\rho) = \pi\eta_R(\rho) = 0$. Hence $\tilde{\pi}(\rho) \in \operatorname{Ker} \eta_A = I_A$. Because the given exact sequence is a split extension, it follows from 2.1.2 that $\tilde{\sigma}(I_A) \subseteq I_R$. But then we get $\sum_{j\in J} c_j w_j \in I_R$, which contradicts the minimality of ρ. □

We recall that an algebra is ***monomial*** if it admits a presentation such that the binding ideal is generated by monomials. ***String algebras*** are special types of monomial algebras for which we refer to [**20**]. ***Gentle algebra*** are special types of string algebras, see[**12**] Chapter X. For ***special biserial algebras***, we refer to [**30**].

COROLLARY 2.2.4. *Let R be a split extension of A by a nilpotent bimodule E, with a presentation respecting A. If R is monomial, string, gentle or special biserial, then so is A.*

PROOF. In each case, the defining conditions on the bound quiver of R remain satisfied if one cuts arrows so that the conditions of 2.2.3 are satisfied. □

As an interesting particular case, if R is a trivial extension of A (by either the minimal injective cogenerator $D({}_AA_A)$ or the relation bimodule $\mathrm{Ext}^2_A(DA\,,\,A)$) and R is monomial, string, gentle or special biserial, then so is A.

The reader will connect the notion of consistent cut of relation with that of admissible cut of an algebra, introduced in [**24**] in the case of selfinjective trivial extensions and in [**16**] in the case of cluster tilted algebras, see 2.2.5(b) below, and also [**2**].

EXAMPLES 2.2.5. (a) We show that one-point extensions may be viewed as split extensions. Let B be an algebra, and M a B-module, then

$$R = B[M] = \begin{pmatrix} B & 0 \\ M & \mathbf{k} \end{pmatrix} = \left\{ \begin{pmatrix} b & 0 \\ m & \lambda \end{pmatrix} \;\middle|\; b \in B, m \in M, \lambda \in \mathbf{k} \right\}$$

becomes an algebra when equipped with the ordinary matrix addition and the multiplication induced from the B-module structure of M. It is called the ***one-point extension*** of R by M, see [**27**]. The quiver Q_R equals Q_B plus an extra point x, called the extension point, which is a source in Q_R.

Cutting all arrows having x as a source is certainly a consistent cut. Therefore R is a split extension of $A = B \times \mathbf{k}$ by the bimodule E such that $E_A = M$ while $D({}_AE) = S_x{}^{\dim_{\mathbf{k}} M}$.

(b) Let Q be a quiver with neither loops nor cycles of length two. A full subquiver of Q is a ***chordless cycle*** if it is induced by a set of points $\{x_1, \ldots, x_p\}$ such that the only edges on it are precisely the x_i —— x_{i+1}, where we set $x_{p+1} = x_1$, see [**17**]. A quiver Q is called ***cyclically oriented*** if each chordless cycle is an oriented cycle, see [**18**].

Let R be a cluster tilted algebra with a cyclically oriented quiver, for instance a representation-finite cluster tilted algebra, then Q_R is consistently cut by exactly one arrow from each branch of a relation if and only if the resulting algebra A is an admissible cut of R, that is, R is the trivial extension of A by its relation bimodule. This indeed follows easily from [**18**](4.2)(3.4)(4.7).

(c) The following example, due to M. I. Platzeck (private communication) shows that in 2.2.3(b), the condition that η_R respects A is necessary.

Assume char $\mathbf{k} \neq 2$ and let R be given by the quiver

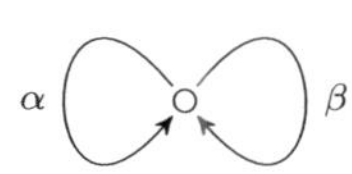

bound by $\alpha^2 = 0$, $\alpha\beta = \beta\alpha$, $\beta^2 = 0$. Let E be the ideal generated by α, then R is the split extension of $A = R/E$ given by the quiver

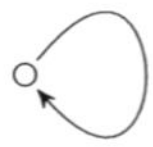

bound by $\beta^2 = 0$. Here, the given presentation of R respects A.

Let now $\alpha' = \alpha$, $\beta' = \alpha + \beta$. Then R is given by the quiver

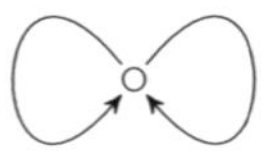

bound by $\alpha'^2 = 0$, $\beta'^2 = 2\beta'\alpha'$, $\alpha'\beta' = \beta'\alpha'$. Taking E' as the ideal generated by α', we get $R/E' \cong A$, as before. However, the second relation is not consistently cut. This presentation of R does not respect A.

3. Modules over split-by-nilpotent extensions

3.1. The change of rings functors. Let R be a split extension of A by a nilpotent bimodule E. There is an obvious embedding of $\operatorname{mod} A$ into $\operatorname{mod} R$, but the latter is in general much larger than the former. For instance, the path algebra of the Kronecker quiver

$$1 \circ \overset{\alpha}{\underset{\beta}{\leftleftarrows}} \circ\ 2$$

is a split extension of the path algebra of the quiver

$$1 \circ \xleftarrow{\ \alpha\ } \circ\ 2.$$

The first is representation-infinite while the second has only 3 isoclasses of indecomposable modules.

Because A is a quotient of R, we have the classical change of rings functors, see [**21**]:

(a) The ***induction*** functor $- \otimes_A R_R \colon \operatorname{mod} A \longrightarrow \operatorname{mod} R$. Modules in its image are called ***induced***.

(b) The ***restriction*** functor $- \otimes_R A_A \colon \operatorname{mod} R \longrightarrow \operatorname{mod} A$. Modules in its image are called ***restricted***.

(c) The ***coinduction*** functor $\operatorname{Hom}_A({}_RR_A\,, -) \colon \operatorname{mod} A \longrightarrow \operatorname{mod} R$. Modules in its image are called ***coinduced***.

(d) The ***corestriction*** functor $\operatorname{Hom}_R({}_AA_R\,, -) \colon \operatorname{mod} R \longrightarrow \operatorname{mod} A$. Modules in its image are called ***corestricted***.

We also have obvious functors:

(e) the ***forgetful*** functor $- \otimes_R R_A$ or $\operatorname{Hom}_R({}_AR_R\,, -) \colon \operatorname{mod} R \longrightarrow \operatorname{mod} A$

(f) the ***embedding*** functor $- \otimes_A A_R$ or $\operatorname{Hom}_A({}_RA_A\,, -) \colon \operatorname{mod} A \longrightarrow \operatorname{mod} R$.

Besides the usual adjunction relations, we have the following lemma.

LEMMA 3.1.1. *We have isomorphisms of functors*

(a) $- \otimes_A R \otimes_R A_A \cong \mathrm{id}_{\mathrm{mod}\, A}$,
(b) $\mathrm{Hom}_R({}_A A_R\, , \mathrm{Hom}_A({}_R R_A\, , -)) \cong \mathrm{id}_{\mathrm{mod}\, A}$.

PROOF. (a) is obvious and (b) follows from the isomorphisms of functors

$$\mathrm{Hom}_R({}_A A_R\, , \mathrm{Hom}_A({}_R R_A\, , -)) \cong \mathrm{Hom}_A({}_A A \otimes_R R_A\, , -) \cong \mathrm{Hom}_A(A\, , -).$$

□

In the next corollary, we use for the first time a notation that we follow until the end of these notes. Because we deal with modules over two algebras, in order to avoid confusion, we denote A-modules by $L, M, N, \ldots$ and R-modules by $X, Y, Z, \ldots$

COROLLARY 3.1.2. *The following conditions are equivalent for two A-modules L and M:*

(a) $L \cong M$ *(b)* $L \otimes_A R \cong M \otimes_A R$ *(c)* $\mathrm{Hom}_A(R\, , L) \cong \mathrm{Hom}_A(R\, , M)$. □

LEMMA 3.1.3. (a) *An R-module X is projective if and only if:*
i) *$X \otimes_R A$ is projective in* $\mathrm{mod}\, A$, *and*
ii) *$X \otimes_R A \otimes_A R \cong X$ in* $\mathrm{mod}\, R$.
Moreover, in this case, X is indecomposable if and only if so is $X \otimes_R A$.
(b) *An R-module Y is injective if and only if:*
i) $\mathrm{Hom}_R(A\, , Y)$ *is injective in* $\mathrm{mod}\, A$, *and*
ii) $\mathrm{Hom}_A({}_R R_A\, , \mathrm{Hom}_A(A\, , Y)) \cong Y$ *in* $\mathrm{mod}\, R$.
Moreover, in this case, Y is indecomposable if and only if so is $\mathrm{Hom}_R(A\, , Y)$.

PROOF. (a) Let $e \in R$ be an idempotent such that $X = eR$. Then $X \otimes_R A = eR \otimes_R A = eA$ is projective in $\mathrm{mod}\, A$. Also, $X \otimes_R A \otimes_A R \cong eA \otimes_A R \cong eR = X$. So X satisfies i) and ii). Conversely, if X satisfies i) and ii), there exists an idempotent e such that $X \otimes_R A = eA$. But then ii) gives $X \cong X \otimes_R A \otimes_A R \cong eA \otimes_A R \cong eR$. This establishes the first statement.

Assume that X is decomposable, say $X = X_1 \oplus X_2$ with X_1, X_2 nonzero, but that $X \otimes_R A = (X_1 \otimes_R A) \oplus (X_2 \otimes_R A)$ is indecomposable. Then one of the summands is zero, say $X_1 \otimes_R A = 0$. But then $X_1 \cong X_1 \otimes_R A \otimes_A R = 0$, a contradiction. Therefore X is indecomposable. Similarly, X indecomposable implies $X \otimes_R A$ indecomposable. □

Thus, there exists a bijection between isoclasses of indecomposable projective A- and R-modules given by $eA \mapsto eR$, where e is a primitive idempotent. Also, there exists a similar bijection for the injectives.

Because $A = R/E$, the category $\mathrm{mod}\, A$ may be identified with the full subcategory of $\mathrm{mod}\, R$ of the modules X such that $XE = 0$. Given any R-module X, there exists a largest R-submodule of X which is annihilated by E, that is, which is an A-module. This is $K_X = \{x \in X \mid xE = 0\}$.

LEMMA 3.1.4. *Let X be an R-module. We have functorial isomorphisms:*

(a) $X \otimes_R A \cong X/XE$,
(b) $\mathrm{Hom}_R(A\, , X) \cong K_X$.

PROOF. (a) Applying $X \otimes_R -$ to the exact sequence of R-R-bimodules

$$0 \longrightarrow E \xrightarrow{\iota} R \xrightarrow{\pi} A \longrightarrow 0 \qquad (*)$$

yields a commutative diagram with exact rows in $\operatorname{mod} R$

$$\begin{array}{ccccccc} & & X \otimes_R E & \longrightarrow & X \otimes_R R & \longrightarrow & X \otimes_R A & \longrightarrow & 0 \\ & & \downarrow{\scriptstyle \mu'} & & \downarrow{\scriptstyle \mu} & & \downarrow{\scriptstyle \mu''} & & \\ 0 & \longrightarrow & XE & \xrightarrow{i} & X & \xrightarrow{p} & X/XE & \longrightarrow & 0 \end{array}$$

where i, p are respectively the inclusion and projection, μ, μ' are the multiplication maps $x \otimes r \longmapsto xr$ and μ'' is induced by passing to cokernels.

Clearly, μ' is surjective, because of the definition of XE. It is well-known that μ is an isomorphism. Therefore the snake lemma implies that μ'' is injective and also surjective. So it is an isomorphism.

(b) Let $f\colon \operatorname{Hom}_R(R, X) \longrightarrow X$, $u \longmapsto u(1)$ be the well-known functorial isomorphism. For every $u \in \operatorname{Hom}_R(A, X)$, we have

$$f \operatorname{Hom}_A(\pi, X)(u) = f(u\pi) = u\pi(1).$$

So, for $x \in E$, we have $u\pi(1)x = u\pi(x) = 0$. Therefore, the image of the composition $f \operatorname{Hom}_A(\pi, X)$ lies in K_X. That is, there exists $f'\colon \operatorname{Hom}_R(A, X) \longrightarrow K_X$ making the following square commutative

$$\begin{array}{ccc} \operatorname{Hom}_R(A, X) & \xrightarrow{\operatorname{Hom}_A(\pi, X)} & \operatorname{Hom}_R(R, X) \\ {\scriptstyle f'}\downarrow & & \downarrow{\scriptstyle f} \\ K_X & \xhookrightarrow{\quad j \quad} & X \end{array}$$

where j is the inclusion. Applying $\operatorname{Hom}_R(-, X)$ to the exact sequence $(*)$ above shows that $\operatorname{Hom}_R(\pi, X)$ is injective. Therefore so is $f \operatorname{Hom}_R(\pi, X)$ and so is f'.

We prove that f' is surjective. Let $x \in K_X$. Because $x \in X$, there exists $u_x \in \operatorname{Hom}_R(R, X)$ such that $x = u_x(1)$. But then $u_x(E) = u_x(1)E = xE = 0$ hence there exists $v_x\colon A \longrightarrow X$ such that $u_x = v_x\pi$. Then $x = u_x(1) = v_x\pi(1) = f'(v_x)$ and so f' is surjective. Therefore it is an isomorphism. □

3.2. Projective covers and injective envelopes. For the notions of superfluous epimorphisms and essential monomorphisms, we refer the reader, for example, to [**3**].

LEMMA 3.2.1 ([**11**]1.1). *Let X be an R-module.*

(a) *The canonical epimorphism $p_X\colon X \longrightarrow X/XE$ is superfluous.*
(b) *The canonical monomorphism $j_X : K_X \hookrightarrow X$ is essential.*

PROOF. (a) Because of Nakayama's lemma, the canonical epimorphism $f\colon X \longrightarrow X/X \cdot \operatorname{rad} R$ is superfluous. Because $E \subseteq \operatorname{rad} R$, there exists an epimorphism $g\colon X/XE \longrightarrow X/X \cdot \operatorname{rad} R$ such that $f = gp_X$. Assume h is such that $p_X h$ is an epimorphism. Then so is $fh = gp_X h$. Because f is superfluous, h is an epimorphism.

(b) Let Y be a nonzero submodule of X. Because E is nilpotent, there exists $s \geq 1$ such that $YE^{s-1} \neq 0$ but $YE^s = 0$. Let $y \in YE^{s-1}$ be nonzero. Then $yE = 0$ so that $y \in K_X$. Therefore $K_X \cap Y \neq 0$ and we are done. □

COROLLARY 3.2.2. [**11**]*1.2 Let M be an A-module.*

(a) *There is a bijection between the isoclasses of indecomposable summands of M in* $\operatorname{mod} A$ *and $M \otimes_A R$ in* $\operatorname{mod} R$, *given by $N \mapsto N \otimes_A R$.*
(b) *There is a bijection between the isoclasses of indecomposable summands of M in* $\operatorname{mod} A$ *and* $\operatorname{Hom}_A(R, M)$ *in* $\operatorname{mod} R$, *given by $N \mapsto \operatorname{Hom}_A(R, N)$.*

PROOF. (a) Suppose N is indecomposable in $\operatorname{mod} A$ but $N \otimes_A R = X_1 \oplus X_2$ in $\operatorname{mod} R$. Then $N \cong N \otimes_A R \otimes_R A \cong (X_1 \otimes_R A) \oplus (X_2 \otimes_R A)$. Because N is indecomposable, $X_1 \otimes_R A$, say, is zero. So $X_1/X_1E = 0$. But p_{X_1} is superfluous so $X_1 = 0$. Thus $N \otimes_A R$ is indecomposable. The rest of the proof is an application of 3.1.2. □

LEMMA 3.2.3. [**11**]*1.3 Let M be an A-module.*

(a) *If $f \colon P \longrightarrow M$ is a projective cover in* $\operatorname{mod} A$, *then $f \otimes_A R \colon P \otimes_A R \longrightarrow M \otimes_A R$ is a projective cover in* $\operatorname{mod} R$.
(b) *If $g \colon M \longrightarrow I$ is an injective envelope in* $\operatorname{mod} A$, *then* $\operatorname{Hom}_A(R, g) \colon \operatorname{Hom}_A(R, M) \longrightarrow \operatorname{Hom}_A(R, I)$ *is an injective envelope in* $\operatorname{mod} R$.

PROOF. (a) Clearly, $P \otimes_A R$ is projective in $\operatorname{mod} R$ and $f \otimes_A R$ is an epimorphism. Consider the commutative square:

$$\begin{array}{ccc} P \otimes_A R & \xrightarrow{f \otimes_A R} & M \otimes_A R \\ \downarrow{\scriptstyle p_{P \otimes_A R}} & & \downarrow{\scriptstyle p_{M \otimes_A R}} \\ P & \xrightarrow{f} & M \end{array}$$

where we have used that $(M \otimes_A R)/(M \otimes_A R)E \cong M \otimes_A R \otimes_R A \cong M$ and similarly for P. It suffices to prove that $f \otimes_A R$ is superfluous. Let h be such that $(f \otimes_A R)h$ is an epimorphism. Then $p_{M \otimes_A R}(f \otimes_A R)h = f p_{P \otimes_A R} h$ is an epimorphism. Because both f and $p_{P \otimes_A R}$ are superfluous, h is an epimorphism. □

We have a similar result when passing from $\operatorname{mod} R$ to $\operatorname{mod} A$.

LEMMA 3.2.4. [**14**]*3.1 Let X be an R-module.*

(a) *If $f \colon \tilde{P} \longrightarrow X$ is a projective cover in* $\operatorname{mod} R$, *then $f \otimes_R A \colon \tilde{P} \otimes_R A \longrightarrow X \otimes_R A$ is a projective cover in* $\operatorname{mod} A$.
(b) *If $g \colon X \longrightarrow \tilde{I}$ is an injective envelope in* $\operatorname{mod} R$, *then* $\operatorname{Hom}_R(A, g) \colon \operatorname{Hom}_R(A, X) \longrightarrow \operatorname{Hom}_R\left(A, \tilde{I}\right)$ *is an injective envelope in* $\operatorname{mod} A$.

PROOF. (a) First $\tilde{P} \otimes_R A$ is projective in mod A, see 3.1.3, and $f \otimes_R A$ is an epimorphism. Next,

$$\begin{aligned}\operatorname{top}(\tilde{P} \otimes_R A) = \operatorname{top}(\tilde{P}/\tilde{P}E) &\cong \frac{\tilde{P}/\tilde{P}E}{\tilde{P}/\tilde{P}E \cdot \operatorname{rad} A} \cong \frac{\tilde{P}/\tilde{P}E}{\tilde{P}/\tilde{P}E \cdot \operatorname{rad}(R/E)} \\ &\cong \frac{\tilde{P}/\tilde{P}E}{(\tilde{P} \cdot \operatorname{rad} R)/\tilde{P}E} \cong \frac{\tilde{P}}{\tilde{P} \cdot \operatorname{rad} A} \cong \frac{X}{X \cdot \operatorname{rad} R} \cong \frac{X/XE}{X/XE \cdot (\operatorname{rad} R)/E} \\ &\cong \operatorname{top}(X \otimes_R A)\end{aligned}$$

where the last isomorphism comes from 3.1.4. □

3.3. Presentations. We now compute minimal projective presentations and injective copresentations of R-modules

COROLLARY 3.3.1 ([**11**]1.3). *Let M be an A-module.*

(a) *If* $P_1 \xrightarrow{f_1} P_0 \xrightarrow{f_0} M \longrightarrow 0$ *is a projective presentation in* mod A, *then* $P_1 \otimes_A R \xrightarrow{f_1 \otimes_A R} P_0 \otimes_A R \xrightarrow{f_0 \otimes_A R} M \otimes_A R \longrightarrow 0$ *is a projective presentation in* mod R. *Further, if the first is minimal, then so is the second.*

(b) *If* $0 \longrightarrow M \xrightarrow{g^0} I^0 \xrightarrow{g^1} I^1$ *is an injective copresentation in* mod A, *then* $0 \longrightarrow \operatorname{Hom}_A(R, M) \xrightarrow{\operatorname{Hom}_A(R, g^0)} \operatorname{Hom}_A(R, I^0) \xrightarrow{\operatorname{Hom}_A(R, g^1)} \operatorname{Hom}_A(R, I^1)$ *is an injective copresentation in* mod R. *Further, if the first is minimal, then so is the second.*

PROOF. (a) The first statement is clear. If the given projective presentation of M is minimal, then, because of 3.2.3, $f_0 \otimes_A R \colon P_0 \otimes_A R \longrightarrow M \otimes_A R$ is a projective cover in mod R. Because $f_1 \colon P_1 \longrightarrow f_1(P_1)$ is a projective cover in mod A, so is $f_1 \otimes_A R \colon P_1 \otimes_A R \longrightarrow f_1(P_1) \otimes_A R \cong (f_1 \otimes_A R)(P_1 \otimes_A R) \cong \operatorname{Ker}(f_0 \otimes_A R)$ in mod R. □

Clearly, if $\tilde{P}_1 \longrightarrow \tilde{P}_0 \longrightarrow X \longrightarrow 0$ is a projective presentation of X in mod R, then $\tilde{P}_1 \otimes_R A \longrightarrow \tilde{P}_0 \otimes_R A \longrightarrow X \otimes_R A \longrightarrow 0$ is a projective presentation in mod A.

But here, the minimality of the first presentation does not imply that of the second.

EXAMPLES 3.3.2. Let A be given by the quiver

$$1 \circ \xleftarrow{\ \alpha\ } \circ\ 2$$

and R by the quiver

$$1 \circ \underset{\alpha}{\overset{\beta}{\rightleftarrows}} \circ\ 2$$

bound by $\alpha\beta\alpha = 0$, $\beta\alpha\beta = 0$. The simple R-module S_1 has a minimal projective presentation

$$e_2 R \longrightarrow e_1 R \longrightarrow S_1 \longrightarrow 0.$$

Applying $- \otimes_R A$ yields a projective presentation

$$e_2 A \longrightarrow e_1 A \longrightarrow S_1 \otimes_R A \longrightarrow 0.$$

But $\operatorname{Hom}_A(e_2A\,,\,e_1A)=0$, hence $S_1\otimes_R A\cong e_1A$ and the previous presentation is not minimal.

We need, for later purposes, to compute the minimal projective presentation of an A-module, considered as an R-module under the embedding $\operatorname{mod}A\hookrightarrow\operatorname{mod}R$.

LEMMA 3.3.3. *Let M be an A-module.*

(a) *If $P_1\xrightarrow{f_1}P_0\xrightarrow{f_0}M\longrightarrow 0$ is a minimal projective presentation of M in* mod A *and P the projective cover of $P_0\otimes_A E_A$ in* mod A*, then there exists a direct summand P' of P such that*

$$(P_1\oplus P')\otimes_A R\longrightarrow P_0\otimes_A R\longrightarrow M\longrightarrow 0$$

is a minimal projective presentation in mod R.

(b) *If $0\longrightarrow M\xrightarrow{g^0}I^0\xrightarrow{g^1}I^1$ is a minimal injective copresentation of M in* mod A*, and I the injective envelope of* $\operatorname{Hom}_A(E\,,\,I^0)$ *in* mod A*, then there exists a direct summand I' of I such that*

$$0\longrightarrow M\longrightarrow\operatorname{Hom}_A(R\,,\,I^0)\longrightarrow\operatorname{Hom}_A(R\,,\,I^1\oplus I')$$

is a minimal injective copresentation in mod R.

PROOF. (a) Let $p_{M\otimes_A R}\colon M\otimes_A R\longrightarrow M$ be the canonical surjection. Because of 3.2.1, it is a superfluous epimorphism. Because of 3.2.3, so is $f_0\otimes_A R\colon P_0\otimes_A R\longrightarrow M\otimes_A R$. Then their composition $p_{M\otimes_A R}(f_0\otimes_A R)$ is a superfluous epimorphism, hence it is a projective cover in mod R.

As A-modules, we have $P_0\otimes_A R\cong P_0\otimes_A(A\oplus E)\cong P_0\oplus(P_0\otimes_A E)$ and similarly $(M\otimes_A R)_A\cong M\oplus(M\otimes_A E)$. The morphism $f_0\otimes_A R$ then takes the form $\begin{pmatrix} f_0 & 0\\ 0 & f_0\otimes_A E\end{pmatrix}$. Because $p_{M\otimes_A R}\colon x\otimes(a,e)\longmapsto xa$, for $x\in M$ and $(a\,,\,e)\in R$, we get $p_{M\otimes_A R}(f_0\otimes_A R)=(f_0\,,\,0)$.

Let $\tilde{P}_1$ be the projective cover of $\operatorname{Ker}\big(p_{M\otimes_A R}(f_0\otimes_A R)\big)=\Omega^1_R M$. Because $p_{M\otimes_A R}(f_0\otimes_A R)=(f_0,0)$, then $P_0\otimes_A E$ is actually a direct summand of $\Omega^1_R M$, when the latter is viewed as A-module. In fact, $\Omega^1_R M\cong\Omega^1_A M\oplus(P_0\otimes_A E)$ in mod A. The projective cover of $\Omega^1_A M$ in mod A is P_1, while that of $P_0\otimes_A E$ is P. Then, we have a commutative diagram in mod R with exact rows

$$\begin{array}{ccccccc}
P_1\otimes_A R & \xrightarrow{f_1\otimes_A R} & P_0\otimes_A R & \xrightarrow{f_0\otimes_A R} & M\otimes_A R & \longrightarrow & 0\\
{\scriptstyle\binom{1}{0}}\downarrow & & \| & & \downarrow{\scriptstyle p_{M\otimes_A R}} & & \\
(P_1\oplus P)\otimes_A R & \xrightarrow{(f_1\otimes_A R\,,\,\tilde{f})} & P_0\otimes_A R & \longrightarrow & M & \longrightarrow & 0
\end{array}$$

where $\tilde{f}\colon P\otimes_A R\longrightarrow P_0\otimes_A R$ is the composition of the embedding $P_0\otimes_A E_R\longrightarrow P_0\otimes_A R_R$, induced from the embedding ${}_AE_R\subseteq{}_AR_R$ because of the projectivity of P_0, with the projective cover $P_A\longrightarrow P_0\otimes_A E$ in mod A.

The lower row in the preceding diagram is a projective presentation in mod R, but is not necessarily minimal. Assume P'' is a direct summand of $P_1\oplus P$ such that we have a minimal projective presentation in mod R

$$P''\otimes_A R\longrightarrow P_0\otimes_A R\longrightarrow M\longrightarrow 0.$$

Because M is an A-module, it is annihilated by E when viewed as R-module. Hence $M \cong M \otimes_R A$ because of 3.1.4. Applying $- \otimes_R A$ to the previous presentation yields a commutative diagram with exact rows in $\operatorname{mod} A$

$$\begin{array}{ccccccc}
P'' & \longrightarrow & P_0 & \xrightarrow{f_0} & M & \longrightarrow & 0 \\
 & & \| & & \| & & \\
P_1 & \xrightarrow{f_1} & P_0 & \xrightarrow{f_0} & M & \longrightarrow & 0.
\end{array}$$

Because P_1 is the projective cover of $\Omega^1_A M$, there exists an epimorphism $P'' \longrightarrow P_1$ making the diagram commute. Therefore $P'' = P_1 \oplus P'$ and we have a minimal projective presentation in $\operatorname{mod} R$

$$(P_1 \oplus P') \otimes_A R \xrightarrow{(f_1 \otimes_A R,\, \tilde{f}')} P_0 \otimes_A R \xrightarrow{p_{M \otimes_A R}(f_0 \otimes_A R)} M \longrightarrow 0$$

where $\tilde{f}'$ is the restriction of $\tilde{f}$ to $P' \otimes_A R$. □

EXAMPLE 3.3.4. Let A be given by the quiver

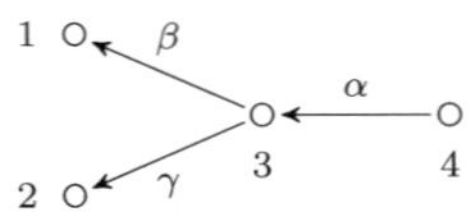

bound by $\alpha\beta = 0$, $\alpha\gamma = 0$, and R be given by the quiver

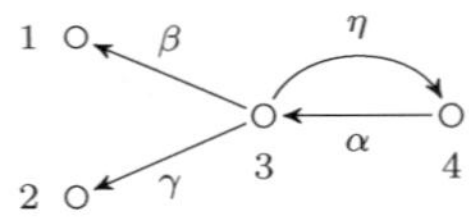

bound by $\alpha\beta = 0$, $\alpha\gamma = 0$, $\eta\alpha\eta\alpha = 0$. Then R is the split extension of A by the nilpotent bimodule E generated by η. The indecomposable (injective) module $M_A = {\scriptstyle\begin{smallmatrix}3\\2\end{smallmatrix}}$ has the minimal projective presentation in $\operatorname{mod} A$

$$0 \longrightarrow e_1A = 1 \longrightarrow e_3A = \begin{smallmatrix} & 3 & \\ 2 & & 1\end{smallmatrix} \longrightarrow M \longrightarrow 0.$$

Because of 3.3.1, a minimal projective presentation for $M \otimes_A R$ is given by

$$e_1R = 1 \xrightarrow{f} e_3R = \begin{smallmatrix} & & 3 & \\ 1 & 2 & & 4 \\ & & & 3 \\ & & & 4\end{smallmatrix} \longrightarrow M \otimes_A R \longrightarrow 0.$$

Then, $M \otimes_A R \cong \begin{smallmatrix} & 3 & \\ 2 & & 4 \\ & & 3 \\ & & 4\end{smallmatrix}$ and also f is a monomorphism, so that $\operatorname{pd}(M \otimes_A R) \leq 1$. Considering $M \otimes_A R$ as an A-module, we get $(M \otimes_A R)_A \cong \begin{smallmatrix}3\\2\end{smallmatrix} \oplus \begin{smallmatrix}4\\3\end{smallmatrix} \oplus 4$. In particular $M \otimes_A E_A = \begin{smallmatrix}4\\3\end{smallmatrix} \oplus 4$ has as projective cover $P = \left(\begin{smallmatrix}4\\3\end{smallmatrix}\right)^2$. Therefore there exists a projective presentation in $\operatorname{mod} R$

$$(e_4R)^2 \oplus e_1R = \begin{pmatrix} 4\\3\\4\\3\\4 \end{pmatrix}^2 \oplus 1 \longrightarrow e_3R = \begin{smallmatrix} & & 3 & \\ 1 & 2 & & 4 \\ & & & 3 \\ & & & 4\end{smallmatrix} \longrightarrow M = \begin{smallmatrix}3\\2\end{smallmatrix} \longrightarrow 0.$$

It is not minimal, but letting $P' = e_4A = \begin{smallmatrix}4\\3\end{smallmatrix}$, we get a minimal projective presentation in $\operatorname{mod} R$

$$e_4R \oplus e_1R \longrightarrow e_3R \longrightarrow M \longrightarrow 0.$$

3.4. Homological dimension one. Working with homological dimension one is easier than with other dimensions, due to its connection with the Auslander-Reiten translation, see [**12**](IV.2.7).

LEMMA 3.4.1 ([**11**]2.1). *For any A-module M, we have*

(a) $\tau_R(M \otimes_A R) \cong \mathrm{Hom}_A({}_RR_A\,,\, \tau_A M)$

(b) $\tau_R^{-1}\,\mathrm{Hom}_A({}_RR_A\,,\, M) \cong (\tau_A^{-1}M) \otimes_A R.$

PROOF. (a) A minimal projective presentation

$$P_1 \longrightarrow P_0 \longrightarrow M \longrightarrow 0$$

in $\mathrm{mod}\, A$ induces, because of 3.3.1, a minimal projective presentation

$$P_1 \otimes_A R \longrightarrow P_0 \otimes_A R \longrightarrow M \otimes_A R \longrightarrow 0.$$

We deduce a commutative diagram with exact rows in $\mathrm{mod}\, R^{\mathrm{op}}$

$$\begin{array}{ccccccc}
\mathrm{Hom}_R(P_0 \otimes_A R\,,\, R) & \longrightarrow & \mathrm{Hom}_R(P_1 \otimes_A R\,,\, R) & \longrightarrow & \mathrm{Tr}(M \otimes_A R) & \longrightarrow & 0 \\
\cong \big\downarrow f & & \cong \big\downarrow g & & \big\downarrow h & & \\
R \otimes_A \mathrm{Hom}_A(P_0\,,\, A) & \longrightarrow & R \otimes_A \mathrm{Hom}_A(P_1\,,\, A) & \longrightarrow & R \otimes_A \mathrm{Tr}\, M & \longrightarrow & 0
\end{array}$$

where the functorial isomorphisms f, g are defined as follows: if e is an idempotent, then $\mathrm{Hom}_R(eA \otimes_A R\,,\, R) \cong \mathrm{Hom}_R(eR\,,\, R) \cong Re \cong R \otimes_A Ae \cong R \otimes_A \mathrm{Hom}_A(eA\,,\, A)$. Then h is deduced by passing to the cokernels and so is an isomorphism. We thus have

$$\begin{aligned}
\tau_R(M \otimes_A R) &\cong \mathrm{D}\mathrm{Tr}(M \otimes_A R) \cong \mathrm{D}(R \otimes_A \mathrm{Tr}\, M) \\
&\cong \mathrm{Hom}_A(R\,,\, \mathrm{D}\mathrm{Tr}\, M) \cong \mathrm{Hom}_A(R\,,\, \tau_A M).
\end{aligned}$$

□

COROLLARY 3.4.2 ([**11**]2.2). *For any A-module M, we have*

(a) $\mathrm{pd}(M \otimes_A R) \leq 1$ *if and only if* $\mathrm{pd}\, M_A \leq 1$ *and* $\mathrm{Hom}_A(\mathrm{D}E\,,\, \tau_A M) = 0$

(b) $\mathrm{id}\,\mathrm{Hom}_A(R\,,\, M) \leq 1$ *if and only if* $\mathrm{id}\, M_A \leq 1$ *and* $\mathrm{Hom}_A\big(\tau_A^{-1}M\,,\, E\big) = 0.$

PROOF. (a) Because of [**12**](IV.2.7), $\mathrm{pd}(M \otimes_A R) \leq 1$ if and only if $\mathrm{Hom}_R(\mathrm{D}R\,,\, \tau_R(M \otimes_A R)) = 0$. Now, we have

$$\begin{aligned}
\mathrm{Hom}_R(\mathrm{D}R\,,\, \tau_R(M \otimes_A R)) &\cong \mathrm{Hom}_R(\mathrm{D}R\,,\, \mathrm{Hom}_A(R\,,\, \tau_A M)) \\
&\cong \mathrm{Hom}_A(\mathrm{D}R \otimes_R R_A\,,\, \tau_A M) \\
&\cong \mathrm{Hom}_A(\mathrm{D}R_A\,,\, \tau_A M) \\
&\cong \mathrm{Hom}_A(\mathrm{D}A\,,\, \tau_A M) \oplus \mathrm{Hom}_A(\mathrm{D}E\,,\, \tau_A M).
\end{aligned}$$

The result follows from another application of [**12**](IV.2.7). □

EXAMPLE 3.4.3. We give an example showing that both conditions are necessary. Let A be the path algebra of the quiver

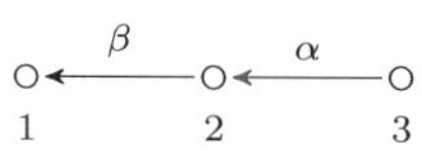

and R be given by the quiver

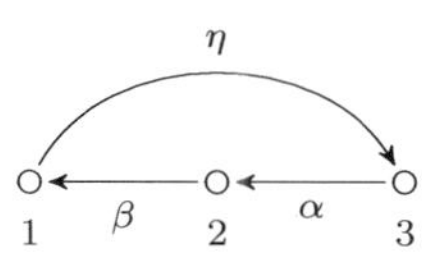

bound by $\alpha\beta\eta = 0$. In this case, one easily sees that $E_A = \left(\begin{smallmatrix}3\\2\\1\end{smallmatrix}\right)^2$, $(\mathrm{D}E)_A = (\begin{smallmatrix}2\\1\end{smallmatrix})^3$. Let $M = \begin{smallmatrix}3\\2\end{smallmatrix}$. We have a minimal projective presentation of M in $\operatorname{mod} A$

$$0 \longrightarrow e_1A \longrightarrow e_3A \longrightarrow M \longrightarrow 0.$$

Applying $- \otimes_A R$ yields a minimal projective presentation in $\operatorname{mod} R$

$$e_1R \xrightarrow{f} e_3R \longrightarrow M \otimes_A R \longrightarrow 0.$$

Therefore, $M \otimes_A R \cong \begin{smallmatrix}3\\2\end{smallmatrix}$. The projective dimension of $\begin{smallmatrix}3\\2\end{smallmatrix}$ in $\operatorname{mod} R$ equals 2. Indeed $\operatorname{Ker} f = e_3R$, so that $\operatorname{pd} M_A \le 1$ but $\operatorname{pd}(M \otimes_A R) > 1$. This shows that the second condition of the corollary is necessary. Actually, $\tau_A(\begin{smallmatrix}3\\2\end{smallmatrix}) = \begin{smallmatrix}2\\1\end{smallmatrix}$ so $\operatorname{Hom}_A(\mathrm{D}E\,, \tau_A M) = 0$.

LEMMA 3.4.4 ([**14**]2.3). *Let M be an A-module.*

(a) *If* $\operatorname{pd} M_R \le 1$, *then* $\operatorname{pd} M_A \le 1$.
(b) *If* $\operatorname{id} M_R \le 1$, *then* $\operatorname{id} M_A \le 1$.

PROOF. (a) Because of 3.4.2, it suffices to prove that $\operatorname{pd}(M \otimes_A R) \le 1$. Let $P_1 \xrightarrow{f_1} P_0 \xrightarrow{f_0} M \longrightarrow 0$ be a minimal projective presentation. Because of 3.3.1, and with its notation, there exists a commutative diagram with exact rows

$$\begin{array}{ccccccc}
P_1 \otimes_A R & \xrightarrow{f_1 \otimes_A R} & P_0 \otimes_A R & \xrightarrow{f_0 \otimes_A R} & M \otimes_A R & \longrightarrow & 0 \\
\downarrow{\scriptstyle \binom{1}{0}} & & \| & & \downarrow & & \\
(P_1 \oplus P') \otimes_A R & \xrightarrow{(f_1 \otimes_A R, \tilde{f}')} & P_0 \otimes_A R & \longrightarrow & M & \longrightarrow & 0.
\end{array}$$

Because $\operatorname{pd} M_R \le 1$, the morphism $(f_1 \otimes_A R, \tilde{f}')$ is injective. Because so is $\binom{1}{0}$, the morphism $f_1 \otimes_A R$ is injective. Therefore $\operatorname{pd}(M \otimes_A R) \le 1$, thus establishing our claim. □

An easy application of this lemma is the following: if R is hereditary, then so is A. But we have a much stronger result, due to Suarez.

THEOREM 3.4.5 ([**33**]3.2)(3.5). *Let R be a split extension of A by a nilpotent bimodule E. Then* $\operatorname{gl.dim.} A \le \operatorname{gl.dim.} R \le \operatorname{gl.dim.} A + \operatorname{pd} A_R$. □

We can also apply 3.4.4 to the study of the left and right parts of an algebra. Recall from [**25**] that, if C is an algebra, the ***left part*** $\mathscr{L}_C$ of $\operatorname{mod} C$ is the full subcategory of $\operatorname{ind} C$ consisting of those indecomposable modules U such that, if there exist V indecomposable and a sequence of nonzero morphisms between indecomposable C-modules

$$V = V_0 \longrightarrow V_1 \longrightarrow \cdots \longrightarrow V_m = U$$

then $\operatorname{pd} V_C \le 1$. One defines dually the ***right part*** $\mathscr{R}_C$ of $\operatorname{mod} C$.

LEMMA 3.4.6 ([**14**]2.4). *Let M be an indecomposable A-module.*

(a) *If $M \otimes_A R$ belongs to $\mathscr{L}_R$, then M belongs to $\mathscr{L}_A$.*
(b) *If $\operatorname{Hom}_A(R, M)$ belongs to $\mathscr{R}_R$, then M belongs to $\mathscr{R}_A$.*
(c) *If $M \otimes_A R$ belongs to $\mathscr{R}_R$, then M belongs to $\mathscr{R}_A$.*
(d) *If $\operatorname{Hom}_A(R, M)$ belongs to $\mathscr{L}_R$, then M belongs to $\mathscr{L}_A$.*

PROOF. (a) Let $L = L_0 \xrightarrow{f_1} L_1 \longrightarrow \cdots \xrightarrow{f_m} L_m = M$ be a sequence of nonzero morphisms between indecomposable A-modules. For each i, $L_i \otimes_A R$ is indecomposable and $f_i \otimes_A R$ is nonzero. So we have a sequence of nonzero morphisms between indecomposable R-modules

$$L \otimes_A R = L_0 \otimes_A R \xrightarrow{f_1 \otimes_A R} L_1 \otimes_A R \longrightarrow \cdots \xrightarrow{f_m \otimes_A R} L_m \otimes_A R = M \otimes_A R.$$

Because $M \otimes_A R \in \mathscr{L}_R$, then $\operatorname{pd}(L \otimes_A R) \leq 1$. Because of 3.4.2, we get $\operatorname{pd} L_A \leq 1$.

(c) We have isomorphisms of $\mathbf{k}$-vector spaces

$$\begin{aligned}\operatorname{Hom}_R(M \otimes_A R, \operatorname{Hom}_A({}_R R_A, M)) &\cong \operatorname{Hom}_A(M \otimes_A R \otimes_R R_A, M)\\ &\cong \operatorname{Hom}_A(M \otimes_A R_A, M)\\ &\cong \operatorname{Hom}_A(M \otimes_A (A \oplus E), M)\\ &\cong \operatorname{Hom}_A(M, M) \oplus \operatorname{Hom}_A(M \otimes_A E, M).\end{aligned}$$

Because $\operatorname{Hom}_A(M, M) \neq 0$, there exists a nonzero morphism $M \otimes_A R \longrightarrow \operatorname{Hom}_A(R, M)$. Now $M \otimes_A R \in \mathscr{R}_R$, which is closed under successors. Hence $\operatorname{Hom}_A(R, M) \in \mathscr{R}_R$. Applying (b), which is proved exactly as (a), we get $M \in \mathscr{R}_A$, as required. □

We now consider different classes of algebras. An algebra C is called ***laura*** if $\mathscr{L}_C \cup \mathscr{R}_C$ is cofinite in $\operatorname{ind} C$, see [**9**] or [**29**]. It is ***left glued*** if $\mathscr{L}_C$ is cofinite in $\operatorname{ind} C$, see [**8**]. ***Right glued*** algebras are defined similarly. An algebra C is ***weakly shod*** if the length of any path from an indecomposable not in $\mathscr{L}_C$ to one not in $\mathscr{R}_C$ is bounded, see [**23**]. It is ***shod*** if every indecomposable has projective dimension or injective dimension at most one, see [**22**]. It is ***quasi-tilted*** if $C_C \in \mathscr{L}_C$, see [**25**]. For ***tilted*** algebras, we refer to [**12**], Chapter VIII. The algebra C is ***right ada*** if $C_C \in \operatorname{add}(\mathscr{L}_C \cup \mathscr{R}_C)$ and ***left ada*** if $\mathrm{D}C_C \in \operatorname{add}(\mathscr{L}_C \cup \mathscr{R}_C)$, see [**1**]. Finally, C is ***ada*** if it is both right and left ada, see [**7**].

THEOREM 3.4.7 ([**14**]2.5 [**35**]1.10 [**1**]3.6 [**7**]2.9). *Let R be a split extension of A by the nilpotent bimodule E.*

(a) *If R is laura, then so is A.*
(b) *If R is right or left glued, then so is A.*
(c) *If R is weakly shod, then so is A.*
(d) *If R is shod, then so is A.*
(e) *If R is quasi-tilted, then so is A.*
(f) *If R is tilted, then so is A.*
(g) *If R is right or left ada, then so is A.*
(h) *If R is ada, then so is A.*

PROOF. (a) Because of 3.4.6, if an indecomposable A-module M does not lie in $\mathscr{L}_A \cup \mathscr{R}_A$, then $M \otimes_A R \notin \mathscr{L}_R \cup \mathscr{R}_R$. Because R is laura, $\mathscr{L}_R \cup \mathscr{R}_R$ is cofinite in $\operatorname{ind} R$. Therefore, $\mathscr{L}_A \cup \mathscr{R}_A$ is cofinite in $\operatorname{ind} A$.

(b) is proved in the same way.

(c) Let $M_0 \notin \mathscr{L}_A$, $M_t \notin \mathscr{R}_A$ be indecomposable A-modules. As seen in 3.4.6, a sequence of nonzero morphisms between indecomposable R-modules $M_0 \longrightarrow M_1 \longrightarrow \cdots \longrightarrow M_t$ induces a sequence of nonzero morphisms between indecomposable R-modules

$M_0 \otimes_A R \longrightarrow M_1 \otimes_A R \longrightarrow \cdots \longrightarrow M_t \otimes_A R$. Because R is weakly shod, t is bounded.

(d) Let M be an indecomposable A-module. Because R is shod, $\operatorname{pd} M_R \leq 1$ or $\operatorname{id} M_R \leq 1$. Then 3.4.4 gives $\operatorname{pd} M_A \leq 1$ or $\operatorname{id} M_A \leq 1$.

(e) Let P be an indecomposable projective A-module. Then $P \otimes_A R$ is an indecomposable projective R-module. Because R is quasi-tilted, $P \otimes_A R \in \mathscr{L}_R$. Because of 3.4.6, $P \in \mathscr{L}_A$.

(f) We refer the reader to [**35**].

(g) Let P be an indecomposable projective A-module. If R is right ada, then $P \otimes_A R \in \mathscr{L}_R \cup \mathscr{R}_R$. Because of 3.4.6, $P \in \mathscr{L}_A$.

(h) Follows from (g). □

3.5. Almost split sequences. We now look for a criterion allowing to verify when an almost split sequence in $\operatorname{mod} A$ embeds as an almost split sequence in $\operatorname{mod} R$.

LEMMA 3.5.1 ([**13**]1.1). *Let M be an indecomposable A-module.*

(a) *Let P_0 be a projective cover of M and P a projective cover of $P_0 \otimes_A E$ in* $\operatorname{mod} A$. *Then there exist a direct summand P' of P and an exact sequence in* $\operatorname{mod} A$

$$0 \longrightarrow \tau_A M \oplus \operatorname{Hom}_A(E, \tau_A M) \longrightarrow \tau_R M \longrightarrow P' \otimes_A \mathrm{D}R \longrightarrow \operatorname{Ker}(p_{M \otimes_A R} \otimes_A \mathrm{D}R) \longrightarrow 0.$$

(b) *Let I_0 be an injective envelope of M and I an injective envelope of* $\operatorname{Hom}_A(E\,, I^0)$ *in* $\operatorname{mod} A$. *Then there exist a direct summand I' of I and an exact sequence in* $\operatorname{mod} A$

$$0 \to \operatorname{Coker} \operatorname{Hom}_A(\mathrm{D}R, j_{\operatorname{Hom}_A(R,M)}) \to \operatorname{Hom}_A(\mathrm{D}R, I') \to \tau_R^{-1} M \to \tau_A^{-1} M \oplus (\tau_A^{-1} M \otimes_A E) \to 0.$$

PROOF. (a) Let $P_1 \xrightarrow{f_1} P_0 \xrightarrow{f_0} M \longrightarrow 0$ be a minimal projective presentation in $\operatorname{mod} A$. Because of 3.3.1, we have a minimal projective presentation

$$P_1 \otimes_A R \xrightarrow{f_1 \otimes_A R} P_0 \otimes_A R \xrightarrow{f_0 \otimes_A R} M \otimes_A R \to 0$$

in $\operatorname{mod} R$. Because of 3.3.3, there exists a direct summand P' of P such that we have a commutative diagram with exact rows in $\operatorname{mod} R$

$$\begin{array}{ccccccc}
P_1 \otimes_A R & \xrightarrow{f_1 \otimes_A R} & P_0 \otimes_A R & \xrightarrow{f_0 \otimes_A R} & M \otimes_A R & \longrightarrow & 0 \\
\downarrow{\scriptstyle \binom{1}{0}} & & \| & & \downarrow{\scriptstyle p_{M \otimes_A R}} & & \\
(P_1 \oplus P') \otimes_A R & \xrightarrow{(f_1 \otimes_A R,\, \tilde{f}')} & P_0 \otimes_A R & \longrightarrow & M & \longrightarrow & 0
\end{array}$$

where the lower sequence is a minimal projective presentation.

Applying the Nakayama functor $- \otimes_R \mathrm{D}R$ yields another commutative diagram with exact rows

$$\begin{array}{ccccccccc}
0 \to & \tau_R(M \otimes_A R) & \xrightarrow{j} & P_1 \otimes_A \mathrm{D}R & \xrightarrow{f_1 \otimes_A \mathrm{D}R} & P_0 \otimes_A \mathrm{D}R & \xrightarrow{f_0 \otimes_A \mathrm{D}R} & M \otimes_A \mathrm{D}R & \to 0 \\
& \downarrow{\scriptstyle u} & & \downarrow{\scriptstyle \binom{1}{0}} & & \| & & \downarrow{\scriptstyle p_{M \otimes R} \otimes_A \mathrm{D}R} & \\
0 \longrightarrow & \tau_R M & \longrightarrow & (P_1 \oplus P') \otimes_A \mathrm{D}R & \longrightarrow & P_0 \otimes_A \mathrm{D}R & \longrightarrow & M \otimes_R \mathrm{D}R & \to 0
\end{array}$$

where j is the inclusion and u is induced by passing to kernels. Because $\binom{1}{0}j$ is injective, so is u.

This diagram induces the two commutative diagrams

$$\begin{array}{ccccccccc} 0 & \longrightarrow & \tau_R(M\otimes_A R) & \longrightarrow & P_1\otimes_A DR & \longrightarrow & X & \longrightarrow & 0 \\ & & \downarrow{\scriptstyle u} & & \downarrow & & \downarrow{\scriptstyle v} & & \\ 0 & \longrightarrow & \tau_R M & \longrightarrow & (P_1\oplus P')\otimes_A DR & \longrightarrow & Y & \longrightarrow & 0 \end{array}$$

$$\begin{array}{ccccccccc} 0 & \longrightarrow & X & \longrightarrow & P_0\otimes_A DR & \longrightarrow & M\otimes_A DR & \longrightarrow & 0 \\ & & \downarrow{\scriptstyle v} & & \| & & \downarrow{\scriptstyle p_{M\otimes R}\otimes_A DR} & & \\ 0 & \longrightarrow & Y & \longrightarrow & P_0\otimes_A DR & \longrightarrow & M\otimes_R DR & \longrightarrow & 0 \end{array}$$

where v is the induced morphism. The snake lemma applied to the second diagram yields v injective and $\operatorname{Coker} v \cong \operatorname{Ker}(p_{M\otimes_A R}\otimes_A DR)$, and to the first diagram an exact sequence

$$0 \longrightarrow \operatorname{Coker} u \longrightarrow P'\otimes_A DR \longrightarrow \operatorname{Coker} v \longrightarrow 0.$$

This latter sequence splices with the exact sequence

$$0 \longrightarrow \tau_R(M\otimes_A R) \xrightarrow{u} \tau_R M \longrightarrow \operatorname{Coker} u \longrightarrow 0$$

to give an exact sequence

$$0 \to \tau_R(M\otimes_A R) \to \tau_R M \to P'\otimes_A DR \to \operatorname{Ker}(p_{M\otimes_A R}\otimes_A DR) \to 0.$$

Finally, as A-modules, we have

$$\tau_R(M\otimes_A R) \cong \operatorname{Hom}_A(R,\tau_A M) \cong \tau_A M \oplus \operatorname{Hom}_A(E,\tau_A M). \qquad \square$$

THEOREM 3.5.2 ([**13**]2.1). *Let M be an indecomposable A-module.*

(a) *If M is nonprojective, then the following conditions are equivalent:*
   i) *the almost split sequences ending with M in* $\operatorname{mod} A$ *and* $\operatorname{mod} R$ *coincide*
   ii) $\tau_A M \cong \tau_R M$
   iii) $\operatorname{Hom}_A(E,\tau_A M)=0$ *and* $M\otimes_A E=0$.

(b) *If M is noninjective, then the following conditions are equivalent:*
   i) *the almost split sequences starting with M in* $\operatorname{mod} A$ *and* $\operatorname{mod} R$ *coincide*
   ii) $\tau_A^{-1}M \cong \tau_R^{-1}M$
   iii) $\operatorname{Hom}_A(E,M)=0$ *and* $\tau_A^{-1}M\otimes_A E=0$.

PROOF. (a) i) implies ii). This is trivial.

ii) implies i). Let $0\to\tau_R M\to X\to M\to 0$ be almost split in $\operatorname{mod} R$. Because $\tau_R M\cong\tau_A M$, the whole sequence lies in $\operatorname{mod} A$. It does not split in $\operatorname{mod} A$, because otherwise it would split in $\operatorname{mod} R$. Let $h\colon L\to M$ be a nonretraction in $\operatorname{mod} A$. Then h is a nonretraction in $\operatorname{mod} R$. Therefore, there exists $h'\colon L\to X$ such that $h=gh'$. Because L, X are A-modules, h' is a morphism in $\operatorname{mod} A$.

ii) implies iii). Because of 3.5.1, there exists a monomorphism $u\colon \tau_A M \oplus \mathrm{Hom}_A(E\,,\,\tau_A M) \longrightarrow \tau_R M$ in $\mathrm{mod}\,A$. Then $\tau_R M \cong \tau_A M$ forces $\mathrm{Hom}_A(E\,,\,\tau_A M) = 0$. Moreover, u is an isomorphism $\tau_R(M \otimes_A R) \cong \tau_R M$. But then $M \otimes_A R \cong M$ and so $M \otimes_A E = 0$.

iii) implies ii). Because $M \otimes_A E = 0$, we have $M \otimes_A R \cong M$, therefore

$$\Omega^1_R(M \otimes_A R) \cong \Omega^1_R M \cong \Omega^1_A M \oplus (P_0 \otimes_A E)$$

using the notation and the proof of 3.3.3. Let $\tilde{P}$ be a projective cover of $\Omega^1_R(M \otimes_A R)$ in $\mathrm{mod}\,A$. Then $\tilde{P} \otimes_A R$ is a projective cover of $\Omega^1_R(M \otimes_A R)$ in $\mathrm{mod}\,R$, because of 3.3.3. Now $\Omega^1_R(M \otimes_A R) \cong \Omega^1_A M$ yields $\tilde{P} \otimes_A R \cong P_1 \otimes_A R$ hence $\tilde{P} = P_1$. Therefore the tops of $\Omega^1_R(M \otimes_A R)$ and $\Omega^1_A M$ are equal in $\mathrm{mod}\,A$, and hence $P_0 \otimes_A E = 0$. But then its projective cover P is zero and so $P' = 0$. Then we have $\tau_R M \cong \tau_A M \oplus \mathrm{Hom}_A(E\,,\,\tau_A M)$. Thus, $\mathrm{Hom}_A(E\,,\,\tau_A M) = 0$ implies $\tau_R M \cong \tau_A M$, as desired. □

The following corollary, due to Hoshino [**26**], played an important rôle in the classification of the representation-finite selfinjective algebras.

COROLLARY 3.5.3 ([**13**]2.3). *Assume $E = {}_A\mathrm{D}A_A$ and let M be an indecomposable A-module.*

(a) *If M is nonprojective, then $\tau_A M \cong \tau_R M$ if and only if* $\mathrm{pd}\, M_A \le 1$, $\mathrm{id}\, \tau_A M \le 1$.

(b) *If M is noninjective, then $\tau_A^{-1} M \cong \tau_R^{-1} M$ if and only if* $\mathrm{pd}\, \tau_A^{-1} M \le 1$, $\mathrm{id}\, M_A \le 1$.

PROOF. (a) We have $\mathrm{pd}\, M_A \le 1$ if and only if $\mathrm{Hom}_A(\mathrm{D}A\,,\,\tau_A M) = 0$, see [**12**](IV.2.7), and $\mathrm{id}\,\tau_A M \le 1$ if and only if $M \otimes_A \mathrm{D}A \cong \mathrm{D}\mathrm{Hom}_A(M\,,\,A) = 0$. □.

Assume now that A is a tilted algebra and $E = \mathrm{Ext}^2_A(\mathrm{D}A\,,\,A)$ so that $R = A \ltimes E$ is cluster tilted. It is shown in [**4**] that any complete slice in $\mathrm{mod}\,A$ embeds in $\mathrm{mod}\,R$ as what is called a local slice, a result extended in [**6**] to algebras B such that there exist surjective morphisms of algebras $R \longrightarrow B \longrightarrow A$. The decisive step was the proof that, if M is an indecomposable lying on a complete slice in $\mathrm{mod}\,A$, then, if M is nonprojective in $\mathrm{mod}\,A$, we have $\tau_A M \cong \tau_R M$ and if it is noninjective, then $\tau_A^{-1} M \cong \tau_R^{-1} M$, see [**6**](3.2.1).

In [**34**](5.9), Treffinger obtained necessary and sufficient conditions for a τ-slice in $\mathrm{mod}\,A$ to embed as a τ-slice in $\mathrm{mod}\,R$.

EXAMPLE 3.5.4. Let A be given by the quiver

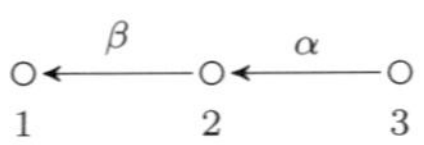

bound by $\alpha\beta = 0$, and R be given by the quiver

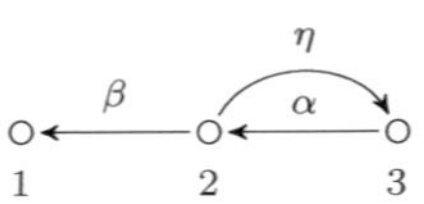

bound by $\alpha\beta = 0$, $\eta\alpha = 0$. Here we find $E_A = (\,3\,)^2$, while $(\mathrm{D}E)_A = {}^{3}_{2}$. Consider first the simple module $S_2 = 2$. We have $\tau_A S_2 = 1$ hence $\mathrm{Hom}_A(E\,,\,\tau_A S_2) = 0$. On the other hand, $S_2 \otimes_A E \cong \mathrm{DHom}_A(S_2\,,\,\mathrm{D}E) = \mathrm{DHom}_A(2\,,\,{}^{3}_{2}) \neq 0$. Therefore the almost split sequences ending with S_2 in mod A and mod R do not coincide. In fact, a quick calculation shows that the first is $0 \longrightarrow 1 \longrightarrow {}^{2}_{1} \longrightarrow 2 \longrightarrow 0$

while the second is $0 \longrightarrow {}_{1}{}^{2}{}_{3} \longrightarrow {}^{2}_{3} \oplus {}^{2}_{1} \longrightarrow 2 \longrightarrow 0$.

On the other hand, looking at $S_3 = 3$ we have $\tau_A S_3 = 2$, so that $\mathrm{Hom}_A(E\,,\,\tau_A S_3) = 0$. Also $S_3 \otimes_A E \cong \mathrm{DHom}_A(S_3\,,\,\mathrm{D}E) = \mathrm{DHom}_A(3\,,\,{}^{3}_{2}) = 0$. Therefore, the almost split sequences ending in S_3 in mod A and mod R coincide.

4. Tilting modules

4.1. Extendable tilting modules. For tilting theory, we refer the reader to [**12**] Chapter IV. Let, as usual, R be a split extension of A by a nilpotent bimodule E.

THEOREM 4.1.1 ([**11**]2.3). *Let T be an A-module, then*

(a) *$T \otimes_A R$ is a partial tilting (or tilting) R-module if and only if T is a partial tilting (or tilting, respectively) A-module,* $\mathrm{Hom}_A(T \otimes_A E\,,\,\tau_A T) = 0$ *and* $\mathrm{Hom}_A(\mathrm{D}E\,,\,\tau_A T) = 0$;

(b) $\mathrm{Hom}_A(R\,,\,T)$ *is a partial cotilting (or cotilting) R-module if and only if T is a partial cotilting (or cotilting, respectively) A-module,* $\mathrm{Hom}_A\big(\tau_A^{-1}T\,,\,\mathrm{Hom}_A(E\,,\,T)\big) = 0$ *and* $\mathrm{Hom}_A\big(\tau_A^{-1}T\,,\,E\big) = 0$.

PROOF. (a) Because of 3.2.2, the number of isoclasses of indecomposable summands of T equals that of $T \otimes_A R$. Also, because of 2.1.2, the ranks of the Grothendieck groups of A and R are equal. Therefore, it suffices to prove the statement about partial tilting modules.

We have isomorphisms of vector spaces

$$\begin{aligned}\mathrm{Hom}_R(T \otimes_A R\,,\,\tau_R(T \otimes_A R)) &\cong \mathrm{Hom}_R(T \otimes_A R\,,\,\mathrm{Hom}_A(R\,,\,\tau_A T))\\ &\cong \mathrm{Hom}_A(T \otimes_A R \otimes_R R\,,\,\tau_A T)\\ &\cong \mathrm{Hom}_A(T \otimes_A R\,,\,\tau_A T)\\ &\cong \mathrm{Hom}_A(T\,,\,\tau_A T) \oplus \mathrm{Hom}_A(T \otimes_A E\,,\,\tau_A T).\end{aligned}$$

If T is a partial tilting module then $\mathrm{pd}\, T_A \leq 1$ implies $\mathrm{Hom}_A(T\,,\,\tau_A T) \cong \mathrm{DExt}^1_A(T\,,\,T) = 0$. Further, $\mathrm{Hom}_A(\mathrm{D}E\,,\,\tau_A T) = 0$ implies $\mathrm{pd}(T \otimes_A R) \leq 1$ because of 3.4.2. Therefore $\mathrm{Hom}_A(T \otimes_A E\,,\,\tau_A T) = 0$ implies

$$\mathrm{Ext}^1_R(T \otimes_A R\,,\,T \otimes_A R) \cong \mathrm{DHom}_R(T \otimes_A R\,,\,\tau_R(T \otimes_A R)) = 0$$

and so $T \otimes_A R$ is a partial tilting R-module.

Conversely, if $T \otimes_A R$ is a partial tilting R-module, 3.4.2 gives $\mathrm{pd}\, T_A \leq 1$ and $\mathrm{Hom}_A(\mathrm{D}E\,,\,\tau_A T) = 0$. Moreover $\mathrm{Hom}_R(T \otimes_A R\,,\,\tau_R(T \otimes_A R)) = 0$ yields $\mathrm{Hom}_A(T \otimes_A E\,,\,\tau_A T) = 0$ and $\mathrm{Ext}^1_A(T\,,\,T) \cong \mathrm{DHom}_A(T\,,\,\tau_A T) = 0$, so T_A is a partial tilting module. □

DEFINITION 4.1.2. (a) A partial tilting (or tilting) A-module is called ***extendable*** if $T \otimes_A R$ is a partial tilting (or tilting, respectively) R-module.

(b) A partial cotilting (or cotilting) A-module is called ***coextendable*** if $\mathrm{Hom}_A(R\,,\,T)$ is a partial cotilting (or cotilting, respectively) R-module.

One reason for looking at this class of (co)tilting modules is that they preserve the splitting character of the algebra.

PROPOSITION 4.1.3 ([**11**]2.5). (a) *If T is an extendable tilting A-module, then $S = \mathrm{End}(T \otimes_A R)$ is the split extension of $B = \mathrm{End}\, T_A$ by the nilpotent bimodule ${}_BW_B = \mathrm{Hom}_A({}_BT_A\,,\, {}_BT \otimes_A E)$.*

(b) *If T is a coextendable cotilting A-module, then $S = \mathrm{End}\,\mathrm{Hom}_A(R\,,\,T)$ is the split extension of $B = \mathrm{End}\, T_A$ by the nilpotent bimodule ${}_BW_B = \mathrm{Hom}_A(\mathrm{Hom}_A(E\,,\,T)\,,\,T)$.*

PROOF. (a) We have vector space isomorphisms

$$\begin{aligned} S = \mathrm{Hom}_R(T \otimes_A R\,,\, T \otimes_A R) &\cong \mathrm{Hom}_A(T\,,\, \mathrm{Hom}_R({}_AR_A\,,\, T \otimes_A R)) \\ &\cong \mathrm{Hom}_A(T\,,\, T \otimes_A R_A) \cong \mathrm{Hom}_A(T\,,\,T) \oplus \mathrm{Hom}_A(T\,,\, T \otimes_A E). \end{aligned}$$

We thus have an exact sequence $0 \longrightarrow W \longrightarrow S \overset{\varphi}{\longrightarrow} B \longrightarrow 0$ where φ is an algebra morphism, and the ideal structure of W is induced from its B-B-bimodule structure. There remains to prove that W is nilpotent. The multiplication in W is that of S and, for any $w \in W$, its image is contained in $T \otimes_A E$. Because E is nilpotent, there exists $s \geq 0$ such that, for any sequence $w_1, \ldots, w_s$ of elements of W, the image of $w_1 \cdots w_s$ lies in $T \otimes_A E^s = 0$. Therefore $W^s = 0$. $\square$

The proof shows that the nilpotency index of W in S does not exceed that of E in R. Thus, if R is a trivial extension of A by E, then S is a trivial extension of B by W.

EXAMPLE 4.1.4. Let A be the path algebra of the quiver

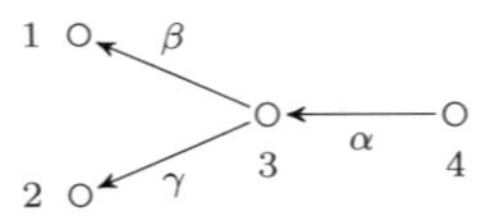

and R be given by the quiver

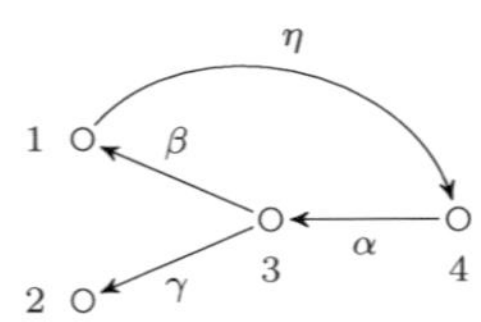

bound by $\beta\eta = 0$, $\eta\alpha\beta = 0$, $\eta\alpha\gamma = 0$. It is easily seen that the A-module

$$T = e_1A \oplus e_4A \oplus D(Ae_1) \oplus D(Ae_4) = 1 \oplus \begin{smallmatrix}4\\3\\1\ 2\end{smallmatrix} \oplus \begin{smallmatrix}4\\3\\1\end{smallmatrix} \oplus 4$$

is tilting. We claim it is extendable. We first observe that $E_A = \begin{smallmatrix}4\\3\end{smallmatrix}$, $(DE)_A = (1)^2$. In particular DE is generated by T so that $\mathrm{Hom}_A(DE\,,\, \tau_AT) \cong D\mathrm{Ext}^1_A(T\,,\, DE) = 0$. We now compute $T \otimes_A R$. We have $e_1A \otimes_A R \cong e_1R = \begin{smallmatrix}1\\4\\3\end{smallmatrix}$ and $e_4A \otimes_A R \cong e_4R = \begin{smallmatrix}4\\3\\1\ 2\end{smallmatrix}$. Also the minimal projective presentations

$$0 \longrightarrow e_2A \longrightarrow e_4A \longrightarrow \begin{smallmatrix}4\\3\\1\end{smallmatrix} \longrightarrow 0$$

$$0 \longrightarrow e_3A \longrightarrow e_4A \longrightarrow 4 \longrightarrow 0$$

induce respectively the minimal projective presentations

$$e_2R \longrightarrow e_4R \longrightarrow \begin{smallmatrix}4\\3\\1\end{smallmatrix} \otimes_A R \longrightarrow 0$$

$$e_3R \longrightarrow e_4R \longrightarrow 4 \otimes_A R \longrightarrow 0.$$

Therefore $\begin{smallmatrix}4\\3\\1\end{smallmatrix} \otimes_A R \cong \begin{smallmatrix}4\\3\\1\end{smallmatrix}$ and $4 \otimes_A R \cong 4$ and $T \otimes_A R = \begin{smallmatrix}1\\4\\3\end{smallmatrix} \oplus \begin{smallmatrix}&4&\\&3&\\2&&1\end{smallmatrix} \oplus \begin{smallmatrix}4\\3\\1\end{smallmatrix} \oplus 4$ so that $T \otimes_A E = \begin{smallmatrix}4\\3\end{smallmatrix}$, which is generated by T. Therefore $\mathrm{Hom}_A(T \otimes_A E\,,\, \tau_A T) = 0$ and T is extendable.

The algebra $\operatorname{End} T$ is given by the quiver

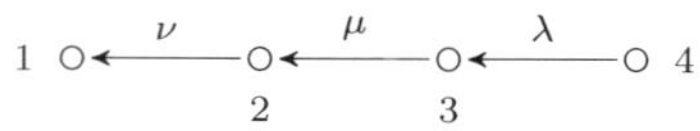

bound by $\lambda\mu\nu = 0$, while $\operatorname{End}(T \otimes_A R)$ is given by the quiver

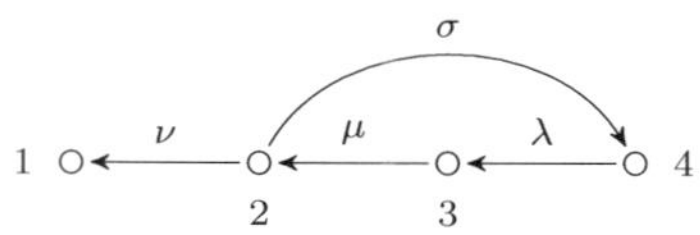

bound by $\lambda\mu\nu = 0$, $\lambda\mu\sigma = 0$, $\sigma\lambda = 0$. It is the split extension of $\operatorname{End} T$ by the bimodule generated by σ.

On the other hand, the tilting A-module $T' = \begin{smallmatrix}&4&\\&3&\\2&&1\end{smallmatrix} \oplus \begin{smallmatrix}4\\3\\2\end{smallmatrix} \oplus \begin{smallmatrix}4\\3\\1\end{smallmatrix} \oplus 4$ is not extendable, because $\mathrm{D}E$ is not generated by T' and then $\mathrm{Ext}^1_A(T'\,,\, \mathrm{D}E) \neq 0$.

Another example of extendable partial tilting module can be found in [**28**] where the authors study cluster tilted algebras from the point of view of induced and coinduced modules. They prove in [**28**](4.9) that, if A is a tilted algebra, $E = \mathrm{Ext}^2_A(\mathrm{D}A\,,\, A)$ and $R = A \ltimes E$, then $(\mathrm{D}E)_A$ is an extendable partial tilting module, and E_A is a coextendable partial cotilting module.

4.2. Induced torsion pairs. For a module M, the notations $\operatorname{Gen} M$ and $\operatorname{Cogen} M$ stand respectively for the class of modules generated and cogenerated by M.

Associated with a tilting A-module T is a torsion pair $(\mathcal{T}(T_A), \mathcal{F}(T_A))$ in $\operatorname{mod} A$ defined by

$$\begin{aligned}\mathcal{T}(T_A) &= \big\{M_A \bigm| \mathrm{Ext}^1_A(T\,,\, M) = 0\big\} = \operatorname{Gen} T\\ \mathcal{F}(T_A) &= \{M_A \mid \mathrm{Hom}_A(T\,,\, M) = 0\} = \operatorname{Cogen}(\tau_A T).\end{aligned}$$

Similarly, associated with a cotilting A-module T is a torsion pair $(\mathcal{T}'(T_A), \mathcal{F}'(T_A))$ given by

$$\begin{aligned}\mathcal{T}'(T_A) &= \{M_A \mid \mathrm{Hom}_A(T\,,\, M) = 0\} = \operatorname{Gen}(\tau_A^{-1} T)\\ \mathcal{F}'(T_A) &= \big\{M_A \bigm| \mathrm{Ext}^1_A(T\,,\, M) = 0\big\} = \operatorname{Cogen} T.\end{aligned}$$

PROPOSITION 4.2.1. (a) *If T is an extendable tilting A-module then*
$X_R \in \mathcal{T}(T \otimes_A R)$ if and only if $X_A \in \mathcal{T}(T)$
$X_R \in \mathcal{F}(T \otimes_A R)$ if and only if $X_A \in \mathcal{F}(T)$.

(b) *If T is a coextendable cotilting A-module then*
$X_R \in \mathcal{T}'(\mathrm{Hom}_A(R\,,\, T))$ if and only if $X_A \in \mathcal{T}'(T)$
$X_R \in \mathcal{F}'(\mathrm{Hom}_A(R\,,\, T))$ if and only if $X_A \in \mathcal{F}'(T)$.

PROOF. (a) The statement follows from the vector space isomorphisms

$$\begin{aligned}\operatorname{Ext}^1_R(T\otimes_A R, X) &\cong D\operatorname{Hom}_R(X\,,\,\tau_R(T\otimes_A R)) \cong D\operatorname{Hom}_R(X\,,\,\operatorname{Hom}_A(R\,,\,\tau_A T))\\ &\cong D\operatorname{Hom}_A(X\otimes_R R_A\,,\,\tau_A T) \cong D\operatorname{Hom}_A(X_A\,,\,\tau_A T) \cong \operatorname{Ext}^1_A(T, X)\end{aligned}$$

and $\operatorname{Hom}_R(T\otimes_A R,\, X) \cong \operatorname{Hom}_A(T\,,\,\operatorname{Hom}_R({}_AR_R\,,\,X)) \cong \operatorname{Hom}_A(T\,,\,X_A)$. □

COROLLARY 4.2.2. (a) *If T is an extendable tilting module, then*

i) $\mathcal{T}(T)\otimes_A R \subseteq \mathcal{T}(T\otimes_A R)$ *always, and* $\mathcal{T}(T)\otimes_A R \supseteq \mathcal{T}(T\otimes_A R)$ *if and only if* $\operatorname{Im}(-\otimes_A R) \supseteq \mathcal{T}(T\otimes_A R)$.

ii) $\mathcal{F}(T)\otimes_A R \subseteq \mathcal{F}(T\otimes_A R)$ *if and only if* $\mathcal{F}(T)\otimes_A E \subseteq \mathcal{F}(T)$, $\mathcal{F}(T)\otimes_A R \supseteq \mathcal{F}(T\otimes_A R)$ *if and only if* $\operatorname{Im}(-\otimes_A R) \supseteq \mathcal{F}(T\otimes_A R)$.

(b) *If T is a coextendable cotilting module, then*

i) $\operatorname{Hom}_A(R, \mathcal{F}'(T)) \subseteq \mathcal{F}'(\operatorname{Hom}_A(R, T))$ *always, and* $\operatorname{Hom}_A(R, \mathcal{F}'(T)) \supseteq \mathcal{F}'(\operatorname{Hom}_A(R, T))$ *if and only if* $\operatorname{Im}\operatorname{Hom}_A(R, -) \supseteq \mathcal{F}'\big(\operatorname{Hom}_A(R, T)\big)$.

ii) $\operatorname{Hom}_A(R, \mathcal{T}'(T)) \subseteq \mathcal{T}'(\operatorname{Hom}_A(R, T))$ *if and only if* $\operatorname{Hom}_A(E, \mathcal{T}'(T)) \subseteq \mathcal{T}'(T)$, $\operatorname{Hom}_A(R, \mathcal{T}'(T)) \supseteq \mathcal{T}'(\operatorname{Hom}_A(R, T)))$ *if and only if* $\operatorname{Im}\operatorname{Hom}_A(R, -) \supseteq \mathcal{T}'\big(\operatorname{Hom}_A(R, T)\big)$.

PROOF. (a) i) Let $M \in \mathcal{T}(T)$. In order to show that $M\otimes_A R \in \mathcal{T}(T\otimes_A R)$, we need, because of 4.2.1, to prove that $M\otimes_A R_A = M\oplus(M\otimes_A E)$ lies in $\mathcal{T}(T)$. We know that $M \in \mathcal{T}(T)$. But M generated by T implies $M\otimes_A E$ generated by $T\otimes_A E$ and the latter is generated by T, because of 4.1.1. This establishes the first statement.

The necessity part of the second statement is clear, so we prove sufficiency. Let $X \in \mathcal{T}(T\otimes_A R)$. The hypothesis says that there exists M_A such that $X \cong M\otimes_A R$. It suffices to prove that $M \in \mathcal{T}(T)$. But this follows from the facts that $X_A \in \mathcal{T}(T)$ and $X_A \cong M\oplus(M\otimes_A E)$.

ii) Let $N \in \mathcal{F}(T)$. We have $N\otimes_A R \in \mathcal{F}(T\otimes_A R)$ if and only if $N\otimes_A R_A = N\oplus(N\otimes_A E) \in \mathcal{F}(T)$ if and only if $N\otimes_A E \in \mathcal{F}(T)$. This implies the first statement. The second one is proved as the corresponding one for $\mathcal{T}(T)$. □

COROLLARY 4.2.3. (a) *Let T be an extendable tilting module, then*

i) *If* $(\mathcal{T}(T\otimes_A R), \mathcal{F}(T\otimes_A R))$ *splits in* $\operatorname{mod} R$, *then* $(\mathcal{T}(T), \mathcal{F}(T))$ *splits in* $\operatorname{mod} A$.

ii) *If* $(\mathcal{T}(T), \mathcal{F}(T))$ *splits in* $\operatorname{mod} A$ *and* $\mathcal{F}(T\otimes_A R) \subseteq \operatorname{Im}(-\otimes_A R)$, *then* $(\mathcal{T}(T\otimes_A R), \mathcal{F}(T\otimes_A R))$ *splits in* $\operatorname{mod} R$.

(b) *Let T be a coextendable cotilting module, then*

i) *If* $(\mathcal{T}'(\operatorname{Hom}_A(R, T)), \mathcal{F}'(\operatorname{Hom}_A(R, T)))$ *splits in* $\operatorname{mod} R$, *then* $(\mathcal{T}'(T), \mathcal{F}'(T))$ *splits in* $\operatorname{mod} A$.

ii) *If* $(\mathcal{T}'(T), \mathcal{F}'(T))$ *splits in* $\operatorname{mod} A$ *and* $\mathcal{T}'(\operatorname{Hom}_A(R, T)) \subseteq \operatorname{Im}\operatorname{Hom}_A(R, -)$, *then* $(\mathcal{T}'(\operatorname{Hom}_A(R, T)), \mathcal{F}'(\operatorname{Hom}_A(R, T)))$ *splits in* $\operatorname{mod} R$.

PROOF. (a) i) Let $M \in \mathcal{T}(T)$, $N \in \mathcal{F}(T)$. We claim that $\operatorname{Ext}^1_A(N\,, M) = 0$. Because of 4.2.1, we have $M_R \in \mathcal{T}(T\otimes_A R)$, $N_R \in \mathcal{F}(T\otimes_A R)$. But then $\operatorname{Ext}^1_R(N\,, M) = 0$. This implies $\operatorname{Ext}^1_A(N\,, M) = 0$.

ii) Let $X \in \mathscr{T}(T\otimes_A R)$, $Y \in \mathscr{F}(T\otimes_A R)$. We claim that $\mathrm{Ext}^1_R(X, Y) = 0$. Because of the hypothesis and 4.2.2, there exists $N \in \mathscr{F}(T)$ such that $Y \cong N \otimes_A R$. Also, $X_R \in \mathscr{T}(T \otimes_A R)$ implies that $X_A \in \mathscr{T}(T)$, that is, $X \in \mathrm{Gen}(T)$. This implies that $X \otimes_A E \in \mathrm{Gen}(T \otimes_A E)$. Because $T \otimes_A E \in \mathrm{Gen}\, T$, see 4.1.1, we get $X \otimes_A E \in \mathscr{T}(T)$. Therefore, $X \otimes_A R_A \cong X \oplus (X \otimes_A E) \in \mathscr{T}(T)$. Hence

$$\begin{aligned}\mathrm{Ext}^1_R(Y,X) &\cong \mathrm{D}\overline{\mathrm{Hom}}_R(X,\tau_R Y) \subseteq \mathrm{DHom}_R(X, \tau_R Y) \cong \mathrm{DHom}_R(X, \tau_R(N\otimes_A R)) \\ &\cong \mathrm{DHom}_R(X, \mathrm{Hom}_A(R, \tau_A N)) \cong \mathrm{DHom}_A(X\otimes_R R_A, \tau_A N) \\ &\cong \mathrm{DHom}_A(X, \tau_A N) \oplus \mathrm{DHom}_A(X\otimes_A E, \tau_A N) = 0\end{aligned}$$

because $N \in \mathscr{F}(T)$ and $(\mathscr{T}(T), \mathscr{F}(T))$ split imply $\tau_A N \in \mathscr{F}(T)$. □

EXAMPLE 4.2.4. Let $e \in A$ be an idempotent such that eA is simple projective noninjective, and E is a nilpotent bimodule such that $eE = Ee = 0$. Then the APR-tilting module $T = \tau_A^{-1}(eA) \oplus (1-e)A$ is extendable.

Indeed, we must show that $\mathrm{D}E$ and $T \otimes_A E$ are generated by T. Now $\mathrm{Ext}^1_A(T, \mathrm{D}E) \cong \mathrm{DHom}_A(\mathrm{D}E, \tau_A T) = \mathrm{DHom}_A(\mathrm{D}E, eA)$ is nonzero if and only if eA is a direct summand of $\mathrm{D}E$. But $\mathrm{Hom}_A(eA, \mathrm{D}E) \cong (\mathrm{D}E)e \cong \mathrm{D}(eE) = 0$. Hence $\mathrm{Ext}^1_A(T, \mathrm{D}E) = 0$ and so $\mathrm{D}E \in \mathrm{Gen}\, T$.

Moreover, there exists an idempotent $e' \in A$ such that we have an almost split sequence

$$0 \longrightarrow eA \longrightarrow e'A \longrightarrow \tau_A^{-1}(eA) \longrightarrow 0. \qquad (*)$$

Applying $-\otimes_A E$ yields $e'E \cong \tau_A^{-1}(eA) \otimes_A E$. Now $e'E \in \mathrm{Gen}\, T$ because $e'Ee = 0$, hence so is $\tau_A^{-1}(eA) \otimes_A E$. Therefore $T \otimes_A E \in \mathrm{Gen}\, T$.

Furthermore, $T \otimes_A R$ is also an APR-tilting module.

Indeed, we first prove that eR is simple projective noninjective in $\mathrm{mod}\, R$. If this is not the case, there exists $\alpha \in (Q_R)_1$ starting at the point corresponding to e. There is no such arrow in Q_A, hence α belongs to E and $\alpha = e\alpha = 0$ gives a contradiction. Next, applying $-\otimes_A R$ to $(*)$ yields an exact sequence

$$0 \longrightarrow \mathrm{Tor}_1^A(\tau_A^{-1}(eA), R) \longrightarrow eR \xrightarrow{\ j\ } e'R \longrightarrow \tau_A^{-1}(eA) \otimes_A R \longrightarrow 0.$$

Because $\mathrm{Tor}_1^A(\tau_A^{-1}(eA), R) \cong \mathrm{DExt}^1_A(\tau_A^{-1}(eA), \mathrm{D}R) = 0$ (for, $\mathrm{D}E \in \mathrm{Gen}\, T$), we deduce that j is a monomorphism. Hence $\tau_A^{-1}(eA) \otimes_A R \cong \tau_R^{-1}(eR)$ and so $T \otimes_A R$ is indeed an APR-tilting module.

Finally, $\mathscr{F}(T \otimes_A R) = \mathrm{add}(eR) = \mathrm{add}(eA \otimes_A R) \subseteq \mathrm{Im}(-\otimes_A R)$ so that the conditions of 4.2.3 are satisfied in this case.

4.3. Restrictions of tilting modules. We consider the reverse problem: given a tilting R-module U, under which conditions is the restricted module $U \otimes_R A$ a tilting A-module (and similarly for cotilting modules)?

LEMMA 4.3.1. (a) *Let U_R be such that $\mathrm{Tor}_1^R(U, A) = 0$ and $0 \to \tilde{P}_1 \xrightarrow{f_1} \tilde{P}_0 \xrightarrow{f_0} U \to 0$ a minimal projective resolution for U, then*

$$0 \longrightarrow \tilde{P}_1 \otimes_R A \xrightarrow{f_1 \otimes_R A} \tilde{P}_0 \otimes_R A \xrightarrow{f_0 \otimes_R A} U \otimes_R A \longrightarrow 0$$

is a minimal projective resolution for $U \otimes_R A$. In particular, $\mathrm{pd}(U \otimes_R A) \leq 1$.

(b) *Let* U_R *be such that* $\operatorname{Ext}^1_R(A, U) = 0$ *and* $0 \to U \xrightarrow{g^0} \tilde{I}^0 \xrightarrow{g^1} \tilde{I}^1 \to 0$ *a minimal injective coresolution for* U, *then*

$$0 \to \operatorname{Hom}_R(A, U) \xrightarrow{\operatorname{Hom}_R(A, g^0)} \operatorname{Hom}_R\left(A, \tilde{I}^0\right) \xrightarrow{\operatorname{Hom}_R(A, g^1)} \operatorname{Hom}_R\left(A, \tilde{I}^1\right) \to 0$$

is a minimal injective coresolution for $\operatorname{Hom}_R(A, U)$. *In particular,* $\operatorname{id}\operatorname{Hom}_R(A, U) \le 1$.

PROOF. (a) Applying $- \otimes_R A$ to the given minimal projective resolution of U_R and using that $\operatorname{Tor}_1^R(U, A) = 0$ yields an exact sequence

$$0 \longrightarrow \tilde{P}_1 \otimes_R A \xrightarrow{f_1 \otimes_R A} \tilde{P}_0 \otimes_R A \xrightarrow{f_0 \otimes_R A} U \otimes_R A \longrightarrow 0.$$

Because $\tilde{P}_0 \otimes_R A$, $\tilde{P}_1 \otimes_R A$ are projective A-modules, this is a projective resolution. In particular, $\operatorname{pd}(U \otimes_R A) \le 1$. Minimality follows from the fact that, because of 3.2.4, $\tilde{P}_0 \otimes_R A$ is a projective cover of $U \otimes_R A$. □

LEMMA 4.3.2. (a) *Let* U_R *be such that* $\operatorname{pd} U \le 1$ *and* $\operatorname{Tor}_1^R(U, A) = 0$, *then*

$$\tau_A(U \otimes_R A) \cong \operatorname{Hom}_R(A, \tau_R U).$$

(b) *Let* U_R *be such that* $\operatorname{id} U \le 1$ *and* $\operatorname{Ext}^1_R(A, U) = 0$, *then*

$$\tau_A^{-1} \operatorname{Hom}_R(A, U) \cong (\tau_R^{-1} U) \otimes_R A.$$

PROOF. (a) Because of 4.3.1, a minimal projective resolution $0 \to \tilde{P}_1 \to \tilde{P}_0 \to U \to 0$ induces a minimal projective resolution in $\operatorname{mod} R$

$$0 \longrightarrow \tilde{P}_1 \otimes_R A \longrightarrow \tilde{P}_0 \otimes_R A \longrightarrow U \otimes_R A \longrightarrow 0.$$

Applying $\operatorname{Hom}_A(-, A)$ yields a commutative diagram with exact rows

$$\begin{array}{ccccc}
\operatorname{Hom}_A\left(\tilde{P}_0 \otimes_R A, A\right) & \longrightarrow & \operatorname{Hom}_A\left(\tilde{P}_1 \otimes_R A, A\right) & \longrightarrow & \operatorname{Tr}(U \otimes_R A) \longrightarrow 0 \\
\| \wr & & \| \wr & & \\
\operatorname{Hom}_A\left(\tilde{P}_0, \operatorname{Hom}_A({}_R A, A)\right) & \longrightarrow & \operatorname{Hom}_A\left(\tilde{P}_1, \operatorname{Hom}_A({}_R A, A)\right) & & \\
\| \wr & & \| \wr & & \\
\operatorname{Hom}_R\left(\tilde{P}_0, A\right) & \longrightarrow & \operatorname{Hom}_R\left(\tilde{P}_1, A_R\right) & \longrightarrow & \operatorname{Ext}^1_R(U, A_R) \longrightarrow 0
\end{array}$$

where the lower row is obtained by applying $\operatorname{Hom}_R(-, A_R)$ to the original minimal projective resolution of U_R. Thus $\operatorname{Tr}(U \otimes_R A) \cong \operatorname{Ext}^1_R(U, A)$ and therefore $\tau_A(U \otimes_R A) \cong D\operatorname{Ext}^1_R(U, A) \cong \operatorname{Hom}_R(A, \tau_R U)$ because $\operatorname{pd} U_R \le 1$. □

THEOREM 4.3.3 ([**14**]3.3). (a) *Let* U_R *be a partial tilting (or tilting) R-module such that* $\operatorname{Tor}_1^R(U, A) = 0$, *then* $U \otimes_R A$ *is a partial tilting (or tilting, respectively) A-module.*

(b) *Let* U_R *be a partial cotilting (or cotilting) R-module such that* $\operatorname{Ext}^1_R(A, U) = 0$, *then* $\operatorname{Hom}_R(A, U)$ *is a partial cotilting (or cotilting, respectively) A-module.*

PROOF. (a) Assume first that U_R is partial tilting and such that $\mathrm{Tor}_1^R(U, A) = 0$. Because of 4.3.1, we have $\mathrm{pd}(U \otimes_R A) \leq 1$. Also we have vector space isomorphisms

$$\begin{aligned} \mathrm{DExt}_A^1(U \otimes_R A, U \otimes_R A) &\cong \mathrm{Hom}_A(U \otimes_R A, \tau_A(U \otimes_R A)) \\ &\cong \mathrm{Hom}_R(U, \mathrm{Hom}_A({}_RA, \tau_A(U \otimes_R A)) \\ &\cong \mathrm{Hom}_R(U, \mathrm{Hom}_A({}_RA, \mathrm{Hom}_R(A, \tau_R U))) \\ &\cong \mathrm{Hom}_R(U, \mathrm{Hom}_R({}_RA \otimes_A A_R, \tau_R U)) \\ &\cong \mathrm{Hom}_R(U, \mathrm{Hom}_R({}_RA_R, \tau_R U)). \end{aligned}$$

Applying $\mathrm{Hom}_R(-, \tau_R U)$ to the exact sequence $0 \longrightarrow {}_RE_R \longrightarrow {}_RR_R \longrightarrow {}_RA_R \longrightarrow 0$ yields a monomorphism

$$0 \longrightarrow \mathrm{Hom}_R({}_RA_R, \tau_R U) \longrightarrow \mathrm{Hom}_R({}_RR_R, \tau_R U) \cong \tau_R U.$$

Applying next $\mathrm{Hom}_R(U, -)$ yields another monomorphism

$$0 \longrightarrow \mathrm{Hom}_R(U, \mathrm{Hom}_R({}_RA_R, \tau_R U)) \longrightarrow \mathrm{Hom}_R(U, \tau_R U) \cong \mathrm{DExt}_R^1(U, U) = 0.$$

Thus $\mathrm{Ext}_A^1(U \otimes_R A, U \otimes_R A) = 0$ and $U \otimes_R A$ is a partial tilting A-module.

If U_R is tilting, then there exists an exact sequence $0 \longrightarrow R_R \longrightarrow U_0 \longrightarrow U_1 \longrightarrow 0$ with $U_0, U_1 \in \mathrm{add}\, U$. Because $\mathrm{Tor}_1^R(U, A) = 0$, applying $- \otimes_R A$ yields an exact sequence

$$0 \longrightarrow A_A \longrightarrow U_0 \otimes_R A \longrightarrow U_1 \otimes_R A \longrightarrow 0.$$

Because $U_0 \otimes_R A$, $U_1 \otimes_R A \in \mathrm{add}(U \otimes_R A)$, this finishes the proof. □

DEFINITION 4.3.4. (a) A partial tilting, or tilting, R-module U is called ***restrictable*** provided $\mathrm{Tor}_1^R(U, A) = 0$ and then $U \otimes_R A$ is called its ***restriction***.

(b) A partial cotilting, or cotilting, R-module U is called ***corestrictable*** provided $\mathrm{Ext}_R^1(A, U) = 0$ and then $\mathrm{Hom}_R(A, U)$ is called its ***corestriction***.

LEMMA 4.3.5. (a) *Let T be an extendable partial tilting (or tilting) A-module, then $T \otimes_A R$ is a restrictable partial tilting (or tilting, respectively) R-module, with restriction T.*

(b) *Let T be a coextendable partial cotilting (or cotilting) A-module, then $\mathrm{Hom}_A(R, T)$ is a corestrictable partial cotilting (or cotilting, respectively) R-module, with corestriction T.*

PROOF. (a) Assume that T is an extendable partial tilting, or tilting, A-module. We claim that $T \otimes_A R$ is restrictable, that is $\mathrm{Tor}_1^R(T \otimes_A R, A) = 0$. Let $0 \longrightarrow P_1 \longrightarrow P_0 \longrightarrow T \longrightarrow 0$ be a minimal projective resolution in $\mathrm{mod}\, A$. Because T is extendable, $\mathrm{pd}(T \otimes_A R) \leq 1$, hence 3.3.1 gives a minimal projective resolution

$$0 \longrightarrow P_1 \otimes_A R \longrightarrow P_0 \otimes_A R \longrightarrow T \otimes_A R \longrightarrow 0.$$

Applying $-\otimes_R$ yields a commutative diagram with exact rows

$$\begin{array}{ccccccccc}
0 \to \mathrm{Tor}_1^R(T\otimes_A R, A) & \to & P_1\otimes_A R\otimes_R A & \to & P_0\otimes_A R\otimes_R A & \to & T\otimes_A R\otimes_R A & \to & 0 \\
 & & \| \wr & & \| \wr & & \| \wr & & \\
0 & \longrightarrow & P_1 & \longrightarrow & P_0 & \longrightarrow & T & \longrightarrow & 0.
\end{array}$$

Thus $\mathrm{Tor}_1^R(T\otimes_A R\,,\,A)=0$ and so $T\otimes_A R$ is restrictable. On the other hand, $T\otimes_A R\otimes_R A\cong T$: the restriction of $T\otimes_A R$ is T. $\square$

THEOREM 4.3.6 ([**14**]3.4). (a) *The functors* $-\otimes_A R$ *and* $-\otimes_R A$ *induce mutually inverse bijections*

$$\{\textit{extendable tilting } A\textit{-modules}\} \underset{-\otimes_R A}{\overset{-\otimes_A R}{\rightleftarrows}} \{\textit{restrictable induced tilting } R\textit{-modules}\}.$$

(b) *The functors* $\mathrm{Hom}_A(R\,,-)$ *and* $\mathrm{Hom}_R(A\,,-)$ *induce mutually inverse bijections*

$$\{\textit{coextendable cotilting } A\textit{-modules}\} \underset{\mathrm{Hom}_R(A\,,-)}{\overset{\mathrm{Hom}_A(R\,,-)}{\rightleftarrows}} \{\textit{corestrictable coinduced } R\textit{-modules}\}.$$

PROOF. (a) If T is an extendable tilting A-module, then $T\otimes_A R$ is induced by definition and it is restrictable because of 4.3.5.

Conversely, if U_R is an induced restrictable tilting module, then $\mathrm{Tor}_1^R(U\,,\,A)=0$. Because of 4.3.5, $U\otimes_R A$ is a tilting A-module. On the other hand, there exists M_A such that $U\cong M\otimes_A R$. But then $U\otimes_R A\cong M\otimes_A R\otimes_R A\cong M\otimes_A A\cong M$ so that $(U\otimes_R A)\otimes_A R\cong M\otimes_A R\cong U$. Thus $U\otimes_R A$ is extendable. $\square$.

The modern guise of tilting theory is τ-tilting theory. In [**33**], Suarez obtained a similar result for (support) τ-tilting modules.

EXAMPLE 4.3.7. There exist restrictable tilting R-modules which are not induced. Let A be the path algebra of the quiver

$$1 \circ \xleftarrow{\ \alpha\ } \circ\ 2$$

and R the path algebra of the Kronecker quiver

$$1 \circ \underset{\beta}{\overset{\alpha}{\leftleftarrows}} \circ\ 2\ .$$

We claim that the APR-tilting module $U_R=\tau_R^{-1}(e_1R)\oplus e_2R$ is restrictable but not induced.

To prove that U is not induced, it suffices to prove that $\tau_R^{-1}(e_1R)=\begin{smallmatrix} & 2 & & 2 & \\ 1 & & 1 & & 1\end{smallmatrix}$ is not induced. Because A has only 3 isoclasses of indecomposable modules of which two are projective, it suffices to compute the R-module induced by the remaining indecomposable $S_2=\,2$. The minimal projective resolution

$$0 \longrightarrow e_1A \longrightarrow e_2A \longrightarrow S_2 \longrightarrow 0$$

in $\mathrm{mod}\,A$ induces one in $\mathrm{mod}\,R$

$$0 \longrightarrow e_1A \longrightarrow e_2A \longrightarrow S_2\otimes_A R \longrightarrow 0.$$

Therefore $S_2\otimes_A R\cong \begin{smallmatrix}2\\1\end{smallmatrix}\not\cong \tau_R^{-1}(e_1R)$.

To prove that U is restrictable, we must show that $\mathrm{Tor}_1^R(U, A) \cong \mathrm{DExt}_R^1(U, \mathrm{D}A) = 0$. But this amounts to showing that $(\mathrm{D}A)_R$ is generated by U in $\mathrm{mod}\, R$. Now $(\mathrm{D}A)_R = {}^2_1 \oplus 2$ and both of its summands are generated by U.

Finally, we compute the restriction of U. We have a minimal projective resolution

$$0 \longrightarrow e_1R \longrightarrow (e_2R)^2 \longrightarrow \tau_R^{-1}(e_1R) \longrightarrow 0.$$

Applying $- \otimes_A R$ yields an exact sequence

$$0 \longrightarrow e_1A \longrightarrow (e_2A)^2 \longrightarrow \tau_R^{-1}(e_1R) \otimes_R A \longrightarrow 0.$$

Therefore $\tau_R^{-1}(e_1R) \otimes_R A \cong {}^2_1 \oplus 2$. Because $e_2R \otimes_R A \cong e_2A = {}^2_1$, we deduce that $U \otimes_R A = ({}^2_1)^2 \oplus 2$.

PROPOSITION 4.3.8 ([**14**]3.5). (a) *Let U be a restrictable tilting R-module, then*

$M_A \in \mathcal{T}(U \otimes_R A)$ *if and only if* $M_R \in \mathcal{T}(U)$,
$M_A \in \mathcal{F}(U \otimes_R A)$ *if and only if* $M_R \in \mathcal{F}(U)$.

(b) *Let U be a corestrictable cotilting R-module, then*

$M_A \in \mathcal{T}'(\mathrm{Hom}_R(A, U))$ *if and only if* $M_R \in \mathcal{T}'(U)$,
$M_A \in \mathcal{F}'(\mathrm{Hom}_R(A, U))$ *if and only if* $M_R \in \mathcal{F}'(U)$.

PROOF. (a) This follows from the vector space isomorphisms

$$\begin{aligned}\mathrm{Ext}_A^1(U \otimes_R A, M) &\cong \mathrm{DHom}_A(M, \tau_A(U \otimes_R A))\\ &\cong \mathrm{DHom}_A(M, \mathrm{Hom}_R(A, \tau_A U))\\ &\cong \mathrm{DHom}_R(M \otimes_A A_R, \tau_A U)\\ &\cong \mathrm{DHom}_R(M_R, \tau_R U)\\ &\cong \mathrm{Ext}_R^1(U, M_R),\end{aligned}$$

and

$$\begin{aligned}\mathrm{Hom}_A(U \otimes_R A, M) &\cong \mathrm{Hom}_R(U, \mathrm{Hom}_A(A, M))\\ &\cong \mathrm{Hom}_R(U, M_R).\end{aligned}$$ □

COROLLARY 4.3.9 ([**14**]3.5). (a) *If U is a restrictable tilting R-module such that $(\mathcal{T}(U), \mathcal{F}(U))$ splits in* $\mathrm{mod}\, R$, *then $(\mathcal{T}(U \otimes_R A), \mathcal{F}(U \otimes_R A))$ splits in* $\mathrm{mod}\, A$.

(b) *If U is a corestrictable cotilting R-module such that $(\mathcal{T}'(U), \mathcal{F}'(U))$ splits in* $\mathrm{mod}\, R$, *then $(\mathcal{T}'(\mathrm{Hom}_R(A, U)), \mathcal{F}'(\mathrm{Hom}_R(A, U)))$ splits in* $\mathrm{mod}\, A$.

PROOF. (a) Let $M \in \mathcal{T}(U \otimes_R A)$, $N \in \mathcal{F}(U \otimes_R A)$. Then $M_R \in \mathcal{T}(U)$, $N \in \mathcal{F}(U)$. Because of the hypothesis, $\mathrm{Ext}_R^1(N, M) = 0$. Therefore $\mathrm{Ext}_A^1(N, M) = 0$. □

REMARK 4.3.10 ([**14**]3.7). It is useful to observe that an R-module U verifies $\mathrm{Tor}_1^R(U, A) = 0$ if and only if the multiplication $U \otimes_R E \longrightarrow UE$, $x \otimes e \mapsto xe$ (for $x \in U$, $e \in E$) is an isomorphism of A-modules. Indeed, applying $U \otimes_R -$ to the exact sequence $0 \longrightarrow {}_RE_A \longrightarrow {}_RR_A \longrightarrow {}_RA_A \longrightarrow 0$ yields a commutative diagram with exact rows

$$\begin{array}{ccccccccc} 0 & \longrightarrow & \mathrm{Tor}_1^R(U\otimes_A R, A) & \longrightarrow & U\otimes_R E & \longrightarrow & U\otimes_R R & \longrightarrow & U\otimes_R A & \longrightarrow & 0 \\ & & & & \downarrow{\scriptstyle \mu'} & & \downarrow{\scriptstyle \mu} & & \downarrow{\scriptstyle \mu''} & & \\ & & 0 & \longrightarrow & UE & \xrightarrow{i} & U & \xrightarrow{p} & U/UE & \longrightarrow & 0 \end{array}$$

where μ, μ' are the multiplication maps and μ'' is induced by passing to cokernels. As seen in 3.1.4, μ'' is an isomorphism and μ is well-known to be so. Moreover, μ' is surjective so that we have an exact sequence in mod A

$$0 \longrightarrow \mathrm{Tor}_1^R(U \otimes_A R,\, A) \longrightarrow U \otimes_R E \xrightarrow{\mu'} UR \longrightarrow 0.$$

The statement follows.

References

[1] Edson Ribeiro Alvares, Ibrahim Assem, Diane Castonguay, and Rosana R. S. Vargas, *Right ADA algebras*, J. Algebra Appl. **16** (2017), no. 11, 1750210, 14, DOI 10.1142/S0219498817502103. MR3708600

[2] A. Alvarez, D. Bravo, E. Fernández, V. Guazzelli, M. Muller, N. Rojas, and S. Trepode, *Trivial extensions, Admissible cuts and HW-reflections*, in preparation.

[3] Frank W. Anderson and Kent R. Fuller, *Rings and categories of modules*, Springer-Verlag, New York-Heidelberg, 1974. Graduate Texts in Mathematics, Vol. 13. MR0417223

[4] Ibrahim Assem, Thomas Brüstle, and Ralf Schiffler, *Cluster-tilted algebras and slices*, J. Algebra **319** (2008), no. 8, 3464–3479, DOI 10.1016/j.jalgebra.2007.12.010. MR2408327

[5] I. Assem, T. Brüstle, and R. Schiffler, *Cluster-tilted algebras as trivial extensions*, Bull. Lond. Math. Soc. **40** (2008), no. 1, 151–162, DOI 10.1112/blms/bdm107. MR2409188

[6] Ibrahim Assem, Juan Carlos Bustamante, Julie Dionne, Patrick Le Meur, and David Smith, *Representation theory of partial relation extensions*, Colloq. Math. **155** (2019), no. 2, 157–186, DOI 10.4064/cm7511-3-2018. MR3897504

[7] Ibrahim Assem, Diane Castonguay, Marcelo Lanzilotta, and Rosana R. S. Vargas, *Algebras determined by their supports*, J. Pure Appl. Algebra **216** (2012), no. 5, 1134–1145, DOI 10.1016/j.jpaa.2011.10.034. MR2875332

[8] Ibrahim Assem and Flávio Ulhoa Coelho, *Glueings of tilted algebras*, J. Pure Appl. Algebra **96** (1994), no. 3, 225–243, DOI 10.1016/0022-4049(94)90100-7. MR1303283

[9] Ibrahim Assem and Flávio U. Coelho, *Two-sided gluings of tilted algebras*, J. Algebra **269** (2003), no. 2, 456–479, DOI 10.1016/S0021-8693(03)00436-8. MR2015287

[10] Ibrahim Assem, Flávio U. Coelho, and Sonia Trepode, *The bound quiver of a split extension*, J. Algebra Appl. **7** (2008), no. 4, 405–423, DOI 10.1142/S0219498808002928. MR2442069

[11] Ibrahim Assem and Nikolaos Marmaridis, *Tilting modules over split-by-nilpotent extensions*, Comm. Algebra **26** (1998), no. 5, 1547–1555, DOI 10.1080/00927879808826219. MR1622428

[12] Ibrahim Assem, Daniel Simson, and Andrzej Skowroński, *Elements of the representation theory of associative algebras. Vol. 1*, London Mathematical Society Student Texts, vol. 65, Cambridge University Press, Cambridge, 2006. Techniques of representation theory. MR2197389

[13] Ibrahim Assem and Dan Zacharia, *Full embeddings of almost split sequences over split-by-nilpotent extensions*, Colloq. Math. **81** (1999), no. 1, 21–31, DOI 10.4064/cm-81-1-21-31. MR1716185

[14] Ibrahim Assem and Dan Zacharia, *On split-by-nilpotent extensions*, Colloq. Math. **98** (2003), no. 2, 259–275, DOI 10.4064/cm98-2-10. MR2033112

[15] Maurice Auslander, Idun Reiten, and Sverre O. Smalø, *Representation theory of Artin algebras*, Cambridge Studies in Advanced Mathematics, vol. 36, Cambridge University Press, Cambridge, 1997. Corrected reprint of the 1995 original. MR1476671

[16] Michael Barot, Elsa Fernández, María Inés Platzeck, Nilda Isabel Pratti, and Sonia Trepode, *From iterated tilted algebras to cluster-tilted algebras*, Adv. Math. **223** (2010), no. 4, 1468–1494, DOI 10.1016/j.aim.2009.10.004. MR2581376

[17] Michael Barot, Christof Geiss, and Andrei Zelevinsky, *Cluster algebras of finite type and positive symmetrizable matrices*, J. London Math. Soc. (2) **73** (2006), no. 3, 545–564, DOI 10.1112/S0024610706022769. MR2241966

[18] Michael Barot and Sonia Trepode, *Cluster tilted algebras with a cyclically oriented quiver*, Comm. Algebra **41** (2013), no. 10, 3613–3628, DOI 10.1080/00927872.2012.673665. MR3169475

[19] K. Bongartz and P. Gabriel, *Covering spaces in representation-theory*, Invent. Math. **65** (1981/82), no. 3, 331–378, DOI 10.1007/BF01396624. MR643558

[20] M. C. R. Butler and Claus Michael Ringel, *Auslander-Reiten sequences with few middle terms and applications to string algebras*, Comm. Algebra **15** (1987), no. 1-2, 145–179, DOI 10.1080/00927878708823416. MR876976

[21] Henri Cartan and Samuel Eilenberg, *Homological algebra*, Princeton University Press, Princeton, N. J., 1956. MR0077480

[22] Flávio Ulhoa Coelho and Marcelo Américo Lanzilotta, *Algebras with small homological dimensions*, Manuscripta Math. **100** (1999), no. 1, 1–11, DOI 10.1007/s002290050191. MR1714397

[23] Flávio U. Coelho and Marcelo A. Lanzilotta, *Weakly shod algebras*, J. Algebra **265** (2003), no. 1, 379–403, DOI 10.1016/S0021-8693(03)00164-9. MR1984916

[24] E. Fernández, *Extensiones triviales y álgebras inclinadas iteradas*, Ph.D. thesis, Universidad Nacional del Sur, Argentina, 1999.

[25] Dieter Happel, Idun Reiten, and Sverre O. Smalø, *Tilting in abelian categories and quasitilted algebras*, Mem. Amer. Math. Soc. **120** (1996), no. 575, viii+ 88, DOI 10.1090/memo/0575. MR1327209

[26] Mitsuo Hoshino, *Trivial extensions of tilted algebras*, Comm. Algebra **10** (1982), no. 18, 1965–1999, DOI 10.1080/00927878208822817. MR674704

[27] Claus Michael Ringel, *Tame algebras and integral quadratic forms*, Lecture Notes in Mathematics, vol. 1099, Springer-Verlag, Berlin, 1984. MR774589

[28] Ralf Schiffler and Khrystyna Serhiyenko, *Induced and coinduced modules over cluster-tilted algebras*, J. Algebra **472** (2017), 226–258, DOI 10.1016/j.jalgebra.2016.10.009. MR3584877

[29] Andrzej Skowroński, *On Artin algebras with almost all indecomposable modules of projective or injective dimension at most one*, Cent. Eur. J. Math. **1** (2003), no. 1, 108–122, DOI 10.2478/BF02475668. MR1953044

[30] Andrzej Skowroński and Josef Waschbüsch, *Representation-finite biserial algebras*, J. Reine Angew. Math. **345** (1983), 172–181. MR717892

[31] Andrzej Skowroński and Kunio Yamagata, *Frobenius algebras. I*, EMS Textbooks in Mathematics, European Mathematical Society (EMS), Zürich, 2011. Basic representation theory. MR2894798

[32] Andrzej Skowrński, *Frobenius Algebras II: Tilted and Hochschild Extension Algebras*, EMS Textbk. Math., vol. 19, Eur. Math. Soc., 2017.

[33] P. Suarez, *Split-by-nilpotent extensions and support τ-tilting modules*, to appear.

[34] Hipolito Treffinger, *τ-tilting theory and τ-slices*, J. Algebra **481** (2017), 362–392, DOI 10.1016/j.jalgebra.2017.03.004. MR3639480

[35] Stephen Zito, *Short proof of a conjecture concerning split-by-nilpotent extensions*, Arch. Math. (Basel) **111** (2018), no. 5, 479–483, DOI 10.1007/s00013-018-1208-7. MR3859429

Département de mathématiques, Université de Sherbrooke, Sherbrooke, Quebec, J1K 2R1, Canada

Contemporary Mathematics
Volume **761**, 2021
https://doi.org/10.1090/conm/761/15307

A note on sequential walks

Ibrahim Assem, María Julia Redondo, and Ralf Schiffler

Dedicated to José Antonio de la Peña on the occasion of his sixtieth birthday

ABSTRACT. This short note is devoted to motivate and clarify the notion of sequential walk that has been previously introduced by the authors. We also give some applications of this concept.

1. Introduction

The class of tilted algebras, introduced by Happel and Ringel [**16**], is among the most ubiquitous classes in the representation theory of algebras. For instance, any cluster-tilted algebra is the trivial extension of a tilted algebra by a particular bimodule [**2**]. Surprisingly enough, it is difficult to check whether a given algebra is tilted or not without a good knowledge of its module category. Indeed, most known criteria revolve around the existence of a combinatorial configuration called a complete slice, see, for instance, [**6**].

It was thus needed to have a handy criterion depending only on the bound quiver of the algebra. The most powerful criterion so far is the existence of so-called sequential walks. Sequential walks have a long history; they first appeared in [**1**], where it was shown that an iterated tilted algebra of type $\mathbb{A}$ is tilted if and only if it has no sequential walk. They surfaced again in [**17**] under the name of "sequential pairs" in the classification of quasi-tilted string algebras, then in [**8**] in the classification of shod string algebras and in [**12**] in the classification of laura string algebras under the name of "double zeros". Their present guise was introduced in [**4**] in the context of non-necessarily monomial algebras. It was proved there that if an algebra contains a sequential walk, then it is not tilted.

This shows that this notion is very natural. Indeed, as we prove here, it follows from simple considerations on the comparison between the shape of the bound quiver and the homological dimensions of some modules.

It was pointed out to the authors that the definition of sequential walk given in [**4**] contained an ambiguity. It is the first purpose of this note to clarify this ambiguity. While doing so, we slightly generalize the definition of sequential walk, and try to illustrate its usefulness for computing homological dimensions.

Our main result is the following.

The first author was supported by NSERC of Canada.

The second author was a researcher from CONICET of Argentina.

The third author was supported by the NSF-grant DMS-1800860, and by the University of Connecticut.

THEOREM 1.1. *Let A be a finite dimensional algebra over an algebraically closed field. If the quiver of A contains a sequential walk, then A is not shod. In particular, A is not quasi-tilted.*

This note is organized as follows. In Section 2, we recall the definitions and known results which are necessary for the proof of our theorem. In Section 3, we give our definition of sequential walk, try to motivate it, then we prove our theorem. Section 4 is then devoted to applications and examples.

2. Preliminaries

2.1. Notation. Throughout, we let $\Bbbk$ denote an algebraically closed field and A a finite dimensional basic $\Bbbk$-algebra. We recall that any basic and connected finite dimensional $\Bbbk$-algebra A can be written in the form $A \cong \Bbbk Q_A/I$ where $\Bbbk Q_A$ is the path algebra of the quiver Q_A of A and I is an ideal generated by finitely many relations. The pair (Q_A, I) is then called a *bound quiver*. A *relation* on a quiver is a linear combination $\rho = \sum_{i=1}^{m} \lambda_i u_i$, where the λ_i are nonzero scalars and the u_i are paths of length at least two having all the same source and the same target, called respectively the *source* and the *target* of the relation ρ. The relation ρ is called *quadratic* if each of the u_i has length two. It is called a *monomial relation* if it is a path, and a *minimal relation* if $m \geq 2$ and, for any nonempty proper subset $J \subset \{1, 2, \ldots, m\}$ one has $\sum_{j \in J} \lambda_j u_j \notin I$.

Relations in a bound quiver (Q, I) are generated by top relations. Indeed, let $\Bbbk Q^+$ be the two-sided ideal of $\Bbbk Q$ consisting of the linear combinations of paths of length at least one and e_x be the primitive idempotent of A corresponding to a point $x \in (Q_A)_0$. Then a relation $\rho \in e_x I e_y$ is called a *top relation* if its residual class in $e_x \left(\frac{I}{\Bbbk Q^+ I + I \Bbbk Q^+} \right) e_y$ is nonzero. They are called top relations because they correspond to nonzero elements of the top of I, considered as a $\Bbbk Q$-$\Bbbk Q$-bimodule. Intuitively one may think of the top relations as being the shortest ones.

For a point x in the ordinary quiver of A, we denote by P_x, I_x, S_x respectively, the indecomposable projective, injective and simple A-modules corresponding to x. The *support* of an A-module M is the full subquiver $\operatorname{Supp} M$ generated by the points $x \in (Q_A)_0$ such that $Me_x \neq 0$. For further definitions and facts, we refer the reader to [**6, 20**].

2.2. Classes of algebras. We need the following classes of algebras.

DEFINITION 2.1. (a) [**11**] An algebra A is *shod* (for <u>s</u>mall <u>ho</u>mological <u>d</u>imensions) if every indecomposable A-module is of projective or injective dimension at most one.

(b) [**15**] An algebra A is *quasi-tilted* if it is shod of global dimension at most two, or, equivalently, if it is the endomorphism algebra of a tilting object in a hereditary, locally finite abelian $\Bbbk$-category.

(c) [**16**] An algebra is *tilted* if it is the endomorphism algebra of a tilting module over a hereditary algebra.

In particular, every tilted algebra is quasi-tilted, and every quasi-tilted algebra is shod.

2.3. Full subcategories. There is a reduction procedure which we shall use in the proof of our main result, called taking full subcategories.

Let $e \in A$ be an idempotent. The finite dimensional $\Bbbk$-algebra eAe is called the *full subcategory* determined by e; indeed, if one considers A as a category whose objects are the elements of a complete set of primitive orthogonal idempotents and the morphism space from e_x to e_y is $e_xA\,e_y$, then a full subcategory is always of the aforementioned form with e equal to a sum of primitive idempotents.

We need the following lemma.

LEMMA 2.2. *Let A be an algebra and $e \in A$ an idempotent, then*

(a) *If A is shod, then so is eAe.*
(b) *If A is quasi-tilted, then so is eAe.*
(c) *If A is tilted, then so is eAe.*

PROOF. (a) is proved in [**18**, 1.2], (b) in [**15**, II.1.15] and (c) in [**14**, III.6.5]. □

2.4. Split-by-nilpotent extensions. Let A be an algebra and E an A-A-bimodule which is finite dimensional as a $\Bbbk$-vector space. We say that E is equipped with an *associative product* if there exists an A-A-bimodule morphism $E \otimes_A E \to E, e \otimes e' \mapsto ee'$, such that $e(e'e'') = (ee')e''$ for all $e, e', e'' \in E$.

DEFINITION 2.3. Let A, E be as before. An algebra R is called a *split extension of A by E* if

$$R = \{(a, e) \mid a \in A, e \in E\}$$

is equipped with the componentwise addition and the multiplication defined by

$$(a, e)(a', e') = (aa', ae' + ea' + ee'\}$$

for $(a, e), (a', e') \in R$. If E is nilpotent with respect to its product, then R is called a *split-by-nilpotent extension.*

It is clear that, if R is a split extension of A by E, then there exists an exact sequence

$$0 \longrightarrow E \longrightarrow R \xrightarrow{\pi} A \longrightarrow 0 \,,$$

where the projection π is an algebra morphism having a section $\sigma\colon A \to R$. Also, E is nilpotent if and only if it is contained in $\operatorname{rad} R$. If an exact sequence as above and a section σ to π are given, then we say that this sequence *realizes* R as a split extension of A by E. If $E^2 = 0$, then the split extension is called a *trivial extension.*

We need the following lemma.

LEMMA 2.4. *Let R be a split extension of an algebra A by a nilpotent bimodule E.*

(a) *If R is shod, then so is A.*
(b) *If R is quasi-tilted, then so is A.*
(c) *If R is tilted, then so is A.*

PROOF. (a) and (b) are proved in [**7**, 2.5] and (c) in [**21**]. □

2.5. Cutting arrows. Let R be a split extension of an algebra A by a nilpotent bimodule E. We recall how one can pass from R to A by dropping arrows from the quiver of R. Let w be a path in the quiver Q_A of A and α an arrow such that there exist subpaths w_1, w_2 of w such that $w = w_1 \alpha w_2$, then we write $\alpha | w$. Also, when we speak of a relation, then we can assume without loss of generality that it is monomial or minimal. Let thus $\rho = \sum_i \lambda_i u_i$ be a relation on Q_A from x to y, say, with the λ_i nonzero scalars and the u_i paths of length at least two from x to y. We say that ρ is *consistently cut* by a set S of arrows (or that S is a *consistent cut*), if, whenever there exist i and $\alpha_i | u_i$ such that $\alpha_i \in S$ then, for any $j \neq i$, there exists $\alpha_j | u_j$ satisfying $\alpha_j \in S$.

THEOREM 2.5. [**3**, 2.5] *Let $\eta_R \colon \Bbbk Q_R \to R$ be a presentation of R, let E be an ideal of R generated by the classes modulo $I_R = \ker \eta_R$ of a set S of arrows, and let $\pi \colon R \to R/E = A$ be the projection. Assume moreover that every relation in I_R is consistently cut by S, then the exact sequence*

$$0 \longrightarrow E \longrightarrow R \xrightarrow{\pi} A \longrightarrow 0$$

realizes R as a split extension of A by E.

2.6. Reduction of an algebra. We now define a notion which we call reduction of an algebra. We say that an algebra B is a *reduction* of an algebra A if there exist an idempotent $e \in A$ and a two-sided ideal E of eAe contained in its radical such that the sequence

$$0 \longrightarrow E \longrightarrow eAe \longrightarrow B \longrightarrow 0$$

realizes eAe as a split extension of B by the nilpotent bimodule E. Thus B is obtained from A by the combination of two consecutive processes: one first drops points of Q_A by passing to the full subcategory eAe, then one drops arrows as explained in Theorem 2.5. One then has the following obvious corollary of lemmata 2.2 and 2.4.

COROLLARY 2.6. *Let B be a reduction of A.*

(a) *If A is shod, then so is B.*
(b) *If A is quasi-tilted, then so is B.*
(c) *If A is tilted, then so is B.*

EXAMPLE 2.7. Taking full subcategories can be used to make a relation quadratic. Indeed, assume that the bound quiver of an algebra A contains a relation $\rho = \sum_{i=1}^m \lambda_i u_i$ with source x and target y

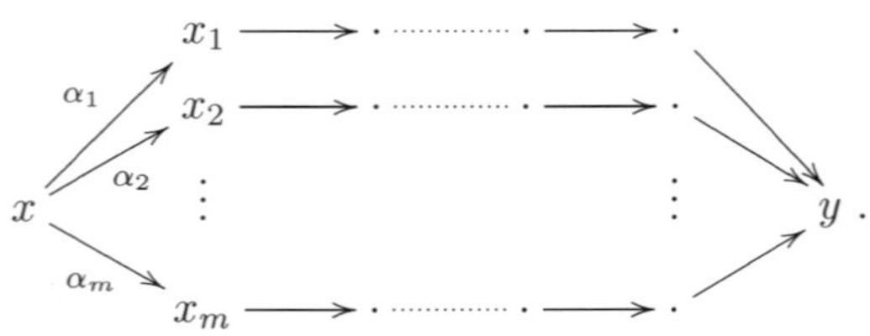

Denote by $x_1, \ldots, x_m$ the immediate successors of x on the paths $u_1, \ldots, u_m$ respectively, and let $e = e_x + \sum_{i=1}^m e_{x_i} + e_y + e'$, where e' is the sum of all primitive idempotents corresponding to points in Q_A not lying on the paths u_i. Then the bound quiver of eAe contains a relation $\rho' = \sum_{i=1}^{m} \lambda_i(\alpha_i\beta_i)$ with source x and target y

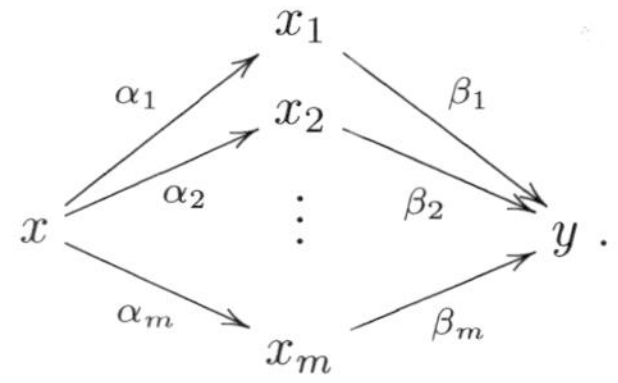

Clearly, ρ' is quadratic. This procedure will be used in the proof of our key lemma 3.9 below.

3. Sequential walks

3.1. The definition. Given an arrow α in a quiver, its formal inverse is denoted by α^{-1}, where we agree that the source of α^{-1} is the target of α and the target of α^{-1} is the source of α. A *walk* w is an expression of the form $w = \alpha_1^{\epsilon_1} \cdots \alpha_t^{\epsilon_t}$, where the α_i are arrows and the $\epsilon_i \in \{+1, -1\}$ are such that the target of $\alpha_i^{\epsilon_i}$ equals the source of $\alpha_{i+1}^{\epsilon_{i+1}}$, for all i. Such a walk is called *reduced* if it contains no expression of one of the forms $\alpha\alpha^{-1}$ or $\alpha^{-1}\alpha$, with α an arrow. It is called a *zigzag walk* if it is of the form $\alpha_1^{\epsilon_1} \cdots \alpha_t^{\epsilon_t}$ with $\epsilon_i \neq \epsilon_{i+1}$ for all i.

We start by recalling the definition of sequential walk as stated in [**4**]. Let w be a nontrivial walk in a bound quiver (Q, I). Assume that one writes $w = uw'v$ where each of u, w', v is a subwalk of w. We say that u, v *point to the same direction* in w if both u and v, or both u^{-1} and v^{-1} are paths in Q.

A reduced walk $w = uw'v$ having u, v pointing to the same direction was called a sequential walk in [**4**] if there is a relation $\rho = \sum_i \lambda_i u_i$ such that $u = u_1$, or $u = u_1^{-1}$, there is a relation $\sigma = \sum_i \mu_i v_i$ such that $v = v_1$, or $v = v_1^{-1}$ and no subpath w_1 of w', or of $(w')^{-1}$ is involved in a relation of the form $\sum \nu_i w_i$.

As mentioned in the introduction, there is an ambiguity in this definition arising from the undefined word "involved". Indeed, the word "involved" can be understood as meaning "is a branch of a relation". But this is not correct as shown in the following example.

EXAMPLE 3.1. Let A be given by the quiver

$$1 \underset{\beta_1}{\overset{\alpha_1}{\rightrightarrows}} 2 \underset{\beta_2}{\overset{\alpha_2}{\rightrightarrows}} 3$$

bound by the relation $\alpha_1\alpha_2 + \beta_1\beta_2 = 0$. Then A is the one-point extension of the Kronecker algebra with quiver

$$2 \underset{\beta_2}{\overset{\alpha_2}{\rightrightarrows}} 3$$

by the indecomposable postprojective module ${}_{3}{}^{2}{}_{3}{}^{2}{}_{3}$. In particular, A is tilted [**19**, 3.5].

Consider the reduced walk $w = (\alpha_1\alpha_2)(\beta_2^{-1}\alpha_1^{-1})(\beta_1\beta_2)$ in the quiver of A, with $u = \alpha_1\alpha_2$, $v = \beta_1\beta_2$ and $w' = \beta_2^{-1}\alpha_1^{-1}$. Then neither w' nor $(w')^{-1} = \alpha_1\beta_2$ is a branch of any relation while u and v satisfy the conditions of the definition of [**4**]. If one understands "involved" as meaning "is a branch of a relation", then w would be a sequential walk and we would get a counterexample to [**4**, 2.4].

We propose the following definition.

DEFINITION 3.2. Let w be a reduced walk in a bound quiver (Q, I), then w is called a *sequential walk* if the following hold.

(a) $w = uw'v$, where u, v point to the same direction, there is a top relation $\rho = \sum_i \lambda_i u_i$ such that $u = u_1$, or $u = u_1^{-1}$, and there is a top relation $\sigma = \sum_i \mu_i v_i$ such that $v = v_1$, or $v = v_1^{-1}$;

(b) no subpath of w', or of $(w')^{-1}$ lies in I, nor is a branch of a relation having a branch which has a point in common with one of the u_i or v_i;

(c) w' itself has no arrows in common with one of the u_i or v_i.

REMARK 3.3. Condition (b) holds for example if no subpath of w' or $(w')^{-1}$ is the branch of a relation.

REMARK 3.4. Our definition of sequential walk is clearly inspired by the definition of sequential pair in [**17**] but it is not identical to it.

Let $A = \Bbbk Q/I$ be a string algebra. In particular A is monomial. A sequential pair of monomial relations is a reduced walk w that contains exactly two zero-relations and these two zero-relations point to the same direction in w. Our definition differs from [**17**] in the case of string algebras. For example, the algebra given by the quiver

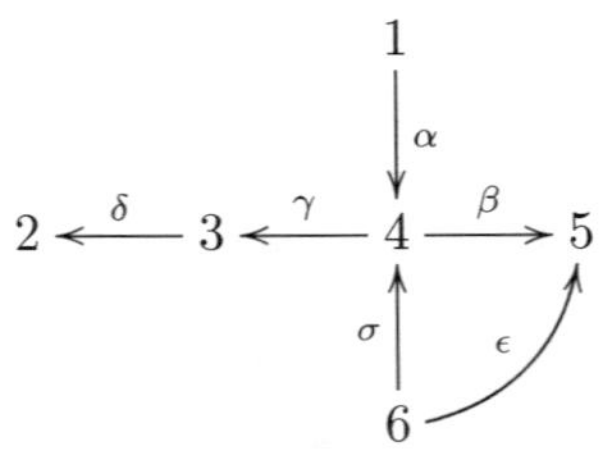

bound by the relations $\alpha\beta = \gamma\delta = \sigma\beta = 0$ contains the sequential pair $\alpha\beta\epsilon^{-1}\sigma\gamma\delta$, where $u = \alpha\beta$, $w' = \epsilon^{-1}\sigma$ and $v = \gamma\delta$, but this is not a sequential walk because the target of σ is a point on u but not an endpoint of u.

Our definition differs from [**17**] also because sequential walks do not detect overlapping relations. For example, the algebra given by the quiver

$$1 \xrightarrow{\alpha} 2 \xrightarrow{\beta} 3 \xrightarrow{\gamma} 4$$

bound by the relations $\alpha\beta = 0 = \beta\gamma$ does not admit a sequential walk.

REMARK 3.5. As we shall see below, if A is a tree algebra of global dimension two, then the two notions of sequential walk and sequential pair coincide.

3.2. Why this notion is natural. It is well-known that if $A = \Bbbk Q/I$ is an algebra, and S_x a simple module, then $\operatorname{pd} S_x > 1$ if and only if x is the source of a top relation in (Q, I). This is easily seen for instance by looking at the radical of P_x. The following considerations go in this direction.

LEMMA 3.6. *Let $A = kQ/I$ be an algebra and M an A-module of projective dimension $d > 1$. Then there exists $x \in (\operatorname{Supp} M)_0$ and, for each $i \leq d$, there exists $y_i \in Q_0$ such that $\operatorname{Ext}^i_A(S_x, S_{y_i}) \neq 0$.*

PROOF. We first claim that M has a composition factor S_x such that $\operatorname{pd} S_x \geq d$. Consider the socle series

$$0 \subsetneq \operatorname{soc} M \subsetneq \operatorname{soc}^2 M \subsetneq \cdots \subsetneq \operatorname{soc}^\ell M = M,$$

where ℓ is the Loewy length of M, see for example [**4**, V.I]. If there exists a simple summand of $\operatorname{soc} M$ of projective dimension d, then we are done. Otherwise, there exists $i < \ell$ such that $\operatorname{pd}\operatorname{soc}^i M < d$ and $\operatorname{pd}\operatorname{soc}^{i+1} M = d$. The short exact sequence

$$0 \longrightarrow \operatorname{soc}^i M \longrightarrow \operatorname{soc}^{i+1} M \longrightarrow \frac{\operatorname{soc}^{i+1} M}{\operatorname{soc}^i M} \longrightarrow 0$$

yields an exact sequence of functors

$$\operatorname{Ext}^d_A\left(\frac{\operatorname{soc}^{i+1} M}{\operatorname{soc}^i M}, -\right) \longrightarrow \operatorname{Ext}^d_A(\operatorname{soc}^{i+1} M, -) \longrightarrow \operatorname{Ext}^d_A(\operatorname{soc}^i M, -) = 0.$$

Now $\operatorname{Ext}^d_A(\operatorname{soc}^{i+1} M, -) \neq 0$ implies $\operatorname{Ext}^d_A(\frac{\operatorname{soc}^{i+1} M}{\operatorname{soc}^i M}, -) \neq 0$. Hence, because the module $\frac{\operatorname{soc}^{i+1} M}{\operatorname{soc}^i M}$ is semisimple, there exists a simple composition factor S_x of M such that $\operatorname{pd} S_x \geq d$.

We deduce the result. Because of our claim, there exists a minimal projective resolution

$$P_{d+1} \xrightarrow{f_{d+1}} P_d \xrightarrow{f_d} P_{d-1} \xrightarrow{f_{d-1}} \cdots \xrightarrow{f_1} P_0 \xrightarrow{f_0} S_x \longrightarrow 0$$

with $P_i \neq 0$ for all $i \leq d$. Consider the projection $p_i \colon P_i \to \operatorname{top} P_i$. Then $p_i f_{i+1} = 0$ because of the minimality of the resolution. Moreover, p_i does not factor through f_i because P_i is the projective cover of $\operatorname{Ker} f_{i-1}$. Therefore $\operatorname{Ext}^i_A(S_x, \operatorname{top} P_i) \neq 0$ and there exists a simple composition factor S_{y_i} of $\operatorname{top} P_i$ such that $\operatorname{Ext}^i_A(S_x, S_{y_i}) \neq 0$. □

COROLLARY 3.7. *Let M be a module of projective dimension $d \geq 2$. Then there exists $x \in (\operatorname{Supp} M)_0$ which is the source of a top relation in (Q, I).*

PROOF. Because $d \geq 2$, there exists y_2 such that $\operatorname{Ext}^2_A(S_x, S_{y_2}) \neq 0$. We then apply [**9**]. □

Let thus M be a module with both projective and injective dimension larger than one. Because of Corollary 3.7 and its dual, there exist two points x, y in $\operatorname{Supp} M$ which are respectively the source of a top relation and the target of a top relation. The notion of sequential walk (and all other similar notions like sequential pairs, double zeros etc.) arose from the attempt to connect y to x by a walk.

3.3. A necessary condition for shod. In this subsection, we prove our main result. We start with the following remark.

REMARK 3.8. If the quiver of an algebra A contains a nonzero walk $w = \alpha_1^{\epsilon_1} \cdots \alpha_t^{\epsilon_t}$, then there is a reduction B of A in which w is replaced by a zigzag walk which is moreover a full subcategory of B. Let $x_1, \ldots, x_r$ be all the sources on w and $y_1, \ldots, y_s$ all the sinks, then set $e = \sum_{i=1}^r e_{x_i} + \sum_{j=1}^s e_{y_j} + e'$, where e' is the sum of the primitive idempotents corresponding to points in Q_A not lying on w. In eAe, w is replaced by a new walk w' which is a zigzag walk. However, we may have in eAe new arrows between the x_i and the y_j not corresponding to subpaths of w. Let S be the set of all these arrows, namely those arrows β in Q_{eAe} whose source and target lie on w' but neither β nor β^{-1} lies on w'. Since w' is a zigzag walk, it contains no (branches of) relations, so S is a consistent cut, as defined in 2.5. Let E be the two-sided ideal of eAe generated by S and $B = eAe/E$. Then it follows from Theorem 2.5 that eAe is a split extension of B by E. Moreover, w' is a full subcategory of B.

LEMMA 3.9. *Let A have a sequential walk. Then there exists a reduction B of A containing one of the following (perhaps not full) subquivers.*

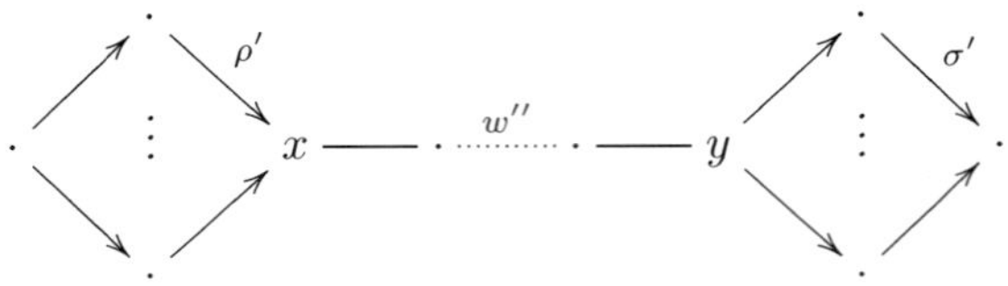

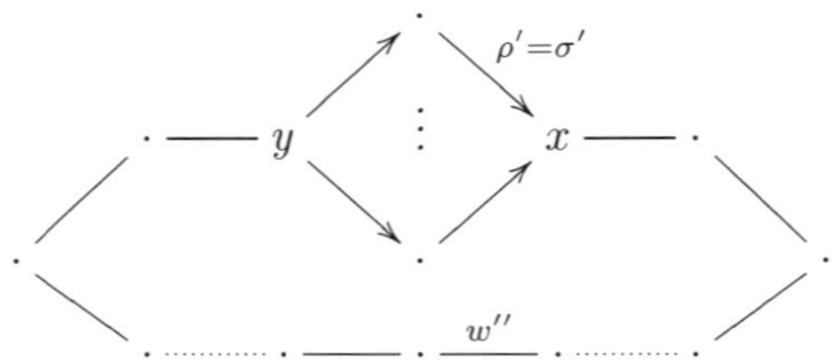

where ρ', σ' are quadratic relations, w'' is a zigzag walk having no point in common with ρ', σ' except x and y.

Moreover, w'' generates a full subcategory of B.

PROOF. Let $w = uw'v$ be a sequential walk in A. Then u, v are branches of top relations $\rho = \sum \lambda_i u_i$ from a to b, say, and $\sigma = \sum \mu_j v_j$ from c to d, say, while w' is a walk from b to c satisfying the conditions of Definition 3.2. Clearly, one may have $\rho = \sigma$. We construct B in the following steps.

(a) We make ρ, σ quadratic, using the recipe of Example 2.7 above. Let e_1 be the sum of the primitive idempotents of A corresponding to:

(1) The sources and targets a, b, c, d of ρ and σ.
(2) All immediate successors of a, c lying on one of the paths u_i, v_j respectively.
(3) All points of Q_A not lying on any of the paths u_i, v_j.

Then $A_1 = e_1 A e_1$ is a full subcategory of A whose quiver contains a sequential pair $w = u'w'v'$ where u', v' are branches of quadratic relations and w' is the same walk as before.

(b) We make w' a zigzag walk, as in Remark 3.8 above. Let $x_1, \ldots, x_r$ and $y_1, \ldots, y_s$ be respectively all the sources and all the sinks on w', and e_2 be the sum of the primitive idempotents corresponding to the x_i, y_j and all the points of Q_{A_1} not on w'. Then $A_2 = e_2 A_1 e_2$ contains a zigzag walk w'' which replaces w', such that there are perhaps additional arrows in Q_{A_2} between points of w'' not corresponding to subpaths of w'.

(c) We eliminate excessive arrows in A_2 making w'' a full subcategory of the resulting algebra. Consider the set S of all arrows α in A_2 whose source and target lies in w'', but such that neither α nor α^{-1} belongs to w''. Since there are no relations on the zigzag walk w'', the set S is a consistent cut. Let E be the two-sided ideal of A_2 generated by S, and let $B = A_2/E$. Then we get an exact sequence

$$0 \longrightarrow E \longrightarrow A_2 \longrightarrow B \longrightarrow 0$$

realizing A_2 as a split extension of B by E, as seen in Theorem 2.5.

It is then clear that the bound quiver of B contains one of the subquivers in the statement, the second one occurring if $\rho = \sigma$. □

We call the reduction B as in the lemma the *standard reduction* corresponding to a given sequential walk.

LEMMA 3.10. *Let A have a sequential walk and B the corresponding standard reduction of A. Then the string module of the zigzag walk w'' in B has projective and injective dimension larger than one.*

PROOF. Let $M = M(w'')$ be the string module corresponding to w'', that is, M is the B-module defined as a representation by

$$M(x) = \begin{cases} \Bbbk & \text{if } x \text{ is a point of} w''; \\ 0 & \text{otherwise}; \end{cases}$$

and

$$M(\alpha) = \begin{cases} \mathrm{id} & \text{if } \alpha \text{ is an arrow of } w''; \\ 0 & \text{otherwise.} \end{cases}$$

Notice that in B, there may be arrows between points in w'' and points on one of the relations ρ', σ'. But because of the definition of the string module, for any such arrow β, we have $M(\beta) = 0$. Because no subpath of w'' is a branch of a relation in B, then M is indeed a B-module. Moreover, M is a string module, see [**10**], and in particular, it is indecomposable.

We now prove that $\operatorname{pd} M_B > 1$. The support of M contains the source y of σ', Therefore the projective cover of M admits a direct summand P_z such that z lies on w'', and either $z = y$ or there is an arrow $z \to y$. It is easily seen that a nonprojective summand of $\Omega^1 S_y$ is a direct summand of $\Omega^1 M$. Because $\Omega^1 S_y$ is not projective, neither is $\Omega^1 M$. This establishes the claim.

Dually, we also have $\operatorname{id} M_B > 1$. □

We are now ready to prove our main result. It generalizes [**4**, 2.4].

THEOREM 3.11. *Let $A = \Bbbk Q/I$ be an algebra having a sequential walk. Then A is not shod. In particular, A is not quasi-tilted.*

PROOF. Let B be the standard reduction of A corresponding to the sequential walk. By Lemma 3.10, there exists an indecomposable B-module that has both injective and projective dimension larger than one. Therefore B is not shod.

Because of Corollary 2.6, neither is A. □

EXAMPLE 3.12. Consider the quiver

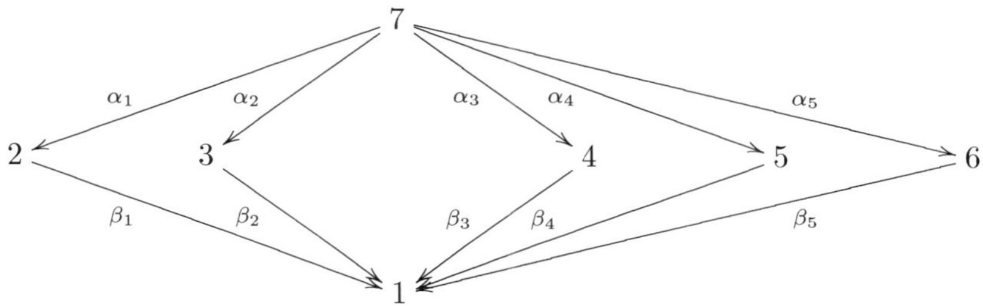

bound by the relations $\alpha_1\beta_1 + \alpha_2\beta_2 = 0$, $\alpha_3\beta_3 + \alpha_4\beta_4 + \alpha_5\beta_5 = 0$. We show how to perform the reduction procedure of Theorem 3.11. Let here

$$w = (\alpha_1\beta_1)\beta_3^{-1}\alpha_3^{-1}(\alpha_2\beta_2),$$

with $u = \alpha_1\beta_1$, $v = \alpha_2\beta_2$ and $w' = \beta_3^{-1}\alpha_3^{-1}$. It is clear that u, v and w' satisfy the conditions of Definition 3.2.

We first eliminate points by taking $e = e_1 + e_2 + e_3 + e_7$. Then eAe is given by the quiver

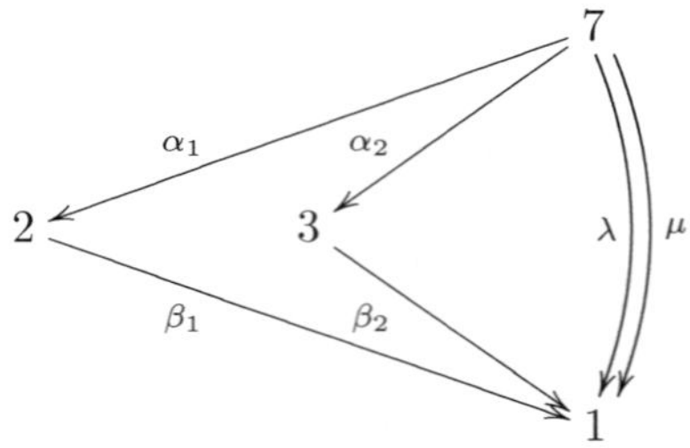

bound by the relation $\alpha_1\beta_1 + \alpha_2\beta_2 = 0$. This is a split extension of the algebra B given by the quiver

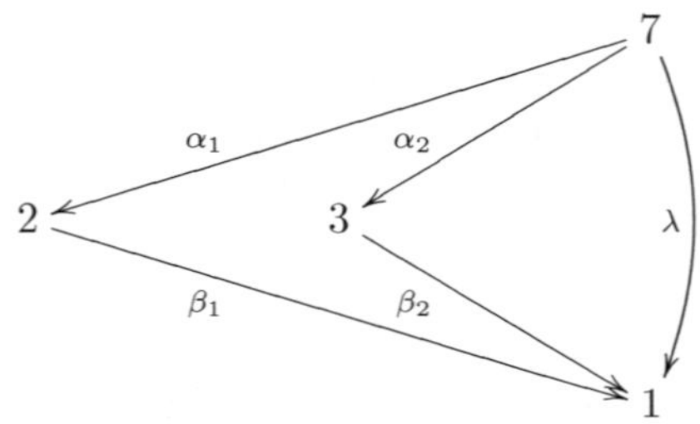

bound by $\alpha_1\beta_1 + \alpha_2\beta_2 = 0$, by the two-sided ideal generated by the arrow μ.

The indecomposable B-module M of the proof is the module $M = \substack{7\\1}$ supported by the arrow λ. Clearly, we have a minimal projective resolution

$$0 \longrightarrow P_1 \longrightarrow P_2 \oplus P_3 \longrightarrow P_7 \longrightarrow M \longrightarrow 0$$

so $\operatorname{pd} M_B = 2$. Notice that $\Omega^1 M = \substack{2\ 3\\1}$ while $\Omega^1 S_7 = \substack{2\ 3\\1} \oplus 1$. Thus $\Omega^1 M$ and $\Omega^1 S_7$ have a common summand but are not equal. Similarly, $\operatorname{id} M_B = 2$. Thus B, and A, are not shod.

The following examples illustrate that the converse of Theorem 3.11 does not hold without additional conditions on the algebra A or the module M. In Section 4, we give examples of such additional conditions.

EXAMPLE 3.13. If an indecomposable module has both projective and injective dimension larger than one, this does not necessarily imply the existence of a sequential walk. Indeed, there exist x, y in the support of the module such that x is the starting point of a top relation v, and y the ending point of a top relation u. Then there exists a walk w' from x to y inside the support of M, but this does not imply that $w = uw'v$ is a sequential walk because it may not be reduced. For instance, let A be the monomial tree algebra given by the quiver

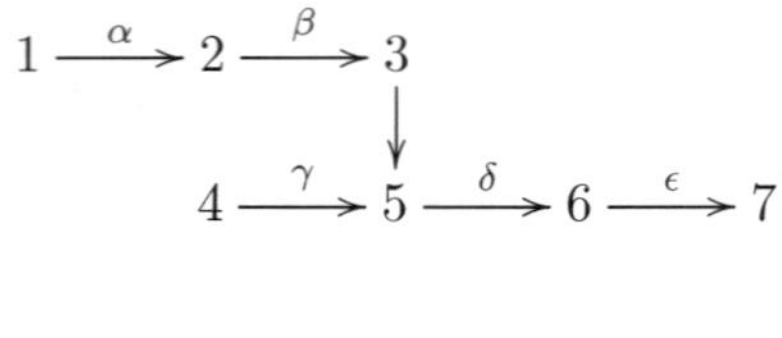

bound by the relations $\alpha\beta = 0 = \gamma\delta\epsilon$, then there is no sequential walk but the module $M = {}^{3}{}_{5}{}^{4}$ has both projective and injective dimension 2.

EXAMPLE 3.14. There exist sincere indecomposable modules of projective and injective dimension 2. Let A be given by the quiver

$$1 \underset{\beta_1}{\overset{\alpha_1}{\rightrightarrows}} 2 \underset{\beta_2}{\overset{\alpha_2}{\rightrightarrows}} 3$$

bound by the relations $\alpha_1\alpha_2 = 0$ and $\beta_1\beta_2 = 0$. Note that A is gentle and in particular tame. The indecomposable module $M = \begin{smallmatrix}1\\2\\3\end{smallmatrix}$ is sincere and of both projective and injective dimension 2.

4. Applications and examples

4.1. The case of global dimension two. Because sequential walks do not detect overlaps, it is natural to think of them in the context of algebras of global dimension two.

PROPOSITION 4.1. *If A is a monomial algebra of global dimension 2 and M is a uniserial A-module whose injective and projective dimensions are both larger than one then there exists a sequential walk in A.*

PROOF. Since M is uniserial, the support of M is of the form

$$z_1 \longrightarrow z_2 \longrightarrow \cdots \qquad \longrightarrow z_\ell\ .$$

By Corollary 3.7 and its dual, there exist $z_i, z_j \in \operatorname{Supp} M$ such that z_j is the source of a top relation v, and z_i the ending point of a top relation u. Since A is monomial of global dimension 2, the two relations u, v do not overlap [**13**]. Thus $i \leq j$. Now let w' be a path $z_i \to \cdots \to z_j$ in $\operatorname{Supp} M$. Then the composition $uw'v$ is a sequential walk in A. □

COROLLARY 4.2. *Let A be a Nakayama algebra of global dimension 2. Then there exists a sequential walk in A if and only if A is not tilted.*

PROOF. Necessity follows from Theorem 3.11. To show sufficiency, suppose A is not quasi-tilted. Then there exists an indecomposable A-module M of both projective and injective dimension 2. Since A is a Nakayama algebra, A is monomial and M is uniserial. Now the result follows from Proposition 4.1 and the fact that quasi-tilted Nakayama algebras are representation-finite, and hence tilted. □

We have the following characterization of projective dimension 2.

PROPOSITION 4.3. *Let M be an indecomposable module over an algebra of global dimension 2 such that one of the sinks in the support of M is the starting point of a top relation. Then the projective dimension of M is two.*

PROOF. Indeed, we have $\operatorname{pd} S_x > 1$ and an exact sequence

$$0 \longrightarrow S_x \longrightarrow M \longrightarrow M/S_x \longrightarrow 0.$$

Hence we have an exact sequence of functors

$$\operatorname{Ext}^2_A(M/S_x, -) \longrightarrow \operatorname{Ext}^2_A(M, -) \longrightarrow \operatorname{Ext}^2_A(S_x, -) \longrightarrow 0,$$

because $\operatorname{Ext}^3_A = 0$. Hence $\operatorname{Ext}^2_A(S_x, -) \neq 0$ implies $\operatorname{Ext}^2_A(M, -) \neq 0$. □

For tree algebras of global dimension two, sequential walks are easy to characterize.

PROPOSITION 4.4. *Let A be a tree algebra of global dimension two. Then the two notions of sequential walk and sequential pair coincide.*

PROOF. Indeed, because A is a tree algebra (not necessarily a string algebra) it is monomial, and the global dimension two means that no two top relations overlap. Because Q is a tree, two points of Q are connected by a unique walk, and so a sequential walk as well as a sequential pair mean a walk of the form $w = uw'v$, where u, v are monomial relations pointing to the same direction while w' is any walk not containing a relation. □

4.2. An application to laura algebras. For our next corollary, we recall a few notions. Given an algebra A, we denote by $\operatorname{ind} A$ a full subcategory of $\operatorname{mod} A$ consisting of exactly one representative from each isomorphism class of indecomposable modules. The *left part* $\mathcal{L}_A$ of $\operatorname{mod} A$ consists of all modules M in $\operatorname{ind} A$ such that, for every L for which there exists a path of nonzero morphisms from L to M, we have $\operatorname{pd} L \leq 1$. The *right part* $\mathcal{R}_A$ of $\operatorname{mod} A$ is defined dually, and an algebra A is called a *laura algebra* if and only if $\mathcal{L}_A \cup \mathcal{R}_A$ is cofinite in $\operatorname{ind} A$.

In [**12**], Dionne has shown that a string algebra is laura if and only if its bound quiver does not contain a combinatorial configuration called intertwined double zero which we now define.

DEFINITION 4.5. [**12**, 2.1.1] Let $A = \Bbbk Q/I$ be a string algebra. A reduced walk w in Q is called an *intertwined double zero* if $w = \rho_1 w_1 w_2 w_3 \rho_2$ where

(a) $\rho_1 = \alpha_1 \dots \alpha_n, \rho_2 = \beta_1 \dots \beta_m$ are monomial relations pointing in the same direction,
(b) neither $\alpha_2 \dots \alpha_n w_1 w_2 w_3 \beta_1 \dots \beta_{m-1}$ nor its inverse contains a monomial relation, and
(c) w_2 is a band.

The next corollary shows that one direction of Dionne's result follows from ours. Observe first that, if $A = \Bbbk Q/I$ is a string algebra, and w is an intertwined double zero then w is a sequential walk in our sense.

COROLLARY 4.6. *Let A be a string algebra having an intertwined double zero. Then A is not laura.*

PROOF. It follows from Definition 4.5 that if $w = \rho_1 w_1 w_2 w_3 \rho_2$ as above is an intertwined double zero, then, for any $n \geq 1$, the reduced walk $w_n = \rho_1 w_1 w_2^n w_3 \rho_2$ is also a sequential walk in the bound quiver of A. Because of Lemma 3.10, we have $\operatorname{pd} M(w_n) > 1$ and $\operatorname{id} M(w_n) > 1$. In particular, the $M(w_n)$ form an infinite family of nonisomorphic indecomposable modules lying neither in $\mathcal{L}_A$ nor in $\mathcal{R}_A$. Thus A is not laura. □

4.3. An application to 2-Calabi-Yau tilted algebras. Let $C = \Bbbk Q/I$ be a quasi-tilted algebra. In particular, its relation extension $\widetilde{C} = \Bbbk\widetilde{Q}/\widetilde{I}$, which is the trivial extension of C by the bimodule $\operatorname{Ext}^2_C(DC, C)$ is a cluster-tilted algebra or a 2-Calabi-Yau tilted algebra of canonical type, see [**5**, 3.1]. A walk $w = \alpha w' \beta$ in $(\widetilde{Q}, \widetilde{I})$ is called a *C-sequential walk* if

(i) w' consists entirely of old arrows, see [**2**] for the terminology "old" vs "new" arrows;

(ii) α, β are new arrows corresponding respectively to old relations $\rho = \sum_i \lambda_i u_i$ and $\sigma = \sum_j \mu_j v_j$;

(iii) For all i, j, the walk $w = u_i w' v_j$ is sequential in (Q, I).

Then we have

COROLLARY 4.7. *Let C be a quasi-tilted algebra. Then the bound quiver of its relation extension $\widetilde{C}$ contains no C-sequential walk.*

4.4. Example. We have seen in Example 3.1 that the algebra A given by the quiver

$$1 \underset{\beta_1}{\overset{\alpha_1}{\rightrightarrows}} 2 \underset{\beta_2}{\overset{\alpha_2}{\rightrightarrows}} 3$$

bound by the relation $\alpha_1\alpha_2 + \beta_1\beta_2 = 0$ is a tilted algebra. Notice that here $w = (\alpha_1\alpha_2)\beta_2^{-1}\alpha_1^{-1}(\beta_1\beta_2)$ is not a sequential walk in the sense of Definition 3.2, since the subpath $(w')^{-1} = \alpha_1\beta_2$ has arrows in common with the branches of the relation $u = \alpha_1\alpha_2$ and $v = \beta_1\beta_2$. In fact, this bound quiver contains no sequential walks.

On the other hand, the algebra A' given by the same quiver but bound by the relation $\alpha_1\alpha_2 = 0$ contains evidently the sequential walk $w = (\alpha_1\alpha_2)\beta_2^{-1}\beta_1^{-1}(\alpha_1\alpha_2)$. In particular, Theorem 3.11 implies that this algebra is not tilted.

It is interesting to note that the relation extensions of both algebras A and A' have the same quiver $\widetilde{Q}$. Therefore the associated cluster categories $\mathcal{C}_A$ and $\mathcal{C}_{A'}$ are categorifications of the same cluster algebra $\mathcal{A}(\widetilde{Q})$.

Acknowledgment

The authors thank Claire Amiot for useful discussions on the topic.

References

[1] Ibrahim Assem, *Tilted algebras of type* $\mathbf{A}_n$, Comm. Algebra **10** (1982), no. 19, 2121–2139, DOI 10.1080/00927878208822826. MR675347

[2] I. Assem, T. Brüstle, and R. Schiffler, *Cluster-tilted algebras as trivial extensions*, Bull. Lond. Math. Soc. **40** (2008), no. 1, 151–162, DOI 10.1112/blms/bdm107. MR2409188

[3] Ibrahim Assem, Flávio U. Coelho, and Sonia Trepode, *The bound quiver of a split extension*, J. Algebra Appl. **7** (2008), no. 4, 405–423, DOI 10.1142/S0219498808002928. MR2442069

[4] Ibrahim Assem, María Julia Redondo, and Ralf Schiffler, *On the first Hochschild cohomology group of a cluster-tilted algebra*, Algebr. Represent. Theory **18** (2015), no. 6, 1547–1576, DOI 10.1007/s10468-015-9551-x. MR3435102

[5] Ibrahim Assem, Ralf Schiffler, and Khrystyna Serhiyenko, *Cluster-tilted and quasi-tilted algebras*, J. Pure Appl. Algebra **221** (2017), no. 9, 2266–2288, DOI 10.1016/j.jpaa.2016.12.008. MR3631717

[6] Ibrahim Assem, Daniel Simson, and Andrzej Skowroński, *Elements of the representation theory of associative algebras. Vol. 1*, London Mathematical Society Student Texts, vol. 65, Cambridge University Press, Cambridge, 2006. Techniques of representation theory. MR2197389

[7] Ibrahim Assem and Dan Zacharia, *On split-by-nilpotent extensions*, Colloq. Math. **98** (2003), no. 2, 259–275, DOI 10.4064/cm98-2-10. MR2033112

[8] Jennifer Bélanger and Cecilia Tosar, *Shod string algebras*, Comm. Algebra **33** (2005), no. 8, 2465–2487, DOI 10.1081/AGB-200065183. MR2159481

[9] Klaus Bongartz, *Algebras and quadratic forms*, J. London Math. Soc. (2) **28** (1983), no. 3, 461–469, DOI 10.1112/jlms/s2-28.3.461. MR724715

[10] M. C. R. Butler and Claus Michael Ringel, *Auslander-Reiten sequences with few middle terms and applications to string algebras*, Comm. Algebra **15** (1987), no. 1-2, 145–179, DOI 10.1080/00927878708823416. MR876976

[11] Flávio Ulhoa Coelho and Marcelo Américo Lanzilotta, *Algebras with small homological dimensions*, Manuscripta Math. **100** (1999), no. 1, 1–11, DOI 10.1007/s002290050191. MR1714397

[12] Julie Dionne, *Algebres de cordes de type laura et conjecture de Skowronski* (French), ProQuest LLC, Ann Arbor, MI, 2008. Thesis (Ph.D.)–Universite de Sherbrooke (Canada). MR2712494

[13] E. L. Green, D. Happel, and D. Zacharia, *Projective resolutions over Artin algebras with zero relations*, Illinois J. Math. **29** (1985), no. 1, 180–190. MR769766

[14] Dieter Happel, *Triangulated categories in the representation theory of finite-dimensional algebras*, London Mathematical Society Lecture Note Series, vol. 119, Cambridge University Press, Cambridge, 1988. MR935124

[15] Dieter Happel, Idun Reiten, and Sverre O. Smalø, *Tilting in abelian categories and quasitilted algebras*, Mem. Amer. Math. Soc. **120** (1996), no. 575, viii+ 88, DOI 10.1090/memo/0575. MR1327209

[16] Dieter Happel and Claus Michael Ringel, *Tilted algebras*, Trans. Amer. Math. Soc. **274** (1982), no. 2, 399–443, DOI 10.2307/1999116. MR675063

[17] François Huard and Shiping Liu, *Tilted string algebras*, J. Pure Appl. Algebra **153** (2000), no. 2, 151–164, DOI 10.1016/S0022-4049(99)00101-2. MR1780740

[18] Mark Kleiner, Andrzej Skowroński, and Dan Zacharia, *On endomorphism algebras with small homological dimensions*, J. Math. Soc. Japan **54** (2002), no. 3, 621–648, DOI 10.2969/jmsj/1191593912. MR1900960

[19] Miki Oryu and Ralf Schiffler, *On one-point extensions of cluster-tilted algebras*, J. Algebra **357** (2012), 168–182, DOI 10.1016/j.jalgebra.2012.02.013. MR2905247

[20] Ralf Schiffler, *Quiver representations*, CMS Books in Mathematics/Ouvrages de Mathématiques de la SMC, Springer, Cham, 2014. MR3308668

[21] Stephen Zito, *Short proof of a conjecture concerning split-by-nilpotent extensions*, Arch. Math. (Basel) **111** (2018), no. 5, 479–483, DOI 10.1007/s00013-018-1208-7. MR3859429

Département de Mathématiques, Université de Sherbrooke, Sherbrooke, Québec, Canada, J1K 2R1

Email address: ibrahim.assem@usherbrooke.ca

Departamento de Matemática, Universidad Nacional del Sur, Av. Alem 1253, 8000 Bahía Blanca, Argentina

Email address: mredondo@criba.edu.ar

Department of Mathematics, University of Connecticut, Storrs, Connectivut 06269-3009

Email address: schiffler@math.uconn.edu

Contemporary Mathematics
Volume **761**, 2021
https://doi.org/10.1090/conm/761/15308

A characterization of representation infinite quiver settings

Grzegorz Bobiński

ABSTRACT. We characterize pairs $(Q, \mathbf{d})$ consisting of a quiver Q and a dimension vector $\mathbf{d}$, such that over a given algebraically closed field k there are infinitely many representations of Q of dimension vector $\mathbf{d}$. We also present an application of this result to the study of algebras with finitely many orbits with respect to the action of (the double product) of the group of units.

1. Introduction

Throughout the paper k is an algebraically closed field. A well-known theorem of Gabriel [**5**] states that the Euclidean quivers are the minimal representation infinite quivers, where a quiver is called representation infinite, if it has infinitely many indecomposable k-representations (up to isomorphism). One also shows that Q is representation infinite if and only if there exists a dimension vector $\mathbf{d}$ such that there are infinitely many representations of Q of dimension vector $\mathbf{d}$ (a stronger and more general version of this observation is a content of the famous second Brauer–Thrall conjecture, first proved by Bautista [**2**]). The main result of the paper can be viewed as a refinement of Gabriel's theorem. Namely, we show that the pairs $(Q, \mathbf{h}_Q)$, where Q is a Euclidean quiver and $\mathbf{h}_Q$ is the associated radical vector, are the minimal pairs $(Q', \mathbf{d})$ consisting of a quiver Q' and a dimension vector $\mathbf{d}$, such that there are infinitely many representations of Q' of dimension vector $\mathbf{d}$ (we call such pairs representation infinite quiver settings).

The proof of the result consists of two steps, which seem to be known before, but apparently have not been joined together. First, we use a result of Skowroński and Zwara [**15**] (see Proposition 3), which says that a pair $(Q, \mathbf{d})$ is a representation infinite quiver setting if and only if there exists a nonzero dimension vector $\mathbf{d}' \leq \mathbf{d}$ such that $q_Q(\mathbf{d}') \leq 0$, where q_Q is the associated Tits form. The second step is Proposition 4, which states that if $q_Q(\mathbf{d}) \leq 0$, for a nonzero dimension vector $\mathbf{d}$, then there exists a Euclidean subquiver Q' of Q (not necessarily full) such that $\mathbf{h}_{Q'} \leq \mathbf{d}'$, where $\mathbf{d}'$ is the restriction of $\mathbf{d}$ to Q'. This latter result should be well-known (at least to the experts), however we could not spot it in the literature. In particular, it seems that it has never been used in the context of quiver representations. Consequently, we include its short proof for completeness.

2010 *Mathematics Subject Classification.* Primary 16G20, 16G10, 14L30.

The author acknowledges the support of the National Science Center grant no. 2015/17/B/ST1/01731.

The problem discussed in the paper has a geometric aspect. Namely, a pair $(Q, \mathbf{d})$ is a representation finite (i.e., not representation infinite) quiver setting if and only if there are only finitely many orbits (with respect to a natural action) in the variety $\operatorname{rep}_Q(\mathbf{d})$ of representations of Q of dimension vector $\mathbf{d}$. In particular, there is a dense orbit in $\operatorname{rep}_Q(\mathbf{d})$, hence for example Schofield's description of the ring of semi-invariants [**13**] applies. Moreover, according to [**15**, Theorem 2] the degeneration order in $\operatorname{rep}_Q(\mathbf{d})$ coincides with the extension order.

Another source of interest in the problem comes from the study of algebras with finitely many orbits (see [**9**]). Here we say that an algebra A has finitely many orbits if there are only finitely many orbits in A with respect to the action of $U(A) \times U(A)$ given by $(u, v) * a := uav^{-1}$, where $U(A)$ is the group of units of A (see Section 3 for some motivation for this problem). As we explain in Section 3, if $\operatorname{rad}^2(A) = 0$, then one may associate to A a quiver setting $(\Delta_A, \mathbf{w}_A)$ in such a way, that A has finitely many orbits if and only if $(\Delta_A, \mathbf{w}_A)$ is representation finite. Consequently, our main theorem gives a criterion for A to have finitely many orbits. We compare our criterion with [**9**, Theorem 10] at the end of the paper.

2. Main result

By a *quiver* Q we mean a finite set Q_0 of *vertices* together with a finite set Q_1 of *arrows* and two maps $s = s_Q, t = t_Q \colon Q_1 \to Q_0$, which assign to each arrow $\alpha \in Q_1$ its *starting vertex* $s\alpha$ and *terminating vertex* $t\alpha$. By a *walk* in a quiver Q we mean a sequence $(x_0, \dots, x_n)$ of vertices such that, for each $1 \leq i \leq n$, x_{i-1} and x_i are connected by an arrow (i.e. there exists an arrow $\alpha \in Q_1$ such that $\{s\alpha, t\alpha\} = \{x_{i-1}, x_i\}$). A quiver Q is *connected* if for any vertices $x, y \in Q_0$ there exists a walk $(x_0, \dots, x_n)$ in Q such that $x_0 = x$ and $x_n = y$. A walk $(x_0, \dots, x_n)$ in Q is called a *cycle* if $n > 0$, $x_{i-1} \neq x_{i+1}$ for each $0 < i < n$, and $x_0 = x_n$ (in particular, $n \neq 2$). We say that a quiver Q has *multiple arrows* if there exist arrows $\alpha \neq \beta$ which connect the same vertices, i.e. $\{s\alpha, t\alpha\} = \{s\beta, t\beta\}$.

By a *dimension vector* for a quiver Q we mean an element of $\mathbb{N}^{Q_0}$. A *representation* of Q of dimension vector $\mathbf{d}$ is a tuple $M = (M_\alpha)_{\alpha \in Q_1}$ of linear maps $M_\alpha \colon k^{\mathbf{d}(s\alpha)} \to k^{\mathbf{d}(t\alpha)}$, $\alpha \in Q_1$. The set of such representations is an affine space and we denote it by $\operatorname{rep}_Q(\mathbf{d})$. Let $\operatorname{GL}(\mathbf{d})$ be the set of tuples $g = (g_x)_{x \in Q_0}$ such that $g_x \colon k^{\mathbf{d}(x)} \to k^{\mathbf{d}(x)}$ is a k-linear automorphism for each $x \in Q_0$. Two representations M and N of dimension vector $\mathbf{d}$ are said to be *isomorphic* if the exists a tuple $g \in \operatorname{GL}(\mathbf{d})$ such that $N_\alpha = g_{t\alpha} M_\alpha g_{s\alpha}^{-1}$ for each $\alpha \in Q_1$. Following [**3**] we call a pair $(Q, \mathbf{d})$ consisting of a quiver Q and a dimension vector $\mathbf{d}$ a *quiver setting*. A quiver setting $(Q, \mathbf{d})$ is called *representation finite* if there are only finitely many (up to isomorphism) representations of Q of dimension vector $\mathbf{d}$. Equivalently, there are finitely many $\operatorname{GL}(\mathbf{d})$-orbits in $\operatorname{rep}_Q(\mathbf{d})$ (with respect to the action induced by the isomorphism formula). A quiver setting which is not representation finite is said to be *representation infinite*.

Let Q and Q' be two quivers. By a *quiver morphism* $\psi \colon Q \to Q'$ we mean a pair of functions $\psi_0 \colon Q_0 \to Q_0'$ and $\psi_1 \colon Q_1 \to Q_1'$ such that $s_{Q'}(\psi_1 \alpha) = \psi_0(s_Q \alpha)$ and $t_{Q'}(\psi_1 \alpha) = \psi_0(t_Q \alpha)$. A quiver morphism $\psi \colon Q \to Q'$ is called *injective* if both ψ_0 and ψ_1 are injective. If this is the case, then Q may be viewed as a subquiver of Q' (where we identify Q with its image via ψ). Note that in the paper the subquivers are not necessarily fully.

If $\psi \colon Q \to Q'$ is a quiver morphism and $\mathbf{d}$ is a dimension vector for Q, then one defines a dimension vector $\psi_* \mathbf{d}$ for Q' via $(\psi_* \mathbf{d})(y) := \sum_{x \in \psi_0^{-1}(y)} \mathbf{d}(x)$, for $y \in Q_0'$. Dually, if $\mathbf{d}'$ is the dimension vector for Q', then $\psi^* \mathbf{d}'$ is a dimension vector for Q given by $(\psi^* \mathbf{d}')(x) := \mathbf{d}'(\psi_0 x)$, for $x \in Q_0$.

Given two quiver settings $(Q, \mathbf{d})$ and $(Q', \mathbf{d}')$ we write $(Q, \mathbf{d}) \leq (Q', \mathbf{d}')$, if there exists an injective quiver morphism $\psi \colon Q \to Q'$ such that $\mathbf{d} \leq \psi^* \mathbf{d}'$. Since ψ is injective, the inequality $\mathbf{d} \leq \psi^* \mathbf{d}'$ is equivalent to the inequality $\psi_* \mathbf{d} \leq \mathbf{d}'$. Thus the condition $(Q, \mathbf{d}) \leq (Q', \mathbf{d}')$ means that Q can be identified with a subquiver of Q' in such a way that $\mathbf{d} \leq \mathbf{d}'|_{Q_0'}$ (with respect to this identification). Obviously, the relation $\leq$ is only a preorder on the class of quiver settings.

Recall that Q is a *Euclidean quiver* if its *underlying graph* (the graph obtained from Q by forgetting the orientations of arrows) is one of the following diagrams:

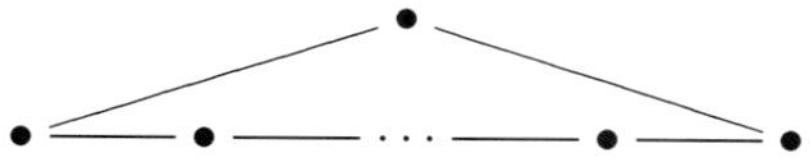

$n + 1$ vertices, $n \geq 0$,

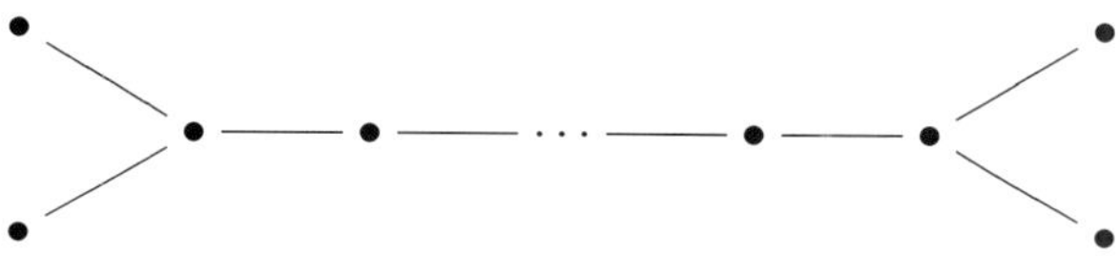

$n + 1$ vertices, $n \geq 4$,

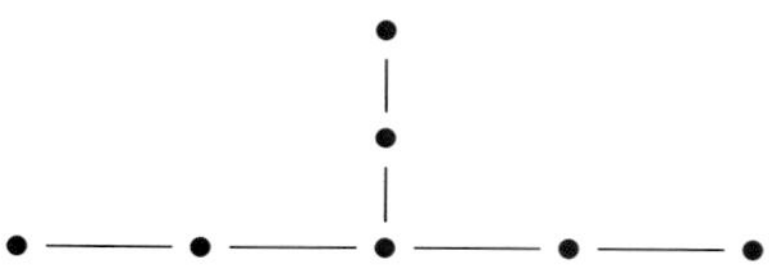

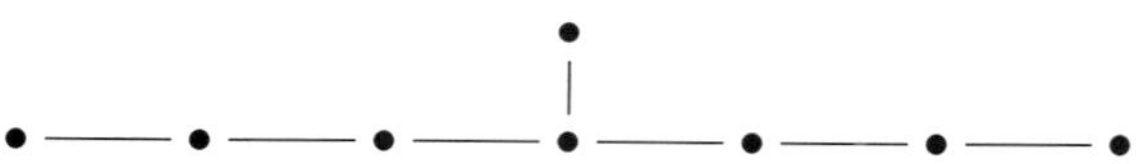

In the above cases we say that Q is a Euclidean quiver of type $\tilde{\mathbb{A}}_n$ ($\tilde{\mathbb{A}}$ for short), $\tilde{\mathbb{D}}_n$ ($\tilde{\mathbb{D}}$ for short), $\tilde{\mathbb{E}}_6$, $\tilde{\mathbb{E}}_7$, and $\tilde{\mathbb{E}}_8$, respectively.

For each Euclidean quiver Q there is a distinguished dimension vector $\mathbf{h}_Q$ defined as follows:

$$\mathbf{h}_Q := \begin{cases} \begin{smallmatrix} & & 1 & & \\ 1 & 1 & \cdots & 1 & 1 \end{smallmatrix} & \text{if } Q \text{ is of type } \tilde{\mathbb{A}}, \\ \begin{smallmatrix} 1 & & & & & & 1 \\ & 2 & 2 & \cdots & 2 & 2 & \\ 1 & & & & & & 1 \end{smallmatrix} & \text{if } Q \text{ is of type } \tilde{\mathbb{D}}, \\ \begin{smallmatrix} & & 1 & & \\ & & 2 & & \\ 1 & 2 & 3 & 2 & 1 \end{smallmatrix} & \text{if } Q \text{ is of type } \tilde{\mathbb{E}}_6, \\ \begin{smallmatrix} & & & 2 & & & \\ 1 & 2 & 3 & 4 & 3 & 2 & 1 \end{smallmatrix} & \text{if } Q \text{ is of type } \tilde{\mathbb{E}}_7, \\ \begin{smallmatrix} & & 3 & & & & & \\ 2 & 4 & 6 & 5 & 4 & 3 & 2 & 1 \end{smallmatrix} & \text{if } Q \text{ is of type } \tilde{\mathbb{E}}_8. \end{cases}$$

The main theorem of the paper is the following.

THEOREM 1. *A quiver setting $(Q, \mathbf{d})$ is representation infinite if and only if there exists a Euclidean quiver Q' such that $(Q', \mathbf{h}_{Q'}) \leq (Q, \mathbf{d})$. In other words, $(Q, \mathbf{d})$ is representation infinite if and only if Q has a Euclidean subquiver Q' such that $\mathbf{h}_{Q'} \leq \mathbf{d}|_{Q_0'}$.*

We have the following reformulation of Theorem 1, which can be seen as a refinement of Gabriel's Theorem.

COROLLARY 2. *The quiver settings $(Q, \mathbf{h}_Q)$, where Q is a Euclidean quiver, are precisely the minimal representation infinite quiver settings.*

An important role in the proof of the above theorem is played by the Tits form q_Q, which is the quadratic form $q_Q \colon \mathbb{Z}^{Q_0} \to \mathbb{Z}$ defined by

$$q_Q(\mathbf{d}) := \sum_{x \in Q_0} \mathbf{d}(x)^2 - \sum_{\alpha \in Q_1} \mathbf{d}(s\alpha)\mathbf{d}(t\alpha) \qquad (\mathbf{d} \in \mathbb{Z}^{Q_0}).$$

The following fact is proved in [**15**, Section 3].

PROPOSITION 3. *A quiver setting $(Q, \mathbf{d})$ is representation infinite if and only if there exists a nonzero dimension vector $\mathbf{d}'$ for Q such that $\mathbf{d}' \leq \mathbf{d}$ and $q_Q(\mathbf{d}') \leq 0$.*

The original proof of Skowroński and Zwara uses a deep theorem due to Kac [**6**, Theorem 2]. Here we present a more elementary proof, which was suggested by William Crawley–Boevey. We need to recall some basic equalities. First, if M is a representation of dimension vector $\mathbf{d}$, then

$$q_Q(\mathbf{d}) = \dim_k \operatorname{End}_Q(M) - \dim_k \operatorname{Ext}^1_Q(M, M)$$

(see for example [**10**, Lemma 2.2]) and

$$\dim \mathrm{GL}(\mathbf{d})M = \dim \mathrm{GL}(\mathbf{d}) - \dim_k \operatorname{End}_Q(M)$$

(see for example [**8**, (2.2)]). Moreover, if $\mathbf{d}$ is dimension vector, then immediately from the definition it follows that

$$q_Q(\mathbf{d}) = \dim \mathrm{GL}(\mathbf{d}) - \dim \operatorname{rep}_Q(\mathbf{d}).$$

In particular, the orbit $\mathrm{GL}(\mathbf{d})M$ is dense in $\operatorname{rep}_Q(\mathbf{d})$ if and only if $\operatorname{Ext}^1_Q(M, M) = 0$. Indeed, the orbit $\mathrm{GL}(\mathbf{d})M$ is dense if and only if $\dim \mathrm{GL}(\mathbf{d})M = \dim \operatorname{rep}_Q(\mathbf{d})$, and

$$\dim \mathrm{GL}(\mathbf{d})M = \dim \operatorname{rep}_Q(\mathbf{d}) - \dim_k \operatorname{Ext}^1_Q(M, M).$$

PROOF OF PROPOSITION 3. Assume first that the quiver setting $(Q, \mathbf{d})$ is representation finite. In this case, one uses an argument due to Tits. Namely, for each nonnzero dimension vector $\mathbf{d}' \leq \mathbf{d}$, there are only finitely many $\mathrm{GL}(\mathbf{d}')$-orbits in $\operatorname{rep}_Q(\mathbf{d}')$. Consequently, there is a dense orbit $\mathrm{GL}(\mathbf{d}')M$ in $\operatorname{rep}_Q(\mathbf{d}')$, hence $\operatorname{Ext}^1_Q(M, M) = 0$, thus $q_Q(\mathbf{d}') = \dim_k \operatorname{End}_Q(M) > 0$.

Now assume that the quiver setting $(Q, \mathbf{d})$ is representation infinite. Then there exists a dimension vector $\mathbf{d}' \leq \mathbf{d}$ such that there are infinitely many pairwise nonisomorphic indecomposable representations of Q of dimension vector $\mathbf{d}'$ (in particular, $\mathbf{d}' \neq 0$). If there exists an indecomposable representation X of Q of dimension vector $\mathbf{d}'$ which is not a *brick* (i.e. $\operatorname{End}_Q(X) \neq k$), then by [**11**, Theorem] there exists a subrepresentation Y of X which is a brick (in particular, indecomposable), such that $\operatorname{Ext}^1_Q(Y, Y) \neq 0$. If $\mathbf{d}''$ is the dimension vector of Y, then we have $\mathbf{d}'' \neq 0$, $\mathbf{d}'' \leq \mathbf{d}' \leq \mathbf{d}$, and

$$q_Q(\mathbf{d}'') = \dim_k \operatorname{End}_Q(Y) - \dim_k \operatorname{Ext}^1_Q(Y, Y) \leq 1 - 1 = 0.$$

Hence we may assume that every indecomposable representation of Q of dimension vector $\mathbf{d}'$ is a brick. If $\operatorname{Ext}^1_Q(X, X) \neq 0$, for such an indecomposable representation X, then, similarly as above, we get $q_Q(\mathbf{d}') \leq 0$. Thus we may assume $\operatorname{Ext}^1_Q(X, X) = 0$, for each indecomposable representation X of Q of dimension vector $\mathbf{d}'$. Consequently, the orbit $\mathrm{GL}(\mathbf{d}')X$ is dense in $\operatorname{rep}_Q(\mathbf{d}')$, for every such a representation X. However, there can be at most one dense orbit in $\operatorname{rep}_Q(\mathbf{d}')$, hence we get a contradiction with the assumption that there are infinitely many pairwise nonisomorphic indecomposable representations of Q of dimension vector $\mathbf{d}'$. □

Now we may prove one implication of Theorem 1.

PROOF OF THEOREM 1, PART I. We first prove that if there exists a Euclidean quiver Q' such that $(Q', \mathbf{h}'_Q) \leq (Q, \mathbf{d})$, then $(Q, \mathbf{d})$ is representation infinite. In fact, it is enough to prove that $(Q, \mathbf{h}_Q)$ is representation infinite for each Euclidean quiver Q. This follows from well-known representation theory of Euclidean quivers (see for example [**12**, Section 3.6]), but we may also refer to Proposition 3, since $q_Q(\mathbf{h}_Q) = 0$ as one easily checks. □

Before we continue the proof, we need some more notation. Let $(-,-) = (-,-)_Q \colon \mathbb{Z}^{Q_0} \times \mathbb{Z}^{Q_0} \to \mathbb{Z}$ be the symmetric bilinear form associated with q_Q, i.e.

$$(\mathbf{d}, \mathbf{d}') := q_Q(\mathbf{d} + \mathbf{d}') - q_Q(\mathbf{d}) - q_Q(\mathbf{d}') \qquad (\mathbf{d}, \mathbf{d}' \in \mathbb{Z}^{Q_0}).$$

In particular,

$$q_Q(\mathbf{d}) = \tfrac{1}{2}(\mathbf{d}, \mathbf{d}) = \tfrac{1}{2} \sum_{x \in Q_0} \mathbf{d}(x)(\mathbf{d}, \mathbf{e}_x),$$

where $\mathbf{e}_x$, $x \in Q_0$, are the standard basis vectors.

PROPOSITION 4. *Let $(Q, \mathbf{d})$ be a quiver setting such that $\mathbf{d}$ is nonzero and $q_Q(\mathbf{d}) \leq 0$. Then there exists a Euclidean subquiver Q' of Q such that $\mathbf{h}_{Q'} \leq \mathbf{d}|_{Q'_0}$.*

PROOF. Obviously we may assume that Q is connected and $\mathbf{d}$ is *sincere*, i.e. $\mathbf{d}(x) \neq 0$ for each $x \in Q_0$. If either there is a *loop* (i.e. an arrow $\alpha \in Q_1$ with $s\alpha = t\alpha$) in Q or Q has multiple arrows, then one easily sees that there is a Euclidean subquiver Q' of Q of type $\tilde{\mathbb{A}}$ such that $\mathbf{h}_{Q'} \leq \mathbf{d}|_{Q'_0}$. Thus for the rest of the proof we assume that Q has neither loops nor multiple arrows. In particular, this implies that $|Q_0| \geq 3$. Indeed, if $Q_0 = \{x\}$, then $q_Q(\mathbf{d}) = \mathbf{d}(x)^2 > 0$, while if $Q_0 = \{x, y\}$, then

$$q_Q(\mathbf{d}) = \mathbf{d}(x)^2 - \mathbf{d}(x)\mathbf{d}(y) + \mathbf{d}(y)^2 = \tfrac{1}{2}[\mathbf{d}(x)^2 + (\mathbf{d}(x) - \mathbf{d}(y))^2 + \mathbf{d}(y)^2] > 0.$$

We observe that we may assume $(\mathbf{d}, \mathbf{e}_x) \leq q_Q(\mathbf{d})$, for each $x \in Q_0$. Indeed, if $(\mathbf{d}, \mathbf{e}_x) \geq q_Q(\mathbf{d}) + 1$ for some $x \in Q_0$, then

$$\begin{aligned} q_Q(\mathbf{d} - \mathbf{e}_x) &= \tfrac{1}{2}(\mathbf{d} - \mathbf{e}_x, \mathbf{d} - \mathbf{e}_x) \\ &= q_Q(\mathbf{d}) - (\mathbf{d}, \mathbf{e}_x) + q_Q(\mathbf{e}_x) \leq q_Q(\mathbf{d}) - (q_Q(\mathbf{d}) + 1) + 1 \leq 0, \end{aligned}$$

hence we may replace $\mathbf{d}$ by $\mathbf{d} - \mathbf{e}_x$ and proceed by induction. Using the above assumption we get

$$q_Q(\mathbf{d}) = \tfrac{1}{2} \sum_{x \in Q_0} \mathbf{d}(x)(\mathbf{d}, \mathbf{e}_x) \leq \tfrac{1}{2} q_Q(\mathbf{d}) \sum_{x \in Q_0} \mathbf{d}(x).$$

Consequently, if $q_Q(\mathbf{d}) < 0$, then $\sum_{x \in Q_0} \mathbf{d}(x) \leq 2$, which contradicts the inequality $|Q_0| \geq 3$, since $\sum_{x \in Q_0} \mathbf{d}(x) \geq |Q_0|$. We conclude $q_Q(\mathbf{d}) = 0$. Using again that $(\mathbf{d}, \mathbf{e}_z) \leq q_Q(\mathbf{d}) = 0$, for each $z \in Q_0$, we get, for $x \in Q_0$,

$$0 = q_Q(\mathbf{d}) \geq (\mathbf{d}, \mathbf{e}_x) \geq \sum_{y \in Q_0} \mathbf{d}(y)(\mathbf{d}, \mathbf{e}_y) = 2q_Q(\mathbf{d}) = 0,$$

i.e., $(\mathbf{d}, \mathbf{e}_x) = 0$, for each $x \in Q_0$. From the famous Vinberg's characterization of Euclidean quivers (see for example [**7**, Corollary 4.3]), this implies that Q is a Euclidean quiver. Moreover, another well-known property of Euclidean quivers (see for example [**1**, Lemma VII.4.2]) implies that $\mathbf{d} = m\mathbf{h}_Q$, for some $m > 0$, thus in particular $\mathbf{d} \geq \mathbf{h}_Q$. □

We finish now the proof of Theorem 1.

Proof of Theorem 1, Part II. Assume that $(Q, \mathbf{d})$ is a representation infinite quiver setting. We know from Proposition 3 that there exists a nonzero dimension vector $\mathbf{d}'$ for Q such that $\mathbf{d}' \leq \mathbf{d}$ and $q_Q(\mathbf{d}') \leq 0$. Consequently, Proposition 4 implies that there exists a Euclidean subquiver Q' of Q such that $\mathbf{h}_{Q'} \leq \mathbf{d}'|_{Q'_0} \leq \mathbf{d}|_{Q'_0}$. □

3. An application to the algebras with finitely many orbits

For a k-algebra A we denote by $U(A)$ the group of units of A. By abuse of notation (and following [**9**]), by an $U(A)$-*orbit in* A we mean an orbit with respect to the action of $U(A) \times U(A)$ on A given by $(u, v) * a := uav^{-1}$. In [**9**] the authors study the algebras with finitely many $U(A)$-orbits. In order to put this study in a wider context, we introduce some notation.

First, for an algebra A, $\mathcal{I}(A)$ and $\mathcal{L}(A)$ denote the lattices of ideals and left ideals, respectively. The group $U(A)$ acts on $\mathcal{L}(A)$ by right translations and we denote by $\mathcal{C}(A)$ the orbit space. Then $\mathcal{C}(A)$ is a semigroup, with the product given by $[I][J] := [IJ]$, where IJ is the linear subspace of A spanned by the products ab, $a \in I$, $b \in J$. The following theorem is one source of motivation for studying algebras with finitely many orbits.

Theorem 5 ([**9**, Theorem 6]). *Consider the following conditions for a finite dimensional k-algebra A.*

(1) *A is of finite representation type.*
(2) *$\mathcal{C}(A)$ is finite.*
(3) *A has finitely many $U(A)$-orbits.*
(4) *$\mathcal{I}(A)$ is a distributive lattice.*

Then (1) $\implies$ (2) $\implies$ (3) $\implies$ (4).

Recall that an algebra A is of *finite representation type*, if there are only finitely many indecomposable A-modules (up to isomorphism). We remark that the implication (3) $\implies$ (4) holds, since we are working over an algebraically closed, hence infinite, field.

In [**9**, Section 3] the authors study the algebras with finitely many orbits such that $\mathrm{rad}^2(A) = 0$, where $\mathrm{rad}(A)$ is the Jacobson radical of A. We discuss now a connection of this case with quiver settings. In particular from now on, we assume all algebras have a vanishing radical square.

Let A be a finite dimensional algebra (with $\operatorname{rad}^2(A) = 0$). Put $\overline{A} := A/\operatorname{rad}(A)$. By Wedderburn–Artin Theorem there exist positive integers $n_1, \ldots, n_l$ such that

$$\overline{A} \simeq \mathbb{M}_{n_1}(k) \times \cdots \times \mathbb{M}_{n_l}(k).$$

Let $\overline{e}_1, \ldots, \overline{e}_l$ be the idempotents corresponding to this decomposition, and $e_1, \ldots, e_l$ their lifts. For $1 \leq i \leq l$, we put $\overline{A}_i := \overline{e}_i \overline{A} \overline{e}_i$.

We associate to A a quiver setting $(\Delta_A, \mathbf{w}_A)$ in the following way. The vertices of Δ_A are the pairs $(0, i)$, $(1, i)$, for $1 \leq i \leq l$. There are only arrows of the form $(1, j) \to (0, i)$ and the number of arrows from $(1, j)$ to $(0, i)$ equals the length $r_{i,j}$ of $e_i \operatorname{rad}(A) e_j$ as an $\overline{A}_i$-$\overline{A}_j$-bimodule. Observe that $\overline{A}_i \otimes_k \overline{A}_j$ is a simple algebra, thus $e_i \operatorname{rad}(A) e_j \simeq S^{r_{i,j}}$, where $S = (\mathbb{M}_{n_i \times n_j}(k))^{r_{i,j}}$ is the unique simple $\overline{A}_i$-$\overline{A}_j$-bimodule. Finally, $\mathbf{w}_A(0, i) := n_i =: \mathbf{w}_A(1, i)$, for each $1 \leq i \leq l$.

We have the following result.

PROPOSITION 6. *Let A be a finite dimensional algebra with $\operatorname{rad}^2(A) = 0$. Then A has finitely many orbits if and only if $(\Delta_A, \mathbf{w}_A)$ is a representation finite quiver setting. Thus A has finitely many $U(A)$-orbits if and only if there is no subquiver Q of Δ_A of Euclidean type such that $\mathbf{h}_Q \leq \mathbf{w}_A|_{Q_0}$.*

PROOF. Recall first from [**9**, Proposition 9] that A has finitely many $U(A)$-orbits if and only if there are finitely many $U(A) \times U(A)$-orbits in $\operatorname{rad}(A)$ (recall that by convention $U(A)$-orbits in A are actually $U(A)$-$U(A)$-orbits, hence one implication is obvious). Moreover, the $U(A) \times U(A)$-orbits in $\operatorname{rad}(A)$ coincide with the $U(\overline{A}) \times U(\overline{A})$-orbits in $\operatorname{rad}(A)$. Since $e_i \operatorname{rad}(A) e_j \simeq (\mathbb{M}_{n_i \times n_j}(k))^{r_{i,j}}$, thus

$$\operatorname{rad}(A) = \prod_{1 \leq i,j \leq l} (\mathbb{M}_{n_i \times n_j}(k))^{r_{i,j}} = \operatorname{rep}_{\Delta_A}(\mathbf{w}_A).$$

Moreover,

$$U(\overline{A}) \times U(\overline{A}) = \prod_{i=1}^{n} \mathrm{GL}_{n_i}(k) \times \prod_{i=1}^{n} \mathrm{GL}_{n_i}(k) = \mathrm{GL}(\mathbf{w}_A),$$

and the action of $U(\overline{A}) \times U(\overline{A})$ on $\operatorname{rad}(A)$ coincides with the action of $\mathrm{GL}(\mathbf{w}_A)$ on $\operatorname{rep}_{\Delta_A}(\mathbf{w}_A)$. This finishes the proof of the former statement. The latter one follows from the former one and Theorem 1. $\square$

Note that if $(\Delta, \mathbf{w}) = (\Delta_A, \mathbf{w}_A)$, for an algebra A, then we have the following:

(1) the vertex set Δ_0 is a disjoint union of two subsets Δ_0' and Δ_0'' such that there is a bijection $\delta \colon \Delta_0'' \to \Delta_0'$;
(2) every arrow starts in Δ_0'' and terminates in Δ_0';
(3) $\mathbf{w}(\delta y) = \mathbf{w}(y)$, for each $y \in \Delta_0''$.

Observe that every setting $(\Delta, \mathbf{w})$ with the above properties is (up to isomorphism) of the form $(\Delta_A, \mathbf{w}_A)$, for some algebra A. Namely, as a vector space A equals $\prod_{x \in \Delta_0'} \mathbb{M}_{\mathbf{w}(x)}(k) \times \prod_{\substack{x \in \Delta_0' \\ y \in \Delta_0''}} (\mathbb{M}_{\mathbf{w}(x) \times \mathbf{w}(y)}(k))^{r_{x,y}}$, where $r_{x,y}$ is the number of arrows from y to x, and the multiplication is given by

$$((M_x)_{x\in\Delta_0'}, (M_{x,y,i})_{\substack{x\in\Delta_0' \\ y\in\Delta_0'' \\ 1\le i\le r_{x,y}}})((N_x)_{x\in\Delta_0'}, (N_{x,y,i})_{\substack{x\in\Delta_0' \\ y\in\Delta_0'' \\ 1\le i\le r_{x,y}}}) \\ := ((M_xN_x)_{x\in\Delta_0'}, (M_xN_{x,y,i} + M_{x,y,i}N_{\delta(y)})_{\substack{x\in\Delta_0' \\ y\in\Delta_0'' \\ 1\le i\le r_{x,y}}}).$$

We compare now Proposition 6 with [**9**, Theorem 10]. First observe that the authors of [**9**] assume that the lattice $\mathcal{I}(A)$ is distributive. This assumption is justified by the implication (3) $\Longrightarrow$ (4) of Theorem 5. Using [**4**, Theorem 1] $\mathcal{I}(A)$ being distributive is equivalent to $r_{i,j} \le 1$, for all i, j. In other words, it means there are no multiple arrows in Δ_A. Recall, that if there are multiple arrows in Δ_A, then one easily finds a subquiver Q of Δ_A of type $\tilde{\mathbb{A}}_1$ such that $\mathbf{h}_Q \le \mathbf{w}_A|_{Q_0}$.

There are two (equivalent) sets of conditions in [**9**, Theorem 10]. We concentrate on the one which is easier to formulate. Namely, we have the following set of conditions:

(1) there are no cycles in Δ_A;
(2) if $\mathbf{w}_A(x) \ge 2$, then there at most three arrows starting and at most three arrows terminating at x;
(3) if $\mathbf{w}_A(s\alpha) \ge 2$ and $\mathbf{w}_A(t\alpha) \ge 2$, for an arrow α of Δ_A, then the number of arrows starting at $s\alpha$ plus the number of arrows terminating at $t\alpha$ is at most 4.

Obviously, the first condition means that $(Q, \mathbf{h}_Q) \not\le (\Delta_A, \mathbf{w}_A)$, where Q is a quiver of type $\tilde{\mathbb{A}}$. In the same way, the second condition excludes the settings $(Q, \mathbf{h}_Q)$, where Q is a quiver of type $\tilde{\mathbb{D}}_4$. Finally, the last condition excludes (up to duality) the settings $(Q, \mathbf{d})$, where Q is the quiver

and $\mathbf{d} := \begin{smallmatrix} 2 & 1 & 1 \\ 1 & 2 & \end{smallmatrix}$. However, the above setting is representation finite. This means that unfortunately [**9**, Theorem 10] is false. Note that, for $(Q, \mathbf{d})$ as above, $(Q, \mathbf{d}) \le (Q', \mathbf{h}_{Q'})$, where Q' is a Euclidean quiver of one of the types $\tilde{\mathbb{D}}_n$, $n > 4$, $\tilde{\mathbb{E}}_6$, $\tilde{\mathbb{E}}_7$, $\tilde{\mathbb{E}}_8$, thus the conditions in [**9**, Theorem 10] are sufficient for A to have finitely many $U(A)$-orbits (but not necessary as pointed out above).

Acknowledgment

We thank Daniel Simson for a hint, which allowed to significantly shorten the proof of Proposition 4. In particular, the proof of the equality $q_Q(\mathbf{d}) = 0$ below follows arguments in the proof of [**14**, Theorem XIV.1.3].

References

[1] Ibrahim Assem, Daniel Simson, and Andrzej Skowroński, *Elements of the representation theory of associative algebras. Vol. 1*, London Mathematical Society Student Texts, vol. 65, Cambridge University Press, Cambridge, 2006. Techniques of representation theory. MR2197389

[2] Raymundo Bautista, *On algebras of strongly unbounded representation type*, Comment. Math. Helv. **60** (1985), no. 3, 392–399, DOI 10.1007/BF02567422. MR814146

[3] Raf Bocklandt, *Symmetric quiver settings with a regular ring of invariants*, Linear Algebra Appl. **365** (2003), 25–43, DOI 10.1016/S0024-3795(02)00407-X. Special issue on linear algebra methods in representation theory. MR1987326

[4] Victor Camillo, *Distributive modules*, J. Algebra **36** (1975), no. 1, 16–25, DOI 10.1016/0021-8693(75)90151-9. MR573061

[5] Peter Gabriel, *Unzerlegbare Darstellungen. I* (German, with English summary), Manuscripta Math. **6** (1972), 71–103; correction, ibid. 6 (1972), 309, DOI 10.1007/BF01298413. MR332887

[6] Victor G. Kac, *Infinite root systems, representations of graphs and invariant theory*, Invent. Math. **56** (1980), no. 1, 57–92, DOI 10.1007/BF01403155. MR557581

[7] Victor G. Kac, *Infinite-dimensional Lie algebras*, 3rd ed., Cambridge University Press, Cambridge, 1990. MR1104219

[8] Hanspeter Kraft and Christine Riedtmann, *Geometry of representations of quivers*, Representations of algebras (Durham, 1985), London Math. Soc. Lecture Note Ser., vol. 116, Cambridge Univ. Press, Cambridge, 1986, pp. 109–145. MR897322

[9] Jan Okniński and Lex E. Renner, *Algebras with finitely many orbits*, J. Algebra **264** (2003), no. 2, 479–495, DOI 10.1016/S0021-8693(03)00129-7. MR1981417

[10] Claus Michael Ringel, *Representations of K-species and bimodules*, J. Algebra **41** (1976), no. 2, 269–302, DOI 10.1016/0021-8693(76)90184-8. MR422350

[11] Claus Michael Ringel, *Bricks in hereditary length categories*, Results Math. **6** (1983), no. 1, 64–70, DOI 10.1007/BF03323325. MR714659

[12] Claus Michael Ringel, *Tame algebras and integral quadratic forms*, Lecture Notes in Mathematics, vol. 1099, Springer-Verlag, Berlin, 1984. MR774589

[13] Aidan Schofield, *Semi-invariants of quivers*, J. London Math. Soc. (2) **43** (1991), no. 3, 385–395, DOI 10.1112/jlms/s2-43.3.385. MR1113382

[14] Daniel Simson and Andrzej Skowroński, *Elements of the representation theory of associative algebras. Vol. 2*, London Mathematical Society Student Texts, vol. 71, Cambridge University Press, Cambridge, 2007. Tubes and concealed algebras of Euclidean type. MR2360503

[15] Andrzej Skowroński and Grzegorz Zwara, *Degenerations in module varieties with finitely many orbits*, Trends in the representation theory of finite-dimensional algebras (Seattle, WA, 1997), Contemp. Math., vol. 229, Amer. Math. Soc., Providence, RI, 1998, pp. 343–356, DOI 10.1090/conm/229/03327. MR1676229

Faculty of Mathematics and Computer Science, Nicolaus Copernicus University, ul. Chopina 12/18, 87-100 Toruń, Poland

Email address: gregbob@mat.umk.pl

Contemporary Mathematics
Volume **761**, 2021
https://doi.org/10.1090/conm/761/15309

Dualities for modules of finite projective dimension

Birge Huisgen-Zimmermann and Manuel Saorín

Dedicated to José Antonio de la Peña on the occasion of his sixtieth birthday

ABSTRACT. The purpose of this article is threefold: The first is to provide an overview of old and recent findings which relate finite dimensional tilting modules over a finite dimensional algebra Λ to dualities defined on resolving subcategories of $\mathcal{P}^{<\infty}(\Lambda\text{-mod})$, the category of finitely generated left Λ-modules of finite projective dimension. The second goal is to supplement the picture available so far by new results, based on the transfer of representation-theoretic information via such partial dualities, and to place the array of contravariant equivalences into a broader context. The third objective is to apply the findings (those on dualities induced by strong tilting modules in particular) to truncated path algebras: A streamlined presentation of existing theory is accompanied by a fully worked example. The pertinent computational methods are developed at the end; they show that the algebras obtained from a truncated path algebra Λ through iterated strong tilting, as well as their $\mathcal{P}^{<\infty}$-categories and $\mathcal{P}^{<\infty}$-approximations, may be constructed from the quiver and Loewy length of Λ.

1. Introduction

During the past four decades, tilting functors have asserted their role as powerful tools for comparing the module categories of finite dimensional algebras. Next to the fact that they often induce "almost-equivalences" on the level of these categories themselves, they give rise to triangle equivalences on the level of the corresponding derived categories. The multi-step trip to this insight started with the reflection functors of Bernstein-Gelfand-Ponomarev which underwent several stages of generalization through work of Auslander-Platzeck-Reiten, Brenner-Butler, Happel-Ringel, Bongartz, Miyashita and others (see [**1**,**4**,**9**,**10**,**22**,**32**]). The breakthrough to equivalences linking the derived category of an algebra to those of its tilted companions was ushered in by Happel [**20**] and Cline-Parshall-Scott [**13**], then fully established by Rickard [**33**]; see also Keller's approach by way of DG-algebras [**30**]. The resulting derived Morita theory has found immediate application to finite dimensional representation theory, e.g., to modular representations of finite groups (see [**31**] for the connection to Broué's conjecture and [**34**] for partial confirmations to date).

By contrast, the present article focuses on *contravariant* equivalences of an equally natural format which are induced by tilting modules. These dualities

---

2010 *Mathematics Subject Classification.* Primary 16G10; Secondary 16D90, 16E05, 16E10, 16G20.

provide a separate set of bridges connecting an algebra Λ to its tilted algebras. (Throughout, Λ denotes a finite dimensional algebra over a field.)

Our notion of a *tilting module* T is Miyashita's generalized version, allowing for arbitrary finite projective dimensions of T, as opposed to restricting to $\operatorname{p\,dim} T \le 1$. (For unexplained italicized terms, we refer to Section 2.) In particular, any tilting (left) module over an algebra Λ belongs to the full subcategory $\mathcal{P}^{<\infty}(\Lambda\text{-mod})$ consisting of the finitely generated left Λ-modules with finite projective dimension. Among the tilting modules, distinguished specimens dubbed *strong* were first singled out by Auslander and Reiten in [**5**]. They are the tilting modules T which are Ext-*injective relative to the objects in* $\mathcal{P}^{<\infty}(\Lambda\text{-mod})$, meaning that $\operatorname{Ext}^i_\Lambda(\mathcal{P}^{<\infty}(\Lambda\text{-mod}), T) = 0$ for all $i \ge 1$; in light of Section 2.C, this description of strong tilting modules is equivalent to the original definition. Due to [**5**], strong tilting modules are unique up to repeats of their indecomposable direct summands; i.e., there is at most one *basic* strong tilting object in Λ-mod, and existence is equivalent to *contravariant finiteness* of $\mathcal{P}^{<\infty}(\Lambda\text{-mod})$. Subsequently, these special tilting modules were re-encountered from a different angle by Happel and Unger in [**23**]: Existence provided, the basic strong tilting object in Λ-mod is the unique smallest element in the poset of all basic tilting objects in Λ-mod. The pertinent partial order was introduced by Riedtmann and Schofield [**35**]; namely, $T_1 \le T_2$ precisely when the *right perpendicular category* $T_1^\perp$ of T_1 is contained in that of T_2; here $T^\perp$ consists of the objects $Y \in \Lambda\text{-mod}$ which satisfy $\operatorname{Ext}^i_\Lambda(T, Y) = 0$ for all $i \ge 1$. (Clearly, this poset always has a largest element, namely the basic projective generator in Λ-mod.) Strong tilting modules were further explored for Auslander-Gorenstein algebras by Iyama and Zhang in [**29**]. In the present context, we will find them to play a particularly prominent role: They are precisely those tilting modules which induce dualities defined on the full category $\mathcal{P}^{<\infty}(\Lambda\text{-mod})$.

The underlying theme of the paper is to exhibit the mongrel status of tilting modules which, by definition, combine properties of projective generators with properties of injective cogenerators. As is well-known, the former are the modules whose covariant Hom-functors induce classical Morita equivalences $\Lambda\text{-mod} \longleftrightarrow \Lambda'\text{-mod}$, while the latter are characterized by the fact that their contravariant Hom-functors induce dualities $\Lambda\text{-mod} \longleftrightarrow \text{mod-}\Lambda'$; in each case Λ' is the opposite of the endomorphism ring of the Λ-module inducing the pertinent functors. Suppose that ${}_\Lambda T_{\widetilde{\Lambda}}$ is a tilting bimodule, where $\widetilde{\Lambda} = \operatorname{End}_\Lambda(T)^{\mathrm{op}}$. Equivalences of suitable subcategories of Λ-mod and $\widetilde{\Lambda}$-mod have been amply studied (see, e.g., [**31**], [**1**], [**36**]). As was first discovered by Miyashita in his pivotal 1986 paper (if not advertised in an explicit format), tilting modules also give rise to dualities linking subcategories of Λ-mod to subcategories of mod-$\widetilde{\Lambda}$. These dualities – they reflect the resemblance of tilting modules to injective cogenerators – have received far less attention than the covariant equivalences. We aim at providing the foundations for a systematic exploration.

We begin by reviewing Miyashita's dualities. They are defined on *resolving subcategories* of $\mathcal{P}^{<\infty}(\Lambda\text{-mod})$ and are *strictly exact* in the following sense: A functor $F : \mathfrak{C} \to \mathfrak{C}'$, co- or contravariant, linking two categories $\mathfrak{C}$ and $\mathfrak{C}'$ of modules over algebras Λ and Λ', is strictly exact if F takes any short exact sequence $0 \to C_1 \to C_2 \to C_3 \to 0$ of Λ-modules with $C_i \in \mathfrak{C}$ to a short exact sequence of Λ'-modules. Miyashita's Theorem has a converse: According to [**25**, Theorem 1], every strictly exact duality defined on a resolving subcategory $\mathfrak{C}$ of $\mathcal{P}^{<\infty}(\Lambda\text{-mod})$

is of the form $\mathrm{Hom}_\Lambda(-,T)$ for a tilting module T, and $\mathfrak{C}$ necessarily has the form $\mathcal{P}^{<\infty}(\Lambda\text{-mod}) \cap {}^\perp T$; here ${}^\perp T$ is the *left perpendicular category* of T. Regarding the sizes of the subcategories coupled by tilting-induced equivalences or dualities, the rule of thumb is this: The larger the tilting module T is in the Riedtmann-Schofield order, the closer its traits are to those of a projective generator; the smaller T is in this partial order, the more its behavior resembles that of an injective cogenerator. The case of a projective generator and that of a strong tilting module constitute the extreme positions on this seesaw.

The main theorems on dualities, including comparisons with classical scenarios and links to work of Auslander, Reiten and others, are assembled in Sections 3, 4. In Section 5, we propose several open problems which address connections between, on one hand, the subcategories of $\mathcal{P}^{<\infty}(\Lambda\text{-mod})$ that carry dualities and naturally associated subcategories of $\mathcal{P}^{<\infty}(\Lambda\text{-Mod})$ on the other. ($\mathcal{P}^{<\infty}(\Lambda\text{-Mod})$ stands for the category of arbitrary left Λ-modules with finite projective dimension.)

In Section 6, we apply the general results on strong tilting to truncated path algebras, i.e., to algebras of the form $\Lambda = \Lambda_L = KQ/\langle \text{all paths of length } L+1\rangle$ for some quiver Q and some positive integer L. We review the findings of [**25**] without proofs; in parallel we analyze an example that illustrates each step. In a nutshell: If Λ is truncated, $\mathcal{P}^{<\infty}(\Lambda\text{-mod})$ is always contravariantly finite. This guarantees a basic strong tilting module ${}_\Lambda T$, which induces a duality $\mathcal{P}^{<\infty}(\Lambda\text{-mod}) \longleftrightarrow \mathcal{P}^{<\infty}(\text{mod-}\widetilde{\Lambda}) \cap {}^\perp(T_{\widetilde{\Lambda}})$. In general, the righthand category is properly contained in $\mathcal{P}^{<\infty}(\text{mod-}\widetilde{\Lambda})$, meaning that T fails to be strong as a tilting module over $\widetilde{\Lambda}$. However, the category $\mathcal{P}^{<\infty}(\text{mod-}\widetilde{\Lambda})$ always has its own basic strong tilting module. This tilting module of the second generation is in fact strong also as a tilting module over its endomorphism ring $\widetilde{\widetilde{\Lambda}}$, whence the process of iterated strong tilting turns periodic: $\Lambda \rightsquigarrow \widetilde{\Lambda} \rightsquigarrow \widetilde{\widetilde{\Lambda}} \rightsquigarrow \widetilde{\Lambda} \rightsquigarrow \cdots$. In each round, the objects in the $\mathcal{P}^{<\infty}$-categories of Λ, $\widetilde{\Lambda}$, $\widetilde{\widetilde{\Lambda}}$ are characterized in terms of their intrinsic structure. Moreover, the strong tilting objects, as well as the $\mathcal{P}^{<\infty}$-approximations arising in the theory, are algorithmically accessible from Q and L. The constructive aspects are supplemented in Sections 7 and 8.

We point to an undercurrent of Section 6: The homological picture resulting from this section reinforces the idea that many results and tools specific to the theory of hereditary algebras may be carried over to truncated path algebras in a fairly natural manner (note that the hereditary algebras are among the truncated ones). The motivation that drives the study of the more general class of algebras is, in essence, the same as in the hereditary case: Namely, to explore the finite dimensional representation theory of a quiver Q per se (not necessarily acyclic in the general truncated case), unmodified by relations; in the presence of oriented cycles, this is achieved by letting the Loewy lengths $L+1$ of the algebras Λ_L grow. Curiously, the process of strongly tilting a truncated path algebra Λ reveals its kinship to a hereditary algebra more clearly than the original category Λ-mod does (see, e.g., Corollary 32).

In Section 2, we recall some background. However, the article is far from being self-contained in that we include proofs only where we do not have references to cite.

2. Background on tilting, strong tilting, and contravariant finiteness

Throughout the first five sections, Λ will denote an arbitrary basic finite dimensional algebra over a field K unless otherwise specified. We fix a full sequence $e_1, \dots, e_n$ of primitive idempotents of Λ. By J we denote the Jacobson radical of Λ and by S_i the simple module $\Lambda e_i / J e_i$. Moreover, we write Λ-Mod and Λ-mod for the categories of arbitrary, resp. finitely generated, left Λ-modules, and Mod-Λ, mod-Λ for the corresponding categories of right Λ-modules.

To fix our terminology, we recall several definitions, some of which have evolved over time.

2.A. $\mathcal{X}$-coresolutions and $\mathcal{X}$-resolutions. Let $\mathcal{X}$ be a subcategory of Λ-Mod. An *$\mathcal{X}$-coresolution* of a left Λ-module M is an exact sequence of the form

$$0 \longrightarrow M \longrightarrow X_0 \longrightarrow X_1 \longrightarrow X_2 \longrightarrow \cdots$$

with all $X_i \in \mathcal{X}$; the *length* of the coresolution is the maximal l with $X_l \neq 0$. The concept of an *$\mathcal{X}$-resolution* of M is dual.

By $\mathfrak{co\text{-}res}^{<\infty}(\mathcal{X})$ and $\mathfrak{Co\text{-}Res}^{<\infty}(\mathcal{X})$ we denote the full subcategories of Λ-mod and Λ-Mod, respectively, which consist of the modules with finite $\mathcal{X}$-coresolutions. The categories $\mathfrak{res}^{<\infty}(\mathcal{X})$ and $\mathfrak{Res}^{<\infty}(\mathcal{X})$ are defined analogously. In case $\mathcal{X} = \text{add}(X)$ or $\mathcal{X} = \text{Add}(X)$ for a module X, we abbreviate further to $\mathfrak{co\text{-}res}^{<\infty}(X)$ and $\mathfrak{res}^{<\infty}(X)$, resp., $\mathfrak{Co\text{-}Res}^{<\infty}(X)$ and $\mathfrak{Res}^{<\infty}(X)$.

2.B. (Strong) tilting modules.

DEFINITION. Following Miyashita [**32**], we call a left Λ-module T a *tilting module* in case (i) T belongs to $\mathcal{P}^{<\infty}(\Lambda\text{-mod})$, (ii) $\text{Ext}^i(T,T) = 0$ for all $i \geq 1$, and (iii) the left regular module ${}_\Lambda\Lambda$ belongs to $\mathfrak{co\text{-}res}^{<\infty}(T)$.

According to [**5**], a tilting module ${}_\Lambda T$ is *strong* if $\mathcal{P}^{<\infty}(\Lambda\text{-mod}) \subseteq \mathfrak{co\text{-}res}^{<\infty}(T)$; due to Proposition 18 in Section 4.B, this definition is equivalent to the description of a strong tilting module given in the introduction (requiring that the functors $\text{Ext}^i_\Lambda(-,T)$ vanish on $\mathcal{P}^{<\infty}(\Lambda\text{-mod})$ for all $i \geq 1$).

The module T is *basic* if it has no indecomposable direct summand of multiplicity ≥ 2.

Given an arbitrary tilting module ${}_\Lambda T$ with $\widetilde{\Lambda} = \text{End}_\Lambda(T)^{\text{op}}$, the properties of the bimodule ${}_\Lambda T_{\widetilde{\Lambda}}$ are known to be left-right symmetric in the following sense: $T_{\widetilde{\Lambda}}$ is in turn a tilting module, and Λ is canonically isomorphic to $\text{End}_{\widetilde{\Lambda}}(T)$, i.e., the K-algebra map $\Lambda \to \text{End}_{\widetilde{\Lambda}}(T)$, $\lambda \mapsto (x \mapsto \lambda x)$, is an isomorphism (see [**32**, Proposition 1.4]); in other words, ${}_\Lambda T_{\widetilde{\Lambda}}$ is a *balanced bimodule*. For the family of covariant equivalences linking suitable subcategories of Λ-mod to subcategories of $\widetilde{\Lambda}$-mod – they originally motivated the exploration of tilting modules – we refer to [**32**, Theorems 1.14 – 1.16], where the current level of generality is attained. See also [**31**] and the *Handbook of Tilting Theory*, [**1**], for numerous additional references.

2.C. Contravariant finiteness of resolving subcategories of Λ-mod. The following concepts were introduced by Auslander and Smalø in [**7**]; for an alternative development of this approximation theory, see [**16, 17**].

DEFINITION. Let $\mathfrak{C}$ be a full subcategory of Λ-mod. Then $\mathfrak{C}$ is said to be *resolving* if $\mathfrak{C}$ contains the projective modules and is closed under extensions and kernels of surjective homomorphisms in Λ-mod.

Now let $\mathfrak{C} \subseteq \Lambda$-mod be a resolving subcategory. A (right) $\mathfrak{C}$-*approximation* of a module $M \in \Lambda$-mod is a homomorphism $\phi : \mathcal{A} \to M$ such that $\mathcal{A} \in \mathfrak{C}$ and every map in $\operatorname{Hom}_\Lambda(\mathfrak{C}, M)$ factors through ϕ; note that any such approximating map ϕ is surjective, since $\mathfrak{C}$ contains the projectives in Λ-mod. Whenever such an approximation of M exists, there is one of minimal K-dimension, $\mathcal{A}(M) = \mathcal{A}_\mathfrak{C}(M)$, which is uniquely determined by M up to isomorphism. In case every $M \in \Lambda$-mod has a $\mathfrak{C}$-approximation, the category $\mathfrak{C}$ is called *contravariantly finite* (in Λ-mod). To explain the terminology, one observes that $\mathfrak{C}$ is contravariantly finite precisely when the restricted contravariant Hom-functor $\operatorname{Hom}(-, M)|_\mathfrak{C}$ is finitely generated in the functor category $\operatorname{Fun}(\mathfrak{C}^{\mathrm{op}}, \mathbf{Ab})$.

Evidently, $\mathcal{P}^{<\infty}(\Lambda\text{-mod})$ is contravariantly finite whenever Λ has either finite global dimension or vanishing left finitistic dimension. But the condition has traction only in cases lying strictly between these extremes. The following sufficient condition for contravariant finiteness of a resolving subcategory $\mathfrak{C}$ was established in [**5**, Proposition 3.7]: If each of the simple left Λ-modules $S_1, \dots, S_n$ has a $\mathfrak{C}$-approximation, then $\mathfrak{C}$ is contravariantly finite. Moreover, by [**5**, Proposition 3.8], for contravariantly finite $\mathfrak{C}$, the modules in $\mathfrak{C}$ are precisely those in $\operatorname{add}\big(\operatorname{filt}(\mathcal{A}_\mathfrak{C}(S_1), \dots, \mathcal{A}_\mathfrak{C}(S_n))\big)$, where $\mathcal{A}_\mathfrak{C}(S_i)$ is the minimal $\mathfrak{C}$-approximation of S_i. (Notation: For $M_1, \dots, M_u \in \Lambda$-mod, the category $\operatorname{filt}(M_1, \dots, M_u)$ consists of the modules permitting finite filtrations with consecutive factors among the M_i.) Clearly, this implies $\sup\{\operatorname{p\,dim} C \mid C \in \mathfrak{C}\} = \sup\{\operatorname{p\,dim} \mathcal{A}_\mathfrak{C}(S_i) \mid 1 \le i \le n\}$.

Of special interest to us will be resolving subcategories of $\mathcal{P}^{<\infty}(\Lambda\text{-mod})$, in particular the category $\mathcal{P}^{<\infty}(\Lambda\text{-mod})$ itself. By dint of the upcoming theory, the following homological assets entailed by contravariant finiteness of $\mathcal{P}^{<\infty}(\Lambda\text{-mod})$ will become relevant (we summarily point to [**2, 5, 7, 26, 28**] for proofs). Suppose for the moment that $\mathcal{P}^{<\infty}(\Lambda\text{-mod})$ is contravariantly finite. Then this category has relative Auslander-Reiten sequences, a bonus that singles it out for an internal representation-theoretic analysis. Moreover, in light of the preceding paragraph, the basic building blocks of the modules in $\mathcal{P}^{<\infty}(\Lambda\text{-mod})$ are the minimal $\mathcal{P}^{<\infty}(\Lambda\text{-mod})$-approximations $\mathcal{A}(S_i)$ of the simples $S_i \in \Lambda$-mod. The same holds for the objects of $\mathcal{P}^{<\infty}(\Lambda\text{-Mod})$, as they turn out to be the direct limits of directed systems in $\mathcal{P}^{<\infty}(\Lambda\text{-mod})$. In particular, the little and big left finitistic dimensions, $\operatorname{fin\,dim} \Lambda$ and $\operatorname{Fin\,dim} \Lambda$, coincide and are equal to the maximum of the projective dimensions of the $\mathcal{A}(S_i)$. It is the key role played by the $\mathcal{A}(S_i)$ that motivates us to pin them down whenever contravariant finiteness of $\mathcal{P}^{<\infty}(\Lambda\text{-mod})$ is confirmed.

While finite dimensional algebras Λ with the property that $\mathcal{P}^{<\infty}(\Lambda\text{-mod})$ is contravariantly finite abound, it may be difficult to decide the status of $\mathcal{P}^{<\infty}(\Lambda\text{-mod})$ for a given algebra Λ. The condition appears to slice through most major classes of algebras which have been delineated by other shared properties (see [**11, 14, 27**] for exceptional classes, and [**21**] for test criteria). One of the classes enjoying this property consists of the truncated path algebras; in Section 6, we will use it to illustrate the theory laid out in Sections 2-5.

2.D. Relative splitting injectivity versus relative Ext**-injectivity, and duals.** Suppose $\mathfrak{C} \subseteq \Lambda$-mod is a full subcategory which is closed under isomorphic objects, direct summands and finite direct sums.

DEFINITION. We call an object Y of $\mathfrak{C}$ *relatively splitting injective in* $\mathfrak{C}$ if every injective morphism in $\operatorname{Hom}_\Lambda(Y, C)$ with $C \in \mathfrak{C}$ splits (compare with [**7**, Section 2]).

Moreover, call Y *relatively* Ext-*injective in* $\mathfrak{C}$ if $\mathrm{Ext}^i_\Lambda(\mathfrak{C}, Y) = 0$ for all $i \geq 1$. The concepts of *relative splitting projectivity* and *relative* Ext-*projectivity* are dual.

REMARKS 1. Suppose that $\mathfrak{C}$ is a resolving subcategory of Λ-mod and hence closed under syzygies. Then every relatively splitting injective object of $\mathfrak{C}$ is relatively Ext-injective in $\mathfrak{C}$. However, the converse fails in general; see Example 5 below. As for the dual attributes: The relative splitting projective objects of $\mathfrak{C}$ coincide with the projectives in Λ-mod, since projective covers of objects in $\mathfrak{C}$ belong to $\mathfrak{C}$. The relative Ext-projective objects of $\mathfrak{C}$ are projective as well. To see this, let $Y \in \mathfrak{C}$ be relatively Ext-projective, and suppose that $f : C \to Y$ with $C \in \mathfrak{C}$ is a surjective homomorphism. Since $\mathrm{Ker}(f) \in \mathfrak{C}$, we have $\mathrm{Ext}^1_\Lambda(Y, \mathrm{Ker}(f)) = 0$, which implies splitness of f.

The following invariant measures the deviation of an object X of $\mathfrak{C}$ from being relatively Ext-injective. It will become relevant in Section 4.

DEFINITION. Let $X \in \mathfrak{C}$. The Ext-*injective dimension* of X *relative to* $\mathfrak{C}$ is

$$\mathrm{Ext.idim}_{\mathfrak{C}}\, X := \inf\{r \in \mathbb{N}_0 \mid \mathrm{Ext}^j_\Lambda(-, X)|_{\mathfrak{C}} = 0 \text{ for all } j \geq r+1\},$$

with the understanding that $\mathrm{Ext.idim}_{\mathfrak{C}}\, X = \infty$ if the specified set is empty. By definition, $\mathrm{Ext.idim}_{\mathfrak{C}}\, X = 0$ precisely when X is a relatively Ext-injective object of $\mathfrak{C}$.

The Ext-*projective dimension* of X *relative to* $\mathfrak{C}$ is defined analogously.

2.E. Well-known connections between tilting and contravariant finiteness. A priori, the notion of a contravariantly finite subcategory of Λ-mod does not indicate the tight connection with tilting theory which will surface in consecutive steps in the sequel.

Let $\mathfrak{C}$ be as in Section 2.D. Call a Λ-module Y a *cogenerator in* $\mathfrak{C}$, if $Y \in \mathfrak{C}$ and every object of $\mathfrak{C}$ embeds into some power Y^k in Λ-mod.

LEMMA 2. *Suppose that $\mathfrak{C}$ is a contravariantly finite resolving subcategory of Λ-mod. Then $\mathfrak{C}$ has a relatively splitting injective cogenerator (obviously unique up to multiplicities of indecomposable direct summands), namely $\mathcal{A}_{\mathfrak{C}}(E)$, the minimal $\mathfrak{C}$-approximation of the basic injective cogenerator E in Λ-mod.*

PROOF. For splitness of any injective homomorphism $\mathcal{A}_{\mathfrak{C}}(E) \hookrightarrow C$ with $C \in \mathfrak{C}$, see e.g. [**7**], or [**14**, Lemma 5.1] for a proof in the current terminology. The cogenerator property of $\mathcal{A}_{\mathfrak{C}}(E)$ for $\mathfrak{C}$ is immediate from the cogenerator property of E in Λ-mod. □

The first link between tilting and contravariant finiteness is provided by the dual of Auslander and Reiten's [**5**, Theorem 5.5(b)]:

THEOREM 3. **Correspondence between basic tilting modules and contravariantly finite subcategories of $\mathcal{P}^{<\infty}(\Lambda\text{-mod})$.** *There is a one-to-one correspondence between the isomorphism classes of basic tilting objects in Λ-mod on one hand and the contravariantly finite subcategories of $\mathcal{P}^{<\infty}(\Lambda\text{-mod})$ on the other. It assigns to any basic tilting module $T \in \Lambda$-mod the subcategory $\mathfrak{co\text{-}res}^{<\infty}({}_\Lambda T)$ of Λ-mod (cf. 2.A).*

A second tie surfaces in

THEOREM 4. **Existence and uniqueness of strong tilting modules.** ([**5**, Proposition 6.3]; use Theorem 5.5, loc. cit., to correct the statement of 6.3.) *The category* Λ-mod *has a strong tilting module if and only if* $\mathcal{P}^{<\infty}(\Lambda\text{-mod})$ *is contravariantly finite in* Λ-mod. *In the positive case, the basic strong tilting module is the direct sum of the distinct (i.e., pairwise nonisomorphic) indecomposable relatively* Ext-*injective objects of* $\mathcal{P}^{<\infty}(\Lambda\text{-mod})$.

In other words, if $\mathfrak{C} = \mathcal{P}^{<\infty}(\Lambda\text{-mod})$ is contravariantly finite and T is the corresponding basic strong tilting module, the modules $X \in \mathfrak{C}$ with $\text{Ext.idim}_{\mathfrak{C}} X = 0$ are precisely the objects of $\text{add}(T)$.

EXAMPLE 5. This example illustrates the fact that relative splitting injectivity in a resolving subcategory $\mathfrak{C} \subseteq \Lambda$-mod is stronger than relative Ext-injectivity in general. Let Λ be the Kronecker algebra. Label the quiver of Λ so that the two equi-directed arrows start in e_1 and end in e_2. Clearly, the subcategory $\mathfrak{C}$ of Λ-mod which consists of all projective left Λ-modules is resolving and contravariantly finite. The basic strong tilting module associated to this category is the left regular module ${}_\Lambda\Lambda$, evidently relatively Ext-injective in $\mathfrak{C}$. On the other hand, the simple projective module Λe_2 fails to be relatively splitting injective in $\mathfrak{C}$. For later use, observe that the minimal $\mathfrak{C}$-approximation of the basic injective cogenerator in Λ-mod is $(\Lambda e_1)^3$.

However, in case $\mathfrak{C} = \mathcal{P}^{<\infty}(\Lambda\text{-mod})$, relative splitting injectivity of an object in $\mathfrak{C}$ amounts to the same as relative Ext-injectivity, since $\mathcal{P}^{<\infty}(\Lambda\text{-mod})$ is closed under cokernels of injective homomorphisms in that case. In fact, we obtain

PROPOSITION 6. **Comparison of strong tilting modules with approximations of injective cogenerators.** [**25**, Suppl. II] *Suppose that* $\mathcal{P}^{<\infty}(\Lambda\text{-mod})$ *is contravariantly finite,* $T \in \Lambda$-mod *a strong tilting module, and* $\mathcal{A}(E)$ *the minimal* $\mathcal{P}^{<\infty}(\Lambda\text{-mod})$*-approximation of the basic injective cogenerator* E *in* Λ-mod. *Then*

$$\text{add}(T) = \text{add}(\mathcal{A}(E)).$$

In Section 4, this proposition will be supplemented as follows: If $T \in \Lambda$-mod is an arbitrary basic tilting module and $\mathfrak{C}$ is the contravariantly finite subcategory $\mathfrak{co\text{-}res}^{<\infty}(T)$ which is associated to T (see Theorem 3), then $\text{add}\big(\mathcal{A}_{\mathfrak{C}}(E)\big) \subseteq \text{add}(T)$. On the other hand, Example 5 shows that the reverse inclusion fails in general.

3. Dualities of subcategories of $\mathcal{P}^{<\infty}$-categories. Consequences

Let Λ' be another finite dimensional K-algebra. For a Λ-Λ'-bimodule ${}_\Lambda Y_{\Lambda'}$, we denote by ${}^\perp({}_\Lambda Y)$ the left perpendicular category of ${}_\Lambda Y$ in Λ-mod, namely the full subcategory of Λ-mod consisting of the Λ-modules X with $\text{Ext}^i_\Lambda(X,Y) = 0$ for all $i \geq 1$. Clearly, ${}^\perp({}_\Lambda Y)$ is a resolving subcategory of Λ-mod, whence so is the intersection $\mathcal{P}^{<\infty}(\Lambda\text{-mod}) \cap {}^\perp({}_\Lambda Y)$. Analogously, one defines the left perpendicular category ${}^\perp(Y_{\Lambda'}) \subseteq \text{mod-}\Lambda'$.

3.A. The main results on dualities. The upcoming fact is readily deduced from Miyashita's Theorem 3.5 in [**32**].

THEOREM AND NOTATION 7. **Miyashita's duality.** *Given a tilting bimodule* ${}_\Lambda T_{\widetilde{\Lambda}}$, *i.e.,* $\widetilde{\Lambda} = \text{End}_\Lambda(T)^{\text{op}}$, *set*

$$\mathfrak{C} = \mathfrak{C}({}_\Lambda T) := \mathcal{P}^{<\infty}(\Lambda\text{-mod}) \cap {}^\perp({}_\Lambda T) \quad \textit{and} \quad \widetilde{\mathfrak{C}} = \widetilde{\mathfrak{C}}(T_{\widetilde{\Lambda}}) := \mathcal{P}^{<\infty}(\text{mod-}\widetilde{\Lambda}) \cap {}^\perp(T_{\widetilde{\Lambda}}).$$

Then the restricted Hom-functors $\operatorname{Hom}_\Lambda(-,T)|_{\mathfrak{C}}$ *and* $\operatorname{Hom}_{\widetilde{\Lambda}}(-,T)|_{\widetilde{\mathfrak{C}}}$ *are inverse dualities* $\mathfrak{C} \longleftrightarrow \widetilde{\mathfrak{C}}$.

By the definition of ${}^\perp({}_\Lambda T)$ and ${}^\perp(T_{\widetilde{\Lambda}})$, the inverse dualities secured by this theorem are *strictly exact*, meaning that they take those sequences $0 \to X_1 \to X_2 \to X_3 \to 0$ (in the specified subcategories of Λ-mod, resp., mod-$\widetilde{\Lambda}$) which are exact in the ambient full module categories to sequences of the same ilk.

Clearly, the inclusion "$\mathcal{P}^{<\infty}(\Lambda\text{-mod}) \subseteq {}^\perp({}_\Lambda T)$" is tantamount to relative Ext-injectivity of the tilting module ${}_\Lambda T$ in $\mathcal{P}^{<\infty}(\Lambda\text{-mod})$. Theorem 7 thus shows that every tilting bimodule ${}_\Lambda T_{\widetilde{\Lambda}}$ which is strong on both sides yields a duality $\mathcal{P}^{<\infty}(\Lambda\text{-mod}) \longleftrightarrow \mathcal{P}^{<\infty}(\text{mod-}\widetilde{\Lambda})$. The converse provided by Corollary 9 below places additional emphasis on the distinguished role of strong tilting modules relative to categories of modules of finite projective dimension.

We supplement Miyashita's theorem by showing that strictly exact dualities between resolving subcategories of $\mathcal{P}^{<\infty}(\Lambda\text{-mod})$ and $\mathcal{P}^{<\infty}(\text{mod-}\Lambda')$ are always afforded by tilting bimodules according to the blueprint of Theorem 7.

The analogy of the upcoming result with classical Morita duality will be further stressed by Proposition 18 and Observation 20 below. (For artinian rings Λ and Λ', a "classical Morita duality" between Λ and Λ' is a contravariant equivalence Λ-mod $\longleftrightarrow$ mod-Λ', well-known to be always induced by a contravariant Hom-functor which is determined by an injective cogenerator in Λ-mod.) The connection with dualities linking the $\mathcal{P}^{<\infty}$-categories of Λ-mod and mod-Λ' is evident in case Λ is a Gorenstein algebra (i.e., the left and right regular modules have finite injective dimension). In that case, the Λ-modules of finite injective dimension coincide with those of finite projective dimension, whence the basic injective cogenerator $E \in \Lambda$-mod is a tilting module, obviously strong. Observe that ${}_\Lambda E$ then gives rise to the standard Morita duality $\operatorname{Hom}_\Lambda(-,E) \cong \operatorname{Hom}_K(-,K) : \Lambda\text{-mod} \longleftrightarrow \text{mod-}\Lambda$, which restricts to a duality $\mathcal{P}^{<\infty}(\Lambda\text{-mod}) \longleftrightarrow \mathcal{P}^{<\infty}(\text{mod-}\Lambda)$. However, beyond this scenario, the dualities encountered in Theorems 7 and 8 are not induced by dualities of classical Morita type in that they do not result from restrictions of dualities defined on Λ-mod.

THEOREM 8. **Correspondence between tilting modules and dualities defined on resolving subcategories of $\mathcal{P}^{<\infty}$-categories.** [**25**, Theorem 1] *Let* Λ *and* Λ' *be finite dimensional algebras, and let* $\mathfrak{C} \subseteq \mathcal{P}^{<\infty}(\Lambda\text{-mod})$ *and* $\mathfrak{C}' \subseteq \mathcal{P}^{<\infty}(\text{mod-}\Lambda')$ *be resolving subcategories of* Λ-mod *and* mod-Λ', *respectively.*

Suppose $\mathfrak{C}$ *is dual to* $\mathfrak{C}'$ *by way of strictly exact contravariant additive functors*

$$F : \mathfrak{C} \longrightarrow \mathfrak{C}' \quad \text{and} \quad F' : \mathfrak{C}' \longrightarrow \mathfrak{C}$$

such that $F' \circ F$ *and* $F \circ F'$ *are isomorphic to the pertinent identity functors.*

Then there exists a tilting bimodule ${}_\Lambda T_{\Lambda'}$ *such that:*

(a) $F \cong \operatorname{Hom}_\Lambda(-,T)\,|_{\mathfrak{C}}$ *and* $F' \cong \operatorname{Hom}_{\Lambda'}(-,T)\,|_{\mathfrak{C}'}$;

(b) $\mathfrak{C} = \mathcal{P}^{<\infty}(\Lambda\text{-mod}) \cap {}^\perp({}_\Lambda T)$, *and* $\mathfrak{C}' = \mathcal{P}^{<\infty}(\text{mod-}\Lambda') \cap {}^\perp(T_{\Lambda'})$; *in other words,* $\mathfrak{C} = \mathfrak{C}({}_\Lambda T)$ *and* $\mathfrak{C}' = \mathfrak{C}(T_{\Lambda'})$;

(c) $\mathfrak{C} = \mathfrak{co\text{-}res}^{<\infty}({}_\Lambda T)$ *and* $\mathfrak{C}' = \mathfrak{co\text{-}res}^{<\infty}(T_{\Lambda'})$, *i.e., the modules in* $\mathfrak{C}$ *are precisely those left* Λ*-modules which have finite coresolutions by objects in* $\operatorname{add}({}_\Lambda T)$, *and the modules in* $\mathfrak{C}'$ *are those right* Λ'*-modules which have finite coresolutions by objects in* $\operatorname{add}(T_{\Lambda'})$.

In particular, the functors F and F' are "minimal dualities", in the sense that they do not induce any duality $\mathfrak{C}_0 \longleftrightarrow \mathfrak{C}'_0$ between proper resolving subcategories $\mathfrak{C}_0$ of $\mathfrak{C}$ and $\mathfrak{C}'_0$ of $\mathfrak{C}'$.

Addendum: *Provided that $\mathfrak{C}$ and $\mathfrak{C}'$ are closed also under cokernels of injective morphisms in the ambient module categories, arbitrary additive dualities $F : \mathfrak{C} \longrightarrow \mathfrak{C}'$ and $F' : \mathfrak{C}' \longrightarrow \mathfrak{C}$ are strictly exact.*

We follow with several consequences. For the first, note that the categories $\mathfrak{C} = \mathcal{P}^{<\infty}(\Lambda\text{-mod})$ and $\mathfrak{C}' = \mathcal{P}^{<\infty}(\text{mod-}\Lambda')$ satisfy the conditions spelled out in the addendum to Theorem 8.

COROLLARY 9. *Let Λ and Λ' be finite dimensional algebras. Then any pair of inverse dualities (F, F')*

$$\mathcal{P}^{<\infty}(\Lambda\text{-mod}) \longleftrightarrow \mathcal{P}^{<\infty}(\text{mod-}\Lambda')$$

is isomorphic to a pair of functors $\big(\mathrm{Hom}_\Lambda(-,T), \mathrm{Hom}_{\Lambda'}(-,T)\big)$ for some tilting bimodule ${}_\Lambda T_{\Lambda'}$ which is strong on both sides.

COROLLARY 10. *Again, let Λ, Λ' be finite dimensional algebras. If there are resolving subcategories of $\mathcal{P}^{<\infty}(\Lambda\text{-mod})$ and $\mathcal{P}^{<\infty}(\text{mod-}\Lambda')$, respectively, which are dual via strictly exact functors, then Λ and Λ' are derived equivalent.*

On combining Theorems 7 and 8 with Theorem 3, we can further sharpen the picture, since these results imply $\mathfrak{C}({}_\Lambda T) = \mathfrak{co}\text{-}\mathfrak{res}^{<\infty}({}_\Lambda T)$ for any tilting module ${}_\Lambda T$.

COROLLARY 11. *The resolving subcategories $\mathfrak{C}$ of $\mathcal{P}^{<\infty}(\Lambda\text{-mod})$ which are strictly dual to resolving subcategories $\mathfrak{C}'$ of $\mathcal{P}^{<\infty}(\text{mod-}\Lambda')$ (for a suitable finite dimensional algebra Λ' depending on $\mathfrak{C}$) are precisely the contravariantly finite ones, i.e., those of the form $\mathfrak{co}\text{-}\mathfrak{res}^{<\infty}({}_\Lambda T)$ for some tilting module ${}_\Lambda T$.*

We stress the special case of finite global dimension.

COROLLARY 12. *Suppose Λ has finite global dimension. Then there is a one-to-one correspondence between the isomorphism classes of basic tilting modules in* Λ-mod *and the equivalence classes of strictly exact dualities $\mathfrak{C} \longleftrightarrow \mathfrak{C}'$ which are defined on resolving subcategories $\mathfrak{C}$ of* Λ-mod*; in each case, $\mathfrak{C}'$ is a resolving subcategory of* mod-Λ' *for some algebra Λ' of finite global dimension that depends on $\mathfrak{C}$.*

COMMENT. We do not have an example of inverse dualites F, F' between resolving subcategories $\mathfrak{C} \subseteq \mathcal{P}^{<\infty}(\Lambda\text{-mod})$ and $\mathfrak{C}' \subseteq \mathcal{P}^{<\infty}(\text{mod-}\Lambda')$ for which F or F' fails to be strictly exact. In general, relative epimorphisms in a resolving subcategory $\mathfrak{C}$ of $\mathcal{P}^{<\infty}(\Lambda\text{-mod})$ need not be surjections though. For instance, take Λ to be $KQ/\langle\beta\alpha\rangle$, where Q is the quiver $1 \xrightarrow{\alpha} 2 \xrightarrow{\beta} 3$, and let $\mathfrak{C}$ be the full subcategory of $\mathcal{P}^{<\infty}(\Lambda\text{-mod}) = \Lambda\text{-mod}$ consisting of the projective modules. Then the map $f \in \mathrm{Hom}_\Lambda(\Lambda e_2, \Lambda e_1)$ with $f(e_2) = \alpha$ is a relative epimorphism in $\mathfrak{C}$. Nonetheless, every duality between $\mathfrak{C}$ and some resolving $\mathfrak{C}' \subseteq \text{mod-}\Lambda'$ is easily seen to be strictly exact in this case.

3.B. Cotilting modules and dualities of $\mathcal{I}^{<\infty}$-categories. We briefly address the cotilting situation by spelling out the duals of Theorems 7 and 8. (The corollaries to these theorems have obvious duals as well.) We now focus on a *cotilting* module $C \in \text{mod-}\Lambda$, the K-dual of a tilting object in Λ-mod. The concept of a *strong* cotilting module is in turn dual to that of a strong tilting module (see [**5**, definition preceding Proposition 6.3]), and the attributes "*coresolving*" / "*covariantly finite*" for subcategories of mod-Λ are the obvious duals of "resolving" / "contravariantly finite" (see [**7**] and [**5**]). In explicit terms: A cotilting module C_Λ is strong in case every module in $\mathcal{I}^{<\infty}(\text{mod-}\Lambda)$ (= the subcategory of mod-Λ consisting of the modules of finite injective dimension) has a finite add(C)-resolution; that is, every right Λ-module M of finite injective dimension admits a resolution $0 \to C^{(m)} \to \cdots \to C^{(0)} \to M \to 0$ with $C^{(i)} \in \text{add}(C)$. Parallel to Theorem 4, a strong cotilting module C is a relatively Ext-projective object of $\mathcal{I}^{<\infty}(\text{mod-}\Lambda)$.

By applying the main results of Section 3.A to the tilting module $D(C)$, where $D = \text{Hom}_K(-, K)$, we obtain

THEOREM 13. **Correspondence between cotilting modules and dualities defined on coresolving subcategories of $\mathcal{I}^{<\infty}(\text{mod-}\Lambda)$.**

(A) Twin of Miyashita's theorem*: Suppose $C \in \text{mod-}\Lambda$ is a cotilting module with endomorphism ring $\widehat{\Lambda}$. Then the functors* $\text{Hom}_\Lambda(-, C)$ *and* $\text{Hom}_{\widehat{\Lambda}}(-, C)$ *restrict to mutually inverse dualities*

$$\mathcal{I}^{<\infty}(\text{mod-}\Lambda) \cap (C_\Lambda)^\perp \quad \longleftrightarrow \quad \mathcal{I}^{<\infty}(\widehat{\Lambda}\text{-mod}) \cap ({}_{\widehat{\Lambda}}C)^\perp.$$

Equivalently: The covariant Hom-functors $\text{Hom}_\Lambda(C, -)$ *and* $\text{Hom}_{\widehat{\Lambda}}(C, -)$ *induce inverse equivalences between* $\mathcal{I}^{<\infty}(\text{mod-}\Lambda) \cap (C_\Lambda)^\perp$ *and* $\mathcal{P}^{<\infty}(\widehat{\Lambda}\text{-mod}) \cap {}^\perp(D(C)_{\widehat{\Lambda}})$.

(B) Twin of Theorem 8*: Let $\widehat{\Lambda}$ be another finite dimensional algebra, and suppose that $\mathfrak{C} \subseteq \mathcal{I}^{<\infty}(\text{mod-}\Lambda)$ and $\widehat{\mathfrak{C}} \subseteq \mathcal{I}^{<\infty}(\widehat{\Lambda}\text{-mod})$ are coresolving subcategories of* mod-Λ *and* $\widehat{\Lambda}$-mod, *respectively. Moreover, suppose that $\mathfrak{C}$ is dual to $\widehat{\mathfrak{C}}$ by way of mutually inverse strictly exact contravariant additive functors*

$$F : \mathfrak{C} \longrightarrow \widehat{\mathfrak{C}} \quad \textit{and} \quad \widehat{F} : \widehat{\mathfrak{C}} \longrightarrow \mathfrak{C}.$$

Then there exists a cotilting bimodule ${}_{\widehat{\Lambda}}C_\Lambda$ such that:

(a) $F \cong \text{Hom}_\Lambda(-, C)\,|_{\mathfrak{C}}$ *and* $\widehat{F} \cong \text{Hom}_{\widehat{\Lambda}}(-, C)\,|_{\widehat{\mathfrak{C}}}$*;*
(b) $\mathfrak{C} = \mathcal{I}^{<\infty}(\text{mod-}\Lambda) \cap (C_\Lambda)^\perp = \mathfrak{res}^{<\infty}(C_\Lambda)$*;*
(c) $\widehat{\mathfrak{C}} = \mathcal{I}^{<\infty}(\widehat{\Lambda}\text{-mod}) \cap ({}_{\widehat{\Lambda}}C)^\perp = \mathfrak{res}^{<\infty}({}_{\widehat{\Lambda}}C)$. (See 2.A for notation.)

Addendum: *Provided that $\mathfrak{C}$ and $\widehat{\mathfrak{C}}$ are closed also under kernels of surjective morphisms in the ambient module categories, arbitrary additive dualities $F : \mathfrak{C} \longrightarrow \widehat{\mathfrak{C}}$ and $\widehat{F} : \widehat{\mathfrak{C}} \longrightarrow \mathfrak{C}$ are strictly exact.*

REMARK 14. As a special case, we find that any strong cotilting module C_Λ induces inverse (covariant) equivalences

$$\mathcal{I}^{<\infty}(\text{mod-}\Lambda) \longleftrightarrow \mathcal{P}^{<\infty}(\text{mod-}\widehat{\Lambda}) \cap {}^\perp(D(C)_{\widehat{\Lambda}}).$$

Specializing further, we retrieve Proposition 6.6 of [**5**]: If ${}_{\widehat{\Lambda}}C_\Lambda$ is a cotilting bimodule which is strong on both sides, the functor $\text{Hom}_\Lambda(C, -)$ induces an equivalence $\mathcal{I}^{<\infty}(\text{mod-}\Lambda) \to \mathcal{P}^{<\infty}(\text{mod-}\widehat{\Lambda})$. Indeed, by definition, bilateral strongness makes C relatively Ext-projective in both $\mathcal{I}^{<\infty}(\text{mod-}\Lambda)$ and $\mathcal{I}^{<\infty}(\widehat{\Lambda}\text{-mod})$, whence $D(C)_{\widehat{\Lambda}}$ is relatively Ext-injective in the category $\mathcal{P}^{<\infty}(\text{mod-}\widehat{\Lambda})$. This amounts to the equalities $\mathcal{I}^{<\infty}(\text{mod-}\Lambda) \cap (C_\Lambda)^\perp = \mathcal{I}^{<\infty}(\text{mod-}\Lambda)$ and $\mathcal{P}^{<\infty}(\text{mod-}\widehat{\Lambda}) \cap {}^\perp(D(C)_{\widehat{\Lambda}}) = \mathcal{P}^{<\infty}(\text{mod-}\widehat{\Lambda})$. In light of Theorem 13, the converse of this implication holds as well.

Also note: For any cotilting module C_Λ, the categories $\mathcal{I}^{<\infty}(\text{mod-}\Lambda) \cap (C_\Lambda)^\perp$ and $\mathcal{I}^{<\infty}(\widehat{\Lambda}\text{-mod}) \cap ({}_{\widehat{\Lambda}}C)^\perp$ are covariantly finite in mod-Λ and $\widehat{\Lambda}$-mod, respectively (combine the dual of Theorem 3 with Theorem 13.B).

PROOF OF THEOREM 13 FROM THEOREMS 7, 8. Set $T = D(C)$. For part **(A)**, apply Theorem 7 to the tilting module ${}_\Lambda T$ with $\text{End}_\Lambda(T)^{\text{op}} = \widehat{\Lambda}$, so as to obtain inverse dualites $\text{Hom}_\Lambda(-,T)$ and $\text{Hom}_{\widehat{\Lambda}}(-,T)$

$$\mathcal{P}^{<\infty}(\Lambda\text{-mod}) \cap {}^\perp({}_\Lambda T) \longleftrightarrow \mathcal{P}^{<\infty}(\text{mod-}\widehat{\Lambda}) \cap {}^\perp(T_{\widehat{\Lambda}}).$$

Since D takes $(C_\Lambda)^\perp$ to ${}^\perp({}_\Lambda T)$, and ${}^\perp(T_{\widehat{\Lambda}})$ to $({}_{\widehat{\Lambda}}C)^\perp$, this yields the following commutative diagram.

$$\begin{array}{ccc}
\mathcal{P}^{<\infty}(\Lambda\text{-mod}) \cap {}^\perp({}_\Lambda T) & \xrightarrow{\ \text{Hom}_\Lambda(-,T)\ } & \mathcal{P}^{<\infty}(\text{mod-}\widehat{\Lambda}) \cap {}^\perp(T_{\widehat{\Lambda}}) \\
\Big\uparrow D & \nearrow \text{Hom}_\Lambda(C,-) & \Big\downarrow D \\
\mathcal{I}^{<\infty}(\text{mod-}\Lambda) \cap (C_\Lambda)^\perp & \xrightarrow{\ \text{Hom}_\Lambda(-,C)\ } & \mathcal{I}^{<\infty}(\widehat{\Lambda}\text{-mod}) \cap ({}_{\widehat{\Lambda}}C)^\perp
\end{array}$$

A straightforward computation thus yields the first claim. Part **(B)** is deduced from Theorem 8 by a similar argument. □

3.C. Connection with the Riedtmann-Schofield partial order. As mentioned in the introduction, Riedtmann and Schofield [**35**] introduced a partial order on the set of all basic tilting objects in Λ-mod: $T_1 \le T_2$ if and only if $T_1^\perp \subseteq T_2^\perp$. The orientation of this partial order is in synchrony with the sizes of the domains of the partial equivalences induced by the T_i; in [**32**, Theorem 1.16], this was first spelled out for the current notion of a tilting module. To put it in rough terms: The larger T is in this partial order, the more comprehensive the equivalences induced by T. In this vein, the unique largest tilting object in Λ-mod, namely the basic projective generator for Λ-mod, induces an actual Morita equivalence on Λ-Mod.

We next observe that, as the domains of the covariant equivalences induced by tilting modules grow, the domains of Miyashita's contravariant equivalences shrink in tandem. More precisely:

OBSERVATION 15. For tilting modules $T_1, T_2 \in \Lambda$-mod,

$$T_1 \le T_2 \ \text{ if and only if } \ \mathfrak{C}(T_1) \supseteq \mathfrak{C}(T_2).$$

PROOF. To see this, recall that $\mathfrak{C}(T) = \mathcal{P}^{<\infty}(\Lambda\text{-mod}) \cap {}^\perp T$. We first remark that the proof of [**23**, Theorem 2.1] establishes the equivalence "$T_1 \le T_2 \iff T_1 \in T_2^\perp$". We supplement it by showing that the condition $T_1 \in T_2^\perp$, i.e., $T_2 \in {}^\perp T_1$, implies $\mathfrak{C}(T_2) \subseteq \mathfrak{C}(T_1)$. So let $X \in \mathfrak{C}(T_2)$. In view of $\mathfrak{C}(T_2) = \mathfrak{co\text{-}res}^{<\infty}(T_2)$, this means that X has a finite $\text{add}(T_2)$-coresolution $0 \to X \to U^{(0)} \to \cdots \to U^{(m)} \to 0$ with $U^{(j)} \in \text{add}(T_2)$. Since T_2 is left perpendicular to T_1 by hypothesis, we obtain $\text{Ext}^i_\Lambda(U^{(j)}, T_1) = 0$ for $i \ge 1$. The given coresolution of X therefore yields $\text{Ext}^i_\Lambda(X, T_1) = 0$ for $i \ge 1$. □

This viewpoint, namely that small tilting modules in the Riedtmann-Schofield partial order are closer to injective cogenerators than large ones, is reinforced in Section 4.

4. Tilting modules viewed as relatively Ext-injective cogenerators

4.A. Projective resolutions versus add(T)-coresolutions. Let $\mathfrak{C}$ be a resolving contravariantly finite subcategory of $\mathcal{P}^{<\infty}(\Lambda\text{-mod})$. As we glean from Theorems 3, 8 and Corollary 11, there exists a basic tilting module ${}_\Lambda T$ such that $\mathfrak{C} = \mathfrak{C}({}_\Lambda T) = \mathfrak{co\text{-}res}^{<\infty}({}_\Lambda T)$. By definition, the objects of $\mathfrak{C}({}_\Lambda T)$ have finite add$({}_\Lambda T)$-coresolutions, whence they have nonnegative add(T)-codimension in the following sense: For $X \in \mathfrak{C}$, we denote by $\operatorname{codim}_{\operatorname{add}(T)}(X)$ the minimal length of an add(T)-coresolution $0 \to X \to T^{(0)} \to \cdots \to T^{(m)} \to 0$. Again, we set $\widetilde{\Lambda} = \operatorname{End}_\Lambda(T)^{\text{op}}$ and $\widetilde{\mathfrak{C}} = \mathfrak{C}(T_{\widetilde{\Lambda}})$.

LEMMA AND TERMINOLOGY 16. *Keep the above notation.*

(a) *The functor* $F = \operatorname{Hom}_\Lambda(-, T)|_{\mathfrak{C}} : \mathfrak{C} \to \widetilde{\mathfrak{C}}$ *takes the exact sequences* $0 \to X \to T^{(0)} \to \cdots \to T^{(m)} \to 0$ *and* $0 \to P^{(r)} \to \cdots \to P^{(0)} \to X \to 0$, *with* $X \in \mathfrak{C}$, $T^{(i)} \in \operatorname{add}({}_\Lambda T)$ *and* $P^{(j)} \in \operatorname{add}({}_\Lambda\Lambda)$, *to projective resolutions, resp.,* add$(T_{\widetilde{\Lambda}})$*-coresolutions, of* $F(X)$ *in* $\widetilde{\mathfrak{C}}$. *These resolutions have the same lengths as the corresponding ones in* $\mathfrak{C}$. *Thus*

$$\operatorname{p\,dim}_\Lambda X = \operatorname{codim}_{\operatorname{add}(T_{\widetilde{\Lambda}})} \operatorname{Hom}_\Lambda(X, T)$$
$$\operatorname{codim}_{\operatorname{add}({}_\Lambda T)} X = \operatorname{p\,dim}_{\widetilde{\Lambda}} \operatorname{Hom}_\Lambda(X, T).$$

Analogous statements hold for the functor $\operatorname{Hom}_{\widetilde{\Lambda}}(-, T)|_{\widetilde{\mathfrak{C}}} : \widetilde{\mathfrak{C}} \to \mathfrak{C}$.

(b) *The values of* $\operatorname{codim}_{\operatorname{add}({}_\Lambda T)} X$ *are uniformly bounded on* $\mathfrak{C}$, *and those of* $\operatorname{codim}_{\operatorname{add}(T_{\widetilde{\Lambda}})} Y$ *are uniformly bounded on* $\widetilde{\mathfrak{C}}$.

PROOF. To derive part (a) from Miyashita's duality, it suffices to observe that all consecutive kernels arising in the displayed exact sequences, read from right to left, again belong to $\mathfrak{C}$. As for the uniform boundedness claim in (b): $\mathfrak{C}$ and $\widetilde{\mathfrak{C}}$ are contravariantly finite in Λ-mod, resp., mod-$\widetilde{\Lambda}$ by Theorem 3. Therefore the Λ-, resp., $\widetilde{\Lambda}$-projective dimensions of the objects in $\mathfrak{C}$, resp., $\widetilde{\mathfrak{C}}$, are bounded from above by the maximum of the projective dimensions of the minimal $\mathfrak{C}$-, resp., minimal $\widetilde{\mathfrak{C}}$-approximations, of the simples in Λ-mod, resp., mod-$\widetilde{\Lambda}$; see Section 2. □

To stress the analogy of the dualities induced by tilting modules with the scenarios of more classical dualities of module categories, we translate the add(T)-codimension of an object X of $\mathfrak{C} = \mathfrak{C}({}_\Lambda T)$ into the Ext-injective dimension of X in $\mathfrak{C}$, as introduced in 2.D.

COROLLARY 17. *Retain the above hypotheses and notation.*

Then $\operatorname{codim}_{\operatorname{add}(T)} X = \operatorname{Ext.idim}_{\mathfrak{C}} X$ *for all* $X \in \mathfrak{C}$.

Moreover, the following invariants associated to $\mathfrak{C}$ *coincide (and hence are all finite):*

$$\sup\{\operatorname{p\,dim}_\Lambda X \mid X \in \mathfrak{C}\} = \sup\{\operatorname{Ext.idim}_{\mathfrak{C}} X \mid X \in \mathfrak{C}\} = \sup\{\operatorname{p\,dim}_{\widetilde{\Lambda}} Y \mid Y \in \widetilde{\mathfrak{C}}\}$$
$$= \sup\{\operatorname{Ext.idim}_{\widetilde{\mathfrak{C}}} Y \mid Y \in \widetilde{\mathfrak{C}}\}.$$

PROOF. In light of $\operatorname{Ext.idim}_{\mathfrak{C}} T = 0$, the initial claim follows from a straightforward induction on m, applied to an add$({}_\Lambda T)$-coresolution $0 \to X \to T^{(0)} \to \cdots \to T^{(m)} \to 0$ of minimal length.

As for the string of additional equalities: Let c be the maximum of the finite set $\{\operatorname{p\,dim}_\Lambda X \mid X \in \mathfrak{C}\}$. Since the category $\mathfrak{C}$ is resolving, the Λ-projective dimension of X coincides with the relative Ext-projective dimension of X in $\mathfrak{C}$ (see Remark 1). Therefore, Lemma 16 yields $c = \inf\{j \geq 0 \mid \operatorname{Ext}^{j+1}(X,-)|_{\mathfrak{C}} = 0 \text{ for all } X \in \mathfrak{C}\} = \inf\{j \geq 0 \mid \operatorname{Ext}^{j+1}(-,-)|_{\mathfrak{C}\times\mathfrak{C}} = 0\} = \inf\{j \geq 0 \mid \operatorname{Ext}^{j+1}(-,X)|_{\mathfrak{C}} = 0 \mid \text{ for all } X \in \mathfrak{C}\} = \sup\{\operatorname{Ext.idim}_{\mathfrak{C}} X \mid X \in \mathfrak{C}\}$. The second equality is immediate from Lemma 16(a), the final one is symmetric to the first. □

4.B. The case of a strong tilting module. The results of this section are closely intertwined with the theory developed by Auslander and Reiten in [**5**]. The following characterization of strong tilting modules buttresses the viewpoint that dualities $\mathcal{P}^{<\infty}(\Lambda\text{-mod}) \longleftrightarrow \mathcal{P}^{<\infty}(\text{mod-}\widetilde{\Lambda})$ should be viewed as variants of the standard duality Λ-mod $\longleftrightarrow$ mod-Λ in the special case when Λ is Gorenstein. When viewed from this angle, condition (2) below reveals contravariant finiteness of $\mathcal{P}^{<\infty}(\Lambda\text{-mod})$ to be a weakened Gorenstein condition for Λ.

PROPOSITION 18. (See [**25**, Proposition 4].) *Let Λ be any finite dimensional algebra. For $T \in \Lambda$-mod, the following conditions are equivalent:*

(1) *T is a strong tilting module in the sense of Auslander-Reiten, i.e., T is a tilting module and $\mathcal{P}^{<\infty}(\Lambda\text{-mod}) = \mathfrak{co\text{-}res}^{<\infty}({}_\Lambda T)$.*

(2) *T is a relatively* Ext*-injective cogenerator in the category $\mathcal{P}^{<\infty}(\Lambda\text{-mod})$, and all objects in $\mathcal{P}^{<\infty}(\Lambda\text{-mod})$ have finite relative* Ext*-injective dimension.*

(3) *T is a tilting module which is relatively* Ext*-injective in $\mathcal{P}^{<\infty}(\Lambda\text{-mod})$.*

In case these conditions are satisfied, T is even relatively splitting injective in $\mathcal{P}^{<\infty}(\Lambda\text{-mod})$ (cf. Proposition 6). *Moreover,*

$$\sup\{\operatorname{Ext.idim}_{\mathcal{P}^{<\infty}(\Lambda\text{-mod})} X \mid X \in \mathcal{P}^{<\infty}(\Lambda\text{-mod})\} = \operatorname{fin\,dim}\Lambda = \operatorname{Fin\,dim}\Lambda,$$

where $\operatorname{fin\,dim}\Lambda$ *and* $\operatorname{Fin\,dim}\Lambda$ *denote the* left *little and big finitistic dimensions of Λ. This number is smaller than or equal to the right finitistic dimension of $\widetilde{\Lambda}$.*

COROLLARY 19. (See [**25**, Corollary 5] and [**14**, Section 8].) *Suppose that ${}_\Lambda T_{\widetilde{\Lambda}}$ is a tilting bimodule which is strong on both sides. Then the coinciding left finitistic dimensions of Λ are equal to the coinciding right finitistic dimensions of $\widetilde{\Lambda}$.*

4.C. Arbitrary tilting modules. In order to emphasize the conceptual connection between "tilting dualities" and standard dualities, we extend Proposition 18 so as to fit arbitrary tilting modules into an analogous pattern. The description is immediate from Theorems 7, 8 and the results of Section 4.A.

OBSERVATION 20. Suppose $T \in \mathcal{P}^{<\infty}(\Lambda\text{-mod})$ is a basic tilting module. Moreover, let $\mathfrak{C} = \mathfrak{C}({}_\Lambda T)$ and $\widetilde{\mathfrak{C}} = \mathfrak{C}(T_{\widetilde{\Lambda}})$, where $\widetilde{\Lambda} = \operatorname{End}_\Lambda(T)^{\text{op}}$. Then T is a relatively Ext-injective cogenerator in $\mathfrak{C}$, and $\operatorname{Ext.idim}_{\mathfrak{C}} X < \infty$ for all $X \in \mathfrak{C}$. Moreover, the following coinciding invariants are finite:

$$\begin{aligned}\sup\{\operatorname{p\,dim}_\Lambda X \mid X \in \mathfrak{C}\} &= \sup\{\operatorname{Ext.idim}_{\mathfrak{C}} X \mid X \in \mathfrak{C}\} = \\ \sup\{\operatorname{p\,dim}_{\widetilde{\Lambda}} Y \mid Y \in \widetilde{\mathfrak{C}}\} &= \sup\{\operatorname{Ext.idim}_{\widetilde{\mathfrak{C}}} Y \mid Y \in \widetilde{\mathfrak{C}}\}.\end{aligned}$$

As for relative splitting injectivity of T in $\mathfrak{C}$, the upcoming observation will show that the number of those indecomposable direct summands of T which are splitting injective in $\mathfrak{C}$ equals the number of non-isomorphic direct summands of the minimal approximation $\mathcal{A}_{\mathfrak{C}}(E)$ of any injective cogenerator $E \in \Lambda$-mod. Indeed, on combining Observation 20 with Lemma 2, we encounter the announced extension of Proposition 6:

COROLLARY 21. *Let $T \in \Lambda$-mod be a basic tilting module, and let $\mathfrak{C}$ be the contravariantly finite category $\mathfrak{C}({}_\Lambda T)$. Then the minimal $\mathfrak{C}$-approximation $\mathcal{A}_\mathfrak{C}(E)$ of the basic injective cogenerator E in Λ-mod belongs to* $\mathrm{add}(T)$*; in other words, every indecomposable direct summand of $\mathcal{A}_\mathfrak{C}(E)$ is also a direct summand of T. Thus T^k is a $\mathfrak{C}$-approximation of E for some $k \geq 1$ depending on the multiplicities of the indecomposable direct summands of $\mathcal{A}_\mathfrak{C}(E)$.*

In case $\mathfrak{C}$ is closed under arbitrary cokernels of monomorphisms, equality holds: $\mathrm{add}(T) = \mathrm{add}\big(\mathcal{A}_\mathfrak{C}(E)\big)$.

PROOF. Observation 20 guarantees an embedding $\mathcal{A}_\mathfrak{C}(E) \subseteq T^k$ for some k. Since, by Lemma 2, $\mathcal{A}_\mathfrak{C}(E)$ is relatively splitting injective in $\mathfrak{C}$, any such inclusion splits. □

Corollary 21 is another indication of the intermediate position between a projective generator and an injective cogenerator which is held by a tilting module: Any tilting module T comes paired with a homomorphism $\phi(T) : T \to E$ such that $\phi(T)^k : T^k \to E$ is a $\mathfrak{C}({}_\Lambda T)$-approximation of E.

We conclude the section with some open problems triggered by Observation 20 and Corollary 21.

PROBLEM 1. Compare $\operatorname{fin dim} \Lambda$ with

$$\sup\{\operatorname{fin dim} \mathfrak{C}({}_\Lambda T) \mid T \in \Lambda\text{-mod} \text{ is a tilting module}\}.$$

PROBLEM 2. Keep the notation of Corollary 21. For which basic tilting modules ${}_\Lambda T$ is the category $\mathfrak{C}({}_\Lambda T)$ closed under cokernels of injective homomorphisms? This problem leads to the following more probing one: Let $\alpha(T)$ be the number of isomorphism classes of indecomposable objects in $\mathrm{add}\big(\mathcal{A}_{\mathfrak{C}(T)}(E)\big)$; we know that this is the number of indecomposable direct summands of T which are relatively splitting injective in $\mathfrak{C}({}_\Lambda T)$ (as opposed to only being relatively Ext-injective). Clearly, $\alpha(T) \leq \operatorname{rank} K_0(\Lambda)$, with equality holding when T is strong. The invariant $\alpha(T)$ gauges the "degree of relative injectivity" of T or, alternatively, the completeness of $\mathfrak{C}(T)$ with respect to cokernels of embeddings $T \hookrightarrow C$, for $C \in \mathfrak{C}$. Obviously, this measure is far coarser than the Riedtmann-Schofield partial order. We include a simple instance in which $\mathfrak{C}({}_\Lambda T_1) \subsetneqq \mathfrak{C}({}_\Lambda T_2)$ does not imply $\alpha(T_1) < \alpha(T_2)$. Pick up the notation of Example 5 for the Kronecker algebra Λ, and denote by R_2 the indecomposable preprojective module with top dimension 2. For the tilting module $T_1 = {}_\Lambda\Lambda$, we obtained $\alpha(T_1) = 1$. For the alternate choice $T_2 = \Lambda e_1 \oplus R_2$, we find $\mathfrak{C}(T_2) = \mathcal{P}^{<\infty}(\Lambda\text{-mod}) \cap {}^\perp T_2 \supsetneqq \mathfrak{C}(T_1) = \mathrm{add}({}_\Lambda\Lambda)$; on the other hand, the minimal $\mathfrak{C}(T_2)$-approximation of the basic injective cogenerator equals $(R_2)^2$, whence $\alpha(T_1) = \alpha(T_2) = 1$.

Given an algebra Λ, determine the values among $\{1, \dots, \operatorname{rank} K_0(\Lambda)\}$ which are attained by the function $\alpha(T)$, as T traces the tilting objects in Λ-mod. May "gaps" of size ≥ 2 occur? Comment: If the left finitistic dimension of Λ is zero, the only value attained by $\alpha(T)$ is $\alpha({}_\Lambda\Lambda) = \operatorname{rank} K_0(\Lambda)$.

5. The big companion categories of $\mathfrak{co\text{-}res}^{<\infty} T$. A further open problem

Throughout, $T \in \Lambda$-mod stands for a basic tilting module. Again, $\mathfrak{C} = \mathfrak{C}({}_\Lambda T) = \mathfrak{co\text{-}res}^{<\infty}({}_\Lambda T)$ and $\widetilde{\mathfrak{C}} = \mathfrak{C}(T_{\widetilde{\Lambda}}) = \mathfrak{co\text{-}res}^{<\infty}(T_{\widetilde{\Lambda}})$, where $\widetilde{\Lambda} = \mathrm{End}_\Lambda(T)^{\mathrm{op}}$ and $\mathfrak{C}({}_\Lambda T) = \mathcal{P}^{<\infty}(\Lambda\text{-mod}) \cap {}^\perp({}_\Lambda T)$. Recall from Section 2.A that $\mathfrak{Co\text{-}Res}^{<\infty}({}_\Lambda T)$ denotes the full subcategory of Λ-Mod which consists of the modules with finite coresolutions by objects in $\mathrm{Add}(T)$. Evidently, $\mathfrak{Co}\text{-}\mathfrak{Res}^{<\infty}({}_\Lambda T)$ contains $\mathfrak{C}({}_\Lambda T)$, $\mathrm{Add}({}_\Lambda\Lambda)$ and $\mathrm{Add}({}_\Lambda T)$.

PROBLEM 3. How do the categories $\mathfrak{Co}\text{-}\mathfrak{Res}^{<\infty}({}_\Lambda T)$ and $\mathcal{P}^{<\infty}(\Lambda\text{-Mod}) \cap {}^\perp({}_\Lambda T)$ compare?

- By definition, $\mathfrak{Co}\text{-}\mathfrak{Res}^{<\infty}({}_\Lambda T) \subseteq \mathcal{P}^{<\infty}(\Lambda\text{-Mod}) \cap {}^\perp({}_\Lambda T)$.
- In case ${}_\Lambda T$ is a strong tilting module, equality holds by [**26**]. Indeed, in this case $\mathcal{P}^{<\infty}(\Lambda\text{-Mod})$ consists of the direct limits of objects in $\mathcal{P}^{<\infty}(\Lambda\text{-mod})$; all of these belong to ${}^\perp({}_\Lambda T)$, since pure injectivity of T guarantees that the functors $\mathrm{Ext}^i_\Lambda(-,T)$ take direct limits to inverse limits (see [**3**]).

Our interest in the category $\mathfrak{Co}\text{-}\mathfrak{Res}^{<\infty}({}_\Lambda T)$ is buttressed by the following observation:

OBSERVATION 22. The functors $F = \mathrm{Hom}_\Lambda(-,T)$ and $\widetilde{F} = \mathrm{Hom}_{\widetilde{\Lambda}}(-,T)$ induce functors

$$\mathcal{P}^{<\infty}(\Lambda\text{-Mod}) \cap {}^\perp({}_\Lambda T) \longrightarrow \mathfrak{Co}\text{-}\mathfrak{Res}^{<\infty}(T_{\widetilde{\Lambda}}), \text{ and}$$
$$\mathcal{P}^{<\infty}(\text{Mod-}\widetilde{\Lambda}) \cap {}^\perp(T_{\widetilde{\Lambda}}) \longrightarrow \mathfrak{Co}\text{-}\mathfrak{Res}^{<\infty}({}_\Lambda T),$$

respectively. In particular, F and $\widetilde{F}$ induce functors

$$\mathfrak{Co}\text{-}\mathfrak{Res}^{<\infty}({}_\Lambda T) \rightleftarrows \mathfrak{Co}\text{-}\mathfrak{Res}^{<\infty}(T_{\widetilde{\Lambda}}).$$

Obviously these are not dualities in the infinite dimensional setting, but they induce Miyashita's dualities on finite dimensional modules. Moreover, the displayed restrictions of F and $\widetilde{F}$ are strictly exact on the indicated subcategories of Λ-Mod and Mod-$\widetilde{\Lambda}$.

PROOF. We only show the first claim, the second being symmetric and the following one a consequence. For $M \in \mathcal{P}^{<\infty}(\Lambda\text{-Mod}) \cap {}^\perp({}_\Lambda T)$, consider a finite projective resolution

$$(\dagger) \qquad 0 \to P_r \to \cdots \to P_0 \to M \to 0.$$

Write $P_j \cong \bigoplus_{1\le i\le n}(\Lambda e_i)^{(A_{ji})}$. Since the syzygies of the resolution ($\dagger$) again belong to ${}^\perp({}_\Lambda T)$, application of the functor F yields an exact sequence ($\ddagger$) $0 \to F(M) \to \prod_{1\le i\le n} T_i^{A_{0i}} \to \cdots \to \prod_{1\le i\le n} T_i^{A_{ri}} \to 0$, where the $T_i = \mathrm{Hom}_\Lambda(\Lambda e_i, T)$ are objects of $\mathrm{Add}(T_{\widetilde{\Lambda}})$. Since $T_{\widetilde{\Lambda}}$ is endofinite, direct products of objects in $\mathrm{Add}(T_{\widetilde{\Lambda}})$ again belong to this category, which shows ($\ddagger$) to be a finite $\mathrm{Add}(T_{\widetilde{\Lambda}})$-coresolution of $F(M)$. The final comment is obvious in light of the inclusion $\mathfrak{Co}\text{-}\mathfrak{Res}^{<\infty}({}_\Lambda T) \subseteq {}^\perp({}_\Lambda T)$. □

To help tackle Problem 3, we give an alternative description of $\mathfrak{Co}\text{-}\mathfrak{Res}^{<\infty}({}_\Lambda T)$. We thank the referee for communicating Proposition 24 below, the proof of which follows her/his guidelines. Recall that a *cotorsion pair* in Λ-Mod is a pair $(\mathcal{A}, \mathcal{B})$ of subcategories such that $\mathcal{B}$ consists of the Λ-modules B with $\mathrm{Ext}^1_\Lambda(-,B)|_\mathcal{A} = 0$ and $\mathcal{A}$ consists of the Λ-modules A with $\mathrm{Ext}^1_\Lambda(A,-)|_\mathcal{B} = 0$. Such a cotorsion pair is called *hereditary* when $\mathrm{Ext}^i_\Lambda(A,B) = 0$, for all $i > 0$ and all $A \in \mathcal{A}$ and $B \in \mathcal{B}$, which is equivalent to saying that $\mathcal{A}$ is closed under kernels of epimorphisms, and in turn equivalent to $\mathcal{B}$ being closed under cokernels of monomorphisms (see [**18**, Lemma 5.24]). We refer the reader to [**18**] for the terminology concerning cotorsion pairs which will be used below. For a tilting module $T \in \Lambda$-mod, define

$\mathcal{B}_T = \bigcap_{i>0} \mathrm{Ker}(\mathrm{Ext}^i_\Lambda(T,-))$. This yields a hereditary cotorsion pair $(\mathcal{A}_T, \mathcal{B}_T) := ({}^\perp\mathcal{B}_T, \mathcal{B}_T)$ in Λ-Mod to which we will refer as the *tilting cotorsion pair associated to* T. We alert the reader to the fact that this last concept applies more generally to any tilting module, not necessarily finite dimensional, over any ring, as is the case for all results from [**18**] cited in the remainder of this section.

The key point towards our next result is the fact that $\mathrm{Add}(T)$-coresolutions of the objects in $\mathfrak{Co}\text{-}\mathfrak{Res}^{<\infty}(T)$ may be uniformly bounded.

LEMMA 23. *Let $T \in \Lambda$-mod be a tilting module and $(\mathcal{A}_T, \mathcal{B}_T)$ the associated tilting cotorsion pair in Λ-Mod. Then $\mathfrak{Co}\text{-}\mathfrak{Res}^{<\infty}(T) = \mathcal{A}_T$, and the latter category consists of the Λ-modules A that admit exact sequences of the form*

$$0 \to A \to T_0 \to \cdots \to T_k \to 0,$$

with $k \le n$ and all T_i in $\mathrm{Add}(T)$.

PROOF. We know from [**18**, Proposition 13.13] that $\mathcal{A}_T$ is the class of Λ-modules A with $\mathrm{Add}(T)$-coresolutions of lengths $\le n$. Hence $\mathcal{A}_T \subseteq \mathfrak{Co}\text{-}\mathfrak{Res}^{<\infty}(T)$. To check the reverse inclusion, let $X \in \mathfrak{Co}\text{-}\mathfrak{Res}^{<\infty}(T)$ and consider an exact sequence

$$0 \to X \to T_0 \to \cdots \to T_k \to 0,$$

with $k \ge 0$ and T_i in $\mathrm{Add}(T)$. Denoting by Z_i the kernel of the map $T_i \to T_{i+1}$, with the convention that $Z_k = T_k$, we obtain $Z_i \in \mathcal{A}_T$ for $i = k, k-1, \dots, 0$ since $\mathcal{A}_T$ is closed under kernels of epimorphisms. In particular we find $Z_0 = X \in \mathcal{A}_T$. □

This yields the announced description of $\mathfrak{Co}\text{-}\mathfrak{Res}^{<\infty}({}_\Lambda T)$ as follows.

PROPOSITION 24. *If $T \in \Lambda$-mod is a tilting module, then*

$$\mathfrak{Co}\text{-}\mathfrak{Res}^{<\infty}({}_\Lambda T) = \varinjlim \mathfrak{C}({}_\Lambda T).$$

PROOF. Bearing in mind that the Λ-module T is endofinite, it is Σ-*pure-split* in the terminology of [**18**, p.328]. Hence, Proposition 13.55 of [op.cit] tells us that $\mathcal{A}_T$ is closed under direct limits in Λ-Mod. This, together with Lemma 23, proves the inclusion "$\supseteq$".

In order to prove the inclusion $\mathfrak{Co}\text{-}\mathfrak{Res}^{<\infty}({}_\Lambda T) \subseteq \varinjlim \mathfrak{C}({}_\Lambda T)$, recall from Lemma 23 that $\mathfrak{Co}\text{-}\mathfrak{Res}^{<\infty}({}_\Lambda T) = \mathcal{A}_T$ is the first component of the associated tilting cotorsion pair. Thus, by [**18**, Theorem 8.40], we know that $\varinjlim (\mathcal{A}_T \cap \Lambda\text{-mod}) = \varinjlim \mathcal{A}_T$. Bearing in mind that T is Σ-pure-split, we conclude $\varinjlim \mathcal{A}_T = \mathcal{A}_T$ by [**18**, Proposition 13.55].

The proof will be complete once we have checked that $\mathcal{A}_T \cap \Lambda\text{-mod} = \mathfrak{C}({}_\Lambda T)$ or, equivalently, that $\mathfrak{Co}\text{-}\mathfrak{Res}^{<\infty}({}_\Lambda T) \cap \Lambda\text{-mod} = \mathfrak{co}\text{-}\mathfrak{res}^{<\infty}({}_\Lambda T)$. It is clear that $\mathfrak{co}\text{-}\mathfrak{res}^{<\infty}({}_\Lambda T) \subseteq \mathfrak{Co}\text{-}\mathfrak{Res}^{<\infty}({}_\Lambda T) \cap \Lambda\text{-mod}$, whence we are left to prove the inclusion $\mathcal{A}_T \cap \Lambda\text{-mod} \subseteq \mathfrak{co}\text{-}\mathfrak{res}^{<\infty}({}_\Lambda T) = \mathfrak{C}({}_\Lambda T) = \mathcal{P}^{<\infty}(\Lambda\text{-mod}) \cap {}^\perp({}_\Lambda T)$. Since $\mathcal{A}_T = {}^\perp({}_\Lambda T^\perp)$ (with orthogonals in Λ-Mod) and $T \in ({}_\Lambda T)^\perp$, we obtain $\mathcal{A}_T \cap \Lambda\text{-mod} \subseteq {}^\perp({}_\Lambda T)$. Therefore it only remains to be checked that $\mathcal{A}_T \cap \Lambda\text{-mod} \subseteq \mathcal{P}^{<\infty}(\Lambda\text{-mod})$. To do so, let $A \in \mathcal{A}_T \cap \Lambda\text{-mod}$, fix an $\mathrm{Add}(T)$-coresolution $0 \to A \to T_0 \to \cdots \to T_k \to 0$, and define Z_i, for $i = k, k-1, \dots, 0$, as in the proof of Lemma 23. Bearing in mind that $\mathcal{P}^{<\infty}(\Lambda\text{-Mod})$ is closed under kernels of epimorphisms, we see that all the Z_i, and in particular $Z_0 = A$, belong to $\mathcal{P}^{<\infty}(\Lambda\text{-Mod})$. But this means that $A \in \mathcal{P}^{<\infty}(\Lambda\text{-mod})$ as desired. □

6. Applications to truncated path algebras

In the sequel, Λ will denote a truncated path algebra, say

$$\Lambda = KQ/\langle \text{all paths of length } L+1\rangle$$

for some quiver Q and positive integer L. We identify the vertices of Q with a full sequence of primitive idempotents $e_1, \dots, e_n$ of Λ. A vertex e of Q (alias a primitive idempotent of Λ) is *precyclic* in case there is an oriented path in Q which starts in e and ends on an oriented cycle; the vertex e is *postcyclic* if the dual holds, and *critical* if it is both pre- and postcyclic. In particular, every vertex that lies on an oriented cycle is critical. Moreover, we say that a simple module S_i is precyclic (postcyclic, critical) if the vertex e_i has the corresponding property.

Due to the key homological difference between precyclic and nonprecyclic vertices of Q, we will fix a convenient indexing: $e_1, \dots, e_m$ are the distinct precyclic vertices, $e_{m+1}, \dots, e_n$ the nonprecyclic ones. Moreover, we will adopt the following notation:

$$\varepsilon = \sum_{\text{nonprecyclic}} e_i = \sum_{m+1 \le i \le n} e_i.$$

We summarily refer to [**15**], [**14**] and [**25**] for proofs of results that are not explicitly referenced. The theory will be illustrated with a fully worked example, which we set up at the outset. Alongside the development of the theory, we will display (without proofs) the results of our computations for this example. The methods for obtaining them will be developed in Section 7, the actual computations carried out in Section 8.

6.A. Reference example. For our graphing conventions, we point to [**19**, Section 2].

Let $\Lambda = KQ/I$, where Q is the following quiver, and I is the ideal of the path algebra KQ which is generated by all paths of length 3.

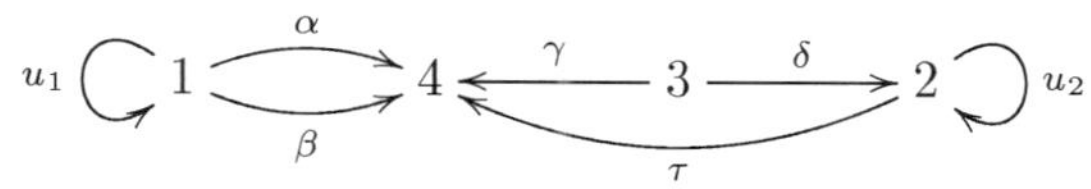

The precyclic vertices are e_1, e_2, e_3 in this case, with e_1, e_2 critical and e_3 a precyclic source. For immediate use in the upcoming section we display the indecomposable projectives and the indecomposable injectives in Λ-mod:

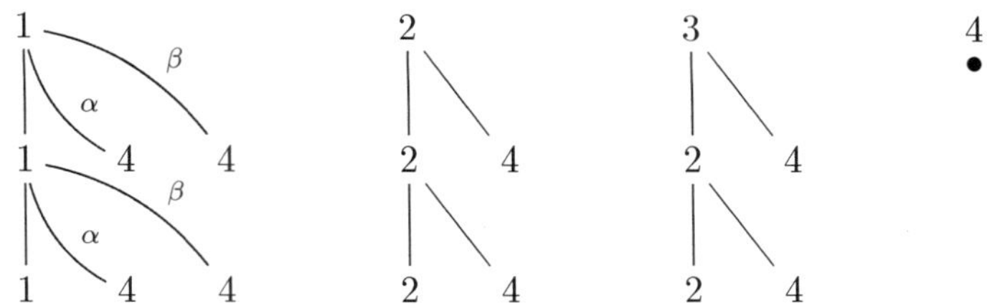

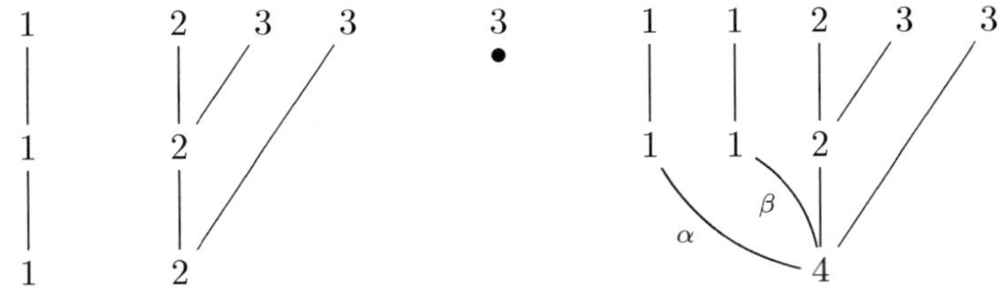

6.B. Characterization of the objects in $\mathcal{P}^{<\infty}(\Lambda\text{-mod})$. From the definition of ε, we obtain $(1-\varepsilon)\Lambda\varepsilon = 0$. Hence, for any $M \in \Lambda\text{-Mod}$, the subspace εM is a Λ-submodule of M. The pivotal feature of the left Λ-modules M of finite projective dimension is the structure of $M/\varepsilon M$.

THEOREM AND NOTATION 25. [**14**, Theorem 3.1] *Let $M \in \Lambda\text{-Mod}$. Then $\operatorname{p\,dim}_\Lambda M < \infty$ if and only if $M/\varepsilon M$ is projective as a module over the algebra $\Lambda/\Lambda\varepsilon\Lambda = \Lambda/\varepsilon\Lambda$.*

In other words, $\operatorname{p\,dim}_\Lambda M < \infty$ precisely when $M/\varepsilon M$ is a direct sum of copies of the modules $\mathcal{A}_i = \Lambda e_i/\varepsilon\Lambda e_i = \Lambda e_i/\varepsilon J e_i$ for the precyclic vertices e_i $(1 \le i \le m)$.

In our **reference example**, the indecomposable projective left $(\Lambda/\Lambda\varepsilon\Lambda)$-modules are $\mathcal{A}_1$, $\mathcal{A}_2$, $\mathcal{A}_3$ determined by the graphs

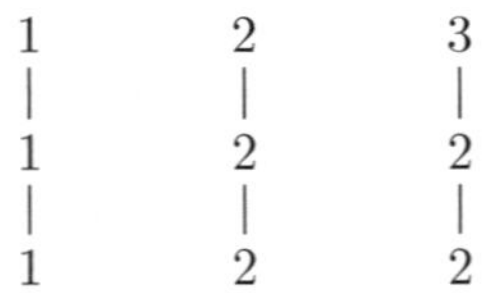

REMARK. The category $(\varepsilon\Lambda\varepsilon)$-mod is always a subcategory of Λ-mod because $\varepsilon\Lambda = \Lambda\varepsilon\Lambda$. It is, in fact, a localizing subcategory, which is contained in $\mathcal{P}^{<\infty}(\Lambda\text{-mod})$ since the quiver of $\varepsilon\Lambda\varepsilon$ is acyclic. By Theorem 25, the quotient category $\big(\mathcal{P}^{<\infty}(\Lambda\text{-mod})/((\varepsilon\Lambda\varepsilon)\text{-mod}\big)$ has finite representation type. However, assuming that Q has at least one postcyclic vertex which fails to be critical, we find: No matter what the representation type of $(\varepsilon\Lambda\varepsilon)$-mod, the extensions $\operatorname{Ext}^1_\Lambda(M/\varepsilon M, \varepsilon M)$ for $M \in \mathcal{P}^{<\infty}(\Lambda\text{-mod})$ always grow sufficiently complex for increasing L' to entail infinite representation type of $\mathcal{P}^{<\infty}\big(\Lambda_{L'}\text{-mod}\big)$ for $L' \gg 0$; here $\Lambda_{L'} = KQ/\langle\text{the paths of length } L'+1\rangle$. The following minimal quiver Q illustrating this was brought to the authors' attention by R. Kinser: Take $\Lambda_{L'} = \mathbb{C}Q/\langle\text{the paths of length } L'+1\rangle$ based on the quiver

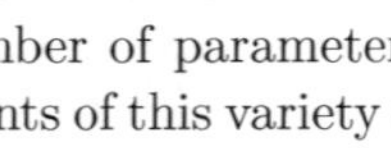

The authors checked that $\mathcal{P}^{<\infty}(\Lambda_{L'}\text{-mod})$ has infinite representation type for $L' \ge 8$ by considering the generic number of parameters of the variety $\mathfrak{Grass}^{S_1^2}_{\mathbf{d}}$ for $\mathbf{d} = \big(2(L'+1), L'\big)$. Note that all points of this variety encode modules of finite projective dimension by Theorem 25; for our notation and for techniques to obtain generic numbers of parameters, we refer to [**19**]. On the side, we mention that $\Lambda_{L'}$ has wild representation type for $L' \ge 4$; see [**24**].

6.C. Contravariant finiteness of $\mathcal{P}^{<\infty}(\Lambda\text{-mod})$ and the basic strong tilting module.

THEOREM 26. **Contravariant finiteness of $\mathcal{P}^{<\infty}(\Lambda\text{-mod})$.** [**14**, Theorem 4.2] *The category $\mathcal{P}^{<\infty}(\Lambda\text{-mod})$ is contravariantly finite, and the minimal $\mathcal{P}^{<\infty}(\Lambda\text{-mod})$-approximation $\mathcal{A}(M)$ of any $M \in \Lambda\text{-mod}$ is $P/\varepsilon\,\mathrm{Ker}(f)$, where $f : P \to M$ is a projective cover of M. In particular, $\mathcal{A}(S_i) = \mathcal{A}_i = \Lambda e_i/\varepsilon J e_i$ for $i \le n$, and the nonprecyclic simple modules $S_{m+1}, \dots, S_n$ coincide with their minimal $\mathcal{P}^{<\infty}(\Lambda\text{-mod})$-approximations.*

Therefore Λ-mod has strong tilting modules in light of Theorem 4. Let T be the basic one. We again write $\widetilde{\Lambda} = \mathrm{End}_\Lambda(T)^{\mathrm{op}}$. The following explicit description of the module ${}_\Lambda T$ permits its construction from quiver and Loewy length of Λ; this description is readily derived from Proposition 6 and Theorem 26.

THEOREM 27. [**14**, Theorem 5.3] *As in Theorem 26, let $\mathcal{A}_i = \mathcal{A}(S_i)$ be the minimal $\mathcal{P}^{<\infty}(\Lambda\text{-mod})$-approximation of S_i, and let $E_i = E(S_i)$ be the injective envelope of S_i. Then the basic strong tilting module in Λ-mod is $T = \bigoplus_{1\le i\le n} T_i$, where $T_i = \mathcal{A}_i$ for $1 \le i \le m$, and $T_i = \mathcal{A}(E_i)$ for $m+1 \le i \le n$.*

Supplement to Theorem 27. [**14**, Theorem 5.3] We provide further structural detail on the T_i in the two cases "e_i precyclic" and "e_i nonprecyclic" (referring to T_i and S_i as "precyclic" or "nonprecyclic" in case e_i has the pertinent property):

• Any precyclic T_i is a local module with a tree graph, all of whose simple composition factors are in turn precyclic.

• Now suppose that T_i is nonprecyclic. Then the minimal $\mathcal{P}^{<\infty}(\Lambda\text{-mod})$-approximation $T_i = \mathcal{A}(E_i) \twoheadrightarrow E_i$ shows E_i to also be a factor module of T_i. The socle of T_i contains precisely one nonprecyclic simple direct summand, namely $S_i = \mathrm{soc}(E_i)$, and every submodule $U \subseteq T_i$ with $U \not\subseteq JT_i$ contains S_i in its socle. Moreover, the factor module $T_i/\varepsilon T_i$ is isomorphic to a direct sum of copies of precyclic T_j's, the multiplicity of any critical T_j as a direct summand of $T_i/\varepsilon T_i$ being the number of paths of length L from e_j to e_i. Finally, the graph of T_i is again a tree (as an undirected graph) and thus determines T_i up to isomorphism.

In our **reference example**, the basic strong tilting module in $\mathcal{P}^{<\infty}(\Lambda\text{-mod})$ is $T = \bigoplus_{i=1}^4 T_i$, where the T_i are pinned down, up to isomorphism, by the following graphs:

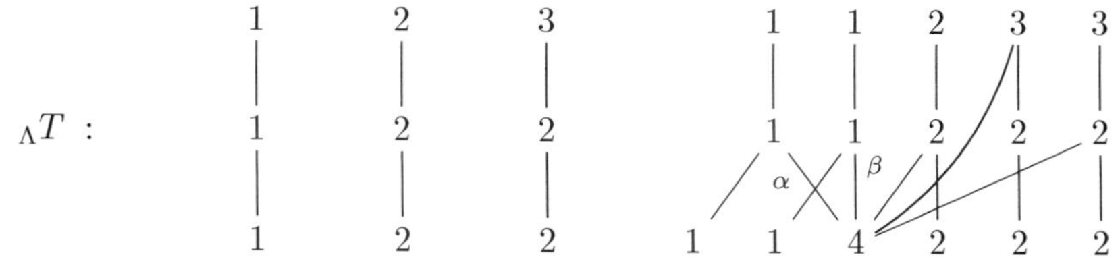

T_1 T_2 T_3 T_4

Here the summands $T_i = \mathcal{A}_i$ for $1 \le i \le 3$ correspond to the precyclic vertices of Q; the only nonprecyclic indecomposable direct summand of T is $T_4 = \mathcal{A}(E_4)$ with $T_4/\varepsilon T_4 \cong T_1^2 \oplus T_2 \oplus T_3^2$.

As was already observed by Auslander and Reiten in [**5**, Proposition 6.5], one-sided strongness of a tilting bimodule does not necessarily carry over to strongness on the opposite side. This asymmetry still occurs over truncated path algebras, our reference example being an instance; note that e_3 is a precyclic source in this example, and apply the upcoming criterion.

Again, we set $\widetilde{\Lambda} = \mathrm{End}_\Lambda(T)^{\mathrm{op}}$. There is a handy criterion for twosided strongness of ${}_\Lambda T_{\widetilde{\Lambda}}$ in the present, truncated, setting:

CRITERION 28. **Left-right symmetry of strongness.** [**14**, Corollary 7.2] *Let ${}_\Lambda T$ be a strong tilting module, and $\widetilde{\Lambda} = \mathrm{End}_\Lambda(T)^{\mathrm{op}}$. Then $T_{\widetilde{\Lambda}}$ is strong also in* mod-$\widetilde{\Lambda}$ *if and only if all precyclic vertices of Q are critical* (*equivalently, if Q does not have a precyclic source*).

6.D. The pivotal endofunctors of the strongly tilted category Mod-$\widetilde{\Lambda}$. In the following, $T = \bigoplus_{1\le i\le n} T_i$ will always stand for the basic strong tilting module in Λ-mod, and $\widetilde{\Lambda}$ for the opposite of its endomorphism ring. Moreover, $\widetilde{J}$ will denote the Jacobson radical of $\widetilde{\Lambda}$. The strongly tilted algebra $\widetilde{\Lambda}$ is in turn a path algebra modulo relations (see [**25**, Observation 10]), but hardly ever truncated unless Q is acyclic. Write $\widetilde{\Lambda} = K\widetilde{Q}/\widetilde{I}$.

Theorem 7 guarantees that the contravariant Hom-functors $\mathrm{Hom}_\Lambda(-,T)$ and $\mathrm{Hom}_{\widetilde{\Lambda}}(-,T)$ induce inverse dualities

$$\mathcal{P}^{<\infty}(\Lambda\text{-mod}) \quad \longleftrightarrow \quad \mathcal{P}^{<\infty}(\text{mod-}\widetilde{\Lambda}) \cap {}^\perp(T_{\widetilde{\Lambda}}).$$

Since the righthand category is properly contained in $\mathcal{P}^{<\infty}(\text{mod-}\widetilde{\Lambda})$ in general, this duality does not a priori permit us to transfer information from $\mathcal{P}^{<\infty}(\text{mod-}\widetilde{\Lambda})$ to $\mathcal{P}^{<\infty}(\Lambda\text{-mod})$, or vice versa. The upcoming analysis of $\mathcal{P}^{<\infty}(\text{mod-}\widetilde{\Lambda})$ through the lens of Q and L additionally hinges on the following **refined partition of the vertices e_i of Q**, and on a synchronized partition of the vertices of $\widetilde{Q}$.

FURTHER CONVENTIONS. We assume the n vertices $e_1, \dots, e_n$ to be ordered as follows:

- $e_1, \dots, e_r$ are the *critical* vertices of Q, i.e., those that are both pre- and postcyclic;
- $e_{r+1}, \dots, e_m$ are the *precyclic vertices which fail to be postcyclic*;
- $e_{m+1}, \dots, e_s$ are the vertices which are *postcyclic, but not precyclic*;
- $e_{s+1}, \dots, e_n$ are the vertices which are *neither pre- nor postcyclic*.

Theorem 27 provided us with a natural bijection $\{e_1, \dots, e_n\} \leftrightarrow \{T_1, \dots, T_n\}$. We supplement it by a bijection $\{T_1, \dots, T_n\} \leftrightarrow \{\text{vertices of } \widetilde{\Lambda}\}$ as follows:

The preferred primitive idempotents $\widetilde{e}_i$ of $\widetilde{\Lambda}$, alias the vertices of $\widetilde{Q}$:

$$\widetilde{e}_i = \iota_i \circ \pi_i : \; T \twoheadrightarrow \; T_i \; \hookrightarrow T,$$

where the ι_i and π_i are the canonical injections and projections, respectively.

Criticality in $\widetilde{\Lambda}$. In our analysis of the homology of $\widetilde{\Lambda}$, the role played by the precyclic vertices $e_1, \dots, e_m$ will be taken over by the critical vertices $\widetilde{e}_1, \dots, \widetilde{e}_r$ of $\widetilde{\Lambda}$, namely those which correspond to the critical idempotents $e_1, \dots, e_r$ of Λ. More precisely: An idempotent $\widetilde{e}_i$ of $\widetilde{\Lambda}$ will be referred to as (*tilted*) *critical*, (*tilted*) *pre-* or *postcyclic* precisely when the corresponding idempotent e_i of Λ has the specified property. In fact, in the sequel we will drop the qualifier "tilted" in reference to criticality (etc.) of the $\widetilde{e}_i$, since we will not incur any danger of ambiguity. (Observe, however, that criticality in this sense is not in line with the position of the vertex $\widetilde{e}_i$ relative to oriented cycles of the quiver $\widetilde{Q}$ of $\widetilde{\Lambda}$; in fact, even for a nonprecyclic vertex e_i of Q, the vertex $\widetilde{e}_i$ of $\widetilde{Q}$ will typically lie on multiple oriented cycles of $\widetilde{Q}$; cf. the reference example below.) We moreover extend the use of the attributes "pre(post)cyclic" and "critical" from the e_i and $\widetilde{e}_i$ to the indecomposable projective modules $\Lambda e_i \in \Lambda$-mod and $\widetilde{e}_i\widetilde{\Lambda} \in$ mod-$\widetilde{\Lambda}$, as well as to the simple modules $S_i = \Lambda e_i/Je_i \in \Lambda$-mod and $\widetilde{S}_i = \widetilde{e}_i\widetilde{\Lambda}/\widetilde{e}_i\widetilde{J} \in$ mod-$\widetilde{\Lambda}$.

Finally, we set $\widetilde{\mu} := \sum_{1\le i\le r} \widetilde{e}_i = \sum_{e_i \text{ critical}} \widetilde{e}_i$.

DEFINITION. **The endofunctors ∇ and Δ, and the critical core.** We introduce three endofunctors, ∇, Δ and $\mathcal{C}$, of Mod-$\widetilde{\Lambda}$ as follows. Let $\widetilde{M} \in$ Mod-$\widetilde{\Lambda}$.

- $\nabla(\widetilde{M}) := \widetilde{M}\,\widetilde{\mu}\,\widetilde{\Lambda}$. Thus $\nabla(\widetilde{M})$ is the unique smallest $\widetilde{\Lambda}$-submodule of $\widetilde{M}$ with the property that $\big(\widetilde{M}/\nabla(\widetilde{M})\big)\,\widetilde{\mu} = 0$. Note that the top of $\nabla(\widetilde{M})$ is a direct sum of critical simples.
- $\Delta(\widetilde{M}) := \{x \in \widetilde{M} \mid x\widetilde{\Lambda}\widetilde{\mu} = 0\} = \operatorname{ann}_{\widetilde{M}}(\widetilde{\Lambda}\widetilde{\mu})$ is the unique largest $\widetilde{\Lambda}$-submodule of $\widetilde{M}$ with $\Delta(\widetilde{M})\,\widetilde{\mu} = 0$. In particular, the socle of $\widetilde{M}/\Delta(\widetilde{M})$ is a sum of critical simples.
- $\widetilde{\Delta} := \Delta(\widetilde{\Lambda}_{\widetilde{\Lambda}})$ is a two-sided ideal of $\widetilde{\Lambda}$, referred to as the *noncritical ideal.* Clearly, $\widetilde{M} \cdot \widetilde{\Delta} \subseteq \Delta(\widetilde{M})$ with equality holding in case $\widetilde{M}$ is projective.
- $\mathcal{C}(\widetilde{M}) := \nabla(\widetilde{M})/\Delta(\nabla(\widetilde{M}))$ is called the *critical core* of $\widetilde{M}$.

By definition, both the top and the socle of $\mathcal{C}(\widetilde{M})$ are direct sums of critical simples in mod-$\widetilde{\Lambda}$. On the other hand, typically $\mathcal{C}(\widetilde{M})\widetilde{\mu} \subsetneqq \mathcal{C}(\widetilde{M})$, i.e., $\mathcal{C}(\widetilde{M})$ has also noncritical composition factors in general (see, e.g., $\widetilde{M} = \widetilde{e}_1\widetilde{\Lambda}$ in the reference example displayed below). The critical cores $\mathcal{C}(\widetilde{e}_i\widetilde{\Lambda})$ will play a crucial role in the sequel.

PROPOSITION 29. **The critical cores of the $\widetilde{e}_i\widetilde{\Lambda}$.** [**25**, Proposition 6]

(a) The critical $\widetilde{e}_i$ $(i \le r)$: Here $\mathcal{C}_i = \mathcal{C}(\widetilde{e}_i\widetilde{\Lambda}) = \widetilde{e}_i\widetilde{\Lambda}/\widetilde{e}_i\widetilde{\Delta}$.

(b) The post- but not precyclic $\widetilde{e}_i$ $(m+1 \le i \le s)$: In this case, $\mathcal{C}(\widetilde{e}_i\widetilde{\Lambda}) = \bigoplus_{k=1}^{r} \mathcal{C}_k^{\mathfrak{m}_{ik}}$, where $\mathfrak{m}_{ik}$ is the number of paths of length L from the vertex e_k of Q to the vertex e_i.

(c) The remaining $\widetilde{e}_i$ $(i \in \{r+1,\dots,m\} \cup \{s+1,\dots,n\})$: Here $\mathcal{C}(\widetilde{e}_i\widetilde{\Lambda}) = 0$.

RETURN TO THE REFERENCE EXAMPLE. As in the general setting, write $\widetilde{\Lambda} = K\widetilde{Q}/\widetilde{I}$. The quiver $\widetilde{Q}^{\mathrm{op}}$ is as follows, and the indecomposable projective right $\widetilde{\Lambda}$-modules are shown in Figure 1 below; we omit labels on edges connecting vertices a and b when there is only a single arrow from a to b.

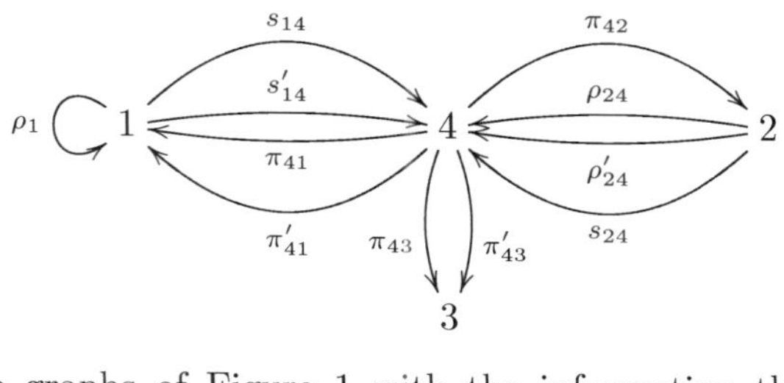

We combine the graphs of Figure 1 with the information that $\widetilde{I}$ is generated by monomial relations and binomial relations of the form $\widetilde{p} - \widetilde{q}$, where $\widetilde{p}$ and $\widetilde{q}$ are paths in $K\widetilde{Q}$. Taking this subsidiary information into account, we find the indecomposable projective modules $\widetilde{e}_i\widetilde{\Lambda}$ to be pinned down up to isomorphism by their graphs.

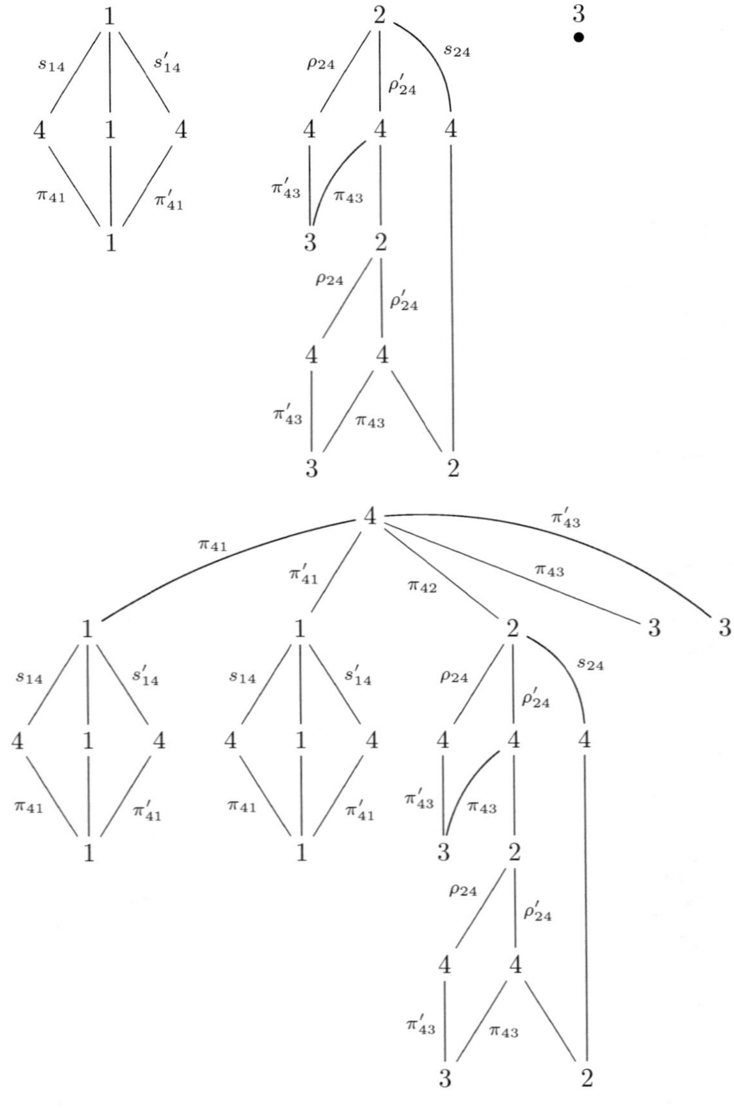

FIGURE 1. The indecomposable projective right $\widetilde{\Lambda}$-modules.

The critical cores of the $\widetilde{e}_i\widetilde{\Lambda}$ are as follows. Namely, $\mathcal{C}(\widetilde{e}_1\widetilde{\Lambda}) = \widetilde{e}_1\widetilde{\Lambda}$. Further, $\mathcal{C}(\widetilde{e}_2\widetilde{\Lambda})$ is the quotient $\widetilde{e}_2\widetilde{\Lambda}/U$, where U is the direct sum of the two copies of the uniserial module with composition factors $(\widetilde{S}_4, \widetilde{S}_3)$ plus the copy of $\widetilde{S}_4$ in the socle of $\widetilde{e}_2\widetilde{\Lambda}$; since there is only a single arrow from $\widetilde{e}_4$ to $\widetilde{e}_2$, the socle of $\widetilde{e}_2\widetilde{\Lambda}$ indeed contains a copy of $\widetilde{S}_4$. Thus $\mathcal{C}(\widetilde{e}_2\widetilde{\Lambda})$ has the graph shown in Figure 2 below. Moreover, $\mathcal{C}(\widetilde{e}_3\widetilde{\Lambda}) = 0$ and $\mathcal{C}(\widetilde{e}_4\widetilde{\Lambda}) \cong \mathcal{C}(\widetilde{e}_1\widetilde{\Lambda})^2 \oplus \mathcal{C}(\widetilde{e}_2\widetilde{\Lambda})$. Observe that the critical core of $\widetilde{e}_4\widetilde{\Lambda}$ is neither a sub- nor a factor module of $\widetilde{e}_4\widetilde{\Lambda}$, but is properly sandwiched between $\nabla(\widetilde{e}_4\widetilde{\Lambda})$ and $\Delta(\nabla(\widetilde{e}_4\widetilde{\Lambda}))$.

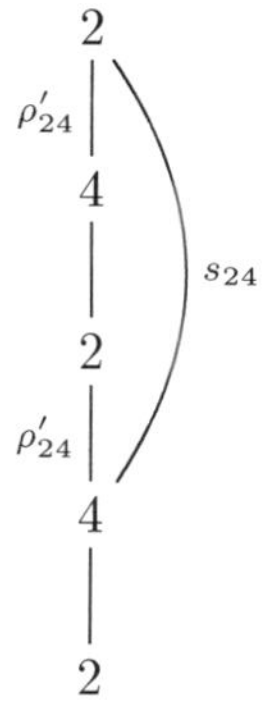

FIGURE 2. $\mathcal{C}(\widetilde{e}_2\widetilde{\Lambda})$.

By Criterion 28, the tilting bimodule ${}_\Lambda T_{\widetilde{\Lambda}}$ fails to be strong in mod-$\widetilde{\Lambda}$, since Q has a precyclic source. We will see however that $\mathcal{P}^{<\infty}(\text{mod-}\widetilde{\Lambda})$ is in turn contravariantly finite and consequently has its own strong tilting module $\widetilde{T} \in \text{mod-}\widetilde{\Lambda}$.

6.E. The structure of the objects in $\mathcal{P}^{<\infty}(\text{mod-}\widetilde{\Lambda})$.

THEOREM 30. [**25**, Theorem 13] *As before, $\mathcal{C}_j = \mathcal{C}(\widetilde{e}_j\widetilde{\Lambda})$ is the critical core of $\widetilde{e}_j\widetilde{\Lambda}$.*

A module $\widetilde{M} \in \text{Mod-}\widetilde{\Lambda}$ has finite projective dimension if and only if its critical core $\mathcal{C}(\widetilde{M})$ is a direct sum of copies of $\mathcal{C}_1, \dots, \mathcal{C}_r$; i.e.,

$$\operatorname{p\,dim}_{\widetilde{\Lambda}} \widetilde{M} < \infty \quad \iff \quad \mathcal{C}(\widetilde{M}) \in \operatorname{Add}(\mathcal{C}_1, \dots, \mathcal{C}_r).$$

Supplement for use in Section 7: $\operatorname{p\,dim}_{\widetilde{\Lambda}} \widetilde{M} < \infty$ *if and only if $\widetilde{M}\widetilde{\mu}$ is projective as a right $\widetilde{\mu}\widetilde{\Lambda}\widetilde{\mu}$-module, if and only if $\widetilde{M}/\Delta(\widetilde{M})$ is projective as a right module over the algebra $\widetilde{\Lambda}/\widetilde{\Delta}$.*

For the notation of Theorem 30, we refer back to 6.D. Regarding the proof of the supplement: The second of the equivalences added on is not explicitly stated in [**25**, Theorem 13], but is addressed in the argument given there.

To derive a first consequence: This criterion permits us to recognize the simple right $\widetilde{\Lambda}$-modules of finite projective dimension. In particular, we find that, outside the case where Q has no precyclic source, mod-$\widetilde{\Lambda}$ has more simples of finite projective dimension than Λ-mod does. More precisely:

COROLLARY 31. [**25**, Corollary 15] *For $j \in \{1, \dots, n\}$, the simple right $\widetilde{\Lambda}$-module $\widetilde{S}_j$ has finite projective dimension if and only if $\widetilde{S}_j$ is noncritical (i.e., $j \in \{r+1, \dots, n\}$).*

Next we address a latent weak heredity property of Λ, which surfaces only on the level of the tilted algebra $\widetilde{\Lambda}$. In our example, the appearance of two copies of $\widetilde{e}_1\widetilde{\Lambda}$ in the radical of $\widetilde{e}_4\widetilde{\Lambda}$, next to a copy of $\widetilde{e}_2\widetilde{\Lambda}$, is not coincidental. Indeed, if $\widetilde{P} = \widetilde{e}_4\widetilde{\Lambda}$, then $\widetilde{P}\widetilde{\mu}\widetilde{\Lambda}$ has top $\widetilde{S}_1^2 \oplus \widetilde{S}_2$ (recall that $\widetilde{\mu}$ is the sum of the critical idempotents $\widetilde{e}_i$, $i \le r$). Hence the following consequence of Theorem 30 predicts this outcome.

Corollary 32. **Weak heredity property of $\widetilde{\Lambda}_{\widetilde{\Lambda}}$.** [**25**, Corollary 16] *For any projective right $\widetilde{\Lambda}$-module $\widetilde{P}$, the submodule $\nabla(\widetilde{P}) = \widetilde{P}\widetilde{\mu}\widetilde{\Lambda}$ is again projective.*

The corollary allows us to obtain a large portion of the structure of $\widetilde{e}_k\widetilde{\Lambda}$ for a post- but not precyclic vertex $\widetilde{e}_k$ from the structure of the critical projectives $\widetilde{e}_i\widetilde{\Lambda}$, via mere inspection of the left Λ-module $T_k \subseteq T$. Namely: The number of arrows in $\widetilde{Q}^{\mathrm{op}}$ from $\widetilde{e}_k$ to a critical $\widetilde{e}_i$ equals the dimension of $e_i(T_k/JT_k)$. Each such arrow, α say, generates a copy $\widetilde{P}(\alpha)$ of $\widetilde{e}_i\widetilde{\Lambda}$ in $\widetilde{e}_k\widetilde{J}$ such that $\widetilde{e}_k\widetilde{J} = \bigoplus_\alpha \widetilde{P}(\alpha) + X$, where α ranges over the arrows starting in $\widetilde{e}_k$ and ending in some critical vertex $\widetilde{e}_i$, and X has only noncritical composition factors. We leave the justification to the reader.

Finally, we compare the "diagnostic homological filtrations" of the modules in Λ-Mod and Mod-$\widetilde{\Lambda}$.

• Each $M \in \Lambda$-Mod has a unique submodule U with the property that all simple composition factors of U have finite projective dimension, while those of M/U have infinite projective dimension (namely $U = \varepsilon M$). This submodule U gives rise to the criterion: $\operatorname{p\,dim}_\Lambda M < \infty$ if and only if M/U is a direct sum of copies of $\mathcal{A}_1, \dots, \mathcal{A}_m$.

• Every module $\widetilde{M} \in$ Mod-$\widetilde{\Lambda}$ contains a unique submodule chain $\widetilde{U} \subseteq \widetilde{V} \subseteq \widetilde{M}$ (namely, $\widetilde{V} = \nabla(\widetilde{M}) \supseteq \widetilde{U} = \Delta(\nabla(\widetilde{M}))$) such that $\widetilde{U}$ and $\widetilde{M}/\widetilde{V}$ have only composition factors of finite projective dimension, while socle and top of $\widetilde{V}/\widetilde{U}$ consist of simples with infinite projective dimension. This chain gives rise to the criterion: $\operatorname{p\,dim}_{\widetilde{\Lambda}} \widetilde{M} < \infty$ if and only if $\widetilde{V}/\widetilde{U}$ is a direct sum of copies of $\mathcal{C}_1, \dots, \mathcal{C}_r$.

Schematically, these filtrations of the objects in Λ-Mod and Mod-$\widetilde{\Lambda}$ look as follows.

	$M \in \Lambda$-Mod	$\widetilde{M} \in$ Mod-$\widetilde{\Lambda}$	
M/U	comp. factors of infinite p dim	comp. factors of finite p dim	$\widetilde{M}/\widetilde{V}$
U	comp. factors of finite p dim	critical core	$\widetilde{V}/\widetilde{U}$
		comp. factors of finite p dim	$\widetilde{U}$

6.F. Contravariant finiteness of $\mathcal{P}^{<\infty}(\text{mod-}\widetilde{\Lambda})$.

Theorem 33. [**25**, Theorem 19] *The category $\mathcal{P}^{<\infty}(\text{mod-}\widetilde{\Lambda})$ is contravariantly finite in* mod-$\widetilde{\Lambda}$. *Moreover, the minimal $\mathcal{P}^{<\infty}(\text{mod-}\widetilde{\Lambda})$-approximations of the simple right $\widetilde{\Lambda}$-modules $\widetilde{S}_i$ can be characterized and constructed from quiver and Loewy length of* Λ (on the basis of Supplement 35 below). *Hence* mod-$\widetilde{\Lambda}$ *has a strong tilting module $\widetilde{T}_{\widetilde{\Lambda}}$.*

By Corollary 31, the noncritical simples $\widetilde{S}_{r+1}, \dots, \widetilde{S}_n$ belong to $\mathcal{P}^{<\infty}(\text{mod-}\widetilde{\Lambda})$, meaning that they coincide with their minimal $\mathcal{P}^{<\infty}(\text{mod-}\widetilde{\Lambda})$-approximations. We describe the minimal approximations of the critical simples $\widetilde{S}_1, \dots, \widetilde{S}_r$. For $k \in \{1, \dots, r\}$, let $\mathcal{C}_k$ again be the critical core of $\widetilde{e}_k\widetilde{\Lambda}$. Fix an injective envelope $\widetilde{E}(\mathcal{C}_k)$ of $\mathcal{C}_k$.

LEMMA 34. [**25**, Lemma 18] *Still* $k \in \{1, \dots, r\}$. *There is a unique submodule* $\widetilde{\mathcal{A}}_k \subseteq \widetilde{E}(\mathcal{C}_k) = \widetilde{E}(\widetilde{e}_k\widetilde{\Lambda}/\widetilde{e}_k\widetilde{\Delta})$ *which is maximal relative to the following two properties:*

(i) *The generator* $a_k := \widetilde{e}_k + \widetilde{e}_k\widetilde{\Delta}$ *of* $\mathcal{C}_k$ *belongs to* $\widetilde{\mathcal{A}}_k \setminus \widetilde{\mathcal{A}}_k\widetilde{J}$;
(ii) $\widetilde{\mathcal{A}}_k$ *is an object of* $\mathcal{P}^{<\infty}(\text{mod-}\widetilde{\Lambda})$.

Observe that existence of $\widetilde{\mathcal{A}}_k$ follows from the fact that $\mathcal{C}_k$ satisfies (i) and (ii).

The following supplement to Theorem 33 specifies the minimal approximations of the simple right $\widetilde{\Lambda}$-modules.

SUPPLEMENT 35. [**25**, Theorem 19] *The minimal* $\mathcal{P}^{<\infty}(\text{mod-}\widetilde{\Lambda})$*-approximation of any critical simple* $\widetilde{S}_k \in \text{mod-}\widetilde{\Lambda}$ *is* $\widetilde{\mathcal{A}}_k$.

RETURN TO THE REFERENCE EXAMPLE. The two critical simple right $\widetilde{\Lambda}$-modules are $\widetilde{S}_i$ for $i = 1, 2$. For $i = 1$, the minimal $\mathcal{P}^{<\infty}(\text{mod-}\widetilde{\Lambda})$-approximation $\widetilde{\mathcal{A}}_1$ of $\widetilde{S}_1$ is as shown below. It is a proper submodule of the injective envelope $\widetilde{E}(\widetilde{S}_1) = \widetilde{T}_1$, with $\widetilde{T}_1$ as displayed in Figure 3 below. On assuming this equality, we visually identify the graph of $\widetilde{\mathcal{A}}_1$ from that of $\widetilde{T}_1$ by using the description in Lemma 34. As for $i = 2$: The minimal $\mathcal{P}^{<\infty}(\text{mod-}\widetilde{\Lambda})$-approximation $\widetilde{\mathcal{A}}_2$ of $\widetilde{S}_2$ equals the critical core $\mathcal{C}(\widetilde{e}_2\widetilde{\Lambda})$, which is shown in Figure 2 above. To back this, we preempt the equality $\widetilde{E}(\widetilde{S}_2) = \widetilde{T}_2$, with $\widetilde{T}_2$ again given in Figure 3. From the graph of $\widetilde{T}_2$, it is then obvious that that $\widetilde{\mathcal{A}}_2 = \widetilde{T}_2\widetilde{J} = \mathcal{C}(\widetilde{e}_2\widetilde{\Lambda})$. (Justifications of the claimed equalities $\widetilde{E}(\widetilde{S}_i) = \widetilde{T}_i$ are to follow in Section 8.)

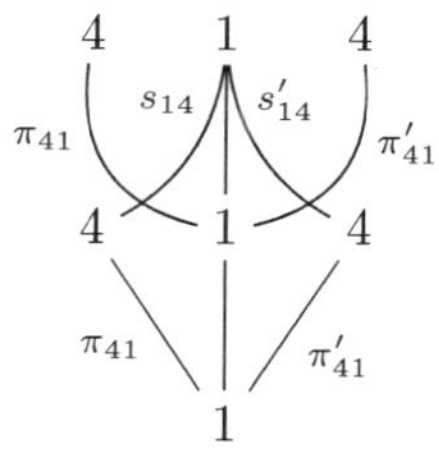

6.G. Iterated strong tilting. For any module $\widetilde{M} \in \text{mod-}\widetilde{\Lambda}$, we fix a $\widetilde{\Lambda}$-injective envelope $\widetilde{E}(\widetilde{M})$ and set $\widetilde{E}_i = \widetilde{E}(\widetilde{S}_i)$. Then $\widetilde{E} := \bigoplus_{1\le i\le n} \widetilde{E}_i$ is the basic injective cogenerator in mod-$\widetilde{\Lambda}$. The minimal $\mathcal{P}^{<\infty}(\text{mod-}\widetilde{\Lambda})$-approximation of $\widetilde{M}$ is denoted by $\widetilde{\mathcal{A}}(\widetilde{M})$.

In light of Theorem 33, $\mathcal{P}^{<\infty}(\text{mod-}\widetilde{\Lambda})$ has a strong tilting module $\widetilde{T}$, and from Proposition 6 we deduce

$$\text{add}(\widetilde{T}_{\widetilde{\Lambda}}) = \text{add}\left(\bigoplus_{i=1}^{n} \widetilde{\mathcal{A}}(\widetilde{E}_i)\right).$$

Now assume $\widetilde{T}$ to be basic. Then $\widetilde{T}_{\widetilde{\Lambda}}$ is the direct sum of one copy of each of the n indecomposable objects in the righthand category.

QUESTION. If we define $\widetilde{\widetilde{\Lambda}}$ to be the endomorphism ring $\text{End}_{\widetilde{\Lambda}}(\widetilde{T})$, is the tilting bimodule ${}_{\widetilde{\widetilde{\Lambda}}}\widetilde{T}_{\widetilde{\Lambda}}$ strong also over $\widetilde{\widetilde{\Lambda}}$?

Towards the positive answer given in the final theorem of this section, one uses the following auxiliary fact, which holds some independent interest. Clearly, the condition that the injective $\widetilde{\Lambda}$-module $\widetilde{E}_i$ embeds into its approximation $\widetilde{\mathcal{A}}(\widetilde{E}_i)$ is equivalent to $\widetilde{E}_i \in \mathcal{P}^{<\infty}(\text{mod-}\widetilde{\Lambda})$. This fails in general. On the other hand, the following weaker statement holds.

LEMMA 36. [**25**, Lemma 20] *The simple right $\widetilde{\Lambda}$-module $\widetilde{S}_i$ is contained in the socle of $\widetilde{\mathcal{A}}(\widetilde{E}_i)$ for each $i \le n$.*

Due to [**5**, Proposition 6.5], strongness of ${}_{\widetilde{\widetilde{\Lambda}}}\widetilde{T}$ is equivalent to the requirement that all simple right $\widetilde{\Lambda}$-modules embed into $\widetilde{T}_{\widetilde{\Lambda}}$. Hence the lemma immediately leads to

THEOREM 37. [**25**, Theorem 21] *Let $\widetilde{T}_{\widetilde{\Lambda}}$ be the basic strong tilting module in $\mathcal{P}^{<\infty}(\text{mod-}\widetilde{\Lambda})$ and $\widetilde{\widetilde{\Lambda}} = \text{End}_{\widetilde{\Lambda}}(\widetilde{T})$. Then the $\widetilde{\widetilde{\Lambda}}$-$\widetilde{\Lambda}$ tilting bimodule $\widetilde{T}$ is strong also as a left $\widetilde{\widetilde{\Lambda}}$-module, and the strongly tilted algebra $\text{End}_{\widetilde{\widetilde{\Lambda}}}(\widetilde{T})^{\text{op}}$ is isomorphic to $\widetilde{\Lambda}$.*

In particular, this shows that the process of iterated strong tilting of Λ-mod becomes stationary. Indeed, combined with Theorem 7, Theorem 37 yields dualities

$$\text{Hom}_{\widetilde{\Lambda}}(-,\widetilde{T}) : \ \mathcal{P}^{<\infty}(\text{mod-}\widetilde{\Lambda}) \quad \longleftrightarrow \quad \mathcal{P}^{<\infty}(\widetilde{\widetilde{\Lambda}}\text{-mod}) \ : \text{Hom}_{\widetilde{\widetilde{\Lambda}}}(-,\widetilde{T}).$$

Clearly, these inverse dualities permit to pivot structural information garnered for the objects of $\mathcal{P}^{<\infty}(\text{mod-}\widetilde{\Lambda})$ to a full complement of information on $\mathcal{P}^{<\infty}(\widetilde{\widetilde{\Lambda}}\text{-mod})$ and vice versa.

RETURN TO THE REFERENCE EXAMPLE. Decompose $\widetilde{T}$ into its indecomposable direct summands, $\widetilde{T} = \bigoplus_{1 \le i \le 4} \widetilde{T}_i$. We display the $\widetilde{T}_i$ in Figure 3 below; for their construction, we refer to the final section.

6.H. Representation type of the categories $\mathcal{P}^{<\infty}(\Lambda\text{-mod})$, $\mathcal{P}^{<\infty}(\text{mod-}\widetilde{\Lambda})$ and $\mathcal{P}^{<\infty}(\widetilde{\widetilde{\Lambda}}\text{-mod})$. The following observation picks up on the comment at the end of Section 6.B. It is an immediate consequence of Theorems 7, 33 and 37.

OBSERVATION 38. There is an injection from the set of (isomorphism classes of) indecomposable left Λ-modules of finite projective dimension to the set of indecomposable right $\widetilde{\Lambda}$-modules of finite projective dimension, and the latter are in one-to-one correspondence with the indecomposable left $\widetilde{\widetilde{\Lambda}}$-modules of finite projective dimension.

In light of the characterization of the objects in $\mathcal{P}^{<\infty}(\Lambda\text{-mod})$ (Theorem 25), it is often manageable to determine the representation type of $\mathcal{P}^{<\infty}(\Lambda\text{-mod})$ from that of $\mathcal{P}^{<\infty}(\varepsilon\Lambda\varepsilon\text{-mod})$ and the extensions in $\text{Ext}^1\big(\Lambda e_i/\varepsilon J e_i, \mathcal{P}^{<\infty}(\varepsilon\Lambda\varepsilon\text{-mod})\big)$ for $i \le m$. In such situations, one obtains an effective handle on the representation types of $\mathcal{P}^{<\infty}(\text{mod-}\widetilde{\Lambda})$ and $\mathcal{P}^{<\infty}(\widetilde{\widetilde{\Lambda}}\text{-mod})$ which allows bypassing the cumbersome computation of quiver and relations of $\widetilde{\Lambda}$ and of the critical cores of the $\widetilde{e}_i\widetilde{\Lambda} \in \text{mod-}\widetilde{\Lambda}$. To illustrate this, we modify the small example given at the end of Section 6.B. If Q is the quiver

$$\omega \circlearrowleft 1 \longrightarrow 2\,,$$

and Λ is the corresponding truncated path algebra of Loewy length 3, then it is not difficult to deduce from Theorem 25 that the category $\mathcal{P}^{<\infty}(\Lambda\text{-mod})$ has finite representation type (note that the factor algebra $\Lambda/\Lambda\omega^2$ is a string algebra of finite type) containing precisely 5 isomorphism classes of indecomposable objects; all of them are local. Given that Q does not have a precyclic source, ${}_\Lambda T_{\widetilde{\Lambda}}$ is strong on both sides by Criterion 28. Hence, by applying $\operatorname{Hom}_\Lambda(-, T)$, one finds that also $\mathcal{P}^{<\infty}(\text{mod-}\widetilde{\Lambda})$ has precisely 5 isomorphism classes of indecomposables, each having a simple socle. It is, in fact, routine to pin them down explicitly. Here $\widetilde{\widetilde{\Lambda}} \cong \Lambda$ because the tilting module T is strong on both sides.

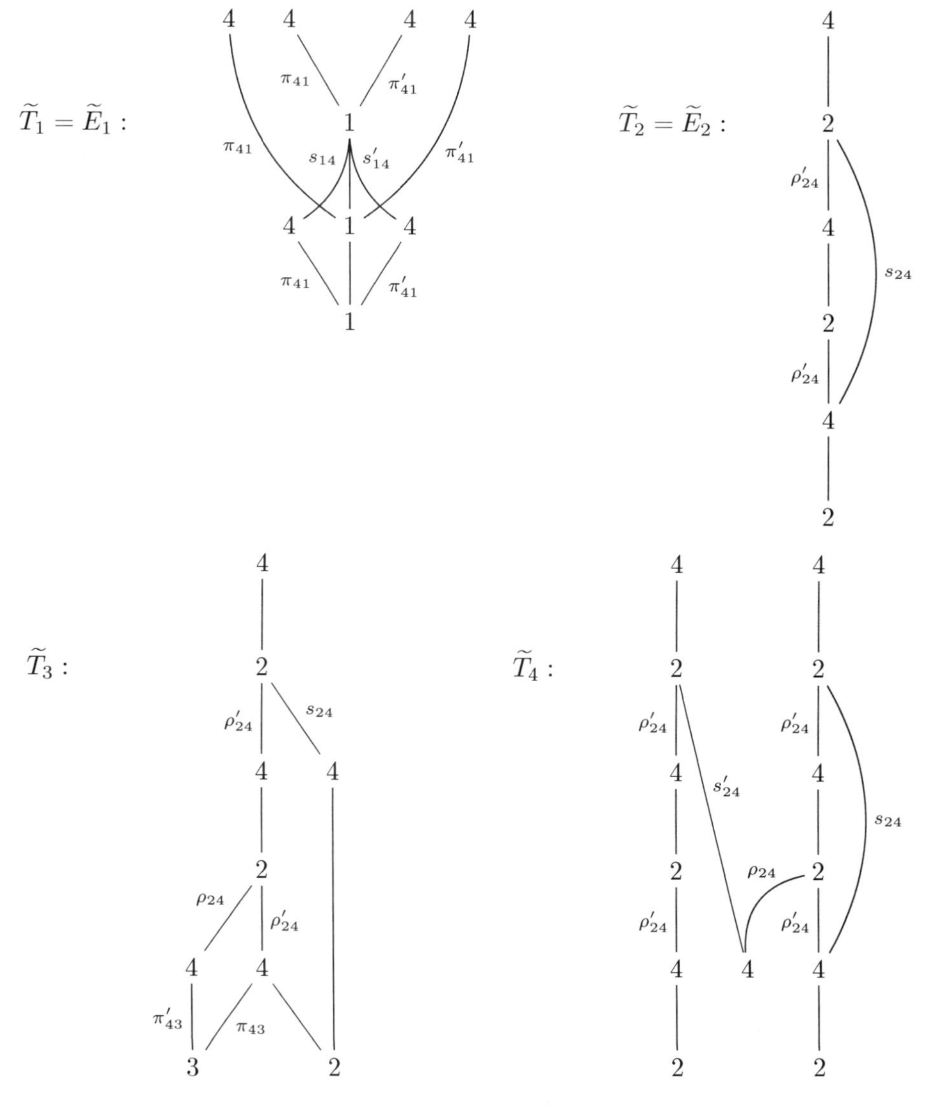

FIGURE 3. $\widetilde{T}_1$, $\widetilde{T}_2$, $\widetilde{T}_3$, $\widetilde{T}_4$.

7. Constructing the minimal $\mathcal{P}^{<\infty}$-approximations in mod-$\widetilde{\Lambda}$

7.A. Results targeting general truncated path algebras. In this section, we assume Λ to be an arbitrary truncated path algebra and $\widetilde{\Lambda} = \operatorname{End}({}_\Lambda T)^{\mathrm{op}}$, where $T \in \Lambda$-mod is the basic strong tilting module. We keep all conventions of Section 6. In particular, we will freely use the endofunctors Δ, ∇ and $\mathcal{C}(-)$ of Mod-$\widetilde{\Lambda}$ which were introduced in 6.D. As before, $\widetilde{\Delta}$ denotes the twosided ideal $\Delta(\widetilde{\Lambda}_{\widetilde{\Lambda}})$ of $\widetilde{\Lambda}$. Motivated by Theorem 30, we will switch back and forth among modules over the algebras $\widetilde{\Lambda}$, $\widetilde{\Lambda}/\widetilde{\Delta}$ and $\widetilde{\mu}\widetilde{\Lambda}\widetilde{\mu}$.

PROPOSITION 39. *Suppose $\widetilde{M} \in$ mod-$\widetilde{\Lambda}$ satisfies $\Delta(\widetilde{M}) = 0$. Then the minimal $\mathcal{P}^{<\infty}$-approximations of $\widetilde{M}$ in* Mod-$\widetilde{\Lambda}$ *and* Mod-$(\widetilde{\Lambda}/\widetilde{\Delta})$ *coincide.*

PROOF. Let $p : \widetilde{X} \longrightarrow \widetilde{M}$ be the minimal $\mathcal{P}^{<\infty}$(mod-$\widetilde{\Lambda}$)-approximation of $\widetilde{M}$. Recall that, for any $\widetilde{Y} \in \mathcal{P}^{<\infty}$(mod-$\widetilde{\Lambda}$), all simple composition factors of the submodule $\Delta(\widetilde{Y})$ are noncritical, whence also $\widetilde{Y}/\Delta(\widetilde{Y})$ belongs to $\mathcal{P}^{<\infty}$(mod-$\widetilde{\Lambda}$) by Corollary 30. By hypothesis, every homomorphism $f \in \operatorname{Hom}_{\widetilde{\Lambda}}(\widetilde{Y}, \widetilde{M})$ induces a homomorphism $\bar{f} : \widetilde{Y}/\Delta(\widetilde{Y}) \to \widetilde{M}$. Thus the approximation property of p shows that all homomorphisms from a module $\widetilde{Y}$ in $\mathcal{P}^{<\infty}$(mod-$\widetilde{\Lambda}$) to $\widetilde{M}$ factor through $\bar{p} : \widetilde{X}/\Delta(\widetilde{X}) \to \widetilde{M}$. Minimality of p consequently yields $\Delta(\widetilde{X}) = 0$, which makes $\widetilde{X}$ a module over $\widetilde{\Lambda}/\widetilde{\Delta}$; indeed, $\dim \widetilde{X}$ is minimal among the K-dimensions of the $\mathcal{P}^{<\infty}$(mod-$\widetilde{\Lambda}$)-approximations of $\widetilde{M}$. It is readily checked that, in this situation, finiteness of $\operatorname{p\,dim}_{\widetilde{\Lambda}} \widetilde{X}$ amounts to the same as finiteness of $\operatorname{p\,dim}_{\widetilde{\Lambda}/\widetilde{\Delta}} \widetilde{X}$, which shows p to also be a minimal $\mathcal{P}^{<\infty}$(mod-$(\widetilde{\Lambda}/\widetilde{\Delta})$-approximation of $\widetilde{M}$. □

The following result says that $\mathcal{P}^{<\infty}$(mod-$\widetilde{\Lambda}$)-approximations of arbitrary $\widetilde{\Lambda}$-modules can be constructed from approximations of $\widetilde{\Lambda}$-modules $\widetilde{N}$ with $\Delta(\widetilde{N}) = 0$.

PROPOSITION 40. *Let $\widetilde{M} \in$ mod-$\widetilde{\Lambda}$, and let $\pi : \widetilde{M} \longrightarrow \widetilde{M}/\Delta(\widetilde{M})$ be the canonical projection. If $p : \widetilde{Y} \longrightarrow \widetilde{M}/\Delta(\widetilde{M})$ is the minimal $\mathcal{P}^{<\infty}$(mod-$\widetilde{\Lambda}$)-approximation of $\widetilde{M}/\Delta(\widetilde{M})$ then the map $q : \widetilde{X} \longrightarrow \widetilde{M}$ arising in the following (bi)cartesian square is the minimal $\mathcal{P}^{<\infty}$(mod-$\widetilde{\Lambda}$)-approximation of $\widetilde{M}$:*

$$\begin{array}{ccccccccc}
0 & \longrightarrow & \Delta(\widetilde{M}) & \longrightarrow & \widetilde{X} & \xrightarrow{\ \widetilde{\pi}\ } & \widetilde{Y} & \longrightarrow & 0 \\
 & & \| & & \big\downarrow{\scriptstyle q} & \blacksquare \quad \bullet & \big\downarrow{\scriptstyle p} & & \\
0 & \longrightarrow & \Delta(\widetilde{M}) & \longrightarrow & \widetilde{M} & \xrightarrow{\ \pi\ } & \widetilde{M}/\Delta(\widetilde{M}) & \longrightarrow & 0
\end{array}$$

PROOF. Suppose $\widetilde{Z} \in \mathcal{P}^{<\infty}$(mod-$\widetilde{\Lambda}$) and $f \in \operatorname{Hom}_{\widetilde{\Lambda}}(\widetilde{Z}, \widetilde{M})$. Then $\pi \circ f$ factors through p, i.e., there is a morphism $g : \widetilde{Z} \longrightarrow \widetilde{Y}$ such that $p \circ g = \pi \circ f$. The universal property of pullbacks then yields a unique morphism $h : \widetilde{Z} \longrightarrow \widetilde{X}$ such that $\widetilde{\pi} \circ h = g$ and $q \circ h = f$. In particular, f factors through q, and hence q is a $\mathcal{P}^{<\infty}$-approximation of $\widetilde{M}$.

For the minimality of q, note that, by Proposition 39, we know that $\Delta(\widetilde{Y}) = 0$. Looking at the upper row in the commutative diagram of the statement, we conclude that $\Delta(\widetilde{X}) \cong \Delta(\widetilde{M})$ and $\widetilde{Y} \cong \widetilde{X}/\Delta(\widetilde{X})$. It is obvious that, after these identifications, we may view $\widetilde{\pi}$ as the canonical projection $\widetilde{X} \twoheadrightarrow \widetilde{X}/\Delta(\widetilde{X})$ and p as the map $\widetilde{q} : \widetilde{X}/\Delta(\widetilde{X}) \longrightarrow \widetilde{M}/\Delta(\widetilde{M})$ induced by q.

Now suppose $u : \widetilde{X} \longrightarrow \widetilde{X}$ is an endomorphism such that $q \circ u = q$. Then clearly $\widetilde{q} \circ \widetilde{u} = \widetilde{q}$, where $\widetilde{u} : \widetilde{X}/\Delta(\widetilde{X}) \longrightarrow \widetilde{X}/\Delta(\widetilde{X})$ is the endomorphism induced by u. By the right minimality of $\widetilde{q} \cong p$, the latter equality implies that $\widetilde{u}$ is an isomorphism. We deduce that u is an isomorphism: Since $q_{|\Delta(\widetilde{X})} : \Delta(\widetilde{X}) \xrightarrow{\cong} \Delta(\widetilde{M})$ is an isomorphism, we find $u_{|\Delta(\widetilde{X})} : \Delta(\widetilde{X}) \longrightarrow \Delta(\widetilde{X})$ to be an isomorphism. When combined with the fact that $\widetilde{u}$ is an isomorphism, this shows that, indeed, u is an isomorphism. □

The task of calculating minimal $\mathcal{P}^{<\infty}$-approximations of $\widetilde{\Lambda}$-modules is now reduced to calculating them for those $\widetilde{\Lambda}$-modules $\widetilde{F}$ which satisfy $\Delta(\widetilde{F}) = 0$. To tackle it, we pick up on Supplement 34, where the minimal approximations of the critical simples in mod-$\widetilde{\Lambda}$ are described.

DEFINITION. Let $\widetilde{F}$ be a finitely generated right $\widetilde{\Lambda}$-module such that $\Delta(\widetilde{F}) = 0$, and let $q : \widetilde{P} \longrightarrow \nabla(\widetilde{F})$ be the projective cover of $\nabla(\widetilde{F})$ in Mod-$(\widetilde{\Lambda}/\widetilde{\Delta})$. A Δ-*extension of* q will be a pair $(\widetilde{Y}, \varphi)$ satisfying the following conditions:

(1) $\widetilde{Y}$ is a submodule of a fixed injective envelope $\widetilde{E}(\widetilde{P})$ of $\widetilde{P}$ in mod-$\widetilde{\Lambda}$ such that $\widetilde{Y}$ contains $\widetilde{P}$ and $\widetilde{Y}/\widetilde{P}$ has only noncritical composition factors.

(2) $\varphi : \widetilde{Y} \longrightarrow \widetilde{F}$ is a morphism in mod-$\widetilde{\Lambda}$ such that the restriction $\varphi_{|\widetilde{P}}$ is the composition $\widetilde{P} \xrightarrow{q} \nabla(\widetilde{F}) \overset{j}{\hookrightarrow} \widetilde{F}$, where $j = j_{\widetilde{F}}$ is the inclusion.

We equip the set of Δ-extensions of $q : \widetilde{P} \to \nabla(\widetilde{F})$ (in a fixed injective envelope $\widetilde{E}(\widetilde{P})$) with the partial order of pairs induced by inclusion of the first components and prove:

PROPOSITION 41. *Let* $\widetilde{F} \in$ mod-$\widetilde{\Lambda}$ *with* $\Delta(\widetilde{F}) = 0$, *and let* $q : \widetilde{P} \longrightarrow \nabla(\widetilde{F})$ *be the projective cover of* $\nabla(\widetilde{F})$ *in* mod-$(\widetilde{\Lambda}/\widetilde{\Delta})$. *The poset of* Δ-*extensions of* q *in* $\widetilde{E}(\widetilde{P})$ *is upward directed and has a maximum. If* $(\widetilde{X}, p)$ *is the largest element, then* $p : \widetilde{X} \longrightarrow \widetilde{F}$ *is the minimal* $\mathcal{P}^{<\infty}$(mod-$\widetilde{\Lambda}$)-*approximation of* $\widetilde{F}$.

PROOF. Let us denote by $\mathcal{S}$ the set of Δ-extensions of q. Existence of a maximum in $\mathcal{S}$ will follow once we have proved that this set is directed: Indeed, finiteness of $\dim \widetilde{E}(\widetilde{P})$ will then yield a unique maximal element in $\mathcal{S}$.

So let $(\widetilde{Y}_1, \varphi_1)$ and $(\widetilde{Y}_2, \varphi_2)$ be elements of $\mathcal{S}$. We check that the restrictions of φ_1 and φ_2 to $\widetilde{Y}_1 \cap \widetilde{Y}_2$ are equal. Indeed, by condition (1) of the above definition, all composition factors of $(\widetilde{Y}_1 \cap \widetilde{Y}_2)/\widetilde{P}$ are noncritical. Since $(\varphi_1 - \varphi_2)(\widetilde{P}) = 0$ by condition (2), this shows $(\varphi_1 - \varphi_2)(\widetilde{Y}_1 \cap \widetilde{Y}_2) \subseteq \Delta(\widetilde{F}) = 0$. Consequently, we obtain a well-defined homomorphism $\varphi : \widetilde{Y}_1 + \widetilde{Y}_2 \longrightarrow \widetilde{F}$ which restricts to the φ_i on the summands $\widetilde{Y}_i$. In other words, the pair $(\widetilde{Y}_1 + \widetilde{Y}_2, \varphi)$ is an element of $\mathcal{S}$ which majorizes both of the original pairs.

Note moreover that, for any pair $(\widetilde{Y}, \varphi) \in \mathcal{S}$, we have $\nabla(\widetilde{Y}) = \widetilde{P}$; this is due to condition (1) and the fact that the top of $\widetilde{P}$ is a sum of critical simples.

Now let $p : \widetilde{X} \twoheadrightarrow \widetilde{F}$ be the minimal $\mathcal{P}^{<\infty}$(mod-$\widetilde{\Lambda}$)-approximation of $\widetilde{F}$. We will show that $(\widetilde{X}, p)$ is isomorphic to a Δ-extension of q, i.e., we will prove existence of an element $(\widetilde{X}', p') \in \mathcal{S}$, together with an isomorphism $\xi : \widetilde{X} \xrightarrow{\cong} \widetilde{X}'$, such that $p' \circ \xi = p$.

To verify this claim, we will in a first step reduce the problem to finding an element $(\widetilde{Y},\varphi) \in \mathcal{S}$, together with a morphism $u : \widetilde{X} \longrightarrow \widetilde{Y}$, such that $\varphi \circ u = p$. Indeed, suppose we are in possession of $(\widetilde{Y},\varphi) \in \mathcal{S}$ and u as specified. Minimality of p then shows u to be a split monomorphism giving rise to a decomposition $\widetilde{Y} = \mathrm{Im}(u) \oplus \widetilde{Y}'$ with $\varphi_{|\widetilde{Y}'} = 0$. This clearly entails that $(\widetilde{X},p)$ is isomorphic to $(\mathrm{Im}(u), \varphi_{|\mathrm{Im}(u)})$ via the isomorphism $u : \widetilde{X} \xrightarrow{\cong} \mathrm{Im}(u)$. To ascertain that the latter pair belongs to $\mathcal{S}$, we observe that $\widetilde{P} = \nabla(\widetilde{Y}) = \nabla(\mathrm{Im}(u)) \oplus \nabla(\widetilde{Y}')$. Since $\varphi_{|\widetilde{P}}$ induces the projective cover $q : \widetilde{P} \longrightarrow \nabla(\widetilde{F})$ in mod-$(\widetilde{\Lambda}/\widetilde{\Delta})$, we obtain a matrix decomposition $q \cong \begin{pmatrix}\varphi_{|\nabla(\mathrm{Im}(u))} & \varphi_{|\widetilde{Y}'}\end{pmatrix} : \widetilde{P} = \nabla(\widetilde{Y}) = \nabla(\mathrm{Im}(u)) \oplus \nabla(\widetilde{Y}') \longrightarrow \nabla(\widetilde{F})$. In light of $\varphi_{|\widetilde{Y}'} = 0$, minimality of q thus implies $\nabla(\widetilde{Y}') = 0$, i.e., $\widetilde{Y}'$ is a submodule of $\widetilde{E}(\widetilde{P})$ which has only noncritical composition factors. From $\Delta(\widetilde{P}) = 0$, we therefore deduce that $\widetilde{P} \cap \widetilde{Y}' = 0$, which yields $\widetilde{Y}' = 0$. Hence $(\mathrm{Im}(u), \varphi_{|\mathrm{Im}(u)})$ indeed belongs to $\mathcal{S}$, and the first step is complete.

Let us prove the existence of $(\widetilde{Y},\varphi)$ and u. By Proposition 39, we have $\Delta(\widetilde{X}) = 0$, i.e., $\nabla(\widetilde{X})$ is the critical core of $\widetilde{X}$ and thus projective as a right $\widetilde{\Lambda}/\widetilde{\Delta}$-module by Theorem 30. Hence the induced map $p_{|\nabla(\widetilde{X})} : \nabla(\widetilde{X}) \longrightarrow \nabla(\widetilde{F})$ factors through q, which yields a morphism $g : \nabla(\widetilde{X}) \longrightarrow \widetilde{P}$ with $q \circ g = p_{|\nabla(\widetilde{X})}$. We thus obtain a commutative diagram as shown below, where the upper square is the pushout of g and the inclusion $j_{\widetilde{X}} : \nabla(\widetilde{X}) \hookrightarrow \widetilde{X}$:

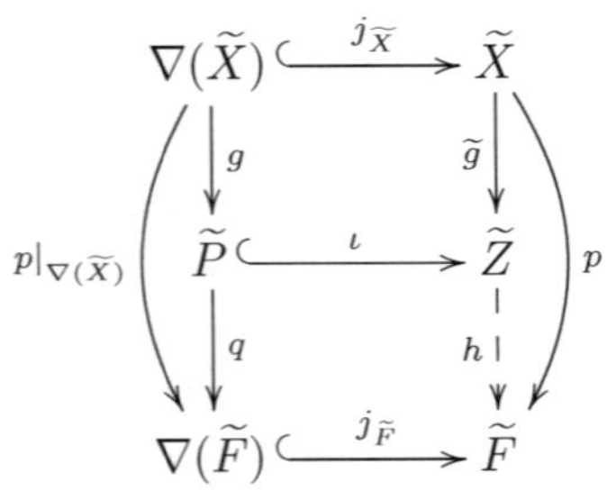

A homomorphism h as indicated by the dotted arrow exists by the pushout property. In view of $\Delta(\widetilde{F}) = 0$, we moreover have a factorization $h : \widetilde{Z} \xrightarrow{\rho} \widetilde{Z}/\Delta(\widetilde{Z}) \xrightarrow{\bar{h}} \widetilde{F}$, where ρ is the canonical map. We set $\widetilde{Y} := \widetilde{Z}/\Delta(\widetilde{Z})$. It remains to be checked that the composition $\widetilde{P} \xhookrightarrow{\iota} \widetilde{Z} \xrightarrow{\rho} \widetilde{Y}$ is an essential monomorphism whose cokernel has only noncritical composition factors. Once this is in place, we know that, up to isomorphism, the monomorphism $\rho \circ \iota$ may be wiewed as an embedding of $\widetilde{Y}$ into $\widetilde{E}(\widetilde{P})$ and that the choice $\varphi = \bar{h}$ establishes the status of $(\widetilde{X},p)$ as a copy of an element of $\mathcal{S}$.

To verify the final auxiliary claim, note that $\mathrm{Ker}(\rho\circ\iota) = \widetilde{P} \cap \Delta(\widetilde{Z}) \subseteq \Delta(\widetilde{P}) = 0$. As for essentiality of the image of $\rho \circ \iota$, the above diagram yields epimorphisms $\widetilde{X}/\nabla(\widetilde{X}) \cong \mathrm{Coker}(j_{\widetilde{X}}) \twoheadrightarrow \mathrm{Coker}(\iota) \twoheadrightarrow \mathrm{Coker}(\rho\circ\iota)$, and hence all composition factors of $\mathrm{Coker}(\rho\circ\iota)$ are noncritical. Given any submodule $\widetilde{Y}'$ of $\widetilde{Y}$ with $\mathrm{Im}(\rho\circ\iota) \cap \widetilde{Y}' = 0$, we infer that $\widetilde{Y}'$ is free of critical composition factors as well, i.e., $\widetilde{Y}' \subseteq \Delta(\widetilde{Y}) = 0$. Thus $\rho \circ \iota$ is indeed an essential monomorphism.

We have shown that $(\widetilde{X}, p) \cong (\widetilde{X}', p')$ with $(\widetilde{X}', p') \in \mathcal{S}$. It is now straightforward to verify that the latter pair is, in fact, maximal in $\mathcal{S}$. We leave the detail to the reader. $\square$

7.B. The critical corner $\widetilde{\mu}\widetilde{\Lambda}\widetilde{\mu}$ of a truncated path algebra. The following general observation can be found in [**25**, Section 6, 1st paragraph]. Let $\mu = \sum_{1 \le i \le r} e_i$ be the sum of the critical idempotents of Λ, and (as before) let $\widetilde{\mu} = \sum_{1 \le i \le r} \widetilde{e}_i$ be the corresponding sum in $\widetilde{\Lambda}$.

LEMMA 42. *The algebras $\mu\Lambda\mu$ and $\widetilde{\mu}\widetilde{\Lambda}\widetilde{\mu}$ are isomorphic. Any representative of the isomorphism class of these algebras will be called the* critical corner *of Λ or $\widetilde{\Lambda}$.*

We are interested in functorial connections between the category Mod-$\widetilde{\Lambda}$ and the module category of the critical corner $\widetilde{\mu}\widetilde{\Lambda}\widetilde{\mu}$ of $\widetilde{\Lambda}$. Recall that the functor H : Mod-$\widetilde{\Lambda} \longrightarrow$ Mod-$\widetilde{\mu}\widetilde{\Lambda}\widetilde{\mu}$ which takes $\widetilde{M}$ to $\widetilde{M}\widetilde{\mu}$ is naturally isomorphic to the functor $\mathrm{Hom}_{\widetilde{\Lambda}}(\widetilde{\mu}\widetilde{\Lambda}, -)$: Mod-$\widetilde{\Lambda} \longrightarrow$ Mod-$\mathrm{End}_{\widetilde{\Lambda}}(\widetilde{\mu}\widetilde{\Lambda})$. As such it is exact and has a left adjoint, namely $G = - \otimes_{\widetilde{\mu}\widetilde{\Lambda}\widetilde{\mu}} \widetilde{\mu}\widetilde{\Lambda}$: Mod-$\widetilde{\mu}\widetilde{\Lambda}\widetilde{\mu} \longrightarrow$ Mod-$\widetilde{\Lambda}$. The unit of this adjunction $\eta : 1_{\text{Mod-}\widetilde{\mu}\widetilde{\Lambda}\widetilde{\mu}} \longrightarrow H \circ G$ is readily seen to be a natural equivalence. These comments lead to the following conclusion:

PROPOSITION 43. *Suppose that e_i is a critical idempotent of Λ (equivalently, $\widetilde{e}_i$ is critical in $\widetilde{\Lambda}$). If $\widetilde{E}_i = \widetilde{E}(\widetilde{S}_i)$ is the corresponding indecomposable injective right $\widetilde{\Lambda}$-module, then $\widetilde{E}_i\widetilde{\mu}$ is an indecomposable injective right $\widetilde{\mu}\widetilde{\Lambda}\widetilde{\mu}$-module.*

In particular: If the critical corner $\mu\Lambda\mu$ is a self-injective algebra, then the projective dimension of $\widetilde{E}_i$ is finite.

PROOF. Let H and G be the functors defined in the previous paragraph. Let $j : \widetilde{V} \to \widetilde{W}$ be a monomorphism in mod-$\widetilde{\mu}\widetilde{\Lambda}\widetilde{\mu}$, and denote by $\widetilde{U}$ the kernel of $G(j)$. In a first step, we show that $\mathrm{Hom}_{\widetilde{\Lambda}}(\widetilde{U}, \widetilde{E}_i) = 0$. Indeed, by applying H to the exact sequence

$$(\dagger) \qquad 0 \longrightarrow \widetilde{U} \longrightarrow G(\widetilde{V}) \xrightarrow{G(j)} G(\widetilde{W}),$$

and keeping in mind that $(H \circ G)(j) \cong j$ is a monomorphism, we obtain $\widetilde{U}\widetilde{\mu} = H(\widetilde{U}) = 0$; in other words, $\widetilde{U}$ is devoid of critical composition factors. Since $\widetilde{E}_i$ has a critical socle, this confirms the equality $\mathrm{Hom}_{\widetilde{\Lambda}}(\widetilde{U}, \widetilde{E}_i) = 0$.

To infer that $j^* := \mathrm{Hom}_{\widetilde{\Lambda}}(j, \widetilde{E}_i\widetilde{\mu})$ is surjective, we use the adjointness of G and H, to obtain the commutative square

$$\begin{array}{ccc}
\mathrm{Hom}_{\widetilde{\Lambda}}(G(\widetilde{W}), \widetilde{E}_i) & \xrightarrow{\mathrm{Hom}(G(j), \widetilde{E}_i)} & \mathrm{Hom}_{\widetilde{\Lambda}}(G(\widetilde{V}), \widetilde{E}_i) \\
\updownarrow \cong & & \updownarrow \cong \\
\mathrm{Hom}_{\widetilde{\mu}\widetilde{\Lambda}\widetilde{\mu}}(\widetilde{W}, \widetilde{E}_i\widetilde{\mu}) & \xrightarrow{j^*} & \mathrm{Hom}_{\widetilde{\mu}\widetilde{\Lambda}\widetilde{\mu}}(\widetilde{V}, \widetilde{E}_i\widetilde{\mu})
\end{array}$$

In view of the preceding paragraph, injectivity of $\widetilde{E}_i$ as a right $\widetilde{\Lambda}$-module ensures that $\mathrm{Hom}_{\widetilde{\Lambda}}(G(j), \widetilde{E}_i)$ is surjective. Hence so is j^*. This proves the first of our two assertions.

We deduce the second from the first: In case the critical corner of $\widetilde{\Lambda}$ is self-injective, $\widetilde{E}_i\widetilde{\mu}$ is a projective right $\widetilde{\mu}\widetilde{\Lambda}\widetilde{\mu}$-module, and therefore $\operatorname{p\,dim}_{\widetilde{\Lambda}} \widetilde{E}_i < \infty$ by Theorem 30. Indecomposabilty of $\widetilde{E}_i\widetilde{\mu}$ is clear, because $\widetilde{e}_i\widetilde{\mu} = \widetilde{e}_i$, whence the socle of $\widetilde{E}_i\widetilde{\mu}$ is simple. □

REMARKS 44. **(1)** Identifying the quiver Q with its associated path category, i.e., the category whose objects are the vertices of Q and whose morphisms are the paths, we find the full subcategory consisting of the critical vertices to be convex, meaning: if $e_i \to e_j \to e_k$ is a composition of two paths in Q such that e_i and e_k are critical, then e_j is critical as well. As an immediate consequence, one obtains that, whenever the critical corner $\mu\Lambda\mu$ is nontrivial, it is again a truncated path algebra which has the same Loewy length as Λ. Note, however, that passage to the critical corner does not preserve connectedness in general.

(2) Any self-injective truncated path algebra Λ is isomorphic to a finite product of Nakayama algebras (see [**6**] for the definition). In light of Remark 1, this shows: If the critical corner $\mu\Lambda\mu$ of Λ is self-injective, its quiver $\mu Q\mu$ is a disjoint union of simple oriented cycles (= oriented cycles which are not products of strictly shorter cycles). This applies to the critical corner in the reference example of Section 6.A, which will be revisited in depth in the next section.

8. Algorithmic aspects of the reference example

In this section we focus on the algebra $\Lambda = KQ/\langle\text{the paths of length } 3\rangle$ of the reference example in 6.A. Our goal is to verify the claims we made throughout Section 6. Again $\widetilde{\Lambda} = K\widetilde{Q}/\widetilde{I}$ denotes the algebra that results from strongly tilting Λ-mod. First, we will use Theorems 26 and 27 to determine the basic strong tilting module $T = \bigoplus_{1\le i\le 4} T_i \in \Lambda$-mod, and next we will find the quiver Q^{op} of $\widetilde{\Lambda}^{op} = \operatorname{End}_\Lambda(T)$, as well as a set of generators for the ideal of relations. Finally, we will apply the recipes of the previous section towards identifying the basic strong tilting module $\widetilde{T}$ in mod-$\widetilde{\Lambda}$. As already mentioned in Section 6, the precyclic vertices of Q are e_1, e_2, and the postcyclic ones are e_1, e_2, e_4, whence e_1 and e_2 are the critical vertices. The idempotent ε of Λ introduced at the outset of Section 6 equals e_4.

Recall that a *top element* of a left Λ-module M is any element $x \in M \setminus JM$ such that $e_i x = x$ for some $i \le 4$; a *full sequence of top elements* of M is a collection of top elements which are linearly independent modulo JM and generate M.

8.A. The indecomposable direct summands of the strong tilting module ${}_\Lambda T$. According to Theorem 27, we have $T_i = \Lambda e_i/\varepsilon\Lambda e_i = \Lambda e_i/e_4\Lambda e_i$ for $i = 1, 2, 3$, while $T_4 = \mathcal{A}(E_4)$ is the minimal $\mathcal{P}^\infty(\Lambda\text{-mod})$-approximation of $E_4 = E(S_4)$. To calculate T_4 by way of Theorem 26, let $\pi : P = \Lambda e_1^2 \oplus \Lambda e_2 \oplus \Lambda e_3^2 \twoheadrightarrow E_4$ be the projective cover of E_4, which sends the top elements e_i of the indecomposable projective modules Λe_i (see the graphs in Section 6.A) to the obvious full sequence of top elements of E_4 as shown below. To compute $P/\varepsilon \operatorname{Ker}(\pi) = P/e_4 \operatorname{Ker}(\pi)$, we may simplify P to the (nonprojective) module Q shown in Figure 4 below, because π vanishes on the additional copies of S_4 in the socle of P. Clearly, $\mathcal{A}(E_4) \cong Q/e_4\operatorname{Ker}(q)$, and $e_4\operatorname{Ker}(q) \cong S_4^4$ is the direct sum of the simple submodules of Q generated by $(\alpha u_1, -\beta u_1, 0, 0, 0)$, $(0, \beta u_1, -\tau u_2, 0, 0)$, $(0, 0, \tau u_2, -\tau\delta, 0)$ and $(0, 0, 0, \tau\delta, -\gamma)$. We conclude that $T_4 \in \Lambda$-mod is indeed the module displayed in Section 6.C.

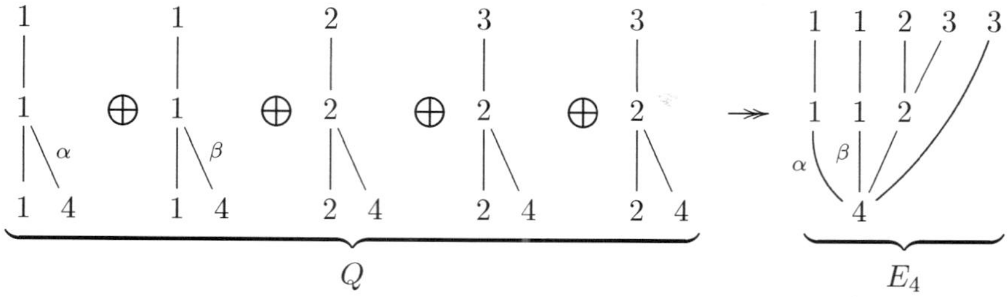

FIGURE 4. The epimorphism $Q \to E_4$.

To facilitate the verification of our subsequent computations, we translate the graphs of the T_i into generators and relators; i.e., we present the T_i as quotients $P_{T_i}/\Omega\, T_i$, where P_{T_i} is the projective cover of T_i suggested by the pertinent graphs (in particular $P_{T_4} = P$ as above), and $\Omega\, T_i$ is the corresponding syzygy of T_i.

PROPOSITION 45. *The following is the list of indecomposable direct summands of the strong tilting left Λ-module T:*

(1) $T_1 = \Lambda e_1/\Omega\, T_1$, *where* $\Omega\, T_1$ *is generated by* αu_1, βu_1, α, β;
(2) $T_2 = \Lambda e_2/\Omega\, T_2$, *where* $\Omega\, T_2$ *is generated by* τu_2 *and* τ;
(3) $T_3 = \Lambda e_3/\Omega\, T_3$, *where* $\Omega\, T_3$ *is generated by* $\tau\delta$ *and* γ;
(4) $T_4 = P_{T_4}/\Omega\, T_4$, *where* $P_{T_4} = \Lambda e_1^2 \oplus \Lambda e_2 \oplus \Lambda e_3^2$ *and* $\Omega\, T_4$ *is the submodule of* P_{T_4} *generated by* $(\beta u_1, 0, 0, 0, 0)$, $(\alpha, 0, 0, 0, 0)$, $(\beta, 0, 0, 0, 0)$, $(0, \alpha u_1, 0, 0, 0)$, $(0, \alpha, 0, 0, 0)$, $(0, \beta, 0, 0, 0)$, $(0, 0, \tau, 0, 0)$, $(0, 0, 0, \gamma, 0)$, $(0, 0, 0, 0, \tau\delta)$, $(\alpha u_1, -\beta u_1, 0, 0, 0)$, $(\alpha u_1, 0, -\tau u_2, 0, 0)$, $(\alpha u_1, 0, 0, -\tau\delta, 0)$, *and* $(\alpha u_1, 0, 0, 0, -\gamma)$.

8.B. Quiver and relations of the endomorphism algebra of ${}_\Lambda T$. We will display the quiver of $\mathrm{End}_\Lambda(T)$, rather than that of $\widetilde{\Lambda}$; in our previous notation, this means that we will determine the quiver $\widetilde{Q}^{\mathrm{op}}$. Passing to the opposite of $\widetilde{\Lambda} = K\widetilde{Q}/\widetilde{I}$ proves convenient towards graphing the indecomposable projective right $\widetilde{\Lambda}$-modules. Evidently, we may think of $\widetilde{Q}^{\mathrm{op}}$ as the quiver having the four vertices $T_1, \dots, T_4$; the set of arrows from T_i to T_j consists of a K-basis for $\mathrm{rad}(T_i, T_j)$ modulo $\mathrm{rad}^2(T_i, T_j)$. To describe our choice of arrows, we organize the non-isomorphisms among the T_i's into three groups, π_{ij}'s, ρ_{ij}'s and s_{ij}'s, with indices i, j pinning down the domain T_i and codomain T_j in each case; when $i = j$ we cut down to a single index. A homomorphism $T_i \to T_j$ belongs to the π-group if its image is not contained in the radical JT_j; it belongs to the ρ-group, if its image is contained in JT_j, but not in J^2T_j, and to the s-group if its image is contained in $J^2T_j = \mathrm{soc}(T_j)$ for each $j \le 4$. The notation P_{T_i} and $\Omega\, T_i$ is carried over from 8.A. Based on these conventions, we next describe a convenient K-basis for the radical of $\mathrm{End}_\Lambda(T)$ in Lemma 46. The proof of this lemma, as well as that of the subsequent one, resides on somewhat cumbersome computations. We leave it to the interested reader to duplicate them.

LEMMA 46. *The Jacobson radical of* $\mathrm{End}({}_\Lambda T)$ *has a K-basis consisting of the morphisms listed below. Regarding the format of the list: Each map $T_i \longrightarrow T_j$ is communicated via a map $f : P_{T_i} \longrightarrow P_{T_j}$ such that $f(\Omega\, T_i) \subseteq \Omega\, T_j$. Moreover, for each of the listed maps, precisely one of the distinguished top elements of its* *domain P_{T_i} is not mapped to zero, whence f is pinned down by its value on this top element:*

(1) $\pi_{41} : T_4 \longrightarrow T_1$, $(e_1, 0, 0, 0, 0) \mapsto e_1$;
(2) $\pi'_{41} : T_4 \longrightarrow T_1$, $(0, e_1, 0, 0, 0) \mapsto e_1$;
(3) $\pi_{42} : T_4 \longrightarrow T_2$, $(0, 0, e_2, 0, 0) \mapsto e_2$;
(4) $\pi_{43} : T_4 \longrightarrow T_3$, $(0, 0, 0, e_3, 0) \mapsto e_3$;
(5) $\pi'_{43} : T_4 \longrightarrow T_3$, $(0, 0, 0, 0, e_3) \mapsto -e_3$;
(6) $\rho_1 : T_1 \longrightarrow T_1$, $e_1 \mapsto u_1$;
(7) $\rho_2 : T_2 \longrightarrow T_2$, $e_2 \mapsto u_2$;
(8) $\rho_{23} : T_2 \longrightarrow T_3$, $e_2 \mapsto \delta$;
(9) $\rho_{24} : T_2 \longrightarrow T_4$, $e_2 \mapsto (0, 0, 0, 0, \delta)$;
(10) $\rho'_{24} : T_2 \longrightarrow T_4$, $e_2 \mapsto (0, 0, u_2, -\delta, 0)$;
(11) $\rho_{41} : T_4 \longrightarrow T_1$, $(e_1, 0, 0, 0, 0) \mapsto u_1$;
(12) $\rho'_{41} : T_4 \longrightarrow T_1$, $(0, e_1, 0, 0, 0) \mapsto u_1$;
(13) $\rho_{42} : T_4 \longrightarrow T_2$, $(0, 0, e_2, 0, 0) \mapsto u_2$;
(14) $\rho_{43} : T_4 \longrightarrow T_3$, $(0, 0, e_2, 0, 0) \mapsto \delta$;
(15) $\rho_4 : T_4 \longrightarrow T_4$, $(0, 0, e_2, 0, 0) \mapsto (0, 0, 0, 0, \delta)$;
(16) $\rho'_4 : T_4 \longrightarrow T_4$, $(0, 0, e_2, 0, 0) \mapsto (0, 0, u_2, -\delta, 0)$
(17) $s_1 : T_1 \longrightarrow T_1$; $e_1 \mapsto u_1^2$;
(18) $s_{14} : T_1 \longrightarrow T_4$, $e_1 \mapsto (u_1^2, 0, 0, 0, 0)$;
(19) $s'_{14} : T_1 \longrightarrow T_4$, $e_1 \mapsto (0, u_1^2, 0, 0, 0)$;
(20) $s_2 : T_2 \longrightarrow T_2$; $e_2 \mapsto u_2^2$;
(21) $s_{23} : T_2 \longrightarrow T_3$, $e_2 \mapsto u_2\delta$;
(22) $s_{24} : T_2 \longrightarrow T_4$, $e_2 \mapsto (0, 0, u_2^2, 0, 0)$;
(23) $s'_{24} : T_2 \longrightarrow T_4$, $e_2 \mapsto (0, 0, 0, u_2\delta, 0)$;
(24) $s''_{24} : T_2 \longrightarrow T_4$, $e_2 \mapsto (0, 0, 0, 0, u_2\delta)$;
(25) $s_{41} : T_4 \longrightarrow T_1$, $(e_1, 0, 0, 0, 0) \mapsto u_1^2$;
(26) $s'_{41} : T_4 \longrightarrow T_1$, $(0, e_1, 0, 0, 0) \mapsto u_1^2$;
(27) $s_{42} : T_4 \longrightarrow T_2$, $(0, 0, e_2, 0, 0) \mapsto u_2^2$;
(28) $s_{43} : T_4 \longrightarrow T_3$, $(0, 0, 0, e_3, 0) \mapsto u_2\delta$;
(29) $s_4 : T_4 \longrightarrow T_4$, $(e_1, 0, 0, 0, 0) \mapsto (u_1^2, 0, 0, 0, 0)$;
(30) $s'_4 : T_4 \longrightarrow T_4$, $(e_1, 0, 0, 0) \mapsto (0, u_1^2, 0, 0, 0)$;
(31) $s''_4 : T_4 \longrightarrow T_4$, $(0, e_1, 0, 0, 0) \mapsto (u_1^2, 0, 0, 0)$;
(32) $s'''_4 : T_4 \longrightarrow T_4$, $(0, e_1, 0, 0, 0) \mapsto (0, u_1^2, 0, 0, 0)$;
(33) $s_4^{iv} : T_4 \longrightarrow T_4$, $(0, 0, e_2, 0, 0) \mapsto (0, 0, u_2^2, 0, 0)$;
(34) $s_4^{v} : T_4 \longrightarrow T_4$, $(0, 0, e_2, 0, 0) \mapsto (0, 0, 0, u_2\delta, 0)$;
(35) $s_4^{vi} : T_4 \longrightarrow T_4$, $(0, 0, e_2, 0, 0) \mapsto (0, 0, 0, 0, u_2\delta)$.

LEMMA 47. *The morphisms π_{41}, π'_{41}, π_{42}, π_{43}, π'_{43}, ρ_1, ρ_{24}, ρ'_{24}, s_{14}, s'_{14} and s_{24} form a basis for* $\operatorname{rad}(\operatorname{End}_\Lambda(T))$ *modulo* $\operatorname{rad}(\operatorname{End}_\Lambda(T))^2$.

THEOREM 48. **(1)** *The quiver $\widetilde{Q}^{\mathrm{op}}$ of* $\operatorname{End}_\Lambda(T)$ *is*

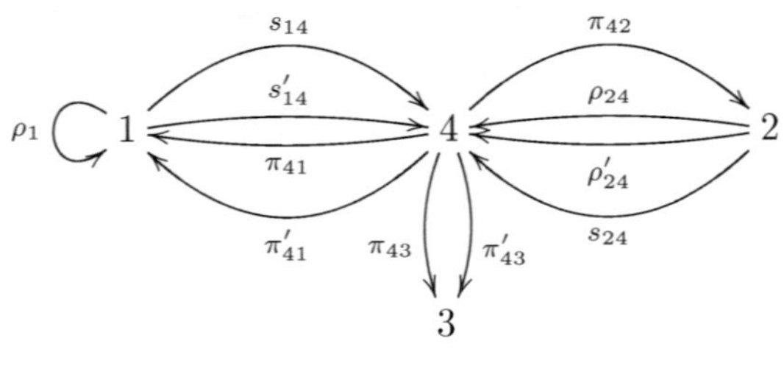

(2) *The following is a set of relations for* $\mathrm{End}_\Lambda(T)$ *in* $K\widetilde{Q}^{\mathrm{op}}$:
- *Monomial relations:* ρ_1^2, $\rho_1 s_{14}$, $\rho_1 s'_{14}$, $s_{14}\pi'_{41}$, $s_{14}\pi_{42}$, $s_{14}\pi_{43}$, $s_{14}\pi'_{43}$, $s'_{14}\pi_{41}$, $s'_{14}\pi_{42}$, $s'_{14}\pi_{43}$, $s'_{14}\pi'_{43}$, $s_{24}\pi_{41}$, $s_{24}\pi'_{41}$, $s_{24}\pi_{43}$, $s_{24}\pi'_{43}$, $s_{24}\pi_{42}\rho_{24}$, $s_{24}\pi_{42}\rho'_{24}$, $s_{24}\pi_{42}s_{24}$, $\rho_{24}\pi_{41}$, $\rho_{24}\pi'_{41}$, $\rho_{24}\pi_{42}$, $\rho_{24}\pi_{43}$, $\rho'_{24}\pi_{41}$, $\rho'_{24}\pi'_{41}$, $\rho'_{24}\pi'_{43}$, $\rho'_{24}\pi_{42}s_{24}$
- *Non-monomial relations:* $s_{14}\pi_{41} - \rho_1^2$, $s'_{14}\pi'_{41} - \rho_1^2$, $s_{24}\pi_{42} - \rho'_{24}\pi_{42}\rho'_{24}\pi_{42}$, $\rho_{24}\pi'_{43} - \rho'_{24}\pi_{43}$.

PROOF. The first assertion is immediate from Lemma 47.

We denote by $\mathcal{R}$ the set of relations listed in (2). Part (1) provides us with a surjective algebra homomorphism $\Psi : K\widetilde{Q}^{\mathrm{op}} \twoheadrightarrow \mathrm{End}({}_\Lambda T)$. The verification that $K\widetilde{Q}^{\mathrm{op}}/\langle\mathcal{R}\rangle = K\widetilde{Q}^{\mathrm{op}}/\mathrm{Ker}(\Psi) \cong \mathrm{End}_\Lambda(T)$ is best carried out by comparing the relations in $\mathcal{R}$ with the graphs of the indecomposable projective left $\mathrm{End}_\Lambda(T)$-modules (= the indecomposable projective right $\widetilde{\Lambda}$-modules) shown in 6.D. $\square$

8.C. The indecomposable direct summands of the strong tilting module $\widetilde{T}_{\widetilde{\Lambda}}$. From Theorems 4 and 33, we know that $\mathcal{P}^{<\infty}(\text{mod-}\widetilde{\Lambda})$ contains a basic strong tiliting module $\widetilde{T} = \bigoplus_{1\le i\le 4}\widetilde{T}_i$. Moreover, Proposition 6 tells us that the $\widetilde{T}_i$ coincide with the indecomposable direct summands of $\bigoplus_{i=1}^4 \mathcal{A}(\widetilde{E}_i)$, where $\mathcal{A}(\widetilde{E}_i)$ is the minimal $\mathcal{P}^{<\infty}(\text{mod-}\widetilde{\Lambda})$-approximation of the indecomposable injective module $\widetilde{E}_i$. Determining $\mathcal{A}(\widetilde{E}_i)$ for $i = 1, 2$ is effortless, since the critical corner $\mu\Lambda\mu = (e_1 + e_2)\Lambda(e_1 + e_2)$ of Λ is a self-injective algebra. Indeed, as a consequence of Proposition 43, we obtain:

COROLLARY 49. *In our reference example,* $\widetilde{T}_1 = \widetilde{E}_1$ *and* $\widetilde{T}_2 = \widetilde{E}_2$. *These modules are pinned down, up to isomorphism, by the graphs shown in Figure* 3.

To construct $\widetilde{T}_3$ and $\widetilde{T}_4$, we follow the recipe developed in the previous section. Since this will require finding the $(\widetilde{\Lambda}/\widetilde{\Delta})$-projective covers of certain $\widetilde{\Lambda}/\widetilde{\Delta}$-modules with tops of the form $\widetilde{S}_1^{m_1} \oplus \widetilde{S}_2^{m_2}$, it will be convenient to have at hand the graphs of the indecomposable projective right $\widetilde{\Lambda}/\widetilde{\Delta}$-modules $\widetilde{e}_i\widetilde{\Lambda}/\widetilde{e}_i\widetilde{\Delta}$ for $i = 1, 2$. We denote these modules by $\hat{P}_1$ and $\hat{P}_2$, respectively. They are graphed in Figure 5 below.

Moreover, it is readily checked that the indecomposable injectives $\widetilde{E}_3$ and $\widetilde{E}_4$ are determined by the graphs in Figure 5. Note that the left-hand graph pins down $\widetilde{E}_3$ up to isomorphism, even though this graph contains a closed edge path; the (unique) scalar involved in the corresponding module comes from the relation $\rho_{24}\pi'_{43} - \rho'_{24}\pi_{43}$.

Clearly, $\widetilde{F}_3 := \widetilde{E}_3/\Delta(\widetilde{E}_3)$ is the uniserial module

$\widetilde{F}_3$

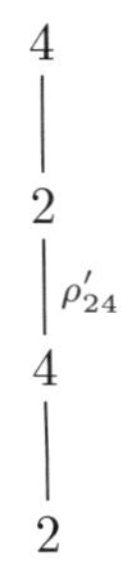

In particular, $\nabla(\widetilde{F}_3)$ is the submodule with composition factors $(\widetilde{S}_2, \widetilde{S}_4, \widetilde{S}_2)$. Its projective cover in $\text{mod-}(\widetilde{\Lambda}/\widetilde{\Delta})$ is the obvious epimorphism $q : \hat{P}_2 \twoheadrightarrow \nabla(\widetilde{F}_3)$. Since

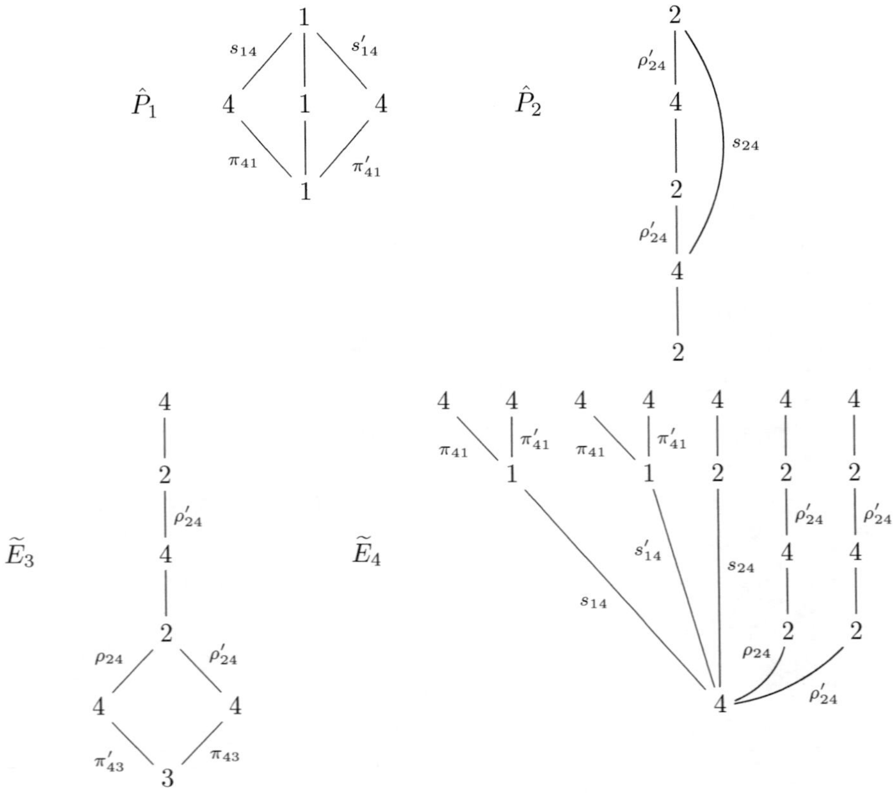

FIGURE 5. $\hat{P}_1$, $\hat{P}_2$, $\widetilde{E}_3$, $\widetilde{E}_4$.

the injective envelope $\widetilde{E}(\hat{P}_2)$ equals $\widetilde{E}_2$ (see Figure 3), we find that $\widetilde{E}_2/\hat{P}_2 \cong \widetilde{S}_4$ has no critical composition factors; moreover, there is a unique morphism $p : \widetilde{E}_2 \longrightarrow \widetilde{F}_3$ with the property that $p_{|\hat{P}_2}$ is the composition $\hat{P}_2 \xrightarrow{q} \nabla(\widetilde{F}_3) \hookrightarrow \widetilde{F}_3 = \widetilde{E}_3/\Delta(\widetilde{E}_3)$. Proposition 41 thus guarantees that p is the minimal $\mathcal{P}^{<\infty}(\text{mod-}\widetilde{\Lambda})$-approximation of $\widetilde{F}_3$. By Proposition 40 the minimal approximation $\rho : \widetilde{\mathcal{A}}(\widetilde{E}_3) \longrightarrow \widetilde{E}_3$ is therefore the pullback of the pair (π_3, p), where $\pi_3 : \widetilde{E}_3 \twoheadrightarrow \widetilde{F}_3 = \widetilde{E}_3/\Delta(\widetilde{E}_3)$ is the canonical map. A routine check now allows us to ascertain that $\mathcal{A}(\widetilde{E}_3)$ is the local module $\widetilde{T}_3$ displayed in Figure 3.

To determine $\widetilde{T}_4$, we compute the minimal $\mathcal{P}^{<\infty}$-approximation $\mathcal{A}(\widetilde{E}_4) \longrightarrow \widetilde{E}_4$, once again guided by Section 7.

Towards identifying the minimal approximation of $\widetilde{E}_4/\Delta(\widetilde{E}_4)$, we find that this quotient decomposes in the form $\widetilde{E}_4/\Delta(\widetilde{E}_4) \cong \widetilde{L}^2 \oplus \widetilde{M} \oplus \widetilde{N}^2$, where $\widetilde{L}$ is given by the diagram 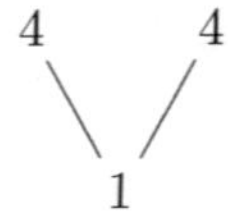, the summand $\widetilde{M}$ is the uniserial module , and $\widetilde{N} = \widetilde{F}_3$ is as introduced above. Clearly, $\nabla(\widetilde{L}) \cong \widetilde{S}_1$ and $\nabla(\widetilde{M}) \cong \widetilde{S}_2$, whence the projective covers of $\nabla(\widetilde{L})$ and $\nabla(\widetilde{M})$ in mod-$(\widetilde{\Lambda}/\widetilde{\Delta})$ are the canonical epimorphisms

$f_1 : \hat{P}_1 = \widetilde{e}_1\widetilde{\Lambda}/\widetilde{e}_1\widetilde{\Delta} \twoheadrightarrow \nabla(\widetilde{L}) \cong \widetilde{S}_1$ and $f_2 : \hat{P}_2 = \widetilde{e}_2\widetilde{\Lambda}/\widetilde{e}_2\widetilde{\Delta} \twoheadrightarrow \nabla(\widetilde{M}) \cong \widetilde{S}_2$. Further, one identifies the injective envelopes of the $\hat{P}_i$ in mod-$\widetilde{\Lambda}$ as being $\widetilde{E}(\widetilde{P}_i) = \widetilde{E}_i$ for $i = 1, 2$. As inspection confirms, all composition factors of the $\widetilde{E}(\hat{P}_i)/\hat{P}_i$ are isomorphic to $\widetilde{S}_4$, whence these quotients are free of critical composition factors. Moreover, the obvious morphisms $q_1 : \widetilde{E}_1 = \widetilde{E}(\hat{P}_1) \longrightarrow \widetilde{L}$ and $q_2 : \widetilde{E}_2 = \widetilde{E}(\hat{P}_2) \longrightarrow \widetilde{M}$ restrict to the maps f_i followed by the respective inclusions $\nabla(\widetilde{L}) \hookrightarrow \widetilde{L}$ and $\nabla(\widetilde{M}) \hookrightarrow \widetilde{M}$. Consequently, another application of Proposition 41 ensures that $q_1 : \widetilde{E}_1 \longrightarrow \widetilde{L}$ and $q_2 : \widetilde{E}_2 \longrightarrow \widetilde{M}$ are the minimal $\mathcal{P}^{<\infty}(\text{mod-}\widetilde{\Lambda})$-approximations of $\widetilde{L}$ and $\widetilde{M}$, respectively. The corresponding information for $\widetilde{N}$ is already available from the computation of $\widetilde{T}_3$. Indeed, that argument shows that the minimal $\mathcal{P}^{<\infty}(\text{mod-}\widetilde{\Lambda})$-approximation of $\widetilde{N}$ is the morphism $p : \widetilde{E}_2 \longrightarrow \widetilde{F}_3 = \widetilde{N}$ specified there. Consequently, the minimal $\mathcal{P}^{<\infty}(\text{mod-}\widetilde{\Lambda})$-approximation of $\widetilde{E}_4/\Delta(\widetilde{E}_4)$ is the induced map $\begin{pmatrix} q_1 & q_1 & q_2 & p & p \end{pmatrix} : (\widetilde{E}_1)^2 \oplus (\widetilde{E}_2)^3 \longrightarrow \widetilde{L}^2 \oplus \widetilde{M} \oplus \widetilde{N}^2 \cong \widetilde{E}_4/\Delta(\widetilde{E}_4)$.

In order to facilitate the application of Proposition 40 towards obtaining the minimal $\mathcal{P}^{<\infty}(\text{mod-}\widetilde{\Lambda})$-approximation of $\widetilde{E}_4$ from that of $\widetilde{E}_4/\Delta(\widetilde{E}_4) \cong \widetilde{L}^2 \oplus \widetilde{M} \oplus \widetilde{N}^2$, we will twice make use of the upcoming "isolation of direct summands" in pullback diagrams.

LEMMA 50. *Let $f \in \operatorname{Hom}_{\widetilde{\Lambda}}(X, N)$, $g \in \operatorname{Hom}_{\widetilde{\Lambda}}(Y, N)$, and $h \in \operatorname{Hom}_{\widetilde{\Lambda}}(U, N)$. Suppose, moreover, that Z is the pullback (viewed as a module) of h and $\begin{pmatrix} f & g \end{pmatrix} \in \operatorname{Hom}_{\widetilde{\Lambda}}(X \oplus Y, N)$. If g factors through h, then $Z \cong Z' \oplus Y$, where Z' is the pullback of h and f.*

PROOF. To introduce our notation, we display the pullback diagram for Z:

$$\begin{array}{ccc} Z & \xrightarrow{\ \widetilde{h}\ } & X \oplus Y \\ {\scriptstyle \widetilde{\chi}}\downarrow & & \downarrow{\scriptstyle \chi = \begin{pmatrix} f & g \end{pmatrix}} \\ U & \xrightarrow{\ h\ } & N \end{array}$$

Moreover, we choose $\eta \in \operatorname{Hom}_{\widetilde{\Lambda}}(Y, U)$ such that $h \circ \eta = g$. To show that $Z \cong Z' \oplus Y$, consider the following square:

$$\begin{array}{ccc} Z' \oplus Y & \xrightarrow{\ \rho\ } & X \oplus Y \\ {\scriptstyle \sigma}\downarrow & & \downarrow{\scriptstyle \chi} \\ U & \xrightarrow{\ h\ } & N \end{array}$$

where ρ and σ are as follows: For $(u, x) \in Z'$, we define $\rho\big((u, x), y\big) = (x, y)$ and $\sigma\big((u, x), y\big) = u + \eta(y)$. In light of the equality $h(u) = f(x)$, a straightforward check shows this diagram to be commutative, whence the universal property of Z yields a unique $\tau \in \operatorname{Hom}_{\widetilde{\Lambda}}(Z' \oplus Y, Z)$ satisfying $\sigma = \widetilde{\chi} \circ \tau$ and $\rho = \widetilde{h} \circ \tau$. Necessarily, τ assigns to an element $\big((u, x), y\big) \in Z' \oplus Y$, the element $\big(u + \eta(y), (x, y)\big) \in Z$.

Clearly, τ is injective. For surjectivity, let $\big(u, (x, y)\big) \in Z$ so that $h(u) = f(x) + g(y)$. Then $\big(u - \eta(y), x\big)$ belongs to Z', and $\tau\big((u - \eta(y), x),\, y\big) = \big(u, (x, y)\big)$. Thus τ is an isomorphism. $\square$

We return to the construction of the minimal $\mathcal{P}^{<\infty}(\text{mod-}\widetilde{\Lambda})$-approximation of $\widetilde{E}_4$. In the first application of the lemma, the role of h is played by $\pi_4 : \widetilde{E}_4 \longrightarrow \widetilde{E}_4/\Delta(\widetilde{E}_4)$, that of f by $\begin{pmatrix} q_2 & p & p \end{pmatrix} : (\widetilde{E}_2)^3 \longrightarrow \widetilde{E}_4/\Delta(\widetilde{E}_4)$, and that of g by $\begin{pmatrix} q_1 & q_1 \end{pmatrix} : (\widetilde{E}_1)^2 \longrightarrow \widetilde{E}_4/\Delta(\widetilde{E}_4)$. In order to bring the lemma to bear, we observe that the induced morphisms $\widetilde{E}_1 \xrightarrow{q_1} \widetilde{L} \xrightarrow{\begin{pmatrix} 1 & 0 & 0 & 0 & 0 \end{pmatrix}} \widetilde{E}_4/\Delta(\widetilde{E}_4)$ and $\widetilde{E}_1 \xrightarrow{q_1} \widetilde{L} \xrightarrow{\begin{pmatrix} 0 & 1 & 0 & 0 & 0 \end{pmatrix}} \widetilde{E}_4/\Delta(\widetilde{E}_4)$ both factor through $\pi_4 : \widetilde{E}_4 \longrightarrow \widetilde{E}_4/\Delta(\widetilde{E}_4)$, whence so does the map $\begin{pmatrix} q_1 & q_1 \end{pmatrix} : (\widetilde{E}_1)^2 \longrightarrow \widetilde{E}_4/\Delta(\widetilde{E}_4)$. Thus we conclude that $\widetilde{\mathcal{A}}(\widetilde{E}_4) = (\widetilde{E}_1)^2 \oplus \hat{T}_4$, where $\hat{T}_4$ is the pullback of π_4 and $\begin{pmatrix} q_2 & p & p \end{pmatrix}$. For notational simplicity, we will write $\varphi = \begin{pmatrix} q_2 & p & p \end{pmatrix}$ in the sequel.

To prepare for another application of Lemma 50, we pass to an alternative decomposition of the domain $(\widetilde{E}_2)^3$ of φ, namely

$$(\widetilde{E}_2)^3 = (\widetilde{E}_2 \oplus 0 \oplus 0) \oplus (0 \oplus \widetilde{E}_2 \oplus 0) \oplus \widetilde{E}',$$

where $\widetilde{E}' \cong \widetilde{E}_2$ is the image of the morphism $\begin{pmatrix} 1 \\ 0 \\ 1 \end{pmatrix} : \widetilde{E}_2 \longrightarrow \widetilde{E}_2 \oplus \widetilde{E}_2 \oplus \widetilde{E}_2$.

Based on the new decomposition of the domain, communicated in the form $(\widetilde{E}_2)^3 = (\widetilde{E}_2)^2 \oplus \widetilde{E}'$, the map φ takes on the form $\varphi = \begin{pmatrix} f & g \end{pmatrix} : (\widetilde{E}_2)^2 \oplus \widetilde{E}' \longrightarrow \widetilde{E}_4/\Delta(\widetilde{E}_4)$, where $f = \begin{pmatrix} q_2 & p \end{pmatrix} : (\widetilde{E}_2)^2 \longrightarrow \widetilde{E}_4/\Delta(\widetilde{E}_4)$ and $g = \varphi_{|\widetilde{E}'} : \widetilde{E}' \longrightarrow \widetilde{E}_4/\Delta(\widetilde{E}_4)$.

We check that the morphism g in turn factors through π_4. To see this, start by noting that $\widetilde{E}_4/\widetilde{E}_4\widetilde{J} \cong \widetilde{S}_4^7$. Tracing the top row of the graph of $\widetilde{E}_4$ from left to right, we number the displayed full sequence of top elements of $\widetilde{E}_4$ by $\widetilde{z}_1, \dots, \widetilde{z}_7$. Next we observe that $\widetilde{E}' \cong \widetilde{E}_2$ is a quotient of $\widetilde{e}_4\widetilde{\Lambda}$; indeed, in view of the graph of $\widetilde{E}_2$ (see Figure 3), $\widetilde{E}' = \widetilde{z}\widetilde{\Lambda}$, where $\widetilde{z} = \widetilde{z}\,\widetilde{e}_4$. We claim that the assignment $\widetilde{z} \mapsto \widetilde{z}_5 + \widetilde{z}_7$ yields a well-defined morphism $\eta : \widetilde{E}' \longrightarrow \widetilde{E}_4$ with the property that $\pi_4 \circ \eta = g$. It suffices to check well-definedness of η; once this is secured, we may conclude that $(\pi_4 \circ \eta)(\widetilde{z}) = \pi_4(\widetilde{z}_5) + \pi_4(\widetilde{z}_7) = g(\widetilde{z})$. In fact, we show that η defines an epimorphism from $\widetilde{E}'$ onto the submodule $\widetilde{X}$ of $\widetilde{E}_4$ which is generated by $\widetilde{z}_5 + \widetilde{z}_7$. From the graph of $\widetilde{E}_4$ it is apparent that the submodule $\widetilde{X}$ may be visualized by way of the diagram in Figure 6 below, which displays the itinerary of the top element $\widetilde{z}_5 + \widetilde{z}_7$ on successive multiplication by arrows. Given that $\widetilde{z}\,\pi_{42}s_{24} = \widetilde{z}\,\pi_{42}\rho'_{24}\pi_{42}\rho'_{24}$, we are thus looking at a copy of $\widetilde{E}_2/\operatorname{soc}(\widetilde{E}_2)$.

The preceding paragraph positions us for a reapplication of Lemma 50, to the effect that $\hat{T}_4 \cong \widetilde{T}'_4 \oplus \widetilde{E}' \cong \widetilde{T}'_4 \oplus \widetilde{E}_2$, where the summand $\widetilde{T}'_4$ is the pullback of $\pi_4 : \widetilde{E}_4 \longrightarrow \widetilde{E}_4/\Delta(\widetilde{E}_4)$ and $f = \begin{pmatrix} q_2 & p \end{pmatrix} : (\widetilde{E}_2)^2 \longrightarrow \widetilde{E}_4/\Delta(\widetilde{E}_4)$. Computation of this pullback finally yields $\widetilde{T}'_4$ to be a copy of the module $\widetilde{T}_4$ displayed in Figure 3. In particular, $\widetilde{T}'_4$ is indecomposable. Since $\widetilde{T}'_4$ is the only indecomposable direct summand of $\widetilde{\mathcal{A}}(\widetilde{E}_4) = (\widetilde{E}_1)^2 \oplus \hat{T}_4 = (\widetilde{E}_1)^2 \oplus \widetilde{T}'_4 \oplus \widetilde{E}_2$ which is not among the $\widetilde{T}_i$ for $i \leq 3$, we conclude that $\widetilde{T}'_4$ supplements $\bigoplus_{1 \leq i \leq 3} \widetilde{T}_i$, so as to yield $\widetilde{T}$. Thus the label $\widetilde{T}_4$ in Section 6 is justified. In summary, we have confirmed the claims with which we concluded 6.G. Namely:

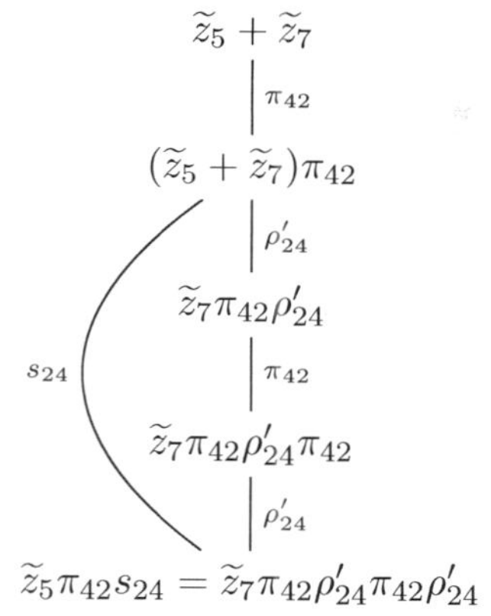

FIGURE 6. Itinerary of $\widetilde{z}_5 + \widetilde{z}_7$.

PROPOSITION 51. *The basic strong tilting object in* mod-$\widetilde{\Lambda}$ *is* $\widetilde{T} = \oplus_{i=1}^{4}\widetilde{T}_i$, *where the* $\widetilde{T}_i$ *are the modules determined (up to isomorphism) by the graphs shown in Figure* 3.

Acknowlegdment

The authors thank the referee for comments. In particular, the present version of Proposition 24 is due to the referee.

References

[1] L. Angeleri Hügel, D. Happel, and H. Krause (eds.), *Handbook of tilting theory*, London Mathematical Society Lecture Note Series, vol. 332, Cambridge University Press, Cambridge, 2007. MR2385175

[2] L. Angeleri-Hügel and J. Trlifaj, *Tilting theory and the finitistic dimension conjectures*, Trans. Amer. Math. Soc. **354** (2002), no. 11, 4345–4358, DOI 10.1090/S0002-9947-02-03066-0. MR1926879

[3] M. Auslander, *Functors and morphisms determined by objects*, Representation theory of algebras (Proc. Conf., Temple Univ., Philadelphia, Pa., 1976), Dekker, New York, 1978, pp. 1–244. Lecture Notes in Pure Appl. Math., Vol. 37. MR0480688

[4] M. Auslander, M. Platzeck, and I. Reiten, *Coxeter functors without diagrams*, Trans. Amer. Math. Soc. **250** (1979), 1–46, DOI 10.2307/1998978. MR530043

[5] M. Auslander and I. Reiten, *Applications of contravariantly finite subcategories*, Adv. Math. **86** (1991), no. 1, 111–152, DOI 10.1016/0001-8708(91)90037-8. MR1097029

[6] M. Auslander, I. Reiten, and S. O. Smalø, *Representation theory of Artin algebras*, Cambridge Studies in Advanced Mathematics, vol. 36, Cambridge University Press, Cambridge, 1997. Corrected reprint of the 1995 original. MR1476671

[7] M. Auslander and S. O. Smalø, *Preprojective modules over Artin algebras*, J. Algebra **66** (1980), no. 1, 61–122, DOI 10.1016/0021-8693(80)90113-1. MR591246

[8] I. N. Bernšteĭn, I. M. Gel′fand, and V. A. Ponomarev, *Coxeter functors, and Gabriel's theorem* (Russian), Uspehi Mat. Nauk **28** (1973), no. 2(170), 19–33. MR0393065

[9] K. Bongartz, *Tilted algebras*, Representations of algebras (Puebla, 1980), Lecture Notes in Math., vol. 903, Springer, Berlin-New York, 1981, pp. 26–38. MR654701

[10] S. Brenner and M. C. R. Butler, *Generalizations of the Bernstein-Gel′fand-Ponomarev reflection functors*, Representation theory, II (Proc. Second Internat. Conf., Carleton Univ., Ottawa, Ont., 1979), Lecture Notes in Math., vol. 832, Springer, Berlin-New York, 1980, pp. 103–169. MR607151

[11] W. D. Burgess and B. Zimmermann-Huisgen, *Approximating modules by modules of finite projective dimension*, J. Algebra **178** (1995), no. 1, 48–91, DOI 10.1006/jabr.1995.1340. MR1358257

[12] H. Cartan and S. Eilenberg, *Homological algebra*, Princeton University Press, Princeton, N. J., 1956. MR0077480

[13] E. Cline, B. Parshall, and L. Scott, *Derived categories and Morita theory*, J. Algebra **104** (1986), no. 2, 397–409, DOI 10.1016/0021-8693(86)90224-3. MR866784

[14] A. Dugas and B. Huisgen-Zimmermann, *Strongly tilting truncated path algebras*, Manuscripta Math. **134** (2011), no. 1-2, 225–257, DOI 10.1007/s00229-010-0395-8. MR2745262

[15] A. Dugas, B. Huisgen-Zimmermann, and J. Learned, *Truncated path algebras are homologically transparent*, Models, modules and abelian groups, Walter de Gruyter, Berlin, 2008, pp. 445–461, DOI 10.1515/9783110203035.445. MR2513257

[16] E. E. Enochs, *Torsion free covering modules*, Proc. Amer. Math. Soc. **14** (1963), 884–889, DOI 10.2307/2035021. MR168617

[17] E. E. Enochs, *Injective and flat covers, envelopes and resolvents*, Israel J. Math. **39** (1981), no. 3, 189–209, DOI 10.1007/BF02760849. MR636889

[18] R. Göbel and J. Trlifaj, *Approximations and endomorphism algebras of modules. Volume 1–Approximations*, Second revised and extended edition, De Gruyter Expositions in Mathematics, vol. 41, Walter de Gruyter GmbH & Co. KG, Berlin, 2012. MR2985554

[19] K. R. Goodearl and B. Huisgen-Zimmermann, *Closures in varieties of representations and irreducible components*, Algebra Number Theory **12** (2018), no. 2, 379–410, DOI 10.2140/ant.2018.12.379. MR3803707

[20] D. Happel, *On the derived category of a finite-dimensional algebra*, Comment. Math. Helv. **62** (1987), no. 3, 339–389, DOI 10.1007/BF02564452. MR910167

[21] D. Happel and B. Huisgen-Zimmermann, *Viewing finite-dimensional representations through infinite-dimensional ones*, Pacific J. Math. **187** (1999), no. 1, 65–89, DOI 10.2140/pjm.1999.187.65. MR1674321

[22] D. Happel and C. M. Ringel, *Tilted algebras*, Trans. Amer. Math. Soc. **274** (1982), no. 2, 399–443, DOI 10.2307/1999116. MR675063

[23] D. Happel and L. Unger, *On a partial order of tilting modules*, Algebr. Represent. Theory **8** (2005), no. 2, 147–156, DOI 10.1007/s10468-005-3595-2. MR2162278

[24] M. Hoshino and J. Miyachi, *Tame two-point algebras*, Tsukuba J. Math. **12** (1988), no. 1, 65–96, DOI 10.21099/tkbjm/1496160637. MR949903

[25] B. Huisgen-Zimmermann, *Dualities from iterated tilting*, Israel J. Math. (to appear).

[26] B. Huisgen-Zimmermann and S. O. Smalø, *A homological bridge between finite- and infinite-dimensional representations of algebras*, Algebr. Represent. Theory **1** (1998), no. 2, 169–188, DOI 10.1023/A:1009948721602. MR1681738

[27] B. Huisgen-Zimmermann and S. O. Smalø, *The homology of string algebras. I*, J. Reine Angew. Math. **580** (2005), 1–37, DOI 10.1515/crll.2005.2005.580.1. MR2130586

[28] K. Igusa, S. O. Smalø, and G. Todorov, *Finite projectivity and contravariant finiteness*, Proc. Amer. Math. Soc. **109** (1990), no. 4, 937–941, DOI 10.2307/2048121. MR1027094

[29] O. Iyama and X. Zhang, *Tilting modules over Auslander-Gorenstein algebras*, Pacific J. Math. **298** (2019), no. 2, 399–416, DOI 10.2140/pjm.2019.298.399. MR3936022

[30] B. Keller, *Deriving DG categories*, Ann. Sci. École Norm. Sup. (4) **27** (1994), no. 1, 63–102. MR1258406

[31] S. König and A. Zimmermann, *Derived equivalences for group rings*, Lecture Notes in Mathematics, vol. 1685, Springer-Verlag, Berlin, 1998. With contributions by Bernhard Keller, Markus Linckelmann, Jeremy Rickard and Raphaël Rouquier. MR1649837

[32] Y. Miyashita, *Tilting modules of finite projective dimension*, Math. Z. **193** (1986), no. 1, 113–146, DOI 10.1007/BF01163359. MR852914

[33] J. Rickard, *Morita theory for derived categories*, J. London Math. Soc. (2) **39** (1989), no. 3, 436–456, DOI 10.1112/jlms/s2-39.3.436. MR1002456

[34] J. Rickard, *Overview of Broué's abelian defect group conjecture*, `people.maths.bris.ac.uk/~majcr/adgc/adgc.html`.

[35] C. Riedtmann and A. Schofield, *On a simplicial complex associated with tilting modules*, Comment. Math. Helv. **66** (1991), no. 1, 70–78, DOI 10.1007/BF02566636. MR1090165

[36] C. M. Ringel, *Tame algebras and integral quadratic forms*, Lecture Notes in Mathematics, vol. 1099, Springer-Verlag, Berlin, 1984. MR774589

Department of Mathematics, University of California, Santa Barbara, California 93106

Email address: birge@math.ucsb.edu

Departamento de Matemáticas, Universidad de Murcia, Aptdo. 4021 30100 Espinardo, Murcia, Spain

Email address: msaorinc@um.es

Contemporary Mathematics
Volume **761**, 2021
https://doi.org/10.1090/conm/761/15310

Frieze varieties are invariant under Coxeter mutation

Kiyoshi Igusa and Ralf Schiffler

ABSTRACT. We define a generalized version of the frieze variety introduced by Lee, Li, Mills, Seceleanu and the second author. The generalized frieze variety is an algebraic variety determined by an acyclic quiver and a generic specialization of cluster variables in the cluster algebra for this quiver. The original frieze variety is obtained when this specialization is $(1, \ldots, 1)$.

The main result is that a generalized frieze variety is determined by any generic element of any component of that variety. We also show that the "Coxeter mutation" cyclically permutes these components. In particular, this shows that the frieze variety is invariant under the Coxeter mutation at a generic point.

The paper contains many examples which are generated using a new technique which we call an invariant Laurent polynomial. We show that a symmetry of a mutation of a quiver gives such an invariant rational function.

CONTENTS

Introduction
1. Preliminaries
2. Definitions and main result
3. Proof of Theorem 2.3
4. Construction from invariant rational functions
5. Examples
6. Symmetry
7. Questions and answers
Acknowledgments
References

Introduction

At the ARTA VI Conference in Mexico celebrating Jose Antonio de la Peña's 60th birthday, the second author presented his paper [**2**] defining "frieze varieties" of an acyclic quiver and, using a result of Jose Antonio de la Peña [**3, 6**] on eigenvalues of the Coxeter matrix of the quiver, to prove the main result: The dimension of this frieze variety is equal to 0,1, or ≥ 2 if and only if the representation type of

2010 *Mathematics Subject Classification.* Primary 13F60, 16G20.

The second author was supported by the NSF grants DMS-1254567 and DMS-1800860 and by the University of Connecticut.

the quiver is finite, tame, or wild, respectively. In particular, the dimension of the frieze variety is 1 if and only if Q is an extended Dynkin quiver.

After the talk, the two authors discussed properties and examples of frieze varieties throughout the conference. This paper is a report on these discussions.

The basic idea was to generalize the notion of frieze varieties by allowing for arbitrary generic vectors $(b_1, b_2, \ldots, b_n)$ instead of the vector $(1, 1, \ldots, 1)$ as the initial condition for the defining recurrence. We show that, for any generic point $a_* = (a_1, \ldots, a_n)$ in any component of the (generalized) frieze variety, the Coxeter mutations of a_* (2.1) will be contained as a dense subset of the variety. (Theorems 2.3, 3.5)

We also show that the Coxeter mutation cyclically permutes the components of the (generalized) frieze variety (Theorem 3.2). In nice examples, we can use this cyclic permutation to generate all polynomials which define all components of the generalized frieze variety out of a single rational function. (Proposition 4.2)

Finally, we also observe that a symmetry of a mutation of a quiver decreases the dimension of the frieze variety. (Proposition 6.2)

1. Preliminaries

We recall the main result of [**2**]. Let Q be a connected finite quiver without oriented cycles and label the vertices $1, 2, \ldots, n$ such that if there is an arrow $i \to j$ then $i > j$.

DEFINITION 1.1. [**2**]

(1) For every vertex $i \in Q_0$ define positive rational numbers $f_i(t)$ $(t \in \mathbb{Z}_{\geq 0})$ recursively by $f_i(0) = 1$ and

$$f_i(t+1) = \frac{1 + \prod_{j \to i} f_j(t) \prod_{j \leftarrow i} f_j(t+1)}{f_i(t)}.$$

(2) For every $t \geq 0$, define the point $P_t = (f_1(t), \ldots, f_n(t)) \in \mathbb{C}^n$.

(3) The *frieze variety* $X(Q)$ of the quiver Q is the Zariski closure of the set of all points P_t $(t \in \mathbb{Z}_{\geq 0})$.

The main result in [**2**] is the following characterization of the finite–tame–wild trichotomy for acyclic quivers Q in terms of its frieze variety $X(Q)$.

THEOREM 1.2. [**2**] *Let Q be an acyclic quiver.*

(a) *If Q is representation finite then the frieze variety $X(Q)$ is of dimension 0.*

(b) *If Q is tame then the frieze variety $X(Q)$ is of dimension 1.*

(c) *If Q is wild then the frieze variety $X(Q)$ is of dimension at least 2.*

2. Definitions and main result

Let Q be a quiver as in section 1, or an acyclic valued quiver. Let $\mathcal{A}(Q)$ be the skew-symmetric (if Q is a quiver) or the skew-symmetrizable (if Q is a valued quiver) cluster algebra of Q and let $\mathbf{x} = (x_1, \ldots, x_n)$ be the initial cluster in $\mathcal{A}(Q)$. Let μ_k denote the mutation in direction k and let x_k' be the new cluster variable obtained by this mutation, thus $\mu_k(\mathbf{x}) = (x_1, \ldots, x_{k-1}, x_k', x_{k+1}, \ldots, x_n)$. We define the *Coxeter mutation* to be the mutation sequence

$$\mu_* = \mu_n \circ \cdots \circ \mu_2 \circ \mu_1, \tag{2.1}$$

where the order $1, 2, \ldots, n$ of the vertices is as in section 1. It is shown in [**2**] that the point P_t in Definition 1.1 is equal to the specialization of the cluster $\mu_*^t(\mathbf{x})$ at $x_i = 1$.

For an arbitrary point $a_* = (a_1, \cdots, a_n) \in \mathbb{C}^n$ with $a_i \neq 0$ and any $1 \leq k \leq n$, let

$$\mu_k(a_1, \cdots, a_n) = (a_1, \cdots, a_{k-1}, a_k', a_{k+1}, \cdots, a_n) \in \mathbb{C}^n$$

where a_k' is obtained from the cluster variable x_k' by specializing the initial cluster variables $x_i = a_i$, $i = 1, 2, \ldots, n$. For generic a_* we will have $a_k' \neq 0$, so the Coxeter mutation can be repeated. Let $\mu_*(a_*) = \mu_n \circ \cdots \circ \mu_1(a_*)$.

We propose the following generalization of the frieze variety.

DEFINITION 2.1. (1) We say that $a_* \in \mathbb{C}^n$ is a *generic specialization* of $\mathbf{x}$ if all coordinates of $\mu_*^t(a_*)$ are nonzero for all $t \geq 0$.

(2) We refer to the set of all $\mu_*^t(a_*) \in \mathbb{C}^n$ for $t \geq 0$ as the *μ_*-orbit* of a_*.

(3) For any generic specialization a_* of $\mathbf{x}$, the *generalized frieze variety* $X(Q, a_*)$ is defined to be the Zariski closure in $\mathbb{C}^n$ of the μ_*-orbit of a_*.

(4) Let $\widetilde{X}(Q, a_*)$ be $X(Q, a_*)$ with zero dimensional components removed. So, $\widetilde{X}(Q, a_*)$ is empty when $X(Q, a_*)$ is finite, e.g. when Q has finite type. (Theorem 1.2, Remark 3.9.)

REMARK 2.2. (1) By the well-known Laurent Phenomenon proved by Fomin and Zelevinsky in [**1**], the coordinates of $\mu_*^t(a_*)$ for any integer t are given by Laurent polynomials in $a_1, \cdots, a_n$. Therefore, $\mu_*^t(a_*)$ is defined for all t as long as $a_* \in (\mathbb{C}^\times)^n$, i.e. $a_i \neq 0$ for all i.

Moreover, by the positivity theorem proved in [**4**] (for acyclic quivers already in [**5**]), if a_* is a positive real vector, then all $\mu_*^t(a_*)$ are positive real vectors. In particular, every positive real vector is a generic specialization.

(2) There are negative integer vectors which are not generic. For example, the vector $a_* = (-1, -1)$ is not generic for the quiver $Q : 1 \leftarrow 2$ since $\mu_*(-1, -1) = (0, -1)$.

(3) The frieze variety of Q is $X(Q) = X(Q, (1, 1, \cdots, 1))$.

(4) We will see that all components of $X(Q, a_*)$ have the same dimension and, therefore, $\widetilde{X}(Q, a_*) = X(Q, a_*)$ when the set is infinite. (Theorem 3.3)

(5) The Zariski closure of the set of all $\mu_*^t(a_*)$ for all $t \in \mathbb{Z}$ is the union of $X(Q, a_*)$ and $X(Q^{op}, a_*)$. One could speculate that these two varieties should always be equal. In any case, our theorems apply to these varieties separately.

We will show that the frieze variety is invariant under mutation in the following sense.

THEOREM 2.3. *If $a_* \in \mathbb{C}^n$ is a generic point on the frieze variety, then a_* is a generic specialization of $\mathbf{x}$ and $X(Q, a_*) = \widetilde{X}(Q)$. More precisely, for each component X_i of $X(Q)$ of dimension ≥ 1 there is a subset $U_i \subset X_i$ given as a countable intersection of open subsets U_i^d so that, for any a_* in any U_i we have $X(Q, a_*) = \widetilde{X}(Q)$.*

COROLLARY 2.4. *If $a_* \in X(Q)$ is a generic point then $\mu_*(a_*) \in X(Q)$.*

We note that frieze varieties often have nongeneric points. See, e.g., Remark 5.1 and the end of Example 3.8.

3. Proof of Theorem 2.3

We prove a more general result (Theorem 3.5 below) using the following lemmas.

LEMMA 3.1. *If b_* is a generic specialization of* $\mathbf{x}$ *and $a_* \in \widetilde{X}(Q,b_*) \cap (\mathbb{C}^\times)^n$ then*

$$\mu_*(a_*) \in \widetilde{X}(Q,b_*).$$

PROOF. Since $\widetilde{X}(Q,b_*)$ contains all but finitely many elements of $X(Q,b_*)$, $\widetilde{X}(Q,b_*) = X(Q,\mu_*^t(b_*))$ for sufficiently large $t > 0$. Replacing b_* with $\mu_*^t(b_*)$, we may therefore assume that $\widetilde{X}(Q,b_*) = X(Q,b_*)$ contains $\mu_*^t(b_*)$ for all $t \geq 0$.

The variety $X(Q,b_*)$ is given by a finite number of polynomials f_j. For any $a_* \in (\mathbb{C}^\times)^n$, $\mu_*(a_*) \in \mathbb{C}^n$ is well-defined and lies in $X(Q,b_*)$ if and only if $f_j(\mu_*(a_*)) = 0$ for all j. Since the coordinates of $\mu_*(x_*)$ are Laurent polynomials in $x_1,\cdots,x_n$, each $f_j(\mu_*(x_*))$ is also a Laurent polynomial in the x_i. So, there are monomials $g_j(x_*)$ with the property that

$$F_j(x_*) := f_j(\mu_*(x_*))g_j(x_*) \in \mathbb{C}[x_1,\cdots,x_n]. \tag{3.1}$$

The polynomials F_j have the property that, for any $a_* \in (\mathbb{C}^\times)^n$, $\mu_*(a_*) \in X(Q,b_*)$ if and only if $F_j(a_*) = 0$ for all j. Since $\mu_*^t(b_*) \in X(Q,b_*)$ for all $t \geq 0$, this implies that $F_j(\mu_*^t(b_*)) = 0$ for all $t \geq 0$. This implies that $F_j = 0$ on the Zariski closure of this set of points: $X(Q,b_*)$. Now let $a_* \in X(Q,b_*)$, then $F_j(a_*) = 0$ for all j, and by the above argument, $\mu_*(a_*) \in X(Q,b_*) = \widetilde{X}(Q,b_*)$ as claimed. □

In order to state the main result (Theorem 3.5 which will generalize Theorem 2.3), we need to consider the irreducible components of the variety $\widetilde{X}(Q,b_*)$. We will show that the the Coxeter mutation μ_* cyclically permutes these components.

THEOREM 3.2. *Let b_* be any generic specialization of the cluster* $\mathbf{x}$*, for example, $b_* = (1,1,\cdots,1)$. Choose $t_0 \geq 0$ so that $\mu_*^t(b_*) \in \widetilde{X}(Q,b_*)$ for all $t \geq t_0$. Then the components of $\widetilde{X}(Q,b_*) = X(Q,\mu_*^{t_0}(b_*))$ can be numbered $X_1,\cdots,X_m$ with the following properties.*

1. *For each $t \geq t_0$, $\mu_*^t(b_*) \in X_i$ if and only if $t \equiv i$ modulo m. In particular, for each $t \geq t_0$, $\mu_*^t(b_*)$ lies in exactly one X_i.*
2. *X_i is the closure of the set of all $\mu_*^{i+km}(b_*)$ for all integers $k \geq t_0/m$.*
3. *For any $a_* \in X_i \cap (\mathbb{C}^\times)^n$ we have $\mu_*(a_*) \in X_{i+1}$ (or X_1 if $i = m$).*

PROOF. By replacing b_* with $\mu_*^{t_0}(b_*)$ we may assume that $t_0 = 0$ and $\widetilde{X}(Q,b_*) = X(Q,b_*)$.

Let $X_1,\cdots,X_m$ be the components of $X(Q,b_*)$.

Consider the polynomials f_{jk} which define the component X_j. As in (3.1), there are polynomials F_{jk} with the property that, for any $a_* \in (\mathbb{C}^\times)^n$, $\mu_*(a_*) \in X_j$ if and only if $F_{jk}(a_*) = 0$ for all k. Then the polynomials F_{jk} and f_{ip} define a subvariety X_{ij} of X_i which contains all $a_* \in (\mathbb{C}^\times)^n \cap X_i$ so that $\mu_*(a_*) \in X_j$. By Lemma 3.1, $\mu_*(a_*) \in \widetilde{X}(Q,b_*) = \cup X_j$, for all $a_* \in (\mathbb{C}^\times)^n \cap X_i$. Therefore, X_i is the union of the subvarieties X_{ij}. Since X_i is irreducible, $X_i = X_{ij}$ for some j. In fact j is uniquely determined by i, but we don't need to verify this.

The equation $X_i = X_{ij}$ implies that, for any $a_* \in X_i \cap (\mathbb{C}^\times)^n$, $\mu_*(a_*) \in X_j$. For each i, choose one such j. Then $\pi(i) = j$ defines a mapping of the set $\{1,2,\cdots,m\}$

to itself. Statement (3) will follow from this after we show that π is a cyclic permutation.

CLAIM 1. π is a permutation which is transitive, i.e. π is an m-cycle.

PROOF. Since $b_* \in X(Q, b_*)$, it must lie in one of the components; suppose that $b_* \in X_i$. Then $\mu_*(b_*) \in X_{\pi(i)}$ and $\mu_*^t(b_*) \in X_{\pi^t(i)}$. Thus, the μ_*-orbit of b_* lies in the union of all X_j where j is in the π-orbit of i. But the closure of the μ_*-orbit of b_* is the union of all the X_j. So π must be transitive, which also implies π is a permutation and, in particular, an m-cycle.

Since π is an m-cycle, we may reindex the sets X_i so that $\pi(i) = i+1$ for $i < m$ and $\pi(m) = 1$ and so that $b_* \in X_m$. □

CLAIM 2. For each $t \geq 0$, $\mu_*^t(b_*)$ lies in only one X_i: the one where $i \equiv t \bmod m$.

PROOF. Suppose not. Then $\mu_*^t(b_*) \in X_i \cap X_j \subset X_j$ where $j \neq i$. By the division algorithm, any $s \geq t$ can be written as $s = p + km$ where $1 \leq p < m$. If $p \neq i$, then $\mu_*^s(b_*) \in X_p$. If $p = i$ then $\mu_*^s(b_*) \in X_j$. So, the set of all $\mu_*^s(b_*)$ for all $s \geq t$ lies in the union of all X_p for $p \neq i$ which is a contradiction. □

These claims prove Statement (1). Statement (2) follows from the definition of X_i. □

Theorem 3.2 can be strengthened as follows.

THEOREM 3.3. *For all generic specializations b_* of $\mathbf{x}$, all components of $X(Q, b_*)$ have the same dimension. In particular, if $X(Q, b_*)$ is infinite, then $\widetilde{X}(Q, b_*) = X(Q, b_*)$ and $t_0 = 0$ in Theorem 3.2 above.*

PROOF. The Coxeter mutation μ_* and its inverse are given by Laurent polynomials which are rational functions whose denominators are monomials. By Theorem 3.2 μ_* gives a bijection between dense subsets of X_i, X_{i+1} which are disjoint from the coordinate hyperplanes. Therefore, the components X_i of $\widetilde{X}(Q, b_*)$ are birationally equivalent and therefore have the same dimension since the dimension of an irreducible variety is the transcendence degree of its field of rational functions. When $X(Q, b_*)$ is finite, the Coxeter mutation clearly acts as a cyclic permutation of that set. So, assume $X(Q, b_*)$ is infinite.

To see that $\widetilde{X}(Q, b_*) = X(Q, b_*)$ in the infinite case, suppose not. Then there must one point $\mu_*^t(b_*)$ not in $\widetilde{X}(Q, b_*)$ so that $\mu_*^{t+1}(b_*) \in X_i$ for some i. Choose a regular function $f : \mathbb{C}^n \to \mathbb{C}$, i.e. a polynomial in n variables, which is zero on X_{i-1} and nonzero on the point $\mu_*^t(b_*)$. Composing with the rational morphism μ_*^{-1} gives a rational function on X_i whose denominator is a monomial and whose numerator is a polynomial function g. Moreover, g is zero on a dense subset of X_i by Theorem 3.2(2) and thus zero on all of X_i, but g is nonzero on $\mu_*^{t+1}(b_*) \in X_i$, since $g(\mu_*^{t+1}(b_*)) = f(\mu_*^t(b_*)) \neq 0$. This contradiction shows that $\mu_*^t(b_*) \in X_{i-1}$ as claimed. □

LEMMA 3.4. *For every component X_i of $X(Q, b_*)$ and every integer $d > 0$, there is a $p_d > d$ and a dense open subset U_i^d of X_i so that, for every $a_* \in U_i^d$, we have the following.*

(1) *For every $0 \leq t \leq p_d$, the coordinates of $\mu_*^t(a_*)$ are nonzero.*

(2) *Any polynomial of degree* $\le d$ *which is zero on* $\mu_*^t(a_*)$ *for all* $0 \le t \le p_d$ *will also be zero on* $X(Q, b_*)$.

Before we prove this lemma, we will show that it implies the following generalization of Theorem 2.3. We use the notation U_i^∞ for the intersection of U_i^d over all $d > 0$.

THEOREM 3.5. *Let* $b_* \in \mathbb{C}^n$ *be any generic specialization of the cluster* $\mathbf{x}$ *of* Q. *Then, for generic* $a_* \in X(Q, b_*)$, *i.e. for* $a_* \in \bigcup U_i^\infty$, *we have* $X(Q, a_*) = X(Q, b_*)$.

PROOF. By (1) in Lemma 3.4, every element of U_i^∞ is a generic specialization of $\mathbf{x}$. By Lemma 3.1, the entire μ_*-orbit of a_* lies in $\widetilde{X}(Q, b_*) = X(Q, b_*)$. So, $X(Q, a_*) \subset X(Q, b_*)$.

If $X(Q, a_*) \ne X(Q, b_*)$ there must be a polynomial f which is zero on the μ_*-orbit of a_* but nonzero on $X(Q, b_*)$. Let $d = \deg f$. Given that f is zero on the μ_*-orbit of $a_* \in U_i^\infty \subset U_i^d$, we conclude by (2) in Lemma 3.4 that f is zero on $X(Q, b_*)$. This contradiction proves the theorem. □

REMARK 3.6. Theorem 2.3 follows from Theorem 3.5 by choosing $b_* = (1, 1, \dots, 1)$.

It remains to prove the lemma:

PROOF OF LEMMA 3.4. We consider only the nontrivial case when $X(Q, b_*)$ is infinite. By Theorems 3.2 and 3.3, $X(Q, b_*) = X_0 \cup \dots \cup X_{m-1}$ and $t_0 = 0$ in Theorem 3.2. By (1) in Theorem 3.2 we have that $\mu_*^i(b_*) \in X_i$. Since $\widetilde{X}(Q, b_*) = X(Q, b_*)$, a polynomial f will be zero on $X(Q, b_*)$ if and only if $f(\mu_*^t(b_*)) = 0$ for all $t \ge i$. The key point of the proof is to show that, for f of degree $\le d$, we only need to check this condition for $t \le p_d$ for some fixed $p_d > 0$. This is a linear condition on the coefficients of f. Since the rank of a linear system is a lower semi-continuous function, there will be an open subset U_i^d of X_i on which this system is defined (Condition (1)) and has maximum rank. This will be the desired set.

We now construct the linear system. With n, d fixed, consider the polynomial mapping

$$P_d : \mathbb{C}^n \to \mathbb{C}^{\binom{n+d}{n}}$$

which sends $x_* = (x_1, \cdots, x_n) \in \mathbb{C}^n$ to the sequence of all monomials in x_j of degree $\le d$. For example, when $n = 2, d = 3$, we have:

$$P_3(x, y) = (1, x, y, x^2, xy, y^2, x^3, x^2y, xy^2, y^3).$$

Then any polynomial function f on $\mathbb{C}^n$ of degree $\le d$ is given as the composition of P_d with a linear mapping $f_* : \mathbb{C}^{\binom{n+d}{n}} \to \mathbb{C}$.

Let $B_d \subset \mathbb{C}^{\binom{n+d}{n}}$ be the vector space span of all $P_d(\mu_*^t(b_*))$ for all $t \ge m$. Then a polynomial f of degree $\le d$ is zero on $X(Q, \mu_*^m(b_*)) = X(Q, b_*)$ if and only if $f_*(B_d) = 0$. So, $P_d(a_*) \in B_d$ for all $a_* \in X(Q, b_*)$. Let k be the dimension of B_d. Then B_d has a basis consisting of $P_d(\mu_*^{t_j}(b_*))$ for some $m \le t_1 < t_2 < \dots < t_k$. These vectors form a $\binom{n+d}{n} \times k$ matrix of rank k. So, there is some $k \times k$ minor M of this matrix which is nonzero. Since $t_j \ge m \ge i$, the entries of the matrix, being monomials in the coordinates of $\mu_*^{t_j - i}(\mu_*^i(b_*))$ for some j, are given as Laurent polynomials in the coordinates of $\mu_*^i(b_*) \in X_i$. Therefore, for each i, the minor M is a Laurent polynomial in the coordinates of $\mu_*^i(b_*)$. Let F_i be the numerator of this polynomial. Then $F_i(\mu_*^i(b_*)) \ne 0$.

Let $p_d = t_k - i$ and let V_i^d be the subset of X_i of all points a_* so that $\mu_*^t(a_*)$ is defined with nonzero coordinates for $0 \leq t \leq p_d$. Since this is an open condition and $\mu_*^i(b_*) \in V_i^d$, V_i^d is a dense open subset of X_i. By Lemma 3.1, $\mu_*^t(a_*) \in X(Q, b_*)$ for all $0 \leq t \leq p_d$. Therefore, $P_d(\mu_*^t(a_*)) \in B_d$ for all $0 \leq t \leq p_d$. The condition that the vectors $P_d(\mu_*^t(a_*))$, for $0 \leq t \leq p_d$ span B_d is an open condition which holds for $a_* = \mu_*^i(b_*) \in X_i \cap V_i^d$. Therefore, it holds on some Zariski open neighborhood of $\mu_*^i(b_*)$ in $X_i \cap V_i^d$. In fact, this condition will hold on the open subset U_i^d of $X_i \cap V_i^d$ on which $F_i \neq 0$.

Since $U_i^d \subset V_i^d$, then $a_* \in U_i^d$ will satisfy Condition (1). For any polynomial f of degree $\leq d$ which is zero on $\mu_*^t(a_*)$ for all $0 \leq t \leq p_d$, the corresponding linear map f_* will vanish on the vector $P_d(\mu_*^t(a_*))$ for all $0 \leq t \leq p_d$. Since these vectors span B_d, $f_*(B_d) = 0$. This implies that f is zero on the set $X(Q, b_*)$, proving Condition (2) and concluding the proof of Lemma 3.4, Theorems 3.5 and 2.3. □

We illustrate some of the concepts in the proof of Lemma 3.4 with two examples.

EXAMPLE 3.7. Consider the A_2 quiver

$$Q: \quad 1 \longleftarrow 2.$$

Then $X(Q)$ has only five points $\mu_*^i(b_*) = (1,1), (2,3), (2,1), (1,2), (3,2)$ for $i = 0, 1, 2, 3, 4$. Thus $m = 5$ and each X_i consists of one point. For $d = 2$, $\binom{n+d}{n} = 6$. So, the five vectors $P_2(\mu_*^i(b_*))$ do not span $\mathbb{C}^6$. These five vectors are the rows of the following 5×6 matrix. (The proof of Lemma 3.4 uses the transpose of this matrix.)

	1	x	y	x^2	xy	y^2
b_*	1	1	1	1	1	1
$\mu_*(b_*)$	1	2	3	4	6	9
$\mu_*^2(b_*)$	1	2	1	4	2	1
$\mu_*^3(b_*)$	1	1	2	1	2	4
$\mu_*^4(b_*)$	1	3	2	9	6	4

The span of these five vectors is $B_2 \subset \mathbb{C}^6$. This is a hyperplane perpendicular to the vector $(3, -2, -2, 1, -1, 1)$. Dot product with this vector gives a linear map $f_* : \mathbb{C}^6 \to \mathbb{C}$, composing with P_2 gives

$$f(x, y) = f_*(P_2(x, y)) = x^2 - xy + y^2 - 2x - 2y + 3.$$

This is the only quadratic polynomial which vanishes on the frieze variety $X(Q)$. The real points form an ellipse centered at $(2, 2)$ with major axis going from $(1, 1)$ to $(3, 3)$.

Here is another example which explains the minor M and numerator F.

EXAMPLE 3.8. Consider the Kronecker quiver

$$Q: \quad 1 \Longleftarrow 2$$

Consider the frieze variety $X(Q)$. The first three points are $b_* = (1, 1)$, $\mu_*(b_*) = (2, 5)$, $\mu_*^2(b_*) = (13, 34)$.

Take $d = 1$. Then $\binom{n+d}{n} = 3$. In order to span $\mathbb{C}^3$ we need three vectors: $P_1(\mu_*^t(b_*))$ for $t = 0, 1, 2$. These are the rows of the following matrix.

	1	x	y
b_*	1	1	1
$\mu_*(b_*)$	1	2	5
$\mu_*^2(b_*)$	1	13	34

Since this has full rank, the determinant of this matrix (which is -15) is the maximal minor. However we need the minor as a Laurent polynomial in the coordinates of $\mu_*^i(b_*)$. Take $i = 1$ and write $\mu_*^1(b_*) = (y_1, y_2)$. Thus $y_1 = x_1', y_2 = x_2'$. In terms of the cluster y_1, y_2, the 3×3 matrix under consideration is:

$$\begin{bmatrix} 1 & \frac{y_1^4+y_2^2+2y_1^2+1}{y_1y_2^2} & \frac{y_1^2+1}{y_2} \\ 1 & y_1 & y_2 \\ 1 & \frac{y_2^2+1}{y_1} & \frac{y_1^4+2y_2^2+y_1^2+1}{y_1^1 y_2} \end{bmatrix}$$

The determinant of this matrix is the rational function M. The numerator of M is the polynomial $F = y_1^2y_2^2M$. This is a polynomial of degree 8 in y_1, y_2. The reason we use these variables is because we are looking for points $a_* = (a_1, a_2)$ close to $\mu_*^1(b_*) = (2, 5)$. What we have already calculated is: $M(2, 5) = -15$.

Using invariant rational functions, the generalized frieze variety $X(Q, b_*)$ can be given as follows. The rational function

$$h(\mathbf{x}) = \frac{x_1^2 + x_2^2 + 1}{x_1x_2}$$

is equal to itself, as an element of $\mathbb{C}(\mathbf{x})$, on all iterated Coxeter mutations of the cluster $\mathbf{x}$. To see this, write it as:

$$h(x_1, x_2) = \frac{x_1 + x_1'}{x_2} = h(x_1', x_2)$$

which is invariant under Coxeter mutation since μ_1 switches the terms x_1, x_1' and similarly for μ_2. So, it is invariant under $\mu_2 \circ \mu_1$. At $b_* = (1, 1)$ it takes the value $h(1, 1) = 3$. This makes

$$x_1^2 + x_2^2 + 1 = 3x_1x_2 \tag{3.2}$$

at all points in $X(Q)$. Note that equation (3.2) is a specialization of the Markov equation $x_1^2 + x_2^2 + x_3^2 = 3x_1x_2x_3$.

Using this equation we can see that the value of M at any point (y_1, y_2) in $X(Q)$ is equal to -15. So, $F = -15y_1^2y_2^2$. The set U_1^d for $d = 1$ is given by $F(\mu_*^t(a_*)) \neq 0$ for three values of t, namely $t = -1, 0, 1$ since we are thinking of a_* as a specialization of $(y_1, y_2) = (x_1', x_2')$. This makes $U_1^d = V_1^d$ the complement in $X(Q)$ of the 12 points consisting of μ_*^{-t}, for $t = -1, 0, 1$, applied to the points $(0, \pm\sqrt{-1}), (\pm\sqrt{-1}, 0)$.

REMARK 3.9. We observe that Theorem 1.2 does not always hold for the generalized frieze variety. For example, when Q is the Kronecker quiver considered above and $b_* = \left(\frac{\sqrt{2}i}{2}, \frac{\sqrt{2}i}{2}\right)$, $X(Q, b_*)$ consists of only two points, b_* and $-b_*$. However, we believe that, for almost all b_*, the analogue of Theorem 1.2 hold.

4. Construction from invariant rational functions

For any $k \geq 0$, the coordinates of $\mu_*^k(\mathbf{x})$ are Laurent polynomials in $\mathbf{x}$. Furthermore, each coordinate of $\mathbf{x}$ is given as a Laurent polynomial in $\mu_*^k(\mathbf{x})$. So, the set of values of $\mu_*^k(\mathbf{x})$ is not contained in any hypersurface in $\mathbb{C}^n$. So, for any rational function $h(\mathbf{x}) \in \mathbb{C}(\mathbf{x})$ and any $t \geq 0$, we have another rational function $h(\mu_*^t(\mathbf{x})) \in \mathbb{C}(\mathbf{x})$ since the denominator of $h(\mu_*^t(\mathbf{x}))$ cannot be identically zero. Suppose, furthermore, that $h(\mathbf{x})$ is a Laurent polynomial in $\mathbf{x}$ and $a_* \in \mathbb{C}^n$ is a generic specialization of $\mathbf{x}$. Then $h(\mu_*^t(a_*))$ is a well-defined complex number for any $t \geq 0$. This is particularly useful when $h(\mathbf{x})$ is periodic in the following sense.

DEFINITION 4.1. We say that a rational function $h(\mathbf{x})$ is *invariant under* μ_*^k if:

$$h(\mu_*^k(\mathbf{x})) = h(\mathbf{x}) \tag{4.1}$$

as an element of $\mathbb{C}(\mathbf{x})$. If $k > 0$ is minimal and $h(\mathbf{x})$ is Laurent, we say that $h(\mathbf{x})$ is an *invariant Laurent polynomial for Q of period k*. For each $t \geq 0$ we will use the notation:

$$h(\mu_*^t(\mathbf{x})) = \frac{f_t(\mathbf{x})}{g_t(\mathbf{x})} \tag{4.2}$$

Note that $f_t, g_t \in \mathbb{C}[\mathbf{x}]$ depend only on the residue class of t modulo the period k.

PROPOSITION 4.2. *Let $h(\mathbf{x})$ be an invariant Laurent polynomial of period k. Let a_* be a generic specialization of $\mathbf{x}$. For each $t \geq 0$, let $c_t = h(\mu_*^t(a_*))$ and let $f_t(\mathbf{x}), g_t(\mathbf{x})$ be as in (4.2). For $0 \leq t < k$, let*

$$F_{j,t}(\mathbf{x}) := f_t(\mathbf{x}) - c_{t+j}g_t(\mathbf{x})$$

be the numerator of the rational function $h(\mu_*^t(\mathbf{x})) - h(\mu_*^{j+t}(a_*))$, and let X_j be the intersection of the k hypersurfaces given by $F_{j,t}(\mathbf{x}) = 0$, for $0 \leq j < k$.

Then the generalized quiver variety $X(Q, a_*)$ is contained in the union $X_0 \cup X_1 \cup \cdots \cup X_{k-1}$.

PROOF. For any $0 \leq j < k$ and $s \geq 0$, let $b_* = \mu_*^{j+ks}(a_*)$. Then, for any $t \geq 0$, we have $h(\mu_*^t(b_*)) = h(\mu_*^{j+t}(a_*)) = c_{j+t}$. Since $F_{j,t}(\mathbf{x})$ is the numerator of $h(\mu_*^t(\mathbf{x})) - h(\mu_*^{j+t}(a_*))$, we get $F_{j,t}(b_*) = 0$. Therefore, $b_* = \mu_*^{j+ks}(a_*)$ lies in X_j for all $s \geq 0$ and the union of the X_j contains the entire μ_*-orbit of a_*. So, $X(Q, a_*) \subset \bigcup X_j$. □

REMARK 4.3. Thus, the single Laurent polynomial $h(\mathbf{x})$ generates k^2 polynomials $F_{j,t}(\mathbf{x})$ giving k varieties X_j whose union contains $X(Q, a_*)$ and, in many cases, is equal to $X(Q, a_*)$ as shown in several examples below. In these examples, all of the rational functions $h(\mu_*^t(\mathbf{x})) = f_t(\mathbf{x})/g_t(\mathbf{x})$ are Laurent polynomials with positive integer coefficients. This is reflected in the fact that the monomials in the polynomials $F_{j,t}(\mathbf{x}) = f_t(\mathbf{x}) - c_{j+t}g_t(\mathbf{x})$ have the same sign except for one: $-c_{j+t}g_t(\mathbf{x})$. We note that this is not a general phenomenon since, e.g., when Q has finite type, $\mu_*^k(\mathbf{x}) = \mathbf{x}$ for some k and, therefore, every Laurent polynomial will be invariant with period dividing k.

5. Examples

To illustrate Proposition 4.2, we give two examples, both tame, where a single invariant Laurent polynomial $h(\mathbf{x})$ whose period k is one less than the number of vertices of Q gives the complete decomposition of $X(Q)$ as a union of k curves.

5.1. The affine quiver $\tilde{A}_2$. Let Q be the quiver:

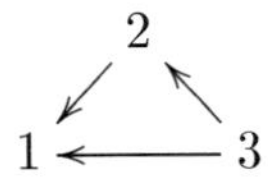

This quiver has the property that $\mu_1 Q \cong Q$ after renumbering the vertices. In terms of the cluster variables (x, y, z), after one mutation, we get back the same quiver with new variables (y, z, x') where $x' = \frac{1}{x}(yz+1)$. For any rational function $h(x, y, z)$ let

$$h'(x, y, z) := h(y, z, x'),$$

where x' is the cluster variable obtained from the cluster (x, y, z) by mutation in x. We also use the notation $h' = h \circ \tilde{\mu}$ where

$$\tilde{\mu}(x, y, z) := (y, z, x'). \tag{5.1}$$

Note that $h^{(3)}(x, y, z) = h'''(x, y, z) = h(\mu_*(x, y, z))$.

For example, let $h(\mathbf{x})$ be the Laurent polynomial

$$h(x, y, z) = \frac{x+z}{y}$$

Then h', h'' are given by

$$h'(x, y, z) = h(y, z, x') = \frac{y + x'}{z} = \frac{xy + yz + 1}{xz}$$

$$h''(x, y, z) = h'(y, z, x') = h(z, x', y') = \frac{z + y'}{x'}$$

Observe that

$$\frac{h}{h''} = \frac{(x+z)x'}{y(z+y')} = \frac{yz + 1 + zx'}{yz + 1 + zx'} = 1$$

and thus $h'' = h$ and, consequently, $h^{(a)} = h$ if a is even, and $h^{(a)} = h'$ if a is odd. Therefore h and h' are invariant under μ_*^2:

$$h(\mu_*^2(x, y, z)) = h^{(6)}(x, y, z) = h(x, y, z),$$

and similarly, $h' \circ \mu_*^2 = h^{(7)} = h'$.

Thus, $h(\mathbf{x})$ is an invariant Laurent polynomial for Q of period 2. So, Proposition 4.2 applies with

$$c_0 = h(1, 1, 1) = 2$$
$$c_1 = h'(1, 1, 1) = 3.$$

So, $X(Q)$ is contained in the union of two curves $X_0 \cup X_1$ where X_0 is given by the polynomial equations $F_{00} = F_{01} = 0$ where

$$F_{00}(x, y, z) = Num(h(x, y, z) - c_0) = x + z - 2y$$

$$F_{01}(x, y, z) = Num(h'(x, y, z) - c_1) = xy + yz + 1 - 3xz$$

and X_1 is given by $F_{11} = F_{12} = 0$ where

$$F_{10}(x, y, z) = Num(h(x, y, z) - c_1) = x + z - 3y$$

$$F_{11}(x, y, z) = Num(h'(x, y, z) - c_0) = xy + yz + 1 - 2xz.$$

From these equations it is easy to verify the observation from [**2**] that X_0 is a nonsingular degree 2 curve. Indeed the equation $F_{00} = 0$ is equivalent to the linear equation $z = 2y - x$ which reduced the second equation to $F_{01}(x, y, z) = F_{01}(x, y, 2y - x) = 0$ which is a nondegenerate quadratic in two variables. Thus

X_0 is a nonsingular curve in $\mathbb{C}^3$ containing the infinite set of points $\mu_*^{2k}(1,1,1)$ for $k \geq 0$. So, it must be the closure of this set. Similarly, the curve X_1 must be the closure of the set of all $\mu_*^{2k+1}(1,1,1)$. We therefore see that the frieze variety $X(Q)$ has two components given by the above four polynomials. These polynomials come from an example worked out in [**2**], but here all four polynomials come from the same Laurent polynomial h.

REMARK 5.1. We note that frieze varieties often have nongeneric points. For example, the first component X_0 of the frieze variety discussed above contains the point $(0, \sqrt{2}i/2, \sqrt{2}i)$. Mutation at the first vertex gives $x' = (yz+1)/x = 0/0$ which is undefined. However, $\tilde{\mu}$ (defined in (5.1) above) sends the 0-component of $X(Q)$ to the 1-component X_1. So, the value of x' can be computed from the linear equation $F_{10}(y,z,x') = 0$:

$$x' = 3z - y = \frac{5}{2}\sqrt{2}i.$$

5.2. The affine quiver $\tilde{A}_n$. More generally, consider the quiver:

Q :
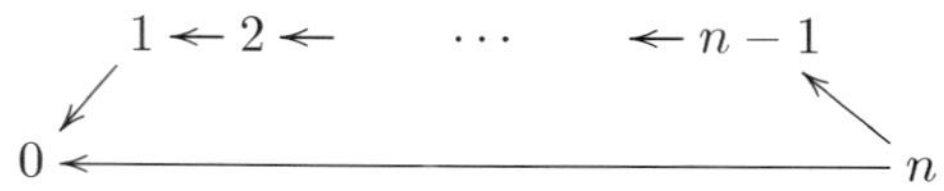

for $n \geq 3$. Let h be the Laurent polynomial:

$$h(x_0, x_1, \cdots, x_n) = \frac{x_{n-2} + x_n}{x_{n-1}}.$$

Mutation gives:

$$h'(x_0, \cdots, x_n) := h(x_1, x_2, \cdots, x_n, x_0) = \frac{x_{n-1} + x_0'}{x_n} = \frac{x_{n-1}x_0 + x_1x_n + 1}{x_0x_n}.$$

Mutating k times for $3 \leq k \leq n$ ($k = 2$ is given in (5.2) below) gives

$$h^{(k)}(x_0, x_1, \cdots, x_n) = h(x_k, x_{k+1}, \cdots, x_n, x_0', \cdots, x_{k-1}') = \frac{x_{k-3}' + x_{k-1}'}{x_{k-2}'}$$

For $k = n$, $h^{(n)} = h$ since the quotient is:

$$\frac{h}{h^{(n)}} = \frac{(x_{n-2} + x_n)x_{n-2}'}{x_{n-1}(x_{n-3}' + x_{n-1}')} = \frac{x_{n-1}x_{n-3}' + 1 + x_nx_{n-2}'}{x_{n-1}x_{n-3}' + 1 + x_nx_{n-2}'} = 1$$

This gives us a cleaner formula for $h^{(k)}$ for $2 \leq k \leq n$ using $2 - n \leq k - n \leq 0$:

$$h^{(k)}(\mathbf{x}) = h^{(k-n)}(\mathbf{x}) = h(x_{k+1}', \cdots, x_n', x_0, x_1, \cdots, x_k) = \frac{x_{k-2} + x_k}{x_{k-1}}. \tag{5.2}$$

The equation $h = h^{(n)}$ also implies that $h^{(nk)} = h$, for all $k \geq 0$. So,

$$h(\mu_*^k(\mathbf{x})) = h^{k(n+1)}(\mathbf{x}) = h^{(k)}(\mathbf{x}).$$

In particular $h(\mu_*^n(\mathbf{x})) = h^{(n)}(\mathbf{x}) = h(\mathbf{x})$. So, $h(\mathbf{x})$ is an invariant Laurent polynomial for Q of period dividing n. To see that the period of h is exactly n we compute c_k:

$$c_k = h^{(k)}(1, 1, \cdots, 1) = h(1, 1, \cdots, 1, 2, 3, \cdots, k+1) = \begin{cases} 3 & \text{if } k = 1; \\ 2 & \text{if } 2 \leq k \leq n. \end{cases} \tag{5.3}$$

Also, $h^{(k)}$ is invariant under μ_*^n, for all $k \geq 0$, since $h^{(k)}(\mu_*^{sn}(\mathbf{x})) = h^{(k+sn)}(\mathbf{x}) = h^{(k)}(\mathbf{x})$. So, for generic a_*, the frieze variety $X(Q, a_*)$ has n components

$$X(Q, a_*) = X_0 \cup X_1 \cup \cdots \cup X_{n-1};$$

the first component X_0, containing the μ_*^n-orbit of $a_* = (1, 1, \cdots, 1)$, is given by the n polynomials $F_{01}, \cdots, F_{0,n-1}, F_{0n} = F_{00}$ where

$$F_{01} = Num(h'(\mathbf{x}) - c_1) = x_0 x_{n-1} + x_1 x_n + 1 - 3x_0 x_n,$$

by Proposition 4.2 and (5.3); and for $2 \leq k \leq n$

$$F_{0k} = Num(h^{(k)}(\mathbf{x}) - c_k) = x_{k-2} + x_k - 2x_{k-1},$$

by Proposition 4.2 ,(5.2) and (5.3).

REMARK 5.2. Note that, for any point in X_0, the equations $F_{0k} = 0$ for $2 \leq k \leq n$ express the coordinates $x_2, x_3, \cdots, x_n$ as linear combinations of x_0, x_1. Thus, points in X_0 are determined by their first two coordinates. Since X_0 contains the 3 points $(1, 1, \cdots), (1, 2, \cdots), (n+1, 2n+3, \cdots)$ and their negatives (where we ignore all but the first two coordinates) and does not contain $(0, 0, \cdots)$, linear algebra in $\mathbb{C}^2$ shows X_0 cannot be a union of two straight lines. So, X_0 is an irreducible degree 2 curve containing the μ_*^n-orbit of the point $a_* = (1, 1, \cdots, 1)$. Since no two distinct curves can have an infinite intersection, X_0 is the Zariski closure of this set.

By Proposition 4.2, X_1 is given by the n polynomial equations $F_{1,t} = 0$ for $0 < t \leq n$

$$F_{1n} = Num(h(\mathbf{x}) - c_1) = x_{n-2} + x_n - 3x_{n-1}$$
$$F_{11} = Num(h'(\mathbf{x}) - c_2) = x_0 x_{n-1} + x_1 x_n + 1 - 2x_0 x_n$$
$$F_{1k} = F_{0k}, \text{ for } k = 2, \cdots, n-1.$$

The other polynomials F_{jt} are similar. As in the case of $\tilde{A}_2$, all polynomials F_{jk} are given by the single Laurent polynomial h and its $n-1$ mutations $h^{(k)}$ for $1 \leq k < n$.

6. Symmetry

One easy observation **[2]** is that, if a permutation σ of $\{1, 2 \cdots, n\}$ leaves the quiver Q invariant, then the frieze variety $X(Q)$ satisfies $x_k = x_{\sigma(k)}$ for all k. In terms of invariant rational functions, $h = x_{\sigma(k)}/x_k$ is invariant under μ_*^2 since μ_* inverts h.

A similar result holds true if a mutation of Q has symmetry. For example,

$$Q: \quad 1 \Longleftarrow 2 \Longleftarrow 3$$

becomes symmetric after one mutation:

$$\mu_1(Q): \quad 1' \Longrightarrow 2 \Longleftarrow 3$$

This implies that

$$h = \frac{x_1'}{x_3} = \frac{x_2^2 + 1}{x_1 x_3}$$

is invariant under μ_*^2. Since $c_0 = h(1, 1, 1) = 2$ and $c_1 = 1/c_0 = 1/2$, the frieze variety of Q is $X(Q) = X_0 \cup X_1$, where X_0 is the hypersurface given by $F_0 = 0$, where

$$F_0 = Num(h(\mathbf{x}) - c_0) = x_2^2 + 1 - 2x_1 x_3$$

and X_1 containing $\mu_*(1,1,1) = (2,5,26)$ is the hypersurface given by $F_1 = 0$, where

$$F_1 = Num\left(h(\mathbf{x}) - \frac{1}{2}\right) = Num\,(2h(\mathbf{x}) - 1) = 2x_2^2 + 2 - x_1x_3.$$

REMARK 6.1. The hypersurface X_0 contains the μ_*^2-orbit of the point $(1,1,\cdots,1)$. Since this set is not contained in any curve by Theorem 1.2, there cannot be a smaller variety containing this set. So, X_0 is the Zariski closure of the μ_*^2 orbit of $(1,1,\cdots,1)$ and similarly for X_1.

More generally we have the following.

PROPOSITION 6.2. *Suppose that i is a sink in the quiver Q and j is a source so that, for any other vertex k, the number n_k of arrows from k to i is equal to the number of arrows from j to k. Then, the frieze variety $X(Q)$ is contained in the union of two hypersurfaces X_0, X_1 given by the equations $F_0 = 0$ and $F_1 = 0$, where*

$$F_0 = 1 - 2x_ix_j + \prod_k x_k^{n_k}$$

$$F_1 = 2 - x_ix_j + 2\prod_k x_k^{n_k}.$$

PROOF. After the mutation $\mu_* = \mu_n \circ \cdots \circ \mu_1$, we will have x_*' where

$$x_i' = \frac{1+\prod x_k^{n_k}}{x_i}, \quad x_j' = \frac{1+\prod (x_k')^{n_k}}{x_j}$$

So, the rational function $h(\mathbf{x}) = x_i'/x_j$ will mutate to

$$h(\mu_*(\mathbf{x})) = \frac{x_i''}{x_j'} = \left(\frac{1+\prod (x_k')^{n_k}}{x_i'}\right)\left(\frac{x_j}{1+\prod (x_k')^{n_k}}\right) = \frac{x_j}{x_i'} = \frac{1}{h(\mathbf{x})}.$$

So, $h(\mathbf{x})$ is μ_*^2 invariant. Since $h(\mathbf{x}) = f(\mathbf{x})/g(\mathbf{x})$ where $f(\mathbf{x}) = 1 + \prod x_k^{n_k}$ and $g(\mathbf{x}) = x_ix_j$, $c_0 = h(1,1,\cdots,1) = 2$, $c_1 = 1/c_0 = \frac{1}{2}$, the numerator of $h(\mathbf{x}) - c_0$ is F_0 and the numerator of$h(\mathbf{x}) - c_1$ is F_1. By Proposition 4.2, the μ_*^2 orbit of $(1,1,\cdots,1)$ satisfies $F_0 = 0$ and the μ_*^2 orbit of $\mu_*(1,1,\cdots,1)$ satisfies $F_1 = 0$. □

7. Questions and answers

We list a few questions from the first version of this paper and short answers to these questions following suggestions by Gordana Todorov. Details will be given in another paper.

(1) Question: In the tame case, does an invariant Laurent polynomial $h(\mathbf{x})$ always exist? Answer: Yes. The cluster character of a regular module in a tube of rank k will be an invariant Laurent polynomial of period k.
(2) Question: Can $h(\mathbf{x})$ be chosen to have positive integer coefficients? Answer: In the tame case yes. In the wild case we also believe the answer is yes since we believe that the only invariant rational functions are the ones given by symmetry of the quiver as in Proposition 6.2.
(3) Question: Can $h(\mathbf{x})$ be chosen such that all iterated Coxeter mutations $h(\mu_*^t(\mathbf{x}))$ have positive integer coefficients? Answer: In the tame case, yes. The answer also seems to be yes in the wild case if, as we suspect, the only invariant rational functions come from symmetry of the quiver

as observed in [**2**] or as given in Proposition 6.2. However we note that, in the latter case, $h(\mu_*(\mathbf{x})) = 1/h(\mathbf{x})$ is not Laurent.

(4) Question: Is the period of the invariant Laurent polynomial always equal to the number of components of $X(Q, a_*)$? Answer: No, a counterexample is given by the following quiver

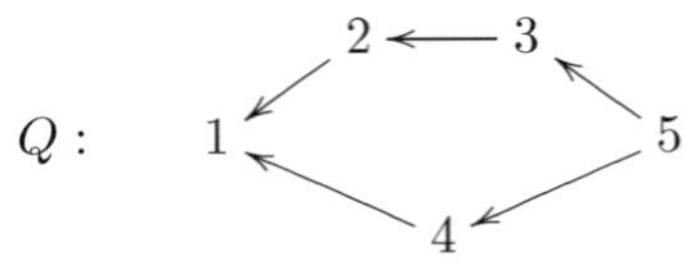

Here there is a tube of rank 3 giving an invariant Laurent polynomial of period 3 and another tube of rank 2 giving an invariant Laurent polynomial of period 2. This suggest that there should be 6 components.

Acknowledgments

The authors wish to thank the organizers of the ARTA VI conference for a very enjoyable and productive event. We also wish to say a special "Happy Birthday" to José Antonio de la Peña and congratulations on his numerous achievements. This paper was also presented by the first author and referenced by the second author at the conference "Cluster Algebras and Representation Theory" held in Kyoto in June 2019. Key observations by Gordana Todorov before and during that conference are added in Section 7 with details to be given in another paper. Also, Salvatore Stella and Alastair King gave us very helpful comments about the first version of this paper. We thank the anonymous referee for pointing out that the results of section 2 hold in the skew-symmetrizable case.

References

[1] Sergey Fomin and Andrei Zelevinsky, *Cluster algebras. I. Foundations*, J. Amer. Math. Soc. **15** (2002), no. 2, 497–529, DOI 10.1090/S0894-0347-01-00385-X. MR1887642

[2] Kyungyong Lee, Li Li, Matthew Mills, Ralf Schiffler, and Alexandra Seceleanu, *Frieze varieties: a characterization of the finite-tame-wild trichotomy for acyclic quivers*, Adv. Math. **367** (2020), 107130, 33, DOI 10.1016/j.aim.2020.107130. MR4080582

[3] José Antonio de la Peña, *Coxeter transformations and the representation theory of algebras*, Finite-dimensional algebras and related topics (Ottawa, ON, 1992), NATO Adv. Sci. Inst. Ser. C Math. Phys. Sci., vol. 424, Kluwer Acad. Publ., Dordrecht, 1994, pp. 223–253. MR1308989

[4] Kyungyong Lee and Ralf Schiffler, *Positivity for cluster algebras*, Ann. of Math. (2) **182** (2015), no. 1, 73–125, DOI 10.4007/annals.2015.182.1.2. MR3374957

[5] Yoshiyuki Kimura and Fan Qin, *Graded quiver varieties, quantum cluster algebras and dual canonical basis*, Adv. Math. **262** (2014), 261–312, DOI 10.1016/j.aim.2014.05.014. MR3228430

[6] Claus Michael Ringel, *The spectral radius of the Coxeter transformations for a generalized Cartan matrix*, Math. Ann. **300** (1994), no. 2, 331–339, DOI 10.1007/BF01450490. MR1299066

Department of Mathematics, Brandeis University, Waltham, Massachusetts 02454
Email address: igusa@brandeis.edu

Department of Mathematics, University of Connecticut, Storrs, Connecticut
Email address: schiffler@math.uconn.edu

Contemporary Mathematics
Volume **761**, 2021
https://doi.org/10.1090/conm/761/15311

On tame strongly simply connected algebras

Stanisław Kasjan and Andrzej Skowroński

Dedicated to José Antonio de la Peña on the occasion of his 60th birthday.

ABSTRACT. In this survey we present the criterion for tameness of strongly simply connected algebras due to Brüstle, de la Peña and Skowroński. We recall relevant concepts of representation theory and discuss some applications and connections to other problems.

1. Introduction

The criterion of tameness of strongly simply connected algebras - Theorem 1.1 below, due to Thomas Brüstle, José Antonio de la Peña and Andrzej Skowroński [**13**] can be considered as a culmination of a research program aimed on better understanding of the concept of representation types of algebras. The purpose of this note is to present the result on the background of classical concepts and results of representation theory, indicating also some less obvious connections to other mathematical ideas. For more information the reader is referred to [**44**].

THEOREM 1.1. [**13**, Main Theorem] *Let A be a strongly simply connected algebra. The following statements are equivalent:*

(1) *A is a tame algebra.*
(2) *The Tits form q_A of A is weakly nonnegative.*

Throughout this article A is an associative algebra of finite dimension over an algebraically closed field K. Usually we assume that A is basic, equivalently, by classical Gabriel's theorem, A is isomorphic to a bound quiver algebra, that is

$$A \cong KQ/I,$$

for some finite quiver Q and an admissible ideal I in the path algebra KQ. We refer to [**2**] for more information about the basic concepts of representation theory of bound quivers and algebras.

The general question of the representation theory of finite dimensional algebras can be formulated as follows: classify (indecomposable) right A -modules of finite K-dimension up to isomorphism. Of course, we are far from satisfactory solution of the problem raised this way.

2010 *Mathematics Subject Classification.* Primary 16G60; Secondary 16G20.

Key words and phrases. Tame representation type, polynomial growth, strongly simply connected algebras, Tits quadratic form.

The second author passed away on October 22, 2020.

Finite dimensional modules can be represented by sequences of matrices and this can be done in several ways. One way is to consider representations of the bound quiver (Q, I) associated to the algebra A, another - based on the idea of analyzing a module through its projective presentation - leads to the concept of Roiter's matrix problems [**50**] or BOCS introduced by Drozd [**19**].

Finding isomorphism classes of A-modules is equivalent to finding "canonical forms" of matrices up to certain admissible operations. Let us remark that matrix problem algorithms allow to obtain a classification of indecomposable A-modules of K-dimension bounded by a number fixed a priori. That does not mean, of course, that a complete classification is always possible.

Let us recall some important facts concerning the easiest situation - the case of representation-finite algebras. By the definition, an algebra A is *representation-finite* (or *of finite representation type*) if it allows only finitely many isomorphism classes of finite dimensional indecomposable A-modules. By famous *1st Brauer-Thrall Conjecture*, which is now a theorem by Roiter [**49**] and Auslander [**1**], A is representation-finite if and only if there is a common bound m_A for the dimensions of indecomposable A-modules. Moreover, if A is not representation-finite (remember that K is algebraically closed), then there exist infinitely many[1] numbers $m \in \mathbb{N}$ such that for each of those m there is infinitely many pairwise non-isomorphic A-modules of K-dimension m. This is the content of the *2nd Brauer-Thrall Conjecture* confirmed by several authors, see [**4**],[**8**], [**12**], [**21**], (also [**38**]). The reader is referred to [**46**] for more on Brauer-Thrall conjectures. Validity of the 2nd Brauer-Thrall conjecture allows to prove that given a natural number d, the class of d -dimensional representation-finite algebras over algebraically closed fields is finitely axiomatizable in a suitable first order language, see [**27**], [**28**], [**29**, Theorem 12.54].

This corresponds to the fact that, given a natural number d, there is only finitely many isomorphism classes of representation-finite algebras having the K-dimension equal d. This is, in turn, a consequence of another deep and difficult result obtained in the 80's of the 20th century: the proof of the existence of a multiplicative basis of any representation-finite algebra over an algebraically closed field [**5**].

Note that arguments of Jensen and Lenzing for the finite axiomatizability are not constructive. However, this can be made more explicit by applying a "numerical version" of 2nd Brauer-Thrall Conjecture, see [**23**, 2.4]. Finally, it is possible to verify in a finite number of steps if given an algebra A is representation-finite or not. In practice, for this purpose we use Bongartz's criterion for representation-finiteness, which shall be discussed later, Auslander-Reiten theory, Galois covering techniques etc. These methods allow, in the representation-finite case, to give a list of representatives of the isomorphism classes of the indecomposable A-modules.

If A is not representation-finite, then there are infinitely many isomorphism classes of indecomposable modules of dimension m for infinitely many numbers m. It may happen that for every m, all but finitely many representatives of those classes can be arranged in finitely many "rational 1-parameter families". In this case we say that A is *of tame representation type* or simply *tame*[2]. We omit precise formulations of the well-known concepts appearing here, the reader is referred to a rich literature on the subject, see for example [**48**], [**51**], [**52**].

[1] By [**58**] it is enough to assume that there exists one such m.

[2] According to our convention a representation-finite algebra is tame.

Otherwise the problem of classification of two non-commuting square matrices up to simultaneous conjugations can be "embedded" into the problem of classifying the indecomposable A-modules, in which case we say that A is *of wild representation type* or *wild.*

This is essentially the content of the famous *Tame-Wild Dichotomy*:

THEOREM 1.2. **[20]**, **[15]** *A is either tame or wild.*

We cannot hope to obtain a complete classification of isomorphism classes of indecomposables in the wild case. Also in the tame case this task can be unreachable. The level of complication is reflected somehow by the stratification of the class of tame algebras into subclasses: domestic, polynomial growth. Let us recall the notions introduced in [**53**], see also [**54**]. See also [**51**, Section 14.4], [**52**, Chapter XIX].

Let A be a tame algebra. Given a natural number m let $\mu_A(m)$ be the lest number of 1-parameter families needed to parameterize (all but finite number) indecomposables of dimension m.

- If $\mu_A(m)$ is bounded, then A is *finite growth.* By the results of [**16**] an algebra is of finite growth if and only if it is *domestic* in the sense of [**47**].
- If $\mu_A(m) \leq m^N$ for some N and arbitrary m, then A is said to be of *polynomial growth.*

We have remarked that, given an algebra A, it is possible to verify wether A is representation-finite or not and it is related with finite axiomatizability of representation-finiteness. This fact is also connected with Gabriel's result that "finite representation type is open" [**22**], more precisely: representation-finite algebras of fixed dimension d induce a Zariski-open subset in the variety of d-dimensional associative algebras. Indeed, an analysis of the proof of Geiss's result on degenerations of wild algebras [**23**] shows that the openness of the varieties of representation-finite algebras can be derived from the existence of finite number of axioms of representation-finiteness.

Let us collect some known analogies for tameness. First of all, the class of tame algebras (of fixed dimension, over algebraically closed fields) is axiomatizable [**30**]. This is related with the observation that, if A is wild, then we can verify it in finite number of steps[3]. However we do not know if tame is finitely axiomatizable, that means that it is possible, that there is an algebra A such that we will never be sure if A is tame or not. If we knew that tame is finite axiomatizable, we would prove that "tame is open" [**30**], by the same kind of arguments that we mentioned above with respect to the openness of finite representation type.

Theorem 1.1 provides us with a criterion for tameness valid for strongly simply connected algebras. The condition equivalent to tameness is expressed in terms of Tits quadratic form.

[3]By applying, for example, matrix problems techniques: we classify canonical forms of indecomposable modules of dimension m, starting from $m = 1$ and increasing m until we meet 2-parameter family of pairwise non-isomorphic indecomposable m-dimensional modules.

2. Tits quadratic form

Let A be a basic finite dimensional K-algebra of the form KQ/I, where $Q = (Q_0, Q_1, s, t)$ is a finite acyclic quiver[4] and I is an admissible[5] ideal of the path algebra KQ. Recall that by a *relation* in the path algebra KQ we mean a linear combination of paths having the same sink and the same source. Let R be a minimal set of relations generating the ideal I. R decomposes into the disjoint union $\bigcup_{i,j\in Q_0} R_{i,j}$, where $R_{i,j}$ denotes the set of the elements starting at the vertex i and ending at j. Although R is not uniquely determined by I, the numbers $r_{i,j} := |R_{i,j}|$ do not depend on the set R [**6**].

The *Tits quadratic form* of the algebra A is the integer quadratic form

$$q_A : \mathbb{Z}^{Q_0} \to \mathbb{Z}$$

defined by the formula

$$q_A(z) = \sum_{i\in Q_0} z_i^2 - \sum_{i\to j\in Q_1} z_i z_j + \sum_{i,j\in Q_0} r_{i,j} z_i z_j \tag{2.1}$$

for $z = (z_i)_{i\in Q_0} \in \mathbb{Z}^{Q_0}$.

Let us also recall another quadratic form associated with an algebra: the *Euler quadratic form*. For this purpose denote by S_i the simple A -module corresponding to the vertex $i \in Q_0$ and assume that A has finite global dimension. The quadratic form

$$\chi_A : \mathbb{Z}^{Q_0} \to \mathbb{Z},$$

defined by the formula

$$\chi_A(z) = \sum_{s=0}^{\infty} \sum_{i,j\in Q_0} (-1)^s \dim \operatorname{Ext}_A^s(S_i, S_j) z_i z_j. \tag{2.2}$$

is called the *Euler quadratic form* of the algebra A.

The Euler form is often called a homological form, by obvious reasons, whereas the Tits form - the geometric quadratic form of an algebra. This is because the summands of the right hand side of (2.1) have interpretations as (bounds of) dimensions of some geometric objects associated with A and z. Thanks to this interpretation one can prove that:

(1) If A is representation-finite, then q_A is weakly positive, that is, $q_A(z) > 0$ for all non-zero vectors z with all coordinates nonnegative [**6**].
(2) If A is tame, then q_A is weakly nonnegative, that is, $q_A(z) \geq 0$ for all vectors z with all coordinates nonnegative [**41**].

The assertion (1) is the easier part of the Bongartz's criterion for the representation-finiteness, we shall discuss the converse later. The converse of (2) (valid under suitable conditions on A) is the content of Theorem 1.1. Let us recall the main steps of the proof of (2), following de la Peña.

It is well-known that the A-modules of dimension vector $z = (z_i)_{i\in Q_0}$ can be represented by the tuples $(M_\alpha)_{\alpha\in Q_1} \in \prod \mathbb{M}_{z_{t(\alpha)}\times z_{s(\alpha)}}(K)$ of matrices satisfying the

[4]As usual, Q_0 is the set of vertices, Q_1 is the set of arrow and $s(\alpha)$ and $t(\alpha)$ denote the source and the target of the arrow α.

[5]That means $(KQ_1)^m \subseteq I \subseteq (KQ_1)^2$ for some m.

relations defining the ideal I. These tuples form the *variety of modules* of dimension vector z denoted by $\mathrm{mod}_A(z)$ and, by Krull Theorem, it is easy to observe that

$$\dim \mathrm{mod}_A(z) \geq \sum_{i \to j \in Q_1} z_i z_j - \sum_{i,j \in Q_0} r_{i,j} z_i z_j. \tag{2.3}$$

The isomorphism classes of modules of dimension vector $z = (z_i)_{i \in Q_0}$ can be identified with the orbits of the natural algebraic group action

$$Gl(z) \times \mathrm{mod}_A(z) \to \mathrm{mod}_A(z)$$

of the group $Gl(z) = \prod_{i \in Q_0} Gl_{z_i}(K)$. Clearly

$$\dim Gl(z) = \sum_{i \in Q_0} z_i^2. \tag{2.4}$$

Now, by (2.1), (2.3) and (2.4),

$$q_A(z) \geq \dim Gl(z) - \dim \mathrm{mod}_A(z). \tag{2.5}$$

Now it is time to apply de la Peña's argument: if A is tame, representatives of the indecomposables of fixed dimension can be arranged into finite number of "one-parameter families". Therefore, there is a "$|z|$ -parameter family" of representatives of the modules of dimension vector z, where $|z| = \sum_{i \in Q_0} z_i$. It follows that if A is tame, then

$$\dim Gl(z) + \sum_{i \in Q_0} z_i \geq \dim \mathrm{mod}_A(z). \tag{2.6}$$

Combining (2.5) and (2.6) we obtain

$$q_A(z) \geq \dim Gl(z) - \dim \mathrm{mod}_A(z) \geq - \sum_{i \in Q_0} z_i.$$

As the left hand side of the above inequality depends quadratically on z, whereas the right hand side - linearly, we conclude that $q_A(z) \geq 0$ for every z with nonnegative coordinates and the proof is complete.

As it is mentioned above, the converses of (1) and (2) are not true in general. Let us recall the well-known Bongartz's example showing this.

EXAMPLE 2.1. Let Q be the quiver

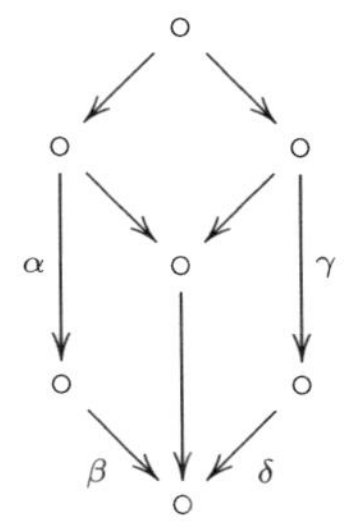

Assume that I_1 is the ideal generated by all commutativity relations, and I_2 the ideal is generated by the commutativity of the upper square and two zero-relations $\alpha\beta$ and $\gamma\delta$. Then $A_1 = KQ/I_1$ is representation-finite, $A_2 = KQ/I_2$ is wild, and the Tits quadratic forms of A_1 and A_2 coincide and are weakly positive.

A lot of effort has been put into proving the converses of (1) and (2) under various assumptions. The most satisfactory results in this direction are the Bongartz's criterion for representation-finiteness and Theorem 1.1 for tameness. The results can be considered as a culmination of a research programme scheduled by Sheila Brenner, who wrote in 1975 about her study on connections between the representation type and definiteness of certain quadratic forms [**10**]: *"This paper is written in the spirit of experimental science. It reports some observed regularities and suggests that there should be a theory to explain them."*

3. Simply connected and strongly simply connected algebras

Let $A = KQ/I$ be as before, assume that the quiver Q is connected and let $m(I)$ be a set of minimal relations generating I. A relation $\sum_{i\in I} \lambda_i u_i \in I$, where $\lambda_i \in K$ and u_i are paths in Q, is *minimal* if $\sum_{i\in J} \lambda_i u_i \notin I$ for any proper subset $J \subseteq I$. We look at Q as a 1-dimensional complex and, whenever two paths u, v are involved into a minimal relation, we attach a 2-cell along the loop uv^{-1}:

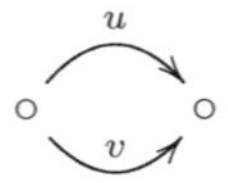

The fundamental group of the obtained complex is called the fundamental group of the bound quiver (Q, I) and it is denoted by $\pi_1(Q, I)$ [**36**]. Let us remark that $\pi_1(Q, I)$ depends not only on KQ/I, it may depend on the particular choice of I.

DEFINITION 3.1. [**3**]. The algebra A is *simply connected* if $\pi_1(Q, I)$ is trivial for every presentation $A \cong KQ/I$.

In the Example 2.1 we have $\pi_1(Q, I_1)$ trivial, whereas $\pi_1(Q, I_2) \cong \mathbb{Z} * \mathbb{Z}$ is a free non-commutative group with 2 free generators.

Combining the results of [**6**], [**7**], [**5**] and [**11**] we derive

THEOREM 3.2. *If A is* **simply connected**, *then A is representation-finite if and only if q_A is weakly positive.*

There is a similar result for algebras having simply connected Galois covering. Results on preservation of (locally) representation-finiteness by Galois coverings, [**9**], are important ingredients in the proof of the general version.

The last sentence of the paper [**6**] is *"There should be a similar result for tame representation type."*

Of course, it is expected that weak nonnegativity of the Tits form implies tameness under suitable assumptions. Actually, there are several results of that kind. Such implication has been proved for instance for:

- hereditary algebras [**37**],
- tilted algebras [**35**],
- quasitilted algebras [**57**].

Observe that it is not enough to assume that A is simply connected, as the following example shows:

EXAMPLE 3.3. Let Q be the quiver

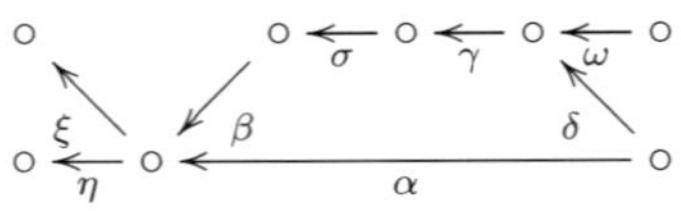

and let I be the ideal generated by the relations:

$$\alpha\xi, \omega\gamma\sigma\beta, \alpha\eta - \delta\gamma\sigma\beta\eta$$

Then $A = KQ/I$ is simply connected, the Tits quadratic form q_A is weakly nonnegative, but A is wild [**13**, Example 1.7].

Thus a strengthening of the concept of simply connectedness is needed.

DEFINITION 3.4. [**55**] An algebra A is **strongly simply connected** if every convex subcategory C of A is simply connected.

It is proved in [**55**] that the condition in the definition above is equivalent to the vanishing of the first Hochschild cohomology groups $H^1(C)$ (with coefficients in the bimodule ${}_CC_C$) for every convex subcategory C of A. By results of Bretscher - Gabriel [**11**] and Martínez - de la Peña [**36**] a triangular representation-finite algebra is simply connected if and only if it is strongly simply connected.

4. Strongly simply connected of polynomial growth

The proof of Theorem 1.1 depends essentially on some older result on polynomial growth algebras.

THEOREM 4.1. *Let A be a strongly simply connected algebra. The following statements are equivalent:*

(1) *A is tame of polynomial growth.*
(2) *The Tits form q_A of A is weakly nonnegative and A does not contain a convex subcategory, which is pg-critical.*

The list of *pg*-critical algebras is given in [**39**].

Representation theoretical properties of tame polynomial growth algebras differ substantially from those of the remaining tame algebras. In particular, we observe characteristic behavior of the Tits and Euler quadratic forms in case of polynomial growth strongly simply connected algebras. Let us recall two theorems:

THEOREM 4.2. [**42**] *Let A be a strongly simply connected algebra. The following conditions are equivalent:*

(1) *A is tame of polynomial growth.*
(2) *For every indecomposable module X of dimension vector z*

$$\chi_A(z) = \dim G(z) - \dim \operatorname{mod}_A(z) \geq 0.$$

In this case every indecomposable module is a smooth point of the module scheme[6].

THEOREM 4.3. [**43**] *Let A be a strongly simply connected algebra and X an indecomposable A-module of dimension vector z. Then:*

(1) $0 \leq q_A(z) \leq 18|Q_0|$.
(2) $0 \leq \chi_A(z) \leq 2 + \#\{$*projective-injective indecomposable A -modules*$\}$.
(3) *If X is a faithful module, then $q_A(z) \leq 2$ and $\chi_A(z) \leq 2$.*

REMARK 4.4. There are also interesting connections between the growth of a tame algebra and certain phenomena on the level of infinite-dimensional modules.

Let us recall that a module is *superdecomposable* if it has no indecomposable direct summands. Here we assume that the field K is countable.

(1) A wild algebra possesses a superdecomposable pure-injective module [**25**].

[6]Let us remark, that here we mean the module scheme, see [**42**] for the details, not the module variety $\operatorname{mod}_A(z)$.

(2) A non-domestic string algebra possesses a superdecomposable pure-injective module [**45**].
(3) An algebra admitting a strongly simply connected Galois covering of non-polynomial growth, in particular, a strongly simply connected algebra of non-polynomial growth, possesses a superdecomposable pure-injective module, under the additional assumption that $char(K) \neq 2$ [**32**], [**33**].
(4) Tubular algebras possess superdecomposable pure-injective modules [**26**].
(5) A strongly simply connected algebra is domestic if and only if does not possess a superdecomposable pure-injective module, under the additional assumption that $char(K) \neq 2$ [**40**].

5. On the proof of the main theorem

We know that the Tits quadratic form of a tame algebra is weakly nonnegative, so we concentrate on the converse implication for strongly simply connected algebras. Very briefly the proof can be sketched as follows. By Theorem 4.1 a strongly simply connected algebra A is tame of polynomial growth if and only if the Tits quadratic form q_A is weakly nonnegative and A does not contain a *pg*-critical convex subcategory. Thus we can restrict our attention to $A = KQ/I$ with q_A weakly nonnegative and containing a *pg*-critical convex subcategory. Without loss of generality we can assume that additionally A has an indecomposable module whose support contains all sources and all sinks of Q . The structure of such algebras can be described quite precisely; they are so called $\mathbb{D}$-algebras in the terminology of [**13**]. Moreover, to A there is associated a *mild and smooth* algebra A^*, which is tame if and only if A is tame. The reader is again referred to [**13**] for the details. So we need to prove that A^* is tame. This final step is done by Geiss's degeneration theorem [**23**], since A^* degenerates to a special biserial algebra, which is tame by [**59**], [**18**], [**14**].

Let us show on an example how to perform degenerations mentioned above. There are two basic, rather known, tricks. First is to degenerate a commutative square to a square with a zero relation:

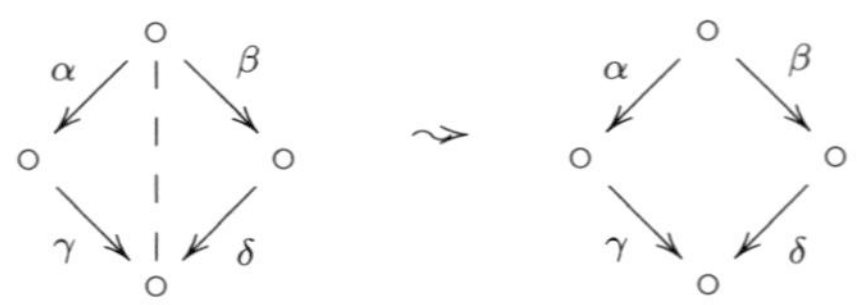

$\alpha\gamma = \beta\delta$ $\qquad\qquad$ $\alpha\gamma = 0$

The degeneration is performed by the following "passage to the limit": $\alpha\gamma - t\beta\delta \to \alpha\gamma$ as $t \to 0$.

Another trick is the following:

$\alpha\gamma = \beta\delta$ $\qquad\qquad$ $\alpha\gamma = 0, \epsilon^2 = 0$

This is based on the observation that the path algebra of the two point quiver without arrows (∘ ∘) is isomorphic to $K \times K \cong K[X]/(X(X-t))$ for $t \neq 0$ and $K[X]/(X(X-t)) \to K(\,\epsilon \circlearrowleft \circ\,)/\epsilon^2$ as $t \to 0$.

Combining this tricks we can degenerate the pg-critical algebra defined by the bound quiver

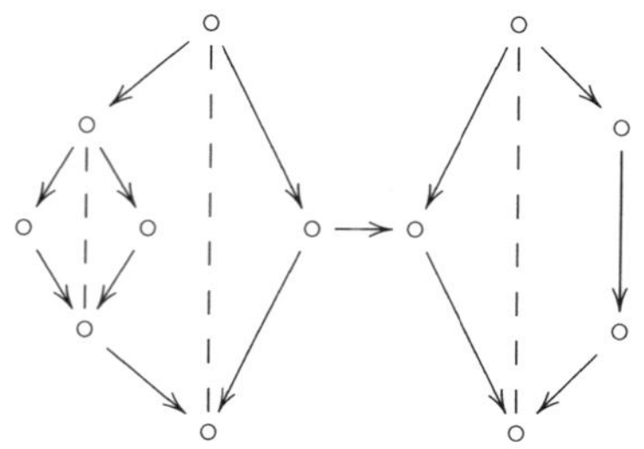

to the special biserial algebra defined by the quiver

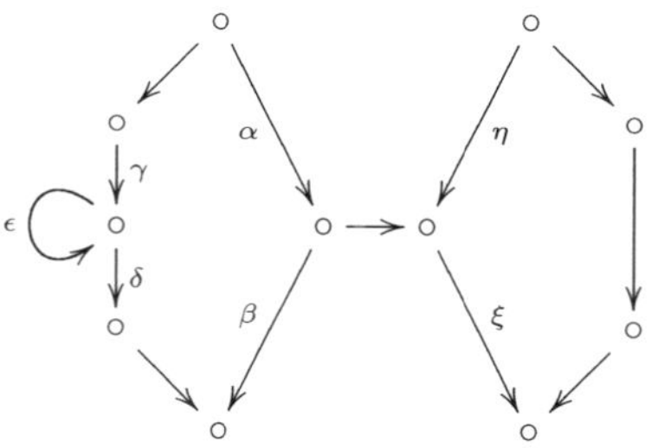

with zero-relations $\alpha\beta = \gamma\delta = \eta\xi = \epsilon^2 = 0$.

6. Some consequences and comments

The main theorem yields a possibility of checking in a finite number of steps wether given a strongly simply connected algebra is tame or not. It is not difficult to observe that also it is possible to check algorithmically wether $A = KQ/I$ is strongly simply connected - here the Hochschild cohomology approach is useful. Thanks to this we obtain

THEOREM 6.1. [**31**] *Given a number d, the class of d-dimensional tame strongly simply connected algebras over algebraically closed fields is finitely axiomatizable.*

A further consequence is

THEOREM 6.2. [**31**] *Given a number d and an algebraically closed field K, the class of d-dimensional tame strongly simply connected K-algebras induces a Zariski-open subset in the variety of the associative d -dimensional K-algebras.*

Let us finish with presenting some general idea of connecting certain properties which seem to be of different origin at the first sight. In the above presentation we tried to highlight connections between some general properties. Namely, we know that:

(1) Finite representation type is open [**22**].
(2) Finite representation type is finitely axiomatizable [**28**].

(3) There is a criterion of representation-finiteness (expressed in terms of the Tits quadratic form) valid for simply connected algebras.
(4) Galois coverings behave nice with respect to representation-finiteness.

Some time ago Jose Antonio de la Peña asked what remains of that in the class of tame algebras? Let us write down that:

(1) We can formulate the conjecture that tame is open.
(2) This conjecture is equivalent to the conjecture that tame is finitely axiomatizable [**30**].
(3) We have quadratic form criterion for tameness of strongly simply connected algebras - Theorem 1.1.
(4) We can formulate the "Galois Covering Preserve Tameness (GCPT) Conjecture": Assume that a torsion-free group G acts freely on the objects of a locally bounded category $\widetilde{A}$. Then $A = \widetilde{A}/G$ is tame if and only if $\widetilde{A}$ is tame.

Let us remark that the "only if" part of GCPT is known by [**17**]. The "if" has been announced by Drozd and Ovsienko. But we are still - excluding Theorem 1.1 - on a conjectural level. However there is an unexpected connection between some of the conjectures.

It is known that we can express the theory of Galois coverings in the language of graded algebras [**24**]. This language is more suitable to our purposes.

Let us consider graded algebras over fields of characteristic p (0 or a prime). We have the following

THEOREM 6.3. [**34**] *If the tame graded algebras of fixed dimension form a finitely axiomatizable class, then GCPT holds for p'-residually finite group G.*

We say that a group G is *p'-residually finite* if for any $a \in G$, $a \neq e$, there exists $H \lhd G$ of finite index not divisible by p such that $a \notin H$.

References

[1] Maurice Auslander, *Applications of morphisms determined by modules*, Representation theory of algebras (Proc. Conf., Temple Univ., Philadelphia, Pa., 1976), Dekker, New York, 1978, pp. 245–327. Lecture Notes in Pure Appl. Math., Vol. 37. MR0491822

[2] I. Assem, D. Simson and A. Skowroński, *Elements of the Representation Theory of Associative Algebras. Vol. 1: Techniques of Representation Theory.* London Mathematical Society Student Texts, Vol. 65. Cambridge University Press, Cambridge, 2006.

[3] Ibrahim Assem and Andrzej Skowroński, *On some classes of simply connected algebras*, Proc. London Math. Soc. (3) **56** (1988), no. 3, 417–450, DOI 10.1112/plms/s3-56.3.417. MR931509

[4] Raymundo Bautista, *On algebras of strongly unbounded representation type*, Comment. Math. Helv. **60** (1985), no. 3, 392–399, DOI 10.1007/BF02567422. MR814146

[5] R. Bautista, P. Gabriel, A. V. Roĭter, and L. Salmerón, *Representation-finite algebras and multiplicative bases*, Invent. Math. **81** (1985), no. 2, 217–285, DOI 10.1007/BF01389052. MR799266

[6] Klaus Bongartz, *Algebras and quadratic forms*, J. London Math. Soc. (2) **28** (1983), no. 3, 461–469, DOI 10.1112/jlms/s2-28.3.461. MR724715

[7] Klaus Bongartz, *A criterion for finite representation type*, Math. Ann. **269** (1984), no. 1, 1–12, DOI 10.1007/BF01455993. MR756773

[8] Klaus Bongartz, *Indecomposables are standard*, Comment. Math. Helv. **60** (1985), no. 3, 400–410, DOI 10.1007/BF02567423. MR814147

[9] K. Bongartz and P. Gabriel, *Covering spaces in representation-theory*, Invent. Math. **65** (1981/82), no. 3, 331–378, DOI 10.1007/BF01396624. MR643558

[10] Sheila Brenner, *Quivers with commutativity conditions and some phenomenology of forms*, Proceedings of the International Conference on Representations of Algebras (Carleton Univ., Ottawa, Ont., 1974), Carleton Univ., Ottawa, Ont., 1974, pp. 25 pp. Carleton Math. Lecture Notes, No. 9. MR0369438

[11] O. Bretscher and P. Gabriel, *The standard form of a representation-finite algebra* (English, with French summary), Bull. Soc. Math. France **111** (1983), no. 1, 21–40. MR710374
€"54, Lecture Notes in Math., Vol. 1177, Springer, Berlin, 1986.

[12] Otto Bretscher and Gordana Todorov, *On a theorem of Nazarova and Roĭter*, Representation theory, I (Ottawa, Ont., 1984), Lecture Notes in Math., vol. 1177, Springer, Berlin, 1986, pp. 50–54, DOI 10.1007/BFb0075257. MR842458

[13] Thomas Brüstle, José Antonio de la Peña, and Andrzej Skowroński, *Tame algebras and Tits quadratic forms*, Adv. Math. **226** (2011), no. 1, 887–951, DOI 10.1016/j.aim.2010.07.007. MR2735778

[14] M. C. R. Butler and Claus Michael Ringel, *Auslander-Reiten sequences with few middle terms and applications to string algebras*, Comm. Algebra **15** (1987), no. 1-2, 145–179, DOI 10.1080/00927878708823416. MR876976

[15] W. W. Crawley-Boevey, *On tame algebras and bocses*, Proc. London Math. Soc. (3) **56** (1988), no. 3, 451–483, DOI 10.1112/plms/s3-56.3.451. MR931510

[16] William Crawley-Boevey, *Tame algebras and generic modules*, Proc. London Math. Soc. (3) **63** (1991), no. 2, 241–265, DOI 10.1112/plms/s3-63.2.241. MR1114510
€"529.

[17] Piotr Dowbor and Andrzej Skowroński, *On Galois coverings of tame algebras*, Arch. Math. (Basel) **44** (1985), no. 6, 522–529, DOI 10.1007/BF01193992. MR797444

[18] Piotr Dowbor and Andrzej Skowroński, *Galois coverings of representation-infinite algebras*, Comment. Math. Helv. **62** (1987), no. 2, 311–337, DOI 10.1007/BF02564450. MR896100

[19] Ju. A. Drozd, *Matrix problems, and categories of matrices* (Russian), Zap. Naučn. Sem. Leningrad. Otdel. Mat. Inst. Steklov. (LOMI) **28** (1972), 144–153. Investigations on the theory of representations. MR0340282

[20] Ju. A. Drozd, *Tame and wild matrix problems* (Russian), Representations and quadratic forms (Russian), Akad. Nauk Ukrain. SSR, Inst. Mat., Kiev, 1979, pp. 39–74, 154. MR600111

[21] Urs Fischbacher, *Une nouvelle preuve d'un théorème de Nazarova et Roiter* (French, with English summary), C. R. Acad. Sci. Paris Sér. I Math. **300** (1985), no. 9, 259–262. MR785064

[22] Peter Gabriel, *Finite representation type is open*, Proceedings of the International Conference on Representations of Algebras (Carleton Univ., Ottawa, Ont., 1974), Carleton Univ., Ottawa, Ont., 1974, pp. 23 pp. Carleton Math. Lecture Notes, No. 9. MR0376769

[23] Christof Geiss, *On degenerations of tame and wild algebras*, Arch. Math. (Basel) **64** (1995), no. 1, 11–16, DOI 10.1007/BF01193544. MR1305654
€"310.

[24] Edward L. Green, *Graphs with relations, coverings and group-graded algebras*, Trans. Amer. Math. Soc. **279** (1983), no. 1, 297–310, DOI 10.2307/1999386. MR704617

[25] Lorna Gregory and Mike Prest, *Representation embeddings, interpretation functors and controlled wild algebras*, J. Lond. Math. Soc. (2) **94** (2016), no. 3, 747–766, DOI 10.1112/jlms/jdw055. MR3614927

[26] Richard Harland and Mike Prest, *Modules with irrational slope over tubular algebras*, Proc. Lond. Math. Soc. (3) **110** (2015), no. 3, 695–720, DOI 10.1112/plms/pdu068. MR3342102

[27] Christian Herrmann, Christian U. Jensen, and Helmut Lenzing, *Applications of model theory to representations of finite-dimensional algebras*, Math. Z. **178** (1981), no. 1, 83–98, DOI 10.1007/BF01218373. MR627096

[28] Christian U. Jensen and Helmut Lenzing, *Homological dimension and representation type of algebras under base field extension*, Manuscripta Math. **39** (1982), no. 1, 1–13, DOI 10.1007/BF01312441. MR672397

[29] Christian U. Jensen and Helmut Lenzing, *Model-theoretic algebra with particular emphasis on fields, rings, modules*, Algebra, Logic and Applications, vol. 2, Gordon and Breach Science Publishers, New York, 1989. MR1057608

[30] Stanisław Kasjan, *On the problem of axiomatization of tame representation type*, Fund. Math. **171** (2002), no. 1, 53–67, DOI 10.4064/fm171-1-3. MR1879883

[31] Stanisław Kasjan, *Tame strongly simply connected algebras form an open scheme*, J. Pure Appl. Algebra **208** (2007), no. 2, 435–443, DOI 10.1016/j.jpaa.2006.01.010. MR2277685

[32] Stanisław Kasjan and Grzegorz Pastuszak, *On the existence of super-decomposable pure-injective modules over strongly simply connected algebras of non-polynomial growth*, Colloq. Math. **136** (2014), no. 2, 179–220, DOI 10.4064/cm136-2-3. MR3257564

[33] Stanisław Kasjan and Grzegorz Pastuszak, *Super-decomposable pure-injective modules over algebras with strongly simply connected Galois coverings*, J. Pure Appl. Algebra **220** (2016), no. 8, 2985–2999, DOI 10.1016/j.jpaa.2016.01.014. MR3471201

[34] Stanisław Kasjan and José Antonio de la Peña, *Galois coverings and the problem of axiomatization of the representation type of algebras*, Extracta Math. **20** (2005), no. 2, 137–150. MR2195198

[35] Otto Kerner, *Tilting wild algebras*, J. London Math. Soc. (2) **39** (1989), no. 1, 29–47, DOI 10.1112/jlms/s2-39.1.29. MR989917

[36] R. Martínez-Villa and J. A. de la Peña, *The universal cover of a quiver with relations*, J. Pure Appl. Algebra **30** (1983), no. 3, 277–292, DOI 10.1016/0022-4049(83)90062-2. MR724038

[37] L. A. Nazarova, *Representations of quivers of infinite type* (Russian), Izv. Akad. Nauk SSSR Ser. Mat. **37** (1973), 752–791. MR0338018

[38] L. A. Nazarova and A. V Roiter, *Categorial matrix problems, and the Brauer-Thrall problem.* (Russian) Izdat. "Naukova Dumka", Kiev, 1973. 100 pp., German translation Mitt. Math. Sem. Giessen Heft 115 (1975).

[39] Rainer Nörenberg and Andrzej Skowroński, *Tame minimal non-polynomial growth simply connected algebras*, Colloq. Math. **73** (1997), no. 2, 301–330, DOI 10.4064/cm-73-2-301-330. MR1446121

[40] Grzegorz Pastuszak, *Strongly simply connected algebras with super-decomposable pure-injective modules*, J. Pure Appl. Algebra **219** (2015), no. 8, 3314–3321, DOI 10.1016/j.jpaa.2014.10.016. MR3320221

[41] J. A. de la Peña, *On the dimension of the module-varieties of tame and wild algebras*, Comm. Algebra **19** (1991), no. 6, 1795–1807, DOI 10.1080/00927879108824229. MR1113958

[42] J. A. de la Peña and A. Skowroński, *Geometric and homological characterizations of polynomial growth strongly simply connected algebras*, Invent. Math. **126** (1996), no. 2, 287–296, DOI 10.1007/s002220050100. MR1411134

[43] J. A. de la Peña and A. Skowroński, *The Tits and Euler forms of a tame algebra*, Math. Ann. **315** (1999), no. 1, 37–59, DOI 10.1007/s002080050317. MR1717543

[44] José Antonio de la Peña and Andrzej Skowroński, *The Tits forms of tame algebras and their roots*, Representations of algebras and related topics, EMS Ser. Congr. Rep., Eur. Math. Soc., Zürich, 2011, pp. 445–499, DOI 10.4171/101-1/10. MR2931903

[45] Gena Puninski, *Superdecomposable pure-injective modules exist over some string algebras*, Proc. Amer. Math. Soc. **132** (2004), no. 7, 1891–1898, DOI 10.1090/S0002-9939-03-07292-7. MR2053958

[46] Claus Michael Ringel, *On algorithms for solving vector space problems. I. Report on the Brauer-Thrall conjectures: Rojter's theorem and the theorem of Nazarova and Rojter*, Representation theory, I (Proc. Workshop, Carleton Univ., Ottawa, Ont., 1979), Lecture Notes in Math., vol. 831, Springer, Berlin, 1980, pp. 104–136. MR607142

[47] Claus Michael Ringel, *On algorithms for solving vector space problems. II. Tame algebras*, Representation theory, I (Proc. Workshop, Carleton Univ., Ottawa, Ont., 1979), Lecture Notes in Math., vol. 831, Springer, Berlin, 1980, pp. 137–287. MR607143

[48] Claus Michael Ringel, *Tame algebras and integral quadratic forms*, Lecture Notes in Mathematics, vol. 1099, Springer-Verlag, Berlin, 1984. MR774589

[49] A. V. Roĭter, *Unboundedness of the dimensions of the indecomposable representations of an algebra which has infinitely many indecomposable representations* (Russian), Izv. Akad. Nauk SSSR Ser. Mat. **32** (1968), 1275–1282. MR0238893

[50] A. V. Rojter, *Matrix problems and representations of BOCSs*, Representation theory, I (Proc. Workshop, Carleton Univ., Ottawa, Ont., 1979), Lecture Notes in Math., vol. 831, Springer, Berlin-New York, 1980, pp. 288–324. MR607144

[51] Daniel Simson, *Linear representations of partially ordered sets and vector space categories*, Algebra, Logic and Applications, vol. 4, Gordon and Breach Science Publishers, Montreux, 1992. MR1241646

[52] Daniel Simson and Andrzej Skowroński, *Elements of the representation theory of associative algebras. Vol. 3*, London Mathematical Society Student Texts, vol. 72, Cambridge University Press, Cambridge, 2007. Representation-infinite tilted algebras. MR2382332

[53] Andrzej Skowroński, *Group algebras of polynomial growth*, Manuscripta Math. **59** (1987), no. 4, 499–516, DOI 10.1007/BF01170851. MR916000

[54] Andrzej Skowroński, *Algebras of polynomial growth*, Topics in algebra, Part 1 (Warsaw, 1988), Banach Center Publ., vol. 26, PWN, Warsaw, 1990, pp. 535–568. MR1171252

[55] Andrzej Skowroński, *Simply connected algebras and Hochschild cohomologies [MR1206961 (94e:16016)]*, Representations of algebras (Ottawa, ON, 1992), CMS Conf. Proc., vol. 14, Amer. Math. Soc., Providence, RI, 1993, pp. 431–447. MR1265301

[56] Andrzej Skowroński, *Simply connected algebras of polynomial growth*, Compositio Math. **109** (1997), no. 1, 99–133, DOI 10.1023/A:1000245728528. MR1473607

[57] Andrzej Skowroński, *Tame quasi-tilted algebras*, J. Algebra **203** (1998), no. 2, 470–490, DOI 10.1006/jabr.1997.7328. MR1622799

[58] Sverre O. Smalø, *The inductive step of the second Brauer-Thrall conjecture*, Canadian J. Math. **32** (1980), no. 2, 342–349, DOI 10.4153/CJM-1980-026-0. MR571928

[59] Burkhard Wald and Josef Waschbüsch, *Tame biserial algebras*, J. Algebra **95** (1985), no. 2, 480–500, DOI 10.1016/0021-8693(85)90119-X. MR801283

Faculty of Mathematics and Computer Science, Nicolaus Copernicus University, Toruń, Poland

Email address: skasjan@mat.umk.pl

Contemporary Mathematics
Volume **761**, 2021
https://doi.org/10.1090/conm/761/15312

A remark on Hochschild cohomology and Koszul duality

Bernhard Keller

Dedicated to the José Antonio de la Peña on the occasion of his sixtieth birthday

ABSTRACT. Applying recent results by Lowen–Van den Bergh we show that Hochschild cohomology is preserved under Koszul–Moore duality as a Gerstenhaber algebra. More precisely, the corresponding Hochschild complexes are linked by a quasi-isomorphism of B_∞-algebras.

1. Introduction

Consider the following statement:

Hochschild cohomology is preserved under Koszul duality.

This is clearly wrong. Indeed, if V is a non zero finite-dimensional vector space over a field k, then the center (=zeroth Hochschild cohomology) of the symmetric algebra SV is SV but the center of the Koszul dual exterior algebra $\Lambda(V^*)$ is finite-dimensional (for a study of the Hochschild cohomology of $\Lambda(V^*)$, cf. [**13**]). To try and save the statement, let us recall that $\Lambda(V^*)$ is in fact the Yoneda algebra

$$\mathsf{Ext}^*_{SV}(k,k)$$

and that, with the zero differential, it is even quasi-isomorphic to the derived endomorphism algebra $\mathsf{RHom}_{SV}(k,k)$. Thus, it is natural to endow $\Lambda(V^*)$ with the grading so that V^* sits in degree 1 and with the zero differential. With this new interpretation of $\Lambda(V^*)$ as a dg (=differential graded) algebra, we find that its zeroth Hochschild cohomology is the *completion* $\widehat{SV}$ of the symmetric algebra at the augmentation ideal. It turns out that in order to get exactly SV, it suffices to replace the dg algebra $\Lambda(V^*)$ with the k-dual dg *coalgebra* ΛV. In fact, we then get an algebra isomorphism

$$HH^*(SV,SV) \xrightarrow{\sim} HH^*(\Lambda V, \Lambda V).$$

Our main results are that this isomorphism generalizes from SV to any augmented dg k-algebra (when we replace ΛV with the Koszul–Moore dual dg coalgebra) and that it lifts to an isomorphism between the corresponding Hochschild cochain complexes in the homotopy category of B_∞-algebras. In particular, the isomorphism in Hochschild cohomology is an isomorphism of Gerstenhaber algebras. The algebra isomorphism for augmented dg algebras follows from the setup of Koszul–Moore

2010 *Mathematics Subject Classification.* Primary 16E40, 18E30.
Key words and phrases. Hochschild cohomology, Koszul duality, Gerstenhaber algebra, B_∞-algebra.

duality, which we recall in section 2. Our first proof of the lift to the B_∞-level is now superseded by recent work of Lowen–Van den Bergh [**9**], which we use in the very short proof of Theorem 3.3.

Previous results relating the Hochschild cohomologies of Koszul(-Moore) dual algebras can be found in [**3**], [**6**] [**4**], [**5**], [**2**] and [**1**] cf. remark 3.4.

2. Reminder on Koszul–Moore duality

We follow Lefèvre-Hasegawa [**8**] and Positselski [**11**], cf. also [**7**] and Appendix A to [**12**]. Let k be a field and A a dg k-algebra. Thus, A is a $\mathbb{Z}$-graded associative algebra with 1

$$A = \bigoplus_{p\in\mathbb{Z}} A^p$$

endowed with a homogeneous linear endomorphism d of degree 1, the differential, such that $d^2 = 0$ and we have the Leibniz rule

$$d(ab) = (da)b + (-1)^p a(db)$$

for all $a \in A^p$ and all $b \in A$. Let $\varepsilon : A \to k$ be an augmentation (a morphism of dg algebras). For example, if V is a vector space (concentrated in degree 0), we can consider $A = SV$ (concentrated in degree 0 with $d = 0$). Dually, let C be a dg coalgebra and $\eta : k \to C$ a co-augmentation. For example, we may consider $C = \Lambda V$, where V is concentrated in degree -1 and $d = 0$. Denote by $\mathsf{Hom}_k(C, A)$ the graded vector space whose nth component is formed by the homogeneous k-linear maps $f : C \to A$ of degree n. We make $\mathsf{Hom}_k(C, A)$ into a differential graded algebra by setting

$$d(f) = d \circ f - (-1)^n f \circ d$$

for f homogeneous of degree n and

$$f * g = \mu \circ (f \otimes g) \circ \Delta$$

for homogeneous f and g, where μ is the multiplication of A and Δ the comultiplication of C. A *twisting cochain* is an element $\tau \in \mathsf{Hom}_k(C, A)$ of degree 1 such that

$$\varepsilon \circ \tau = 0 = \tau \circ \eta \text{ and } d(\tau) + \tau * \tau = 0.$$

For example, with the above notations, the composition of the natural projection and inclusion morphisms

$$\Lambda V \to V \to SV$$

is a twisting cochain. We denote by $\mathsf{Tw}(C, A)$ the set of twisting cochains. From now on, we assume that C is *cocomplete*, i.e. that $\overline{C} = \mathsf{cok}(\eta)$ is the union of the kernels of the maps induced by the iterated comultiplications

$$\Delta^{(n)} : C \to C^{\otimes n}, n \geqslant 2.$$

We denote by Alg the category of augmented dg algebras and by Coalg the category of cocomplete co-augmented dg coalgebras.

PROPOSITION 2.1. a) *The functor* $\mathsf{Tw}(?, A) : \mathsf{Coalg}^{op} \to \mathsf{Set}$ *is representable, i.e. there is an object* $BA \in \mathsf{Coalg}$ *and a functorial bijection*

$$\mathsf{Tw}(C', A) \xrightarrow{\sim} \mathsf{Coalg}(C', BA).$$

b) *The functor* $\mathsf{Tw}(C, ?) : \mathsf{Alg} \to \mathsf{Set}$ *is co-representable, i.e. there is an object* $\Omega A \in \mathsf{Alg}$ *and a functorial bijection*
$$\mathsf{Tw}(C, A') \xrightarrow{\sim} \mathsf{Alg}(\Omega C, A').$$

The dg coalgebra BA is known as the *bar construction* and the dg algebra ΩC as the *cobar construction*. It is not hard to describe BA and ΩC explicitly but we will not need this. Notice that if we denote by DBA the k-dual dg algebra of BA, then we have a canonical isomorphism
$$DBA \xrightarrow{\sim} \mathsf{RHom}_A(k, k)$$
and that the latter is quasi-isomorphic to the Koszul dual $A^!$ (with the generators in differential degree 1) if A is a Koszul algebra concentrated in degree 0. We denote the category of dg right A-modules by $\mathsf{Mod}\, A$. Its localization with respect to all quasi-isomorphisms is the *derived category* $\mathcal{D}A$. A (right) dg comodule M is *cocomplete* if it is the union of the kernels of the maps
$$M \to M \otimes \overline{C}^{n-1}, n \geqslant 2,$$
induced by the iterated comultiplications. We denote by $\mathsf{Com}\, C$ the category of cocomplete right dg C-comodules. It becomes a Frobenius exact category when endowed with the conflations given by the exact sequences whose underlying sequences of graded comodules split. The *category up to homotopy* is the associated stable category. We denote by by $\mathcal{D}C$ the *co-derived category*, i.e. the localization of $\mathsf{Com}\, C$ at the class of all *co-quasi-isomorphisms*. Here, a morphism $s : L \to M$ of dg comodules is a co-quasi-isomorphism if its cone lies in the smallest triangulated subcategory of the category up to homotopy of dg comodules stable under coproducts and containing all totalizations of short exact sequences
$$0 \to X \to Y \to Z \to 0$$
of dg comodules. This definition is due to Positselski [**11**]. One can also characterize the co-quasi-isomorphisms using the cobar construction for dg comodules, cf. [**8**] and below. For the comparison between the two, cf. Appendix A of [**12**].

Let us fix a twisting cochain $\tau : C \to A$. For $M \in \mathsf{Mod}\, A$, let $M \otimes_\tau C$ be the graded comodule $M \otimes C$ endowed with the differential
$$d \otimes \mathbf{1} + \mathbf{1} \otimes d + d_\tau, \text{ where } d_\tau = (\mu \otimes \mathbf{1}) \circ (\mathbf{1} \otimes \tau \otimes \mathbf{1}) \circ (\mathbf{1} \otimes \Delta).$$
For $L \in \mathsf{Com}\, C$, let $L \otimes_\tau A$ be the graded A-module $L \otimes A$ endowed with the differential
$$d \otimes \mathbf{1} + \mathbf{1} \otimes d + d_\tau, \text{ where } d_\tau = (\mathbf{1} \otimes \mu) \circ (\mathbf{1} \otimes \tau \otimes \mathbf{1}) \circ (\Delta \otimes \mathbf{1}).$$

PROPOSITION 2.2. *The pair* $(? \otimes_\tau A, ? \otimes_\tau C)$ *is a pair of adjoint functors between* $\mathsf{Com}\, C$ *and* $\mathsf{Mod}\, A$. *It induces a pair of adjoint functors between* $\mathcal{D}C$ *and* $\mathcal{D}A$.

We define the twisting cochain $\tau : C \to A$ to be *acyclic* if the associated adjoint functors between $\mathcal{D}C$ and $\mathcal{D}A$ are equivalences. In this case, we say that A and C are *Koszul–Moore dual* to each other.

THEOREM 2.3 ([**8**]). *The following are equivalent:*

i) τ *is acyclic;*
ii) τ *induces a quasi-isomorphism* $\Omega C \to A$;
iii) τ *induces a weak equivalence* $C \to BA$ *(i.e. the induced morphism* $\Omega C \to \Omega BA$ *is a quasi-isomorphism);*

iv) *The natural morphism* $A \otimes_\tau C \otimes_\tau A \to A$ *is a quasi-isomorphism.*

For example, it follows from part iv) that SV and ΛV (with the above notations) are Koszul–Moore dual to each other. By part iii), we have Koszul–Moore duality between A and BA and therefore a triangle equivalence

$$\mathcal{D}(BA) \xrightarrow{\sim} \mathcal{D}A.$$

It is remarkable that the dg coalgebra BA determines all of $\mathcal{D}A$ whereas the dual dg algebra $DBA \xrightarrow{\sim} \mathsf{RHom}_A(k,k)$ a priori only determines the thick subcategory of $\mathcal{D}A$ generated by k.

3. The result

We keep the notations of the previous section. In particular, A is an augmented dg algebra and C a co-augmented cocomplete dg coalgebra. Assume that $\tau : C \to A$ is an acylic cochain. Put $A^e = A \otimes A^{op}$ and $C^e = C \otimes C^{op}$ and let

$$\tau^e = \tau \otimes \eta + \eta \otimes \tau : C^e \to A^e.$$

Clearly τ^e is a twisting cochain.

PROPOSITION 3.1. *The twisting cochain* τ^e *is acyclic and the induced equivalence*

$$\mathcal{D}(C^e) \xrightarrow{\sim} \mathcal{D}(A^e)$$

takes the dg bicomodule C *to the dg bimodule* A. *Thus we have an induced isomorphism of graded algebras*

$$HH^*(C) = \mathsf{Ext}^*_{C^e}(C,C) \xrightarrow{\sim} \mathsf{Ext}^*_{A^e}(A,A) = HH^*(A).$$

REMARK 3.2. We see that Koszul–Moore duality preserves Hochschild cohomology as a graded algebra.

PROOF. Let $\varphi_A : A \otimes_\tau C \otimes_\tau A \to A$ be the canonical morphism. Then the canonical morphism

$$\varphi_{A^e} : A^e \otimes_{\tau^e} C^e \otimes_{\tau^e} A^e \to A^e$$

is the composition of the isomorphism

$$(A \otimes A^{op}) \otimes_{\tau^e} (C \otimes C^{op}) \otimes_{\tau^e} (A \otimes A^{op}) \xrightarrow{\sim} (A \otimes_\tau C \otimes_\tau A) \otimes (A^{op} \otimes_\tau C^{op} \otimes_\tau A^{op})$$

with the quasi-isomorphism

$$\varphi_A \otimes \varphi_{A^{op}} : (A \otimes_\tau C \otimes_\tau A) \otimes (A^{op} \otimes_\tau C^{op} \otimes_\tau A^{op}) \to A \otimes A^{op}.$$

Thus τ^e is an ayclic twisting cochain by Theorem 2.3. The induced equivalence takes C to

$$C \otimes_{\tau^e} (A \otimes A^{op}) = A \otimes_\tau C \otimes_\tau A$$

which is quasi-isomorphic to A since τ is acyclic. √

THEOREM 3.3. *The isomorphism of the proposition lifts to an isomorphism in the homotopy category of* B_∞*-algebras between the corresponding Hochschild cochain complexes. In particular, it preserves the Gerstenhaber brackets.*

REMARK 3.4. In [**3**], Buchweitz related the Hochschild cohomology algebras of a Koszul algebra and its Koszul dual. In Theorem 3.5 of [**6**], we showed that for a *Koszul algebra* A (concentrated in differential degree 0), there is a canonical isomorphism in the category of Adams graded B_∞-algebras between the Hochschild complex of A and that of the Koszul dual algebra $A^!$ whose pth graded piece is put into bidegree $(p, -p)$, where the first component is the differential degree and the second component the Adams degree. Here, we get rid of the Koszulity assumption and the Adams grading by using dg coalgebras.

In [**4**], Félix–Menichi–Thomas show that for a simply connected coalgebra C, the Hochschild cohomologies of the dg algebras ΩC and $\mathsf{Hom}_k(C, k)$ are isomorphic as Gerstenhaber algebras. Another proof of this, under less stringent connectedness assumptions, is given in section 3 of Briggs–Gélinas' [**2**].

Herscovich shows in [**5**] that the Tamarkin–Tsygan calculus of an Adams connected (Def. 2.1 of [**10**]) augmented dg algebra and of its Koszul dual are dual to each other. In particular, for this class of dg algebras, Hochschild cohomology of the algebra and its Koszul dual are isomorphic as Gerstenhaber algebras.

For Koszul A_∞–algebras, an isomorphism of the Hochschild cohomologies as weight graded A_∞-algebras (but not as Gerstenhaber algebras) is proved by Berglund–Börjeson in Theorem 3.2 of [**1**].

PROOF. The twisting cochain induces a weak equivalence $BA \to C$. By the argument of [**6**], this yields an isomorphism in the homotopy category of B_∞-algebras between the Hochschild complexes of BA and C (these are sometimes called co-Hochschild complexes). Thus, we may assume that $C = BA$ and τ is the canonical twisting cochain. Let $\widetilde{B}A = A \otimes_\tau C \otimes_\tau A$. Then $\widetilde{B}A$ has a natural structure of dg coalgebra in the category $\mathsf{Mod}(A^e)$ of dg A-A-bimodules endowed with $\otimes_A$. We have a lax monoidal functor

$$F : \mathsf{Mod}(A^e) \to \mathcal{C}(k)$$

taking a bimodule M to $k \otimes_A M \otimes_A k$. The lax structure is given by the morphism

$$F(L \otimes_A M) = k \otimes_A (L \otimes_A M) \otimes_A k = k \otimes_A (L \otimes_A A \otimes_A M) \otimes_A k \to$$

$$k \otimes_A L \otimes_A k \otimes_A M \otimes_A k = (FL) \otimes_k (FM).$$

The functor F sends the A^e-coalgebra $\widetilde{B}A$ to the k-coalgebra $BA = C$. It induces a morphism from the Hochschild complex of $\widetilde{B}A$ to that of C and this is easily seen to be a quasi-isomorphism compatible with the cup product and the brace operations. The claim follows by Theorem 5.1 of [**9**] which states that there is a canonical isomorphism in the homotopy category of B_∞-algebras between the Hochschild complex of $\widetilde{B}A$ and the Hochschild complex of A. It is not hard to check that in homology, it induces the isomorphism of Propostion 3.1. $\surd$

EXAMPLE 3.5. Suppose that k is of characteristic 0 and $\mathfrak{g}$ a finite-dimensional Lie algebra over k. Let $U\mathfrak{g}$ be its enveloping algebra and $\Lambda\mathfrak{g}$ the supersymmetric coalgebra on $\mathfrak{g}$ placed in degree -1 and endowed with the coalgebra differential whose (-2)-component is the bracket $\Lambda^2\mathfrak{g} \to \mathfrak{g}$. Thus, the underlying complex of $\Lambda\mathfrak{g}$ is the Chevalley–Eilenberg complex of $\mathfrak{g}$. Then the map $\tau : \Lambda\mathfrak{g} \to U\mathfrak{g}$ which is the composition of the projection $\Lambda\mathfrak{g} \to \mathfrak{g}$ with the inclusion $\mathfrak{g} \to U\mathfrak{g}$ is an acyclic twisting cochain and we obtain an isomorphism of Gerstenhaber algebras

$$HH^*(\Lambda\mathfrak{g}) \xrightarrow{\sim} HH^*(U\mathfrak{g}).$$

It would be interesting to generalize this example to Lie algebroids.

Acknowledgments

I am grateful to José Antonio de la Peña for his friendship throughout the years and for the co-organization of many successful scientific events. I thank the organizers of the ARTA 7, where this material was first presented. I am indebted to Wendy Lowen for a talk at Trinity College Dublin in May 2019 on the results of [**9**]. Many thanks to Vladimir Dotsenko for reminding me of references [**4**] and [**2**] and to Pedro Tamaroff for pointing out [**1**].

References

[1] Alexander Berglund and Kaj Börjeson, *Koszul A_∞-algebras and free loop space homology*, Proc. Edinb. Math. Soc. (2) **63** (2020), no. 1, 37–65, DOI 10.1017/s0013091519000154. MR4054773

[2] Benjamin Briggs and Vincent Gélinas, *The a-infinity centre of the Yoneda algebra and the characteristic action of Hochschild cohomology on the derived category*, arXiv:1702.00721 [math.RT].

[3] Ragnar-Olaf Buchweitz, *Hochschild cohomology of Koszul algebras*, Talk at the Conference on Representation theory, Canberra, July 2003.

[4] Yves Félix, Luc Menichi, and Jean-Claude Thomas, *Gerstenhaber duality in Hochschild cohomology*, J. Pure Appl. Algebra **199** (2005), no. 1-3, 43–59, DOI 10.1016/j.jpaa.2004.11.004. MR2134291

[5] Estanislao Herscovich, *Hochschild (co)homology of Koszul dual pairs*, J. Noncommut. Geom. **13** (2019), no. 1, 59–85, DOI 10.4171/JNCG/323. MR3941473

[6] Bernhard Keller, *Derived invariance of higher structures on the Hochschild complex*, preprint, 2003, available at the author's home page.

[7] Bernhard Keller, *Koszul duality and coderived categories (after K. Lefèvre)*, unpublished notes, 2003, available at the author's home page.

[8] Kenji Lefèvre-Hasegawa, *Sur les A_∞-catégories*, Thèse de doctorat, Université Denis Diderot – Paris 7, November 2003, arXiv:math.CT/0310337.

[9] Wendy Lowen and Michel Van den Bergh, *The B_∞-structure on the derived endomorphism algebra of the unit in a monoidal category*, arXiv:1907.06026 [math.KT].

[10] Di Ming Lu, John H. Palmieri, Quan Shui Wu, and James J. Zhang, *Koszul equivalences in A_∞-algebras*, New York J. Math. **14** (2008), 325–378. MR2430869

[11] Leonid Positselski, *Two kinds of derived categories, Koszul duality, and comodule-contramodule correspondence*, Mem. Amer. Math. Soc. **212** (2011), no. 996, vi+133, DOI 10.1090/S0065-9266-2010-00631-8. MR2830562

[12] Michel Van den Bergh, *Calabi-Yau algebras and superpotentials*, Selecta Math. (N.S.) **21** (2015), no. 2, 555–603, DOI 10.1007/s00029-014-0166-6. MR3338683

[13] Michael Wong, *A geometric approach to Hochschild cohomology of the exterior algebra*, arXiv:1607.02661 [math.KT].

Université Paris Diderot – Paris 7, UFR de Mathématiques, CNRS, Institut de Mathématiques de Jussieu–Paris Rive Gauche, IMJ-PRG, Bâtiment Sophie Germain, 75205 Paris Cedex 13, France

Email address: bernhard.keller@imj-prg.fr

URL: https://webusers.imj-prg.fr/~bernhard.keller/

Contemporary Mathematics
Volume **761**, 2021
https://doi.org/10.1090/conm/761/15313

Thick subcategories of the stable category of modules over the exterior algebra

Otto Kerner and Dan Zacharia

Dedicated to José Antonio de la Peña for his 60th birthday.

ABSTRACT. We study thick subcategories defined by modules of complexity one in $\underline{\mathrm{mod}}R$, where R is the exterior algebra in $n+1$ indeterminates.

Let R be the exterior algebra in $n+1$ indeterminates, and let $\underline{\mathrm{mod}}R$ denote the stable category of all the finitely generated graded R-modules. This partially semi-expository article is part of a project devoted to the study of the thick subcategories of $\underline{\mathrm{mod}}R$. Note that using the Bernstein-Gelfand-Gelfand correspondence ([**5**]), this is equivalent to studying the thick subcategories of $\mathcal{D}^b(\mathrm{coh}\,\mathbf{P}^n)$-the bounded derived category of coherent sheaves on the projective n-space. This lattice of thick subcategories has been studied before, see for instance the treatment in the $n=1$ in [**22**] using non-crossing partitions ([**20**]), and also in [**12**]. More generally the lattice of thick subcategories of the derived category of a noetherian ring has also been treated extensively in [**4, 8, 19, 25, 31**] to name just a few of the references. While we make use of some of the results in the literature mentioned above, our approach to this problem is different, in the sense that it relies on representation theoretic concepts and methods (Auslander-Reiten theory for instance).

We concentrate in this paper on the study of the thick subcategories of $\underline{\mathrm{mod}}R$ generated by the graded modules of complexity one, which, while having a very nice description, turn out to be quite complicated. Eisenbud has proved in [**9**], that the complexity one graded modules are periodic up to a shift, and so, using the BGG correspondence (see also [**26**]), they correspond to the bounded complexes of coherent sheaves of finite length. It is also easy to see that these modules are weakly Koszul in the sense of [**23**], see also [**18**].

The article is organized as follows: In the first section we review the notation and the background needed. In particular for all non zero linear forms ξ we consider the local linear modules $M_\xi = R/\langle\xi\rangle$ of complexity one. In the second section, we use [**9**] to show that these are the building blocks of the modules of complexity one

2010 *Mathematics Subject Classification.* Primary 16G70; Secondary 16G60, 16E05.

This project was started while both authors were visiting the University of Bielefeld as part of the SFB's "Topologische und spektrale Strukturen in der Darstellungstheorie" program. Most of the results in this paper were obtained in Bielefeld, and also during the authors' visit to the University of Kiel. We both thank Henning Krause and the SFB, and Rolf Farnsteiner for inviting us and for making our stays possible.

(Theorem 2.11), and we study the subcategories $\mathcal{F}(M_\xi)$ consisting of those graded R-modules having a filtration with factors equal to M_ξ. We prove in Theorem 2.8 that these subcategories are wild for $n > 2$ (tame in the case when $n = 2$), and they are all equivalent to the category of finite dimensional modules over the (commutative) power series ring $\Bbbk[[t_1, \cdots, t_n]]$ in $t_1, \ldots, t_n$. We also show that every minimal ideal closed thick category of $\underline{\mathrm{mod}}R$ containing a module of complexity one is generated by a module of the form M_ξ as above. In the third section, we give some criteria for deciding when modules of non maximal complexity have self-extensions. Finally, in the appendix we describe the thick subcategories in the projective line case.

1. Introductory results and background

Let V be an $n+1$ dimensional vector space over an algebraically closed field $\Bbbk$ and let $R = R(V) = \bigwedge V$ be the corresponding exterior algebra. If $\{x_0, \ldots, x_n\}$ is a basis of V, we also write $R = R(x_0, \ldots, x_n)$ and call R the exterior algebra in $x_0, \ldots, x_n$. It is well-known that R is a graded $\Bbbk$-algebra by assigning each indeterminate x_i degree one, and that as a $\Bbbk$-algebra, R is generated by its degree zero and degree one (also called the non zero linear forms) parts. Let $\mathrm{mod}\, R$ be the abelian category of finitely generated *graded* R-modules. For two graded R-modules M and N, $\mathrm{Hom}_R(M, N)$ will *always* denote the space of degree zero homomorphisms from M to N. Also, for two graded R-modules M and N, $\mathrm{Ext}^i_R(M, N)$ will *always* denote the derived functors of the graded Hom between the two modules. If $M = \oplus_j M_j$ is a graded R-module, then its *graded shift* is the graded module $M(i)$ with $M(i)_j = M_{i+j}$ for each $j \in \mathbb{Z}$. By $\underline{\mathrm{mod}}R$ we denote the stable category of finitely generated *graded* R-modules. The objects of $\underline{\mathrm{mod}}R$ are the finitely generated graded R-modules, and $\underline{\mathrm{Hom}}_R(M, N)$ denotes the vector space $\mathrm{Hom}_R(M, N)/\mathcal{P}(M, N)$ where $\mathcal{P}(M, N)$ is the space of the degree zero homomorphisms from M to N factoring through a free R-module. For every finitely generated non projective graded R-modules L and M and homomorphism $L \xrightarrow{u} M$ in $\mathrm{mod}\, R$ we have a pushout diagram ([**16**])

$$\begin{array}{ccccccccc}
0 & \longrightarrow & L & \longrightarrow & I(L) & \longrightarrow & \Omega^{-1}L & \longrightarrow & 0 \\
 & & \downarrow{\scriptstyle u} & & \downarrow & & \| & & \\
0 & \longrightarrow & M & \xrightarrow{v} & N & \xrightarrow{w} & \Omega^{-1}L & \longrightarrow & 0
\end{array}$$

where Ω (respectively Ω^{-1}) denote the syzygy (cosyzygy) functors

$$\Omega, \Omega^{-1} \colon \underline{\mathrm{mod}}R \to \underline{\mathrm{mod}}R$$

and $I(L)$ denotes the injective envelope of L. It is well-known that $\underline{\mathrm{mod}}R$ is a triangulated category where the shift (or suspension) functor is given by the first cosyzygy, and every distinguished triangle is isomorphic in $\underline{\mathrm{mod}}R$ to a triangle of the form

$$L \xrightarrow{u} M \xrightarrow{v} N \xrightarrow{w} \Omega^{-1}L$$

obtained as above from a pushout diagram, where we denote again by u, v, w the induced morphisms in the stable category. So triangle closure in the stable category is induced by extension closure in the category of finitely generated graded R-modules. Note also, that, for every R-modules M and N we have:

$$\underline{\mathrm{Hom}}_R(\Omega M, N) \cong \mathrm{Ext}^1_R(M, N).$$

Recall that a full subcategory $\mathcal{T}$ of $\underline{\text{mod}}R$ is *thick* if it is closed under triangles, direct summands and the suspension functor. We are interested in describing the thick subcategories of $\mathcal{D}^b(\text{coh}\,\mathbf{P}^n)$ or, equivalently (see [**5**]) of $\underline{\text{mod}}R$. In this paper we restrict to thick subcategories consisting of modules of small complexity. We will need some background on the stable graded category of R-modules and we first recall that R is a symmetric Koszul algebra whose Koszul dual is the polynomial algebra S in $n+1$ indeterminants. Let M be a finitely generated graded R-module and let

$$(\mathbb{F}) \cdots \to F^2 \to F^1 \to F^0 \to M \to 0$$

be a minimal graded free resolution of M. The module M is called a *linear module* if each free module F^i in $(\mathbb{F})$ is generated in degree i.

Considering this minimal free resolution $\mathbb{F}$ of M, we let the *i-th Betti number* $\beta_i(M)$ of M be the rank of the free module F_i. Recall that the *complexity* of a finitely generated module M over the exterior algebra is the infimum of the set

$$\{d \in \mathbb{N} | \beta_i(M) \leq ci^{d-1} \text{ for some positive } c \in \mathbb{Q} \text{ and all } i \geq 0\}$$

So an R-module M has *complexity 1* if its Betti numbers $\beta_i(M)$ are bounded.

Note that for a finite dimensional self-injective algebra, the complexity of a module can also be defined by using $\dim_{\Bbbk}\Omega^i M$ instead of $\beta_i(M)$, and the module M has complexity 0 if and only if it is projective. It is well-known that for every module M over the exterior algebra in $n+1$ indeterminates we have $\text{cx}\, M \leq n+1$. The complexity $n+1$ is attained; for instance the trivial module has this complexity.

A very useful characterization of the complexity of an $R(V)$-module M, using *M-regular sequences* is shown in [**1**, Section 3]. It follows directly from this characterization, that the R-modules of complexity smaller than $\dim_{\Bbbk} V$ have even dimension, and that for a subspace $U \subseteq V$, the factor module $R/\langle U\rangle$ has complexity $\dim_{\Bbbk} U$.

In particular, if $\dim_{\Bbbk} U = 1$, then U is a point in the projective space $\mathbf{P}(V)$ and $M_U = R/\langle U\rangle$ is cyclic and has complexity 1.

We will need the following result due to Eisenbud [**9**]:

THEOREM 1.1. *Let M be a finitely generated graded R-module of complexity 1.*
(a) *If M has no indecomposable projective direct summand, then, as R-modules, $M \cong \Omega M$, and as graded modules $M \cong \Omega M(1)$.*
(b) *The module M has a filtration $0 = M_0 \subset M_1 \subset \cdots \subset M_{s-1} \subset M_s = M$ with $M_i/M_{i-1} \cong M_{\xi_i}(j_i)$ for some $\xi_i \in \mathbf{P}(V)$ and some $j_i \in \mathbb{Z}$.* □

This result has no counterpart for modules with complexity bigger then one, as the following example shows:

EXAMPLE 1.2. For $R = R(x, y, z)$ let M be the 4-dimensional R-module with basis $\{e_1, e_2, f_1, f_2\}$ and the following operations of R on M: $e_i z = f_i$, $e_1 x = e_2 y = f_i x = f_i y = f_i z = 0$ and $e_1 y = f_2$, $e_2 x = f_1$. One directly checks that (z) is a maximal M-regular sequence, hence $\text{cx}\, M = 2$ by [**1**]. Any proper indecomposable submodule of M has odd dimension, hence complexity 3. It is not hard to prove that the module M is linear.

We mention the following most probably well-known result, and we include a proof for the convenience of the reader:

LEMMA 1.3. *Let U be a nonzero subspace of V. Then, the module $M = R/\langle U\rangle$ is linear.*

PROOF. The result is clear if $U = V$ so assume that that U is a proper subspace of V. Let $\{x_0, \ldots, x_n\}$ be a basis of V. Without loss of generality we may assume that U is spanned by $\{x_0, \ldots, x_k\}$ for some $k < n$. Note that $R = R(x_0, \ldots, x_k) \boxtimes R(x_{k+1}, \ldots, x_n)$ where $\boxtimes$ denotes the skew tensor product (see [**24, 27**]). Note also that we have an isomorphism of graded R-modules $R/\langle U\rangle \cong \Bbbk \boxtimes R(x_{k+1}, \ldots, x_n)$ and, as skew tensor products of linear modules are linear, the proof is complete. $\square$

EXAMPLE 1.4. Let $R = R(x, y, z)$, and let M be a four dimensional module with Loewy length two, generated in degree 0, and with basis $\{e, ex, ey, ez\}$. Then there are no M-regular elements, hence $\operatorname{cx} M = 3$, by [**1**]. If $M' = M/M\langle\xi\rangle$, where ξ is a no zero linear form, then $\dim_{\Bbbk} M' = 3$ and $\operatorname{cx} M' = 3$.

A graded R module M is called *weakly Koszul* if its quadratic dual $M^!$ is a linear S-module (see [**13, 18**]), where S denotes the Ext-algebra $\operatorname{Ext}^*_R(R/J, R/J)$, with J being the radical of R. Note that $S \cong \Bbbk[y_0, \ldots, y_n]$, where $\{y_0, \ldots, y_n\}$ is a dual basis of $\{x_0, \ldots, x_n\}$. The weakly Koszul modules can be described also in terms of certain types of extensions of shifts of linear modules (see [**23**, p. 679]). First, we recall the following definition:

DEFINITION 1.5. An extension $0 \to L \to M \to N \to 0$ of graded R-modules is called relative if for each $k \geq 0$ we have $MJ^k \cap L = LJ^k$.

For instance, every short exact sequence $0 \to L \to M \to N \to 0$ of linear R-modules is a relative extension. We have the following (see [**23**]):

PROPOSITION 1.6. *Let M be a weakly Koszul R-module and assume that a minimal set of homogeneous generators of M is distributed in degrees $i_0 < i_1 < \ldots < i_p$. Let $L = \langle M_{i_0}\rangle$ be the homogeneous submodule of M generated by the degree i_0 part of M. Then L is a graded shift of a linear module, M/L is a weakly Koszul module generated in degrees $i_1 < \ldots < i_p$, and the extension $0 \to L \to M \to M/L \to 0$ is also a relative extension.*

Using the above proposition, one can use an inductive procedure to show that every weakly Koszul module can be filtered with linear modules (and their degree shifts) using relative extensions. The following proposition was proved in [**13**].

PROPOSITION 1.7. *Let $0 \to L \to M \to N \to 0$ be a relative extension of weakly Koszul modules. Then for each $k \geq 0$ we have relative extensions of weakly Koszul modules $0 \to \Omega^k L \to \Omega^k M \to \Omega^k N \to 0$.*

An immediate consequence is that if M is a weakly Koszul R-module generated in degrees $i_0 < \cdots < i_p$, then ΩM is also weakly Koszul and is generated in degrees $i_0 + 1 < \cdots < i_p + 1$.

An easy inductive argument shows that the following result holds over any selfinjective Koszul algebra :

PROPOSITION 1.8. *Let $0 \to A \to B \to C \to 0$ be a relative extension of weakly Koszul modules. Then $\operatorname{cx} B = \max\{\operatorname{cx} A, \operatorname{cx} C\}$.* $\square$

What makes the weakly Koszul modules relatively ubiquitous over the exterior algebra is the fact that given any finitely generated R-module M, then $\Omega^i M$ is weakly Koszul for some positive integer i (see for instance [**18, 23**]). If M is indecomposable with $\operatorname{cx} M = 1$ then, by Eisenbud's theorem 1.1, $\Omega^r M \cong M(-r)$. Thus Theorem 1.1 also implies the following

COROLLARY 1.9. *All the indecomposable R-modules of complexity one are weakly Koszul.* □

We also have the following ([**23**]):

PROPOSITION 1.10. *Let R be the exterior algebra in $n+1$ indeterminates and $M \in \operatorname{mod} R$ be an indecomposable graded non projective module. Then:*

(1) $\tau M = \Omega^2 M(n+1)$, *where τ denotes the Auslander-Reiten translate in the category of graded R-modules.*
(2) *If $n = 1$ and M is weakly Koszul, then M is linear, or a graded shift of a linear module, and M and τM are both generated in the same degree.*

We start with the following observation:

LEMMA 1.11. *Let M be an indecomposable finitely generated graded module over the exterior algebra in $n+1$ indeterminates, $n \geq 1$. Then:*

(1) *If $\operatorname{cx} M = 1$, then for every $n \geq 1$, $\tau M = M(n-1)$. In particular, if $n = 1$ then $\tau M = M$.*
(2) *Let $L \to M \to N \to \Omega^{-1} L$ be a distinguished triangle in $\underline{\operatorname{mod}} R$. If $\operatorname{cx} L = \operatorname{cx} M = i$, then $\operatorname{cx} N \leq i$. In particular, if $\operatorname{cx} L = \operatorname{cx} M = 1$, then either $\operatorname{cx} N = 1$ or N is free.*

PROOF. We know by Eisenbud's theorem that $M \cong \Omega M(1)$, so also $M \cong \Omega^2 M(2)$, and moreover M must be weakly Koszul. As we saw earlier, $\tau M = \Omega^2 M(n+1)$, so the first part follows immediately. For the second part it suffices to assume that none of the modules involved is free. But then we know that every distinguished triangle $L \to M \to N \to \Omega^{-1} L$ is isomorphic in the stable category of R to a triangle obtained from a pushout diagram that yields a short exact sequence $0 \to M \to N \oplus F \to \Omega^{-1} L \to 0$ for some free R-module F. This completes the proof of the lemma. □

Since indecomposable modules of complexity one are weakly Koszul, see Corollary 1.9, Proposition 1.6 has the following refinement (see also [**15**]):

COROLLARY 1.12. *Let M be an indecomposable module of complexity one over the exterior algebra and assume that M has a minimal set of generators in degrees $i_0 < i_1 < \ldots < i_p$. Let $L = \langle M_{i_0} \rangle$ be the homogeneous submodule of M generated by the degree i_0 part of M and let $N = M/L$. Then both L and N are non projective and have complexity one, and therefore M can be filtered by graded shifts of linear modules of complexity one.*

PROOF. It is obvious that neither L nor N are projective, so by 1.6 and 1.7 we have an induced short exact sequence $0 \to \Omega L \to \Omega M \to \Omega N \to 0$ and it is also clear that the submodule of ΩM generated in degree $i_0 + 1$ is isomorphic to ΩL. So both L and N have also complexity one by 1.8. The corollary follows now from 1.6 and induction on the number of degrees of generation for M. □

Note that the category of graded modules over the exterior algebra is closed under tensor products, where the tensor product is taken over the ground field, and is defined as follows: For two finitely generated graded R-modules M and N, we define $(M \otimes N)_k = \bigoplus_i (M_i \otimes N_{k-i})$ where the R-module structure is induced by

$$(m \otimes n)\xi = (-1)^i m\xi \otimes n + m \otimes n\xi$$

for each $m \in M$, $n \in N_i$, and for each homogeneous element ξ of degree one in R. Moreover, if $0 \to A \to B \to C \to 0$ is a short exact sequence of graded R-module, then for every graded R-module M we have an induced sequence of graded R-modules $0 \to M \otimes A \to M \otimes B \to M \otimes C \to 0$. Following [**3**] we say that a thick subcategory $\mathcal{T}$ has *ideal closure* if for each $M \in \mathcal{T}$ and $X \in \underline{\mathrm{mod}}R$, we have that $M \otimes X \in \mathcal{T}$. Let S denote the unique simple R-module concentrated in degree 0, so that $S(-i)$ is the unique simple module concentrated in degree i. Then it is easy to see that $M \otimes S(-i) \cong M(-i)$. Let $\mathcal{T}_M$ denote the thick subcategory generated by M. We have the following:

PROPOSITION 1.13. *Let R be the exterior algebra in $n+1$ indeterminates, and let M be an indecomposable graded module of complexity 1. Then the thick subcategory $\mathcal{T}_M$ generated by M contains the Auslander-Reiten components containing all the graded shifts of M. In particular, $\mathcal{T}_M$ has ideal closure.*

PROOF. As we have seen above, if M is a finitely generated indecomposable graded module of complexity 1, then $\Omega^{-1}M \cong M(1)$ and $M(-1) \cong \Omega M$, so for each integer i, the graded shifts $M(i) \in \mathcal{T}_M$. Moreover, since for such a module $\tau M \cong M(n-1)$, it follows that $\mathcal{T}_M$ contains also the Auslander-Reiten component containing M and all its graded shifts. To show that $\mathcal{T}_M$ has ideal closure we proceed by induction on the graded length using as a starting point the fact that if L is a semisimple module generated in a single degree, then $M \otimes L$ is isomorphic to a direct sum of graded shifts of M. To be more specific let $L = L_{i_0} \oplus \cdots \oplus L_{i_0+t}$ be a graded R-module and consider the exact sequence $0 \to L_{\geq i_0+1} \to L \to L_{i_0} \to 0$. We have an induced exact sequence of R-modules $0 \to M \otimes L_{\geq i_0+1} \to M \otimes L \to M \otimes L_{i_0} \to 0$. By induction, $M \otimes L_{i_0}$ and $M \otimes L_{\geq i_0+1}$ are in $\mathcal{T}_M$ so $M \otimes L \in \mathcal{T}_M$ too. $\square$

Let $\mathcal{T}$ be a thick subcategory of $\underline{\mathrm{mod}}R$ containing an indecomposable module M of complexity 1, where R is the exterior algebra in $n+1$ indeterminates. For $n = 1$ the module M lies in a homogeneous tube, and if $n > 1$ then M lies in a $\mathbb{Z}A_\infty$ component by [**23**, Theorem 5.7]. So we may always assume that M is quasi-simple.

2. Thick subcategories generated by modules of complexity one

Let $R = R(V)$ denote the exterior algebra in $n+1$ indeterminates $x_0, x_1, \cdots, x_n$. Following [**9**], the complexity one modules of smallest length are cyclic of dimension 2^n, linear and, up to degree shift of the form $R/\langle \xi \rangle$ with $\xi \in \mathbf{P}(V)$.

Let $M = M_\xi$ be such a module. We describe first the thick subcategory $\mathcal{T}_M$ of $\underline{\mathrm{mod}}R$ generated by M. It is clear that $\Omega M(1) = M$ and $\tau M = M(n-1)$. We start by collecting some facts about these modules.

LEMMA 2.1. *Let $\xi \neq \eta$ be two points in $\mathbf{P}(V)$. Then:*

(i) $\mathrm{Hom}_R(M_\xi, M_\eta) = 0$ *and* $\mathrm{Hom}_R(M_\xi, M_\xi) = \underline{\mathrm{Hom}}_R(M_\xi, M_\xi)$ *is one dimensional.*

(ii) $\underline{\mathrm{Hom}}_R(M_\xi, M_\eta(i)) = 0$ *for each integer i. Equivalently,* $\mathrm{Ext}^k_R(M_\xi, M_\eta(i)) = 0$ *for each integer i, and for each non-negative integer k.*

(iii) $\underline{\mathrm{Hom}}_R(M_\xi, M_\xi(i))$ *has dimension $\binom{n}{i}$ if $0 \leq i \leq n$ and is zero otherwise.*

(iv) $\mathrm{Hom}_R(M_\xi(j), M_\eta) = 0 = \mathrm{Hom}_R(M_\xi(j), M_\xi)$ *for all $j > 0$ and for all $j < -n$.*

(v) $\mathrm{Hom}_R(\Omega^i M_\xi, M_\xi) = \underline{\mathrm{Hom}}_R(\Omega^i M_\xi, M_\xi)$ *for all $i \geq 0$.*

PROOF. (i) If there was a nonzero homomorphism $f\colon M_\xi \to M_\eta$, then f would take the degree zero part of M_ξ onto the degree zero part of M_η. Hence f would be onto and a dimension argument would imply that f would have to be an isomorphism. Consequently M_ξ and M_η will have the same annihilator in R contradicting the fact that $\xi \neq \eta$.

Since M_ξ has simple top, the graded Hom space $\mathrm{Hom}_R(M_\xi, M_\xi)$ is one dimensional. It is obvious that no nonzero degree zero homomorphism from M_ξ to M_ξ can factor through a free module so we obtain immediately that $\mathrm{Hom}_R(M_\xi, M_\xi) = \underline{\mathrm{Hom}}_R(M_\xi, M_\xi)$.

(ii) By making a change of basis if necessary we may assume that R is the exterior algebra in $x_0, x_1, \cdots, x_n$ with $x_0 \in \xi$ and $x_n \in \eta$. The result is obvious if either $i \leq 0$ or $i \geq n$, so assume that $0 < i \leq n-1$. Let e_1 span the degree zero part of M_ξ and let e_2 span the degree $-i$ part of $M_\eta(i)$. A graded homomorphism $f\colon M_\xi \to M_\eta(i)$ is given by $f(e_1) = e_2 a$ where a is a homogeneous element of degree i in $x_0, x_1, \cdots, x_{n-1}$, but since e_1 is annihilated by x_0 it turns out that $a = bx_0$ where b is homogenous of degree $i-1$ in $x_1, \cdots, x_{n-1}$. Let $\epsilon\colon M_\xi \to R(1)$ be the inclusion of M_ξ into its injective envelope. Then $\epsilon(e_1) = e_3 x_0$ where e_3 spans the degree -1 part of $R(1)$. We show that f factors through ϵ. A typical graded homomorphism $h\colon R(1) \to M_\eta(i)$ has the form $h(e_3) = e_2 c$ where c is a homogenous element of degree $i-1$ in the indeterminates $x_0, x_1, \cdots, x_{n-1}$. So $h\epsilon(e_1) = e_2 c'$ where c' is a homogenous element of degree i containing x_0 as a factor. It is clear that c can always be chosen so that $h\epsilon = f$ and the proof is complete.

(iii) The result is trivial if i is negative or $i > n$, so assume $0 \leq i \leq n$. Again we may assume without loss of generality $x_0 \in \xi$. Let e_1 span the degree zero part of M_ξ and let e_2 span the degree $-i$ part of $M_\xi(i)$. A degree zero embedding ϵ from M_ξ into its injective envelope $R(1)$ is given by $\epsilon(e_1) = e_3 x_0$ where e_3 spans the degree -1 part of $R(1)$, and a homomorphism $h\colon R(1) \to M_\xi(i)$ is given by $h(e_3) = e_2 a$ where a is a homogenous element of R of degree $i-1$ in $x_1, \cdots, x_n$. Since x_0 annihilates M_ξ and all its graded shifts we get that $\mathrm{Hom}_R(M_\xi, M_\xi(i)) = \underline{\mathrm{Hom}}_R(M_\xi, M_\xi(i))$ and the rest follows immediately.

(iv,v) These also follow immediately. □

As a consequence of part (i) of the above lemma we see that $\mathrm{Ext}^1_R(M_\xi(1), M_\xi)$ is one dimensional since it is isomorphic to $\underline{\mathrm{Hom}}_R(M_\xi, M_\xi)$. In fact up to isomorphism we have a nonsplit exact sequence $0 \to M_\xi \to R(1) \to M_\xi(1) \to 0$ spanning this Ext space. Part (iii) of the lemma implies that $\mathrm{Ext}^1_R(M_\xi, M_\xi(n-1))$ is one dimensional, and then up to isomorphism we have a nonsplit short exact sequence of graded R-modules $0 \to M_\xi(n-1) \to X_\xi \to M_\xi \to 0$, which is an Auslander-Reiten sequence. It is easy to see that the module X_ξ is indecomposable since by [**9**] there are no indecomposable modules of dimension smaller than the dimension of M in this Auslander-Reiten component.

Since for all graded R-modules X, Y and for each $t \geq 1$ we have isomorphisms $\mathrm{Ext}^t_R(X, Y) \cong \underline{\mathrm{Hom}}_R(\Omega^t X, Y) \cong \underline{\mathrm{Hom}}_R(X, \Omega^{-t} Y)$, we get from part (iii) of Lemma 2.1:

COROLLARY 2.2. *Let $R = R(V)$ and let $\xi \in \mathbf{P}(V)$. Then*

(i) $\mathrm{Ext}^1_R(M_\xi, M_\xi(i)) \neq 0$ *if and only if* $0 \leq i+1 \leq n$.

(ii) $\mathrm{Ext}^1_R(M_\xi, M_\xi)$ *is n-dimensional.* $\mathrm{Ext}^2_R(M_\xi, M_\xi(i)) \neq 0$ *if and only if* $0 \leq i+2 \leq n$.

(iii) $\operatorname{Ext}^1_R(M_\xi, M_\xi(i)) \neq 0$ *and* $\operatorname{Ext}^2_R(M_\xi, M_\xi(i)) = 0$ *if and only if* $i = n - 1$. □

Putting together Theorem 1.1 and part (ii) of 2.1, we obtain:

COROLLARY 2.3. *Let* $R = R(V)$ *and let* X *be an indecomposable finitely generated* R*-module of complexity one. Then there exists a unique* $\xi \in \mathbf{P}(V)$ *such that* X *has a filtration with subfactors* $M_\xi(j_i)$, *for suitable* $j_i \in \mathbb{Z}$. □

Having this result in mind, Corollary 1.12 then implies that for the study of indecomposable modules of complexity one, we should consider linear modules X of complexity one, which have a filtration $0 = X_0 \subset X_1 \subset \cdots \subset X_t = X$ with $X_i/X_{i-1} \cong M_\xi$ for a fixed $\xi \in \mathbf{P}(V)$. Let $M = M_\xi$ and $\mathcal{F}(M)$ be the full subcategory of $\operatorname{mod} R$ consisting of the modules having a filtration with subfactors isomorphic to M. Since M has trivial endomorphism ring, by [**28**] the category $\mathcal{F}(M)$ is a full exact and extension closed abelian subcategory of $\operatorname{mod} R$ and M is its unique (up to isomorphism) simple object. As in the classical case, every semi-simple object in $\mathcal{F}(M)$ is isomorphic to a direct sum of copies of M, any two different simple subobjects of a module in $\mathcal{F}(M)$ have zero intersection. Each object in $\mathcal{F}(M)$ has a finite composition series, hence $\mathcal{F}(M)$ is a length category.

Following Gabriel [**10**, Sect.7] one should then consider the Ext-quiver $\mathcal{Q}$ of this category, since it already gives a lot of informations on $\mathcal{F}(M)$: The quiver $\mathcal{Q}$ has one vertex and $n = \dim_{\mathbb{k}} \operatorname{Ext}^1_R(M, M)$ loops. As usual we will identify the category of finite dimensional $\mathbb{k}$-linear representations of $\mathcal{Q}$ with the category $\operatorname{mod} \mathbb{k}\mathcal{Q}$. Since $\mathcal{Q}$ has n loops, the algebra $\mathbb{k}\mathcal{Q} \cong \mathbb{k}\langle t_1, \dots, t_n\rangle$-the free algebra in $t_1, \dots, t_n$.

By Gabriel ([**10**]) any length-category is equivalent to the category of finite dimensional modules over some pseudo-compact ring A (see [**11**]). We will show that in our situation $A = \mathbb{k}[[t_1, \cdots, t_n]]$.

Obviously all the nonzero objects in $\mathcal{F}(M)$ are linear of complexity one. Since $\mathcal{F}(M)$ is a length category, for each nonzero object $X \in \mathcal{F}(M)$ we may consider its radical, $\operatorname{rad}_{\mathcal{F}} X$- the intersection of all maximal subobjects of X in $\mathcal{F}(M)$. Let us show that $\operatorname{rad}_{\mathcal{F}} X \in \mathcal{F}(M)$. Being an abelian category, $\mathcal{F}(M)$ is closed under finite intersections of subobjects. Let $(Y_j)_{j \in J}$ be a family of subobjects of some $Y \in \mathcal{F}(M)$. Since Y has finite $\mathcal{F}(M)$-length, there exists a finite subfamily $\{Y_1, \dots, Y_t\}$ such that $C = \cap_{1 \le i \le t} Y_i \subset Y_j$ for all $j \in J$ that is, $C = \cap_{j \in J} Y_j$, so $\mathcal{F}(M)$ is closed under arbitrary intersections as well.

Letting $\operatorname{rad}^0_{\mathcal{F}} X = X$ and $\operatorname{rad}^{i+1}_{\mathcal{F}} X = \operatorname{rad}_{\mathcal{F}}(\operatorname{rad}^i_{\mathcal{F}} X)$, we obtain the "Loewy series"

$$X \supset \operatorname{rad}_{\mathcal{F}} X \supset \operatorname{rad}^2_{\mathcal{F}} X \supset \cdots \supset \operatorname{rad}^{r-1}_{\mathcal{F}} X \supset \operatorname{rad}^r_{\mathcal{F}} X = 0.$$

The smallest r with the property that $\operatorname{rad}^r_{\mathcal{F}} X = 0$ is called the *Loewy length* of X, written $\ell\ell(X)$. The factor $X/\operatorname{rad}_{\mathcal{F}} X$ is called the $\mathcal{F}$*-top* of X. If $\mathcal{F}$-top$X$ is simple, then X is called $\mathcal{F}$-local or simply local. All the factors $\operatorname{rad}^i_{\mathcal{F}} X/\operatorname{rad}^{i+1}_{\mathcal{F}} X$ are semisimple in the category $\mathcal{F}(M)$. The largest semi-simple subobject of X in $\mathcal{F}(M)$ is called the $\mathcal{F}$-socle of X.

Our first aim is to give a precise description of this category $\mathcal{F}(M)$.

For each $i \in \mathbb{N}$ define the full subcategories $\mathcal{F}^{(i)} = \{X \in \mathcal{F}(M) | \ell\ell(X) \le i\}$. We have an ascending chain of subcategories

$$0 = \mathcal{F}^{(0)} \subset \mathcal{F}^{(1)} \subset \mathcal{F}^{(2)} \subset \cdots,$$

and $\mathcal{F}(M) = \bigcup_{i \in \mathbb{N}} \mathcal{F}^{(i)}$. The inclusion functors $\mathcal{F}^{(i)} \to \mathcal{F}^{(i+1)}$ are right adjoint to the truncation functors $\mathcal{F}^{(i+1)} \to \mathcal{F}^{(i)}$, defined by $Y \mapsto Y/\mathrm{rad}^i_{\mathcal{F}}Y$. The categories $\mathcal{F}^{(i)}$ are exact abelian subcategories of $\mathcal{F}(M)$, but they are not closed under extensions. For each i, let $\mathcal{L}^{(i)}$ be the set of local objects in $\mathcal{F}^{(i)}$ whose tops equal the unique simple object M. We claim that the length of each object in $\mathcal{L}^{(i)}$ is bounded by $(n^i - 1)/(n-1)$.

The claim is clear for $i = 1$. For $i > 1$ and $X \in \mathcal{L}^{(i)}$, then as $\dim_{\Bbbk} \mathrm{Ext}^1_R(M, M) = n$, we have that $\mathrm{rad}_{\mathcal{F}}X/\mathrm{rad}^2_{\mathcal{F}}X \cong M^r$, with $r \leq n$, so $\mathrm{rad}_{\mathcal{F}}X$ is an epimorphic image of a sum of r local modules $L_1, \ldots, L_r$, all contained in $\mathrm{rad}_{\mathcal{F}}X$. By induction, since for each j, $\ell(L_j) \leq (n^{i-1} - 1)/(n-1)$, we get

$$\ell(X) \leq n(n^{i-1} - 1)/(n-1) + 1 = (n^i - n)/(n-1) + 1 = (n^i - 1)/(n-1)$$

which proves the claim. By [**21**], there exists in $\mathcal{L}^{(i)}$ a unique object $P^{(i)}$ of maximal length with the properties that each object in $\mathcal{L}^{(i)}$ is an epimorphic image of $P^{(i)}$, and that for any $X \in \mathcal{F}^{(i)}$, each epimorphism $X \to P^{(i)}$ splits. It is also easy to check that the object $P^{(i)}$ is projective in $\mathcal{F}^{(i)}$ in the usual sense, meaning that, for each epimorphism $X \to Y$ in $\mathcal{F}^{(i)}$, the induced map $\mathrm{Hom}_R(P^{(i)}, X) \to \mathrm{Hom}_R(P^{(i)}, Y)$ is surjective. The object $P^{(i)}$ is then the minimal projective generator in the category $\mathcal{F}^{(i)}$, therefore by Morita theory the functor $\mathrm{Hom}_R(P^{(i)}, -)$ induces an equivalence $\mathcal{F}^{(i)} \cong \mathrm{mod}\, \mathrm{End}_R(P^{(i)})$.

LEMMA 2.4. *For all $i \in \mathbb{N}$ we have $P^{(i)}/\mathrm{rad}^{i-1}_{\mathcal{F}} P^{(i)} \cong P^{(i-1)}$.*

PROOF. Since $P^{(i)}$ generates all elements in $\mathcal{L}^{(i)}$, it generates also all the elements in $\mathcal{L}^{(i-1)}$. So every object in $\mathcal{L}^{(i-1)}$ is a quotient of $P^{(i)}/\mathrm{rad}^{i-1}_{\mathcal{F}} P^{(i)}$, and, since $P^{(i-1)}$ has maximal composition length in $\mathcal{L}^{(i-1)}$, the result follows. □

This lemma will enable us to construct the objects $P^{(i)}$ and their endomorphism rings inductively. For $i = 1$ the category $\mathcal{F}^{(1)} = \mathrm{add}\, M$ is semisimple, so $P^{(1)} = M$, hence $\mathrm{End}_R M = \Bbbk$. For the inductive step we use universal short exact sequences as follows: Let $X \in \mathcal{F}^{(i)}$ be indecomposable with $\dim_{\Bbbk} \mathrm{Ext}^1_R(X, M) = a$. Let $(\eta_s)_{1 \leq s \leq a}$ be a basis of $\mathrm{Ext}^1_R(X, M)$, say

$$\eta_s : 0 \to M \to E_s \to X \to 0.$$

We have a pullback diagram

$$\begin{array}{ccccccccc}
0 & \longrightarrow & M^a & \longrightarrow & E & \longrightarrow & X & \longrightarrow & 0 \\
 & & \Vert & & \downarrow & & \downarrow{\scriptstyle \Delta} & & \\
0 & \longrightarrow & M^a & \longrightarrow & \oplus_s E_s & \longrightarrow & X^a & \longrightarrow & 0
\end{array}$$

where $\Delta : X \to X^a$ is the diagonal embedding. The top short exact sequence $0 \to M^a \to E \to X \to 0$ is called the *universal short exact sequence* ([**6**]) for $\mathrm{Ext}^1_R(X, \mathrm{add}\, M)$ and the connecting homomorphism $\mathrm{Hom}_R(M^a, M) \to \mathrm{Ext}^1_R(X, M)$ is an isomorphism. Note that E is uniquely determined up to isomorphism.

REMARK 2.5. For each $k > 1$, the $\mathcal{F}$-top of $\mathrm{rad}_{\mathcal{F}} P^{(k)}$ equals M^n. Let $e_i \in \mathrm{rad}_{\mathcal{F}} P^{(k)}$ be such that $e_i + \mathrm{rad}^2_{\mathcal{F}} P^{(k)}$ is a generator of the i-th direct summand in M^n above and let e be the generator of the $\mathcal{F}$-top of $P^{(k)}$. For each $1 \leq i \leq n$, let $\pi_i \colon P^{(k)} \to e_i R$ be the $\mathcal{F}$-projective cover such that $\pi_i(e) = e_i$, and let ϵ_i denote the inclusions $e_iR \to P^{(k)}$. We will view each t_i as the endomorphism $\epsilon_i\pi_i\colon P^{(k)} \to P^{(k)}$. Let now $g\colon P^{(k)} \to P^{(k)}$ be an endomorphism. Since $g(\mathrm{rad}_{\mathcal{F}}P^{(k)}) \subseteq \mathrm{rad}_{\mathcal{F}}P^{(k)}$, g induces a homomorphism

$$\overline{g}\colon \mathcal{F}\text{-top}P^{(k)} \to \mathcal{F}\text{-top}P^{(k)}.$$

If $\overline{g} \neq 0$, then g is an automorphism, as $P^{(k)}$ is a local object in $\mathcal{F}(M)$. Otherwise, the image of g must be contained in $\mathrm{rad}_{\mathcal{F}}P^{(k)} = \sum_{i=1}^{n} t_i(P^{(k)})$.

EXAMPLE 2.6. We give an explicit construction of $P^{(2)}$ that will be used shortly. For each $1 \leq s \leq n$, consider the extensions

$$\eta_s\colon 0 \to M \to E_s \to M \to 0$$

where $M = M_\xi$ and $x_0 \in \xi$. Then, as $R(x_1, \cdots, x_n) = R/\langle x_0\rangle$-modules, the modules E_s and M^2 are isomorphic, and are each generated over R by two generators: e, and e_s, satisfying $ex_0 = e_sx_s$, and $e_sx_0 = 0$. It is clear that the set $\{\eta_s\}_{1\leq s\leq n}$ forms a $\Bbbk$-basis for $\mathrm{Ext}^1_R(M, M)$. We now construct the universal short exact sequence for $\mathrm{Ext}^1_R(M, \mathrm{add}\, M)$, and we get

$$0 \to M^n \to P^{(2)} \to M \to 0$$

where $E = P^{(2)}$ has a minimal generating system $\{e, e_1, \ldots, e_n\}$ in degree 0, subject to the relations $ex_0 = \sum_s e_sx_s$ and $e_sx_0 = 0$ for all $1 \leq s \leq n$. We have seen in the previous remark that we may consider each t_i as being an endomorphism of $P^{(2)}$, taking $P^{(2)}$ inside $\mathrm{rad}_{\mathcal{F}}P^{(2)}$. Since $\mathrm{rad}^2_{\mathcal{F}}P^{(2)} = 0$, we have that $t_j(e) = e_j$ and $t_i(e_j) = 0$ for all i and j. Then, since we are considering only the degree zero homomorphisms, it follows that $t_1, \ldots, t_n$ are linearly independent and form a basis of the vector space of endomorphisms factoring through $\mathrm{rad}_{\mathcal{F}}P^{(2)}$. Clearly $t_it_j = 0$ for all i and j, and this means that $\mathrm{End}_R\, P^{(2)} \cong \Bbbk[t_1, \ldots t_n]/\langle t_1, \ldots t_n\rangle^2$ and is commutative.

Consider the universal short exact sequence $0 \to \mathrm{rad}^d_{\mathcal{F}}P^{(d+1)} \to P^{(d+1)} \to P^{(d)} \to 0$ and let $f \in \mathrm{End}_R\, P^{(d+1)}$. Since $\mathrm{rad}^d_{\mathcal{F}}P^{(d+1)}$ is an f-stable submodule, f induces for all $d > 0$ endomorphisms $f_d \in \mathrm{End}_R\, P^{(d)}$, hence algebra homomorphisms $\Phi_d\colon \mathrm{End}_R\, P^{(d+1)} \to \mathrm{End}_R\, P^{(d)}$, for all $d > 0$.

LEMMA 2.7. *The homomorphisms* $\Phi_d\colon \mathrm{End}_R\, P^{(d+1)} \to \mathrm{End}_R\, P^{(d)}$ *are surjective and the diagrams*

$$\begin{array}{ccc} \mathcal{F}^{(d)} & \hookrightarrow & \mathcal{F}^{(d+1)} \\ \downarrow \cong & & \downarrow \cong \\ \mathrm{mod}\,\mathrm{End}_R\, P^{(d)} & \hookrightarrow & \mathrm{mod}\,\mathrm{End}_R\, P^{(d+1)} \end{array}$$

commute for all the natural numbers d.

PROOF. Since $P^{(d)} \in \mathcal{F}^{(d+1)}$ and $P^{(d+1)}$ is projective in $\mathcal{F}^{(d+1)}$, each endomorphism of $P^{(d)}$ can be lifted to an endomorphism of $P^{(d+1)}$, and this proves the surjectivity of the Φ_ds. The equivalences $\mathcal{F}^{(d)} \to \mathrm{mod}\,\mathrm{End}_R\, P^{(d)}$ for $d \geq 1$ are given by the functors $\mathrm{Hom}_R(P^{(d)}, -)$. But $P^{(d+1)}/\mathrm{rad}^d_{\mathcal{F}}P^{(d+1)} \cong P^{(d)}$, and $\mathrm{rad}^d_{\mathcal{F}}P^{(d+1)}$ is in the kernel of every homomorphism $f\colon P^{(d+1)} \to U$ for $U \in \mathcal{F}^{(d)}$. This implies that $\mathrm{Hom}_R(P^{(d+1)}, -)_{|\mathcal{F}^{(d)}} = \mathrm{Hom}_R(P^{(d)}, -)_{|\mathcal{F}^{(d)}}$. □

We now can formulate the conclusion of these constructions:

THEOREM 2.8. *Let $R = R(x_0 \dots, x_n)$ for some positive integer n, and let $M = M_\xi$ be a local R-module of complexity one and dimension 2^n. Then the category $\mathcal{F}(M)$ is equivalent to the category of finite dimensional modules over the power series ring $\Bbbk[[t_1, \dots, t_n]]$.*

PROOF. We will show that $\operatorname{End}_R P^{(d)} \cong \Bbbk[t_1, \dots, t_n]/\langle t_1, \dots, t_n\rangle^d$. Then, the claim will follow by Lemma 2.7 by taking limits. To prove this result, a detailed study of the modules $P^{(d)}$ is necessary. We assume that $x_0 \in \xi$ and we will show by induction on d that:

(A) $\operatorname{rad}_{\mathcal{F}}^{d-1} P^{(d)}$ is $\mathcal{F}$-semisimple, and is a direct sum of $\binom{n+d-2}{d-1}$ copies of M. Note that $\binom{n+d-2}{d-1}$ is the number of pairwise different monomials of degree $d-1$ in $\Bbbk[t_1, \dots, t_n]$. The generators of these copies of M are all in degree zero. Denoting these generators by $e_{i_1 \dots i_{d-1}}$ with $1 \leq i_j \leq n$, we have for all permutations $\pi \in S_{d-1}$:

$$e_{i_1 \dots i_{d-1}} = e_{i_{\pi(1)} \dots i_{\pi(d-1)}}$$

where we consider the indices $i_1 \dots i_{d-1}$ as "monomials" in the "letters" $1, \dots, n$.

(B) For each generating element $e_{i_1 \dots i_{d-2}} + \operatorname{rad}_{\mathcal{F}}^{d-1} P^{(d)}$ of $\operatorname{rad}_{\mathcal{F}}^{d-2} P^{(d)} / \operatorname{rad}_{\mathcal{F}}^{d-1} P^{(d)}$ we have

$$e_{i_1 \dots i_{d-2}} x_0 = \sum_{r=1}^{n} e_{i_1 \dots i_{d-2} r} x_r.$$

Statement (A) is trivially true for $P^{(1)} = M$, and both (A) and (B) were proved in example 2.6 for $d = 2$. Assume that (A) and (B) hold true for all $P^{(s)}$ with $s \leq d$. Therefore we may choose a minimal system of generators of the $\mathcal{F}$-Loewy layers of $P^{(d)}$: $e, e_i, e_{i,j}, \dots e_{i_1 \dots i_{d-1}}$ with $e \in \mathcal{F}$-top of $P^{(d)}$, and with $e_{i_1 \dots i_s}$ in the $\mathcal{F}$-top of $\operatorname{rad}_{\mathcal{F}}^{s} P^{(d)}$, satisfying (A) and (B) for all $s \leq d-1$.

We construct $P^{(d+1)}$ as middle term of the universal short exact sequence

$$0 \to M^c \to P^{(d+1)} \to P^{(d)} \to 0,$$

where $c = \dim_{\Bbbk} \operatorname{Ext}^1_R(P^{(d)}, M) = \dim_{\Bbbk} \underline{\operatorname{Hom}}_R(P^{(d)}, M(1))$. As an $R(x_1 \dots x_n)$-module, M is free since it is indecomposable of maximal Loewy length. Thus, $P^{(d)}$ is equal to a direct sum of $\sum_{s=0}^{d-1} \binom{n+s-1}{s}$ copies of M and is also a free $R(x_1 \dots x_n)$-module. But $P^{(d)}$ is indecomposable as R-module (by the action of x_0) since it is is a local object in the full length category $\mathcal{F}(M)$. In order to evaluate c, let us describe all the graded homomorphisms from $P^{(d)}$ to $M(1)$, as well as those factoring through a free module. Let $g\colon P^{(d)} \to M(1)$ be an R-linear map. Then, g takes each of the generators $e, e_i, \dots e_{i_1 \dots i_{d-1}}$ of (the linear module) $P^{(d)}$ into an element of degree 0 in $M(1)$.

The cyclic module $M(1)$ is generated in degree -1 and let us denote by f its generator. Then, as a $\Bbbk$-vector space, $M(1)_0 = \oplus_{i=1}^n \Bbbk f x_i$. Hence:

$$\begin{array}{ccc} g(e) & = & f \sum_{r=1}^n a_r x_r \\ g(e_i) & = & f \sum_{r=1}^n a_r^i x_r \\ g(e_{ij}) & = & f \sum_{r=1}^n a_r^{ij} x_r \\ \vdots & \vdots & \vdots \\ g(e_{i_1 \dots i_s}) & = & f \sum_{r=1}^n a_r^{i_1 \dots i_s} x_r \\ \vdots & \vdots & \vdots \end{array}$$

By induction, $e_{i_1\dots i_s} = e_{i_{\pi(1)}\dots i_{\pi(s)}}$ for all $1 \le s \le d-1$, so we also get that $a_r^{i_1\dots i_s} = a_r^{i_{\pi(1)}\dots i_{\pi(s)}}$. In this way, for every choice of the scalars $a_r, a_r^i, \dots a_r^{i_1\dots i_s}$, the map g is a well-defined $R(x_1\dots x_n)$-linear map, since $P^{(d)}$ is free over $R(x_1\dots x_n)$. We want to define g as an R-homomorphism, so the action of x_0 has to be considered separately: Since $M(1)$ is annihilated by x_0, we have $g(px_0) = 0$ for all $p \in P^{(d)}$. This implies that

$$0 = g(ex_0) = \sum_{i=1}^n g(e_i)x_i = f\sum_{i=1}^n\sum_{r=1}^n a_r^i x_r x_i.$$

Since $x_rx_i + x_ix_r = 0$ for all r and i, we get $a_r^i = a_i^r$ for all i and r. This means that the set of admissible n^2-tuples $(a_r^i)_{1\le i,r\le n}$ with $a_r^i = a_i^r$ form a subspace of $\Bbbk^{n^2}$ of dimension $\binom{n+1}{2}$.

Consider now for $1 < s < d-1$ the element $e_{i_1\dots i_s}$. We get

$$0 = g(e_{i_1\dots i_s}x_0) = \sum_{j=1}^n g(e_{i_1\dots i_s j}x_j) = f\sum_{j=1}^n\sum_{r=1}^n a_r^{i_1\dots i_s j}x_r x_j.$$

Again, we have $a_r^{i_1\dots i_s j} = a_j^{i_1\dots i_s r}$ for all r and j. Additionally, we see that $a_r^{i_1\dots i_s i_{s+1}} = a_r^{i_{\pi(1)}\dots i_{\pi(s)} i_{\pi(s+1)}}$ for all $\pi \in S_{s+1}$. Consequently we have $a_{i_{s+2}}^{i_1\dots i_s i_{s+1}} = a_{i_{\pi(s+2)}}^{i_{\pi(1)}\dots i_{\pi(s)} i_{\pi(s+1)}}$ for all $\pi \in S_{s+2}$. Hence the families of coefficents $(a_{i_{s+2}}^{i_1\dots i_s i_{s+1}})$ with these relations form $\Bbbk$-vector spaces of dimensions $\binom{n+s}{s+1}$.

Each admissible choice of these coefficients $a_i, a_i^j, \dots, a_{i_{s+2}}^{i_1\dots i_s i_{s+1}}$ defines a unique R-linear map $g\colon P^{(d)} \to M(1)$. Therefore, the $\Bbbk$-dimension of $\mathrm{Hom}_R(P^{(d)}, M(1))$ is $\sum_{j=1}^d \binom{n+j-1}{j}$.

Consider now the vector space $\mathcal{P}_R(P^{(d)}, M(1))$ of maps $P^{(d)} \to M(1)$ factoring through a projective R-module, hence factoring through $R(1)$-the projective cover of $M(1)$. Let $\rho\colon P^{(d)} \to R(1)$ and $\pi\colon R(1) \to M(1)$. Denote the generator of $R(1)$ by $\hat{f}$, so $\pi(\hat{f}) = f$. Since $\pi\rho \in \mathcal{P}_R(P^{(d)}, M(1))$, we get for ρ, similar to the preceding procedure

$$\begin{array}{ccc}
\rho(e) & = & \hat{f}(\sum_{r=1}^n a_r x_r + a x_0) \\
\rho(e_i) & = & \hat{f}(\sum_{r=1}^n a_r^i x_r + a^i x_0) \\
\rho(e_{ij}) & = & \hat{f}(\sum_{r=1}^n a_r^{ij} x_r + a^{ij} x_0) \\
\vdots & \vdots & \vdots \\
\rho(e_{i_1\dots i_s}) & = & \hat{f}(\sum_{r=1}^n a_r^{i_1\dots i_s} x_r + a^{i_1\dots i_s} x_0) \\
\vdots & \vdots & \vdots
\end{array}$$

The coefficients $a_r^{i_1\dots i_s}$ satisfy the same relations as in the first part of the proof ($a_i^j = a_j^i, \dots$), since $\pi\rho \in \mathcal{P}_R(P^{(d)}, M(1)) \subseteq \mathrm{Hom}_R(P^{(d)}, M(1))$. In this way we have defined an $R(x_1\dots x_n)$-linear map. Since it has to be R-linear, the action of x_0 has to be considered as well. We have

$$\rho(e)x_0 = \hat{f}(\sum_{r=1}^n a_r x_r + a x_0)x_0 = \hat{f}\sum_{r=1}^n a_r x_r x_0$$

since $x_0^2 = 0$. On the other hand, we have:

$$\rho(ex_0) = \rho(\sum_j e_j x_j) = \sum_j \hat{f}(\sum_{r=1}^n a_r^j x_r + a^j x_0)x_j = \hat{f}(\sum_{j,r} a_r^j x_r x_j + a^j x_0 x_j).$$

Since $a_i^j = a_j^i$ we have $\sum_{j,r} a_r^j x_r x_j = 0$ and since we want $\rho(ex_0) = \rho(e)x_0$, we see that $a_j = -a^j$, for $1 \le j \le n$. Similarly, we get for $s < d-1$

$$\begin{aligned}\rho(e_{i_1\dots i_s})x_0 &= \hat{f}(\sum_{1\le r\le n} a_r^{i_1\dots i_s} x_r + a^{i_1\dots i_s}x_0)x_0 \\ &= \hat{f}(\sum_{1\le r\le n} a_r^{i_1\dots i_s} x_r x_0) = \rho(e_{i_1\dots i_s}x_0) = \rho(\sum_j e_{i_1\dots i_{s+1,j}} x_j) \\ &= \sum_j \hat{f}(\sum_{1\le r\le n} a_r^{i_1\dots i_{s+1,j}} x_r + a^{i_1\dots i_{s+1,j}}x_0)x_j = \hat{f}\sum_j a^{i_1\dots i_{s+1,j}} x_0 x_j,\end{aligned}$$

since $\sum_j \sum_r a_r^{i_1\dots i_{s+1,j}} x_r x_j = 0$. This means that $a_r^{i_1\dots i_s} = -a^{i_1\dots i_{s+1,r}}$.

Finally, for $d-1$ we get $e_{i_1\dots i_{d-1}}x_0 = 0$ and therefore

$$0 = \rho(e_{i_1\dots i_{d-1}}x_0) = \hat{f}\sum_{r=1}^n a_r^{i_1\dots i_{d-1}} x_r x_0.$$

Thus, all the coefficients $a_r^{i_1\dots i_{d-1}}$ are zero. Notice that the map $\pi\rho$ is independent of the choice of the coefficient $a \in \Bbbk$, and the coefficients $a^{i_1\dots i_s}$ are determined by the $a_{i_s}^{i_1\dots i_{s-1}}$, hence $\dim_\Bbbk \mathcal{P}_R(P^{(d)}, M(1)) = \sum_{j=1}^{d-1}\binom{n+j-1}{j}$. Consequently

$$\begin{aligned}\dim_\Bbbk \operatorname{Ext}_R^1(P^{(d)}, M) &= \dim_\Bbbk \underline{\operatorname{Hom}}_R((P^{(d)}, M(1)) \\ &= \dim_\Bbbk \operatorname{Hom}_R(P^{(d)}, M(1)) - \dim_\Bbbk \mathcal{P}_R(P^{(d)}, M(1)) \\ &= \binom{n+d-1}{d}.\end{aligned}$$

Therefore $\operatorname{rad}_\mathcal{F}^d P^{(d+1)}$ is a direct sum of $\binom{n+d-1}{d}$ copies of M.

Next, we construct generators $e_{i_1\dots i_d}$ for $\operatorname{rad}_\mathcal{F}^d P^{(d+1)}$ satisfying the conditions (A) and (B). For this, we consider a projective cover P of $\operatorname{rad}_\mathcal{F}^{d-1} P^{(d+1)}$ in $\mathcal{F}^{(2)}$. Since $\operatorname{rad}_\mathcal{F}^{d-1} P^{(d)}$ is a direct sum of $\binom{n+d-2}{d-1}$ copies of M, the projective cover of $\operatorname{rad}_\mathcal{F}^{d-1} P^{(d)}$ in $\mathcal{F}^{(2)}$ is a direct sum of $\binom{n+d-2}{d-1}$ copies of $P^{(2)}$. The $\mathcal{F}$-top of $\operatorname{rad}_\mathcal{F}^{d-1} P^{(d+1)}$ is generated by the elements $e_{i_1\dots i_{d-1}} + \operatorname{rad}_\mathcal{F}^d P^{(d+1)}$. Denote the generators of the $\mathcal{F}$-tops of the copies of $P^{(2)}$ in the projective cover P by $f_{i_1\dots i_{d-1}}$, with the convention that $f_{i_1\dots i_{d-1}} = f_{i_{\pi(1)}\dots i_{\pi(d-1)}}$, for all $\pi \in S_{d-1}$. Let us set $f_{i_1\dots i_{d-1}}x_0 = \sum_j f_{i_1\dots i_{d-1}j}x_j$. Then the elements $f_{i_1\dots i_{d-1}j}$ generate the $\mathcal{F}$-radical of P. Let $\gamma\colon P \to \operatorname{rad}_\mathcal{F}^{d-1} P^{(d+1)}$ be the projective cover with $\gamma(f_{i_1\dots i_{d-1}}) = e_{i_1\dots i_{d-1}}$, and define $\gamma(f_{i_1\dots i_{d-1}j}) = e'_{i_1\dots i_{d-1}j}$. As an immediate consequence, we get that $e_{i_1\dots i_{d-1}}x_0 = \sum_j e'_{i_1\dots i_{d-1}j}x_j$.

The identities $0 = -e_{i_1\dots i_{d-2}}x_0^2 = \sum_r e_{i_1\dots i_{d-2}r}x_0 x_r = \sum_r\sum_s e'_{i_1\dots i_{d-2}rs}x_s x_r$ imply that $e'_{i_1\dots i_{d-2}rs} = e'_{i_1\dots i_{d-2}sr}$. But $e_{i_1\dots i_{d-2}i_{d-1}} = e_{i_{\pi(1)}\dots i_{\pi(d-2)}i_{\pi(d-1)}}$ for $\pi \in S_{d-1}$. Therefore the elements $e_{i_1\dots i_{d-2}i_{d-1}i_d} = e'_{i_1\dots i_{d-2}i_{d-1}i_d}$ satisfy the required properties from (A) and (B).

Finally we have to show that $\operatorname{End}_R P^{(d)} = \Bbbk[t_1,\dots,t_n]/\langle t_1,\dots,t_n\rangle^d$. It is trivial for $d = 1$ and by 2.6, we know that $\operatorname{End}_R P^{(2)} = \Bbbk[t_1,\dots,t_n]/\langle t_1,\dots,t_n\rangle^2$ with the action $t_j(e) = e_j$ and $t_i(e_j) = 0$. The functor

$$\operatorname{Hom}_R(P^{(d)}, -)\colon \mathcal{F}^{(d)} \longrightarrow \operatorname{mod} \operatorname{End}_R P^{(d)}$$

is an equivalence and the image of $P^{(d)}$ under this functor is the local algebra $\operatorname{End}_R P^{(d)}$ with identity 1_d. Let Y be module over $\operatorname{End}_R P^{(d)}$. Then each R-linear map $g\colon \operatorname{End}_R P^{(d)} \to Y$ is uniquely determined by the value $g(1_d)$, and for every $y \in Y$ there exists a unique $\operatorname{End}_R P^{(d)}$-homomorphism $g_y\colon \operatorname{End}_R P^{(d)} \to Y$ with $g_y(1_d) = y$. This means that for each object X in the subcategory $\mathcal{F}^{(d)}$, and for every scalar $x \in X_0$, there exists a unique homomorphism $f_x\colon P^{(d)} \to X$ with $f_x(e) = x$ and every R-homomorphism $f\colon P^{(d)} \to X$ must have this form. Since $P_0^{(d)}$ is the $\Bbbk$-span of the set $\{e, e_i, e_{ij}, \dots e_{i_1\dots i_{d-1}}\}$, there exists exactly one endomorphism $t_i \in \operatorname{End}_R P^{(d)}$ such that $t_i(e) = e_i$.

As t_i is R-linear, $t_i(ex_0) = e_i x_0 = \sum_j e_{ij}x_j$ and $t_i(ex_0) = \sum_j t_i(e_j x_j)$, which means $t_i(e_j) = e_{ij} = t_i t_j(e)$. By (A), $e_{ij} = e_{ji}$, so $t_i t_j = t_j t_i$ and the $\binom{n+1}{2}$ different products are linearly independent. We assume by induction that for all $s < d-1$ we have $e_{i_1\dots i_s} = t_{i_1}\cdots t_{i_s}(e)$, hence $t_{i_1}\cdots t_{i_s} = t_{i_{\pi(1)}}\cdots t_{i_{\pi(s)}}$ for all $\pi \in S_s$. All the $\binom{n+s-1}{s}$ monomials in $t_1, \dots, t_n$ of degree s are linearly independent in $\operatorname{End}_R P^{(d)}$. Consider $t_i(e_{i_1\dots i_{d-3}}x_0) = e_{i_1\dots i_{d-3}i}x_0 = \sum_j e_{i_1\dots i_{d-3}ij}x_j = \sum_j t_i(e_{i_1\dots i_{d-3}j})x_j$. Consequently we get

$$t_i(t_{i_1}\cdots t_{i_{d-3}}t_j(e)) = t_i(e_{i_1\dots i_{d-3}j}) = e_{i_1\dots i_{d-3}ji},$$

and so, $e_{i_1\dots i_{d-1}} = e_{i_{\pi(1)}\dots i_{\pi(d-1)}}$ implies that $t_{i_1}\cdots t_{d-1} = t_{i_{\pi(1)}}\cdots t_{i_{\pi(d-1)}}$ for all permutations $\pi \in S_{d-1}$. Therefore, the $\binom{n+d-2}{d-1}$ monomials in $t_1, \dots t_n$ of degree $d-1$ are linearly independent in $\operatorname{End}_R P^{(d)}$. It is easy to show that $t_i(e_{i_1\dots i_{d-1}}) = 0$ for all i and all $(i_1 \dots i_{d-1})$. This means that the algebra $\operatorname{End}_R P^{(d)}$ is isomorphic to $\Bbbk[t_1, \dots, t_n]/\langle t_1, \dots, t_n\rangle^d$. □

Let $M = M_\xi$ and $\mathcal{F}(\{M(i)\}_{i\in\mathbb{Z}})$ denote the full subcategory of $\operatorname{mod} R$ consisting of those graded modules having a filtration with composition factors of the type $M(i)$, $i \in \mathbb{Z}$. Note that the category $\mathcal{F}(\{M(i)\}_{i\in\mathbb{Z}})$ is not an abelian category if $n > 1$.

One consequence of Theorem 1.1 and Corollaries 1.12 and 2.3 is that every indecomposable module of dimension larger than 2^n in $\mathcal{F}(\{M(i)\}_{i\in\mathbb{Z}})$ is a relative extension of two modules in the same subcategory, hence this subcategory is closed under syzygies and by duality, under cosyzygies. Finally, denote by $\underline{\mathcal{F}}$ the image of $\mathcal{F}$ in $\underline{\operatorname{mod}}R$. We have the following consequence of this discussion:

PROPOSITION 2.9. $\mathcal{T}_M = \underline{\mathcal{F}}(\{M(i)\}_{i\in\mathbb{Z}})$.

PROOF. We only need to show that $\underline{\mathcal{F}}(\{M(i)\}_{i\in\mathbb{Z}})$ is closed under triangles in $\underline{\operatorname{mod}}R$. But every triangle $A \to B \to C \to \Omega^{-1}A$ with A and C in $\underline{\mathcal{F}}(\{M(i)\}_{i\in\mathbb{Z}})$ comes from a short exact sequence $0 \to A \to I(A) \oplus B \to C \oplus I \to 0$ where $I(A)$ is the injective envelope of A, and I is some injective R-module. As $\mathcal{F}(\{M(i)\}_{i\in\mathbb{Z}})$ is closed under extensions, $\underline{\mathcal{F}}(\{M(i)\}_{i\in\mathbb{Z}})$ is closed under triangles, and so it is a thick subcategory of the stable module category containing M. □

In order to prove that the thick category generated by a module of the form M_ξ does not contain a proper non zero thick subcategory, we need to recall a few facts. We may assume without loss of generality that $R = R(\xi, x_1, \cdots x_n)$, and thus the polynomial algebra $S = \Bbbk[z, y_1, \cdots, y_n]$, where $\{z, y_1, \cdots, y_n\}$ is the dual basis of $\{\xi, x_1, \cdots x_n\}$. Consider the point S-module $\Bbbk[z] = S/\langle y_1, \cdots, y_n\rangle$. Under the BGG correspondence

$$\underline{\mathrm{mod}}R \to \frac{D^b(\mathrm{mod}\, R)}{\mathcal{P}} \to \frac{D^b(\mathrm{mod}\, S)}{\mathcal{FD}}$$

where $\mathcal{P}$ denotes the thick subcategory of perfect complexes in $D^b(\mathrm{mod}\, R)$ and $\mathcal{FD}$ is the thick subcategory of $D^b(\mathrm{mod}\, S)$ consisting of complexes of finite dimensional modules, the module M_ξ corresponds to the shifted stalk complex $\Bbbk[z][-n]$ in $D^b(\mathrm{mod}\, S)$. In the following we use some arguments, which are well known to topologists, see e.g. [**8**, **25**, **31**] and we are grateful to Greg Stevenson for very helpful conversations. We will need the notion of support of a triangulated subcategory of $D^b(\mathcal{S})$. If C is a complex in $D^b(\mathcal{S})$, we define its support

$$\mathrm{Supp}(C) = \{\mathfrak{p} \in \mathrm{Spec}^h S \mid \ H^*(C)_\mathfrak{p} \neq 0\},$$

where $\mathrm{Spec}^h S$ denotes the homogeneous spectrum of S consisting of the relevant homogeneous prime ideals in S. Then, the support of a triangulated subcategory $\mathcal{A}$ of $D^b(\mathcal{S})$ is the union of the supports of the objects in $\mathcal{A}$.

THEOREM 2.10. *Let $R = R(V)$ be the exterior algebra in $n+1$ indeterminates and let $\xi \in \mathbf{P}(V)$. Let $M = M_\xi$ be a cyclic R-module of complexity one. Then each nonzero element of $\mathcal{T}_M$ generates $\mathcal{T}_M$.*

PROOF. We need to show that $M \in \mathcal{T}_X$ for each nonzero $X \in \mathcal{T}_M$. Let $\sigma(\mathcal{T}_M)$ and $\sigma(\mathcal{T}_X)$ denote the images of $\mathcal{T}_M$ and of $\mathcal{T}_X$ in $D^b(\mathrm{mod}\, S)$ under the BGG correspondence. Both are thick subcategories closed under tensor products since this was the case with $\mathcal{T}_M$ and $\mathcal{T}_X$. But the support of $\sigma(\mathcal{T}_M)$ in $\mathrm{Spec}^h S$, consists only of the ideal $\langle y_1, \cdots, y_n\rangle$ of height n. Therefore, $\mathrm{Supp}(\sigma(\mathcal{T}_M)) = \mathrm{Supp}(\sigma(\mathcal{T}_X))$ and by [**31**, Theorem 3.15] (see also [**8**]), we have $\mathcal{T}_M = \mathcal{T}_X$. □

Summarizing the results of this section, we have:

THEOREM 2.11. *Let $R = R(V)$ be the exterior algebra in $n+1$ indeterminates and let $\mathcal{T}$ be the thick subcategory of $\underline{\mathrm{mod}}R$ consisting of all the modules of complexity one. Then $\mathcal{T} = \bigsqcup_{\xi\in\mathbf{P}(V)} \mathcal{T}_{M_\xi}$ is the decomposition of $\mathcal{T}$ into its irreducible thick subcategories.* □

3. Modules with self-extensions

It is an open problem whether all the indecomposable modules of non-maximal complexity over the exterior algebra have graded self-extension. We cannot give a complete answer even in the complexity one case. We prove in this section that in the case $n = 2$, every module of complexity one or two, has self-extensions, while in the higher dimensional case, we prove this result for linear modules having non maximal complexity. We will use an argument which seems to be well known to geometers, and we are grateful to Greg Stevenson for explaining us how it can be deduced from Böhning's result [**2**, 2.1.4] : A sheaf without self-extensions over $\mathbf{P}(V)$ is locally free. We start with a preliminary result. We have the following ([**1**, 3.8]):

PROPOSITION 3.1. *Let $R = R(x_0, \cdots, x_n)$ and let $S = \Bbbk[x_0, \cdots, x_n]$ where the base field $\Bbbk$ is algebraically closed. Let M be a finitely generated graded R-module. Then $\mathrm{cx}\, M$ equals the Krull dimension of the graded S-module $\mathcal{E}(M) = \mathrm{Ext}^*_R(M, \Bbbk)$.*

This implies that if $\mathcal{E}(M)$ is torsion free, then M must have maximal complexity. In particular, if the sheafification ([**30**]) of $\mathcal{E}(M)$ is locally free, then the complexity of M must equal $n+1$.

THEOREM 3.2. *Let $R = R(x_0, \cdots, x_n)$ and let M be an indecomposable linear module of complexity less or equal to n. Then $\operatorname{Ext}^1_R(M,M) \neq 0$.*

PROOF. Assume that $\operatorname{Ext}^1_R(M,M) = 0$. By Koszul duality, we have that $\operatorname{Ext}^1_S(X,X) = 0$ where $X = \mathcal{E}(M) = \operatorname{Ext}^*_R(M, \Bbbk)$. Let $\widetilde{X}$ be the sheafification of X. It follows from [**17**] that the sheaf $\widetilde{X}$ has no self extension either, and from the proof of [**2**, 2.1.4] that $\widetilde{X}$ is locally free yielding a contradiction to our preceding remarks. □

REMARK 3.3. It would be interesting to have also a proof of the above result that does not require a geometric argument. We have such a proof only in the case that the complexity of our R-linear module equals 1. This proof is presented in an earlier version of this paper, posted in `arXiv:1701.01149`. More precisely, we prove that an indecomposable linear module of complexity one over the exterior algebra in $n+1$ indeterminates has self-extensions.

Since we know the structure of all indecomposable modules of complexity one, see Theorem 1.1 and Corollary 2.3, we can use Mukai's lemma ([**29**, 1.4]) to get a slight generalization:

PROPOSITION 3.4. *Let R be the exterior algebra in $n+1$ indeterminates and let X be an indecomposable module of complexity one minimally generated in degrees $i_0 < i_1 < \ldots < i_{p-1} < i_p$ and assume either $i_1 = i_0 + (n-1)$ or $i_p = i_{p-1} + (n-1)$. Then $\operatorname{Ext}^1_R(X,X) \neq 0$.*

PROOF. Let $i_1 = i_0 + (n-1)$ and let $L = \langle X_{i_0} \rangle$ be the submodule of X generated by X_{i_0}.Then L is a degree shift of a linear module of complexity one, hence $\operatorname{Ext}^1_R(L,L) \neq 0$ by the previous proposition. Consider the short exact sequence $0 \to L \to X \to X/L \to 0$. Then $\operatorname{Hom}_R(L, X/L) = 0$ and $\operatorname{Ext}^2_R(X/L, L) = 0$, hence by [**29**, 1.4] again, $\dim_\Bbbk \operatorname{Ext}^1_R(X,X) \geq \dim_\Bbbk \operatorname{Ext}^1_R(L,L) > 0$. In the second case, $i_p = i_{p-1} + (n-1)$ we use the dual argument. □

COROLLARY 3.5. *Let R be the exterior algebra in $n+1$ indeterminates and let $\mathcal{C}$ be an Auslander-Reiten component in the Auslander-Reiten quiver of* $\operatorname{mod} R$ *containing a shift of a linear module of complexity one. Then all indecomposable modules in $\mathcal{C}$ have self-extensions.*

PROOF. Since $\tau X \cong X(n-1)$ for complexity one modules, any indecomposable module Y in $\mathcal{C}$ has a minimal set of generators in degrees $i_0, i_0 + (n-1), \ldots, i_0 + s(n-1)$ for some integer i_0 and some $s \geq 0$, hence the above proposition applies. □

Using [**29**, 1.4] we get in the two dimensional case:

PROPOSITION 3.6. *Let R be the exterior algebra in 3 indeterminates and let X be an indecomposable module of complexity one or two. Then $\operatorname{Ext}^1_R(X,X) \neq 0$.*

PROOF. If X has complexity one, then X is minimally generated in degrees $i_0, i_0+1, i_0+2, \cdots, i_0+r$, hence it has self extensions by Proposition 3.4. Let X be an indecomposable module such that $\operatorname{cx} X = 2$ and and assume that $\operatorname{Ext}^1_R(X,X) = 0$. By taking syzygies, we may assume without loss of generality that X is weakly

Koszul ([**23**]). Again without loss of generality we may assume that X is minimally generated in degrees $0, 1, \ldots, p$. Then, as in the proof of 3.4 there exists a linear R-module U and a relative extension $0 \to U \to X \to V \to 0$ of weakly Koszul modules of complexity two. Moreover, V is minimally generated in degrees $1, \ldots, p$, and we have $\mathrm{Hom}_R(U, V) = \mathrm{Ext}^2_R(V, U) = 0$. From Mukai's lemma, we infer that $\mathrm{Ext}^1_R(U, U) = 0$. Under Koszul duality and sheafification, U corresponds to a rigid sheaf on the projective plane, so by [**7**], it must be a vector bundle. But the modules corresponding to vector bundles must have maximal complexity by [**1**], hence $\mathrm{cx}\, U = \mathrm{cx}\, X = 3$ and we get a contradiction. □

Appendix A. The projective line case

The thick subcategories of $\underline{\mathrm{mod}}R$ where $R = R(x_0, x_1)$ is the exterior algebra in two indeterminates, were described by Krause in [**22**] using non-crossing partitions. Using the BGG correspondence this yields a description of the thick subcategories of $\mathcal{D}^b(\mathrm{coh}\, \mathbf{P}^1)$. In this appendix we will use an "Auslander-Reiten theoretical approach" to formulate and recapture these results. Note that in the dimension one case, every indecomposable non projective graded R-module that is not simple has Loewy length (therefore graded length) equal to two.

If M is indecomposable with complexity 1, then $\tau M \cong M$, hence M lies in a homogeneous tube Σ, the whole tube is contained in the thick subcategory $\mathcal{T}_M$ and any two indecomposable modules in Σ generate the same thick subcategory. Also for every non zero linear form ξ, the modules M_ξ defined at the beginning of this paper, are uniserial of length two, have complexity one, and are the graded quasi-simple R-modules. Therefore we may assume that the module M is quasi-simple in Σ.

PROPOSITION A.1. *Let M be an indecomposable quasi-simple graded module of complexity 1. Let N be another quasi-simple module of complexity 1 belonging to $\mathcal{T}_M$. Then $N \cong \Omega^{-i} M \cong M(i)$ for some $i \in \mathbb{Z}$.*

PROOF. Using 1.10 we infer that the module M is linear, so it is generated in degree 0. We know that $\mathcal{T}_M$ contains the tube containing M, as well as all the tubes containing each $\Omega^i M = M(-i)$. Since $\mathcal{T}_M$ is connected, there are nonzero maps between the homogeneous tubes contained in $\mathcal{T}_M$. So for every tube in $\mathcal{T}_M$ containing a quasi-simple module L, there is another tube $\mathcal{T}_M$ containing a quasi-simple module N such that we must have $\underline{\mathrm{Hom}}_R(L, N) \neq 0$ or $\underline{\mathrm{Hom}}_R(N, L) \neq 0$. Since L is weakly Koszul it follows from 1.10 that it must be in fact, a (degree) shift of some linear module. Assume that $\underline{\mathrm{Hom}}_R(L, N) \neq 0$. By the preceding remarks and Lemma 1.2., N is either generated in degree 0, or -1. If N is generated in degree 0, since both L and N are uniserial of graded length 2, every nonzero homomorphism from L to N must be an isomorphism, so assume that N is generated in degree -1. But N being also of complexity 1, means that we may write $N = \Omega^{-1} K$ for some linear module K of complexity 1, so we have $\mathrm{Ext}^1_R(L, K) = \underline{\mathrm{Hom}}_R(L, \Omega^{-1} K) \neq 0$. Thus we have a non split extension of linear modules $0 \to K \to X \to L \to 0$ and an induced commutative diagram:

$$\begin{CD}
0 @>>> K @>>> X @>>> L @>>> 0\\
@. @VVfV @VVV @| @.\\
0 @>>> L @>>> Y @>>> L @>>> 0
\end{CD}$$

where the bottom sequence is the Auslander-Reiten sequence ending at L. As before, the homomorphism f is either an isomorphism or zero. But f cannot be zero since this would imply a splitting of the Auslander-Reiten sequence ending at L. So f is an isomorphism, and then $N \cong \Omega^{-1}L$. The case $\underline{\mathrm{Hom}}_R(N, L) \neq 0$ is treated in a similar fashion. $\square$

As an immediate consequence we obtain a description of the thick subcategories generated by indecomposable modules of complexity 1:

PROPOSITION A.2. *Let M be an indecomposable R-module of complexity 1, and let $\mathcal{T}_M$ be the thick subcategory generated by M. Then $\mathcal{T}_M$ is the additive subcategory of $\underline{\mathrm{mod}}^Z R$ generated by $\bigcup_{i\in\mathbb{Z}} \Sigma_i$, where for each integer i, Σ_i denotes the graded Auslander-Reiten component containing the module $M(i)$.*

PROOF. We have seen that $\mathcal{T}_M$ contains the add $\bigcup_{i\in\mathbb{Z}} \Sigma_i$, so we only need to prove that the additive closure of this union is a triangulated subcategory of $\underline{\mathrm{mod}}^{\mathbb{Z}} R$. Let $X \to Y \to Z \to \Omega^{-1}X$ be a triangle where X, Z are in add $\bigcup_{i\in\mathbb{Z}} \Sigma_i$. Obviously Z is also isomorphic to a direct sum of modules of complexity 1. Let C be an indecomposable summand of Y. We may assume that one of $\underline{\mathrm{Hom}}_R(X, C)$ and $\underline{\mathrm{Hom}}_R(C, Z)$ is nonzero. Say we have a non-zero graded map from an indecomposable direct summand X' of X contained in some tube Σ_j to C. As in the proof of the previous lemma, we may assume that both X' and C are quasi-simple and it follows that $C \in \Sigma_j$ or $C \in \Sigma_{j-1}$. $\square$

LEMMA A.3. *Let M be an indecomposable graded module of complexity one, and let $\mathcal{T}$ be a thick subcategory of $\underline{\mathrm{mod}} R$ containing M and a graded simple module. Then $\mathcal{T} = \underline{\mathrm{mod}} R$.*

PROOF. Since $\mathcal{T}_M$ contains all the graded shifts of M, we may assume without loss of generality that M is quasi-simple generated in degree 0. We have an embedding $S(-1) \to M$ and a homomorphism $M \to S$, where S is the unique simple module concentrated in degree zero. From the distinguished triangle

$$S(-1) \to M \to S \to \Omega^{-1}S(-1)$$

we infer that $S(-1) \in \mathcal{T}$ and in this way we get that for each $i \in \mathbb{Z}$, $S(i) \in \mathcal{T}$. Let N be an indecomposable non projective R-module of graded length two. We have a short exact sequence $0 \to S(l)^t \to N \to S(l+1)^s \to 0$ for some positive integers s, t and for an integer l. It follows that $N \in \mathcal{T}$. $\square$

We obtain the following consequence:

COROLLARY A.4. *Let M be an indecomposable graded module of complexity one, and let $\mathcal{T}$ be a thick subcategory of $\underline{\mathrm{mod}} R$ containing M and an indecomposable module belonging to one of the graded transjective components. Then $\mathcal{T} = \underline{\mathrm{mod}} R$.*

PROOF. In view of the previous lemma, it suffices to show that $\mathcal{T}$ contains a graded simple module. Let $X \in \mathcal{T}$ be an indecomposable module of complexity two, and let Γ be the transjective component containing it,

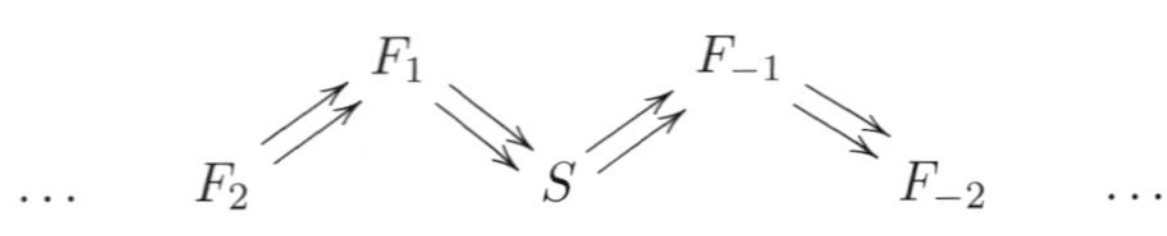

where S is the graded simple module lying in Γ. Assume that $X = F_i$ for some $i \geq 1$. If $i = 2k$ is even, $F_i = \tau^k S = \Omega^{2k} S(2k)$, so the graded simple module $S(2k) \in \mathcal{T}$. If $i = 2k+1$, then $F_i = \tau^k F_1 = \Omega^{2k} F_1(2k)$, implying that $F_1(2k) \in \mathcal{T}$. But $M(2k)$ is also in $\mathcal{T}$ and, as in proofs of the previous lemmas, we have a nonzero homomorphism $f\colon M(2k) \to F_1(2k)$, or a non-zero homomorphism $f\colon F_1(2k) \to M(2k)$. In the first case, since both $F_1(2k)$ and $M(2k)$ are generated in the same degree and have linear resolutions, $\underline{\mathrm{Hom}}_R(M(2k), F_1(2k)) = \mathrm{Hom}_R(M(2k), F_1(2k))$. But $M(2k)$ is quasi-simple, so f is one-to-one and its cokernel is simple. So in this case too, $\mathcal{T}$ contains a graded simple module. The remaining case, and the case when X is of the form F_{-i} for some $i \geq 0$ is treated the same way. □

We turn our attention to the thick subcategories generated by the modules of complexity 2. Let Γ_0 be the transjective component containing the simple module S generated in degree zero. Then for each integer i we have

$$F_{i+1} = \Omega F_i(1) = \Omega^{i+1} S(i+1)$$

where F_0 denotes the simple module S. Let $X \in \Gamma_0$. Then $\mathcal{T}_X$ contains all the syzygies and cosyzygies of X, and it follows that for each integer j, $\mathcal{T}_X$ contains a graded shift of F_j, so it must contain some graded simple module. So when describing $\mathcal{T}_X$ we may assume without loss of generality that X is a graded simple module. We have the following:

LEMMA A.5. *Let S be the graded simple module generated in degree zero. Then $\mathcal{T}_S$ is the additive closure of $\{F_i(-i) | i \in \mathbb{Z}\}$.*

PROOF. $\mathrm{add}\{F_i(-i) | i \in \mathbb{Z}\}$ is obviously closed under syzygies and cosyzygies so to show that it is triangulated we need to show that if two terms of a distinguished triangle are in $\mathrm{add}\{F_i(-i) | i \in \mathbb{Z}\}$, then so is the third. Let X be an indecomposable module of complexity two. Without loss of generality we may clearly assume that $X = F_i = \Omega^i S(i)$ and belongs to the Auslander-Reiten component Γ_0 containing the unique simple module concentrated in degree zero. Then

$$\mathrm{Ext}^1_R(F_i(-i), F_i(-i)) = \mathrm{Ext}^1_R(S, S) = 0.$$

It is also easy to show that for all integers i, j:

$$\underline{\mathrm{Hom}}_R(F_i(-i), F_j(-j)) = \begin{cases} 0 & \text{if } i \neq j. \\ \Bbbk & \text{otherwise.} \end{cases}$$

This clearly implies that each $\mathrm{add}\{F_i(-i) | i \in \mathbb{Z}\}$ is a triangulated subcategory and the proof of the lemma is complete. □

LEMMA A.6. *Let $\mathcal{T}$ be a thick subcategory of $\underline{\mathrm{mod}} R$, containing indecomposable modules $F_i(j)$ and $F_r(s)$ with $i + j \neq r + s$. Then $\mathcal{T} = \underline{\mathrm{mod}} R$.*

PROOF. Assume that $r+s > i+j$. Since for each i, j we have $F_i(j) = \Omega^i S(i+j)$, we infer from $F_i(j) \in \mathcal{T}$ that $S(i+j) \in \mathcal{T}$. Similarly $F_{r+s-(i+j)}(i+j) = \Omega^{r+s} S(i+j)$ is in $\mathcal{T}$. But $\underline{\mathrm{Hom}}_R(F_{r+s-(i+j)}(i+j), S(i+j)) \neq 0$. As in the Kronecker algebra case there exists a short exact sequence

$$0 \to M(i+j) \to F_{r+s-(i+j)}(i+j) \to S(i+j) \to 0,$$

where M is an indecomposable module of complexity 1. From this short exact sequence we get in $\underline{\mathrm{mod}}^Z_R$ a triangle

$$M(i+j) \to F_{r+s-(i+j)}(i+j) \to Si+j) \to \Omega^{-1} M(i+j)$$

Since $F_{r+s-(i+j)}(i+j)$ and $S(i+j)$ are in $\mathcal{T}$, $M(i+j) \in \mathcal{T}$ too. Corollary 2.5 then implies $\mathcal{T} = \underline{\mathrm{mod}}R$. □

We summarize this discussion with the following description of the lattice of thick subcategories of $\underline{\mathrm{mod}}R$:

THEOREM A.7. *Let $R = R(x_0, x_1)$ be the exterior algebra in two indeterminates.*

(1) *The proper thick subcategories of $\underline{\mathrm{mod}}R$ are of the form $\mathcal{T}_\alpha$, where $\phi \neq \alpha \subseteq \mathbf{P}^1$, and* $\mathrm{add}\{F_i(-i+j)$ *for all* $i, j \in \mathbb{Z}\}$.
(2) *The proper thick subcategories of $\underline{\mathrm{mod}}R$ closed under tensor products are only of the form $\mathcal{T}_\alpha$, where $\phi \neq \alpha \subseteq \mathbf{P}^1$.* □

The following picture is useful in illustrating the irreducible thick subcategories consisting of modules of complexity 2.

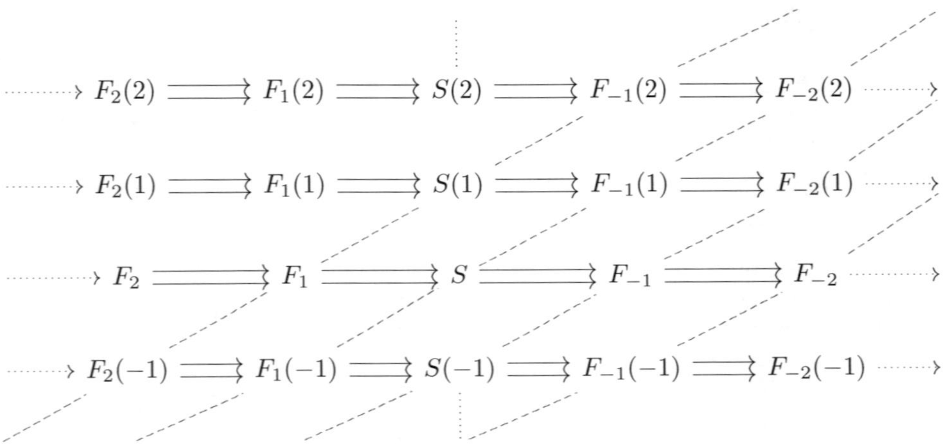

References

[1] Annetta Aramova, Luchezar L. Avramov, and Jürgen Herzog, *Resolutions of monomial ideals and cohomology over exterior algebras*, Trans. Amer. Math. Soc. **352** (2000), no. 2, 579–594, DOI 10.1090/S0002-9947-99-02298-9. MR1603874

[2] Christian Böhning, *Derived categories of coherent sheaves on rational homogeneous manifolds*, Doc. Math. **11** (2006), 261–331. MR2262935

[3] D. J. Benson, Jon F. Carlson, and Jeremy Rickard, *Thick subcategories of the stable module category*, Fund. Math. **153** (1997), no. 1, 59–80, DOI 10.4064/fm-153-1-59-80. MR1450996

[4] David J. Benson, Srikanth Iyengar, and Henning Krause, *Representations of finite groups: local cohomology and support*, Oberwolfach Seminars, vol. 43, Birkhäuser/Springer Basel AG, Basel, 2012. MR2951763

[5] I. N. Bernšteĭn, I. M. Gel′fand, and S. I. Gel′fand, *Algebraic vector bundles on $\mathbf{P}^n$ and problems of linear algebra* (Russian), Funktsional. Anal. i Prilozhen. **12** (1978), no. 3, 66–67. MR509387

[6] Klaus Bongartz, *Tilted algebras*, Representations of algebras (Puebla, 1980), Lecture Notes in Math., vol. 903, Springer, Berlin-New York, 1981, pp. 26–38. MR654701

[7] J.-M. Drezet, *Fibrés exceptionnels et suite spectrale de Beilinson généralisée sur $\mathbf{P}_2(\mathbf{C})$* (French), Math. Ann. **275** (1986), no. 1, 25–48, DOI 10.1007/BF01458581. MR849052

[8] Ivo Dell'Ambrogio and Greg Stevenson, *On the derived category of a graded commutative Noetherian ring*, J. Algebra **373** (2013), 356–376, DOI 10.1016/j.jalgebra.2012.09.038. MR2995031

[9] David Eisenbud, *Periodic resolutions over exterior algebras*, J. Algebra **258** (2002), no. 1, 348–361, DOI 10.1016/S0021-8693(02)00511-2. Special issue in celebration of Claudio Procesi's 60th birthday. MR1958910

[10] Peter Gabriel, *Indecomposable representations. II*, Symposia Mathematica, Vol. XI (Convegno di Algebra Commutativa, INDAM, Rome, 1971), Academic Press, London, 1973, pp. 81–104. MR0340377

[11] Pierre Gabriel, *Des catégories abéliennes* (French), Bull. Soc. Math. France **90** (1962), 323–448. MR232821

[12] A. L. Gorodentsev, S. A. Kuleshov, and A. N. Rudakov, *t-stabilities and t-structures on triangulated categories* (Russian, with Russian summary), Izv. Ross. Akad. Nauk Ser. Mat. **68** (2004), no. 4, 117–150, DOI 10.1070/IM2004v068n04ABEH000497; English transl., Izv. Math. **68** (2004), no. 4, 749–781. MR2084563

[13] Edward L. Green and Roberto Martínez Villa, *Koszul and Yoneda algebras*, Representation theory of algebras (Cocoyoc, 1994), CMS Conf. Proc., vol. 18, Amer. Math. Soc., Providence, RI, 1996, pp. 247–297. MR1388055

[14] E. L. Green, R. Martinez-Villa, I. Reiten, Ø. Solberg, and D. Zacharia, *On modules with linear presentations*, J. Algebra **205** (1998), no. 2, 578–604, DOI 10.1006/jabr.1997.7402. MR1632765

[15] Jinyun Guo, Qiuxian Wu, and Qianhong Wan, *On the tubes of exterior algebras*, Algebra Colloq. **13** (2006), no. 1, 149–162, DOI 10.1142/S1005386706000162. MR2188483

[16] Dieter Happel, *Triangulated categories in the representation theory of finite-dimensional algebras*, London Mathematical Society Lecture Note Series, vol. 119, Cambridge University Press, Cambridge, 1988. MR935124

[17] Dieter Happel and Dan Zacharia, *A note on sheaves without self-extensions on the projective n-space*, Proc. Amer. Math. Soc. **141** (2013), no. 10, 3383–3390, DOI 10.1090/S0002-9939-2013-11305-5. MR3080161

[18] Jürgen Herzog and Srikanth Iyengar, *Koszul modules*, J. Pure Appl. Algebra **201** (2005), no. 1-3, 154–188, DOI 10.1016/j.jpaa.2004.12.037. MR2158753

[19] Michael J. Hopkins, *Global methods in homotopy theory*, Homotopy theory (Durham, 1985), London Math. Soc. Lecture Note Ser., vol. 117, Cambridge Univ. Press, Cambridge, 1987, pp. 73–96. MR932260

[20] Colin Ingalls and Hugh Thomas, *Noncrossing partitions and representations of quivers*, Compos. Math. **145** (2009), no. 6, 1533–1562, DOI 10.1112/S0010437X09004023. MR2575093

[21] Otto Kerner, *Projective covers and injective hulls in abelian length categories*, Rings, modules, algebras, and abelian groups, Lecture Notes in Pure and Appl. Math., vol. 236, Dekker, New York, 2004, pp. 349–352. MR2050723

[22] Henning Krause, *Report on locally finite triangulated categories*, J. K-Theory **9** (2012), no. 3, 421–458, DOI 10.1017/is011011005jkt172. MR2955969

[23] Roberto Martínez-Villa and Dan Zacharia, *Approximations with modules having linear resolutions*, J. Algebra **266** (2003), no. 2, 671–697, DOI 10.1016/S0021-8693(03)00261-8. MR1995131

[24] Roberto Martínez-Villa and Dan Zacharia, *Selfinjective Koszul algebras*, AMA Algebra Montp. Announc. (2003), Paper 5, 5. Théories d'homologie, représentations et algèbres de Hopf. MR2065448

[25] Amnon Neeman, *The chromatic tower for $D(R)$*, Topology **31** (1992), no. 3, 519–532, DOI 10.1016/0040-9383(92)90047-L. With an appendix by Marcel Bökstedt. MR1174255

[26] Christian Okonek, Michael Schneider, and Heinz Spindler, *Vector bundles on complex projective spaces*, Modern Birkhäuser Classics, Birkhäuser/Springer Basel AG, Basel, 2011. Corrected reprint of the 1988 edition; With an appendix by S. I. Gelfand. MR2815674

[27] Alexander Polishchuk and Leonid Positselski, *Quadratic algebras*, University Lecture Series, vol. 37, American Mathematical Society, Providence, RI, 2005. MR2177131

[28] Claus Michael Ringel, *Representations of K-species and bimodules*, J. Algebra **41** (1976), no. 2, 269–302, DOI 10.1016/0021-8693(76)90184-8. MR422350

[29] A. N. Rudakov, *Markov numbers and exceptional bundles on* $\mathbf{P}^2$ (Russian), Izv. Akad. Nauk SSSR Ser. Mat. **52** (1988), no. 1, 100–112, 240, DOI 10.1070/IM1989v032n01ABEH000738; English transl., Math. USSR-Izv. **32** (1989), no. 1, 99–112. MR936525

[30] Jean-Pierre Serre, *Faisceaux algébriques cohérents* (French), Ann. of Math. (2) **61** (1955), 197–278, DOI 10.2307/1969915. MR68874

[31] R. W. Thomason, *The classification of triangulated subcategories*, Compositio Math. **105** (1997), no. 1, 1–27, DOI 10.1023/A:1017932514274. MR1436741

Mathematisches Institut, Heinrich-Heine-Universität, 40225 Düsseldorf, Germany
Email address: kerner@math.uni-duesseldorf.de

Department of Mathematics, Syracuse University, Syracuse, New York 13244
Email address: zacharia@syr.edu

Contemporary Mathematics
Volume **761**, 2021
https://doi.org/10.1090/conm/761/15314

Finite direct sums of cyclic embeddings

Justyna Kosakowska and Markus Schmidmeier

Dedicated to Professor José Antonio de la Peña.

ABSTRACT. In this paper we generalize Kaplansky's combinatorial characterization of the isomorphism types of embeddings of a cyclic subgroup in a finite abelian group given in his 1951 book *"Infinite Abelian Groups"*. For this we introduce partial maps on Littlewood-Richardson tableaux and show that they characterize the isomorphism types of finite direct sums of such cyclic embeddings. We discuss applications to certain embeddings of infinite length, to short exact sequences, to tableaux of special shape, and to endo-submodules.

1. Introduction

In [**1**] Birkhoff stated the problem of classification of subgroups of finite abelian groups. Connected with this problem are results of Kaplansky in [**5**] that present a combinatorial characterization of the isomorphism types of embeddings of a cyclic subgroup in a finite abelian group. This characterization uses sequences of positive integers, see Section 2.1 for details. In this paper we use partial maps on Littlewood-Richardson tableaux (LR-tableaux for short) to generalize this result to finite direct sums of such embeddings. We extend results given in [**9**, Section 2] and present some applications.

Instead of finite abelian groups we work with finitely generated modules over a discrete valuation ring Λ. One of the presented result is Theorem 3.3 saying that a direct sum of cyclic embeddings is uniquely determined by the associated LR-tableau with a partial map satisfying the empty box property, see Section 2.2 for definitions. LR-tableaux were successfully used in [**7**–**11**] to describe algebraic, geometric and combinatorial properties of short exact sequences of Λ-modules. Results obtained there show that relationships between short extact sequences of Λ-modules and LR-tableaux are deep and interesting. This motivates us to collect properties of LR-tableaux of direct sums of cyclic embeddings.

The paper is organized as follows.

In Section 2 we present notation, definitions and basic facts concerning cyclic embeddings. In particular, we recall results of Kaplansky and rewrite them in the language of Littlewood-Richardson tableaux (LR-tableaux).

2010 *Mathematics Subject Classification.* Primary 05E10, 20K27, 47A15.
Key words and phrases. Subgroup embedding, invariant subspace, Littlewood-Richardson tableau, partial map.

In Section 3 we review the correspondence between direct sums of cyclic embeddings and partial maps on LR-tableaux from [**9**, Section 2].

In Section 4 we present four applications of this correspondence. In particular, in Section 4.1, we consider two types of embeddings $A \subset B$, where the ambient space B is not a finite length module. Yet, provided A is p^2-bounded, each such embedding is a direct sum of cyclic embeddings.

Section 4.2 deals with short exact sequences of finite length modules over a PID; we give a combinatorial critereon for the sequence to be a direct sum of cyclic embeddings.

Corollary 4.3 in Section 4.3 states that each LR-tableau of a certain shape can be realized as the LR-tableau of a direct sum of cyclic embeddings. This property is used in [**9, 11**] to study the partition of the variety of short exact sequences of nilpotent linear operators into its irreducible components.

In the final Section 4.4, we describe the endo-submodules of a finite length module over a PID in terms of height sequences and LR-tableaux.

2. Direct sums of cyclic embeddings

Let Λ be a discrete valuation ring with maximal ideal generator p and radical factor field k. By modΛ we denote the category of all (finite length) Λ-modules. It is well known that any finite length Λ-module X is isomorphic to a direct sum

$$X \cong \Lambda/(p^{\alpha_1}) \oplus \Lambda/(p^{\alpha_2}) \oplus \ldots \oplus \Lambda/(p^{\alpha_n}),$$

where $\alpha_1 \geq \alpha_2 \geq \ldots \geq \alpha_n > 0$. The assignment $X \mapsto \alpha = (\alpha_1, \ldots, \alpha_n)$ defines a bijection between the set of isomorphism classes of finite length Λ-modules and the set of all partitions. Given a partition α, we denote by $N_\alpha(\Lambda) = N_\alpha$ the corresponding Λ-module.

Denote by $\mathcal{S}(\Lambda)$ the category of all short exact sequences in modΛ with morphisms given by commutative diagrams. With the componentwise exact structure, $\mathcal{S}(\Lambda)$ is an exact Krull-Remak-Schmidt category. We denote objects in $\mathcal{S}(\Lambda)$ either as short exact sequences $0 \to A \to B \to C \to 0$ of Λ-modules, or as embeddings $(A \subset B)$.

2.1. Cyclic embeddings and poles.

We review Kaplansky's classification of cyclic embeddings.

DEFINITION 2.1. An embedding $(A \subset B)$ is called *cyclic* if the submodule A is cyclic as a Λ-module, that is, if A_Λ is either indecomposable (i.e. of the form $\Lambda/(p^n)$ for some $n \geq 1$) or zero. Examples include the *empty embeddings*, that is, embeddings of the form $(0 \subset B)$. By E_β we denote the empty embedding $(0 \subset N_\beta)$.

A cyclic embedding $(A \subset B)$ is a *pole*, if A is an indecomposable Λ-module (in particular nonzero) and if $(A \subset B)$ has no nonempty direct summand. In Proposition 2.8, we give a detailed description of cyclic embeddings.

For the classification we use height sequences which were originally introduced by Prüfer [**14**] as *Höhenexponenten*. In [**5**, Section 18], the height sequence for an element $a \in B$ is called its *Ulm-sequence*.

DEFINITION 2.2.
- A *height sequence* is a sequence in $\mathbb{N}_0 \cup \{\infty\}$ which is strictly increasing until it reaches ∞ after finitely many steps. A height sequence is *non-empty*, if it has at least one element from $\mathbb{N}_0$.

- Let $B \in \operatorname{mod}\Lambda$. We say an element $a \in B$ has *height* m if $a \in p^m B \setminus p^{m+1}B$. In this case we write $h(a) = m$. By definition, $h(0) = \infty$.
- The *height sequence for* a *in* B is $H_B(a) = (h(p^i a))_{i \in \mathbb{N}_0}$. Usually, we do not write the entries ∞.
- A sequence (m_i) has a *gap* after m_ℓ if $m_{\ell+1} > m_\ell + 1$.

The following result is proved in [**5**, Theorem 25].

PROPOSITION 2.3. *There is a bijection*

$$\{poles\}/_{\cong} \quad \overset{1-1}{\longleftrightarrow} \quad \{\text{finite non-empty strictly increasing sequences in } \mathbb{N}_0\}.$$

EXAMPLE 2.4. Following [**5**], we construct a cyclic embedding $(A \subset B)$ for a given strictly increasing sequence $(m_i)_{0 \le i \le n}$ in $\mathbb{N}_0$. Let $i_1 > i_2 > \cdots > i_s$ be such that (m_i) has gaps exactly after the entries $m_{i_1} > m_{i_2} > \cdots > m_{i_s}$. For $1 \le j \le s$ put $\beta_j = m_{i_j} + 1$ and $\ell_j = m_{i_j} - i_j$, then β and ℓ are strictly decreasing sequences of positive and nonnegative integers, respectively. Let

$$B = N_\beta = \bigoplus_{i=1}^{s} \Lambda/(p^{\beta_i})$$

be generated by elements b_{β_j} of order p^{β_j}. Let $a = \sum_{j=1}^{s} p^{\ell_j} \cdot b_{\beta_j}$ and put $A = (a)$. This yields a cyclic embedding $P((m_i)) = (A \subset B)$.

EXAMPLE 2.5. The height sequence $(1,3,4)$ has gaps after 1 and 4 (so the numbers defined above are $m_0 = 1, m_1 = 3, m_2 = 4, s = 2, i_1 = 2, i_2 = 0, m_{i_1} = 4, m_{i_2} = 1, \beta_1 = 5, \beta_2 = 2, \ell_1 = 2, \ell_2 = 1$), and hence gives rise to the embedding $P((1,3,4)) = ((p^2 b_5 + p b_2) \subset N_{(5,2)})$.

REMARK 2.6. It may be helpful to the reader to visualize cyclic embeddings. The columns of boxes in the picture represent the module $B = N_\beta$, more precisely, the lengths of the columns are the parts of the partition β. The dot or dotted line represents the generator of the submodule $A = (a)$: If a column of boxes corresponds to a summand of B, say generated by b, then a dot in the i-th box from the top represents the element $a = p^{i-1}b$; for each connected line of dots, we take the sum. We will align the columns vertically so that the dotted line becomes horizontal, if this is possible. Here is the picture for the previous example:

$P((1,3,4)):$ 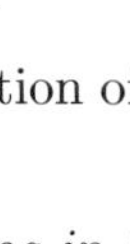

LEMMA 2.7 ([**5**, Lemma 22]). *Suppose the height sequence (m_i) of some element a in B has a gap after m_ℓ. Then $N_{(m_\ell+1)}$ occurs as a direct summand of B.*

PROOF. See [**5**, Lemma 22 and page 29]. □

The following result is a slight generalization of Proposition 2.3. We will use it later.

PROPOSITION 2.8. *Every cyclic embedding in $\mathcal{S}(\Lambda)$ has the form*

$$P((m_i)) \oplus E_\beta$$

for a finite, possibly empty, strictly increasing sequence (m_i) and a possibly empty partition β. More precisely, there is a one-to-one correspondence between the set *of cyclic embeddings up to isomorphy and the set of pairs consisting of a finite, possibly empty, strictly increasing sequence and a partition.*

PROOF. Let $(A \subset B)$ be a cyclic embedding in $\mathcal{S}(\Lambda)$. The category $\mathcal{S}(\Lambda)$ has the Krull-Remak-Schmidt property so $(A \subset B)$ is isomorphic to a direct sum of indecomposable cyclic embeddings. As Λ is a local ring, and as the submodule A is indecomposable or zero, at most one of the summands has a nonzero submodule, hence is a pole, and hence is given uniquely, up to isomorphy, by a finite non-empty strictly increasing sequence in $\mathbb{N}_0$ (Proposition 2.3). (The case where the submodule A is zero is included since the embedding given by the empty height sequence denotes the zero object in $\mathcal{S}(\Lambda)$.) The sum of the remaining indecomposable summands has the form $(0 \subset B')$ where B' is a Λ-module such that $(0 \subset B') \cong E_\beta$ for some partition β. □

REMARK 2.9. Each indecomposable cyclic embedding is either a pole of the form $P((m_i))$ for (m_i) a finite non-empty strictly increasing sequence in $\mathbb{N}_0$ or else an empty embedding of the form $E_{(n)}$ for n a natural number.

2.2. Partitions, tableaux and partial maps. We introduce partial maps on Littlewood-Richardson tableaux and review cyclic embeddings.

DEFINITION 2.10. Let (α, β, γ) be a partition triple. An *LR-tableau* of shape $\beta \setminus \gamma$ and content α is a Young diagram of shape β (the *outer shape* of the tableau) in which the region $\beta \setminus \gamma$ is filled with $(\alpha')_1$ entries 1, $(\alpha')_2$ entries 2, etc., where α' denotes the transpose of α, such that three conditions are satisfied:

- in each row, the entries are weakly increasing;
- in each column, the entries are strictly increasing;
- the lattice permutation property holds, that is, on the right hand side of each column there occur at least as many entries e as there are entries $e+1$ $(e = 1, 2, \ldots)$.

REMARK 2.11. Note that unless the partitions α, β, γ are chosen suitably, LR-tableaux may not exist. In particular, we need $\beta_i \geq \gamma_i$ for all i so that the skew diagram $\beta \setminus \gamma$ is defined. Another necessary condition is that the number $|\alpha|$ of entries equals the number of boxes $|\beta| - |\gamma|$ in the skew diagram. The number of LR-tableaux of a given shape is called the *LR-coefficient;* it is considered a difficult counting problem to determine the LR-coefficient in general, see for example [**3**].

An embedding $(A \subset B)$ with corresponding short exact sequence $0 \to A \to B \to C \to 0$ yields three partitions α, β, γ which describe the isomorphism types of A, B, C, respectively. The Green-Klein Theorem [**6**] states that a partition triple (α, β, γ) can be realized in this way by a short exact sequence if and only if there exists an LR-tableau of shape $\beta \setminus \gamma$ and content α. The tableau of an embedding can be obtained in the following way.

Suppose that the submodule A has Loewy length r. Then the sequence of epimorphisms

$$B = B/p^r A \twoheadrightarrow B/p^{r-1} A \twoheadrightarrow \cdots \twoheadrightarrow B/pA \twoheadrightarrow B/A$$

induces a sequence of inclusions of partitions

$$\beta = \gamma^{(r)} \supset \gamma^{(r-1)} \supset \cdots \supset \gamma^{(1)} \supset \gamma^{(0)} = \gamma,$$

where $N_{\gamma^{(e)}} \simeq B/p^e A$. The partitions define the corresponding LR-tableau $\Gamma = [\gamma^{(0)}, \gamma^{(1)}, \ldots, \gamma^{(r)}]$ as follows. The Young diagram of shape $\beta = \gamma^{(r)}$ is filled in such a way that the region given by $\gamma = \gamma^{(0)}$ remains empty and every box in the skew diagram $\gamma^{(e)} \setminus \gamma^{(e-1)}$ carries the entry $\boxed{e}$. The *shape* of the tableau is the skew diagram $\beta \setminus \gamma$, the *content* is the partition α.

DEFINITION 2.12. The *union* of two tableaux is taken row-wise, so if $\mathrm{E} = [\varepsilon^{(i)}]_{0 \le i \le s}$ and $\mathrm{Z} = [\zeta^{(i)}]_{0 \le i \le t}$ are tableaux then the partitions $\gamma^{(i)}$ for $\Gamma = \mathrm{E} \cup \mathrm{Z}$ are given by taking the union (of ordered multi-sets): $\gamma^{(i)} = \varepsilon^{(i)} \cup \zeta^{(i)}$ where $\varepsilon^{(i)} = \varepsilon^{(s)}$ for $i \ge s$ and $\zeta^{(i)} = \zeta^{(t)}$ for $i \ge t$.

REMARK 2.13. It is easy to check that if embeddings $(A \subset B)$, $(C \subset D)$ have tableaux Γ_1, Γ_2, respectively, then the direct sum $(A \subset B) \oplus (C \subset D)$ has tableau $\Gamma = \Gamma_1 \cup \Gamma_2$.

DEFINITION 2.14. *Column tableaux* (which we call *columns* if the context is clear) are tableaux which have only one column, and where the entries in this column are subsequent natural numbers (not necessarily starting at 1). Each column is one of the following: For $1 \le e \le f$, we denote by $C(e,f)_n$ the column tableau of height n with entries $e, \ldots, f$. Formally, $C(e,f)_n = [\gamma^{(0)}, \ldots, \gamma^{(f)}]$ where $\gamma^{(0)} = \cdots = \gamma^{(e-1)} = (n-f+e-1), \gamma^{(e)} = (n-f+e), \ldots, \gamma^{(f)} = (n)$. By $C(1,0)_n$ we denote the empty column of length n. Note that a column tableau $C(e,f)_n$ may contain empty boxes, this is the case when $n > f-e+1$ holds. Also, the column tableau satisfies the lattice permutation property, and hence is an LR-tableau, if and only if $e = 1$.

EXAMPLE 2.15. The tableau for the embedding $P(1,3,4)$ in Example 2.5 is a union of two column tableaux:

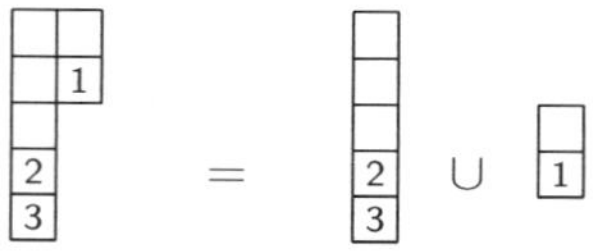

Note that in a union of column tableaux, the vertical columns need no longer form column tableaux.

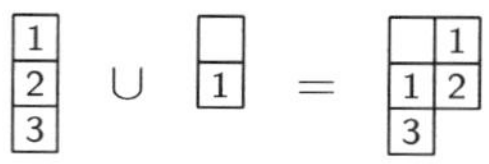

DEFINITION 2.16. A *partial map* g on an LR-tableau Γ assigns to each box $\boxed{e}$ with $e > 1$ a box $\boxed{d}$ with entry $d = e - 1$ such that

(1) g is one-to-one,
(2) for each box b, the row of $g(b)$ is above the row of b.

An *orbit* of a partial map g is a sequence $\boxed{e}, g(\boxed{e}), g^2(\boxed{e}), \ldots, g^{e-1}(\boxed{e})$ given by a box $\boxed{e}$ which is not in the image of g. Thus, the orbits of g are in one-to-one correspondence with the boxes $\boxed{e}$ not in the image of g (the correspondence is given by taking the first box in the orbit), and in one-to-one correspondence with the boxes that have entry 1 (where the correspondence is given by taking the last box).

REMARK 2.17. Given an LR-tableau Γ, the existence of at least one partial map follows from the lattice permutation property.

Given a partial map g on an LR-tableau Γ, a *jump* in row r is a box b in this row with the property that either b has entry $\boxed{1}$ or that $g(b)$ occurs in row s with $s < r - 1$. We say a partial map g on Γ has the *empty box property (EBP)* if for each row r, there are at least as many columns in Γ of exactly $r - 1$ empty boxes as there are jumps in row r.

EXAMPLE 2.18. Consider the following LR-tableau

Γ : 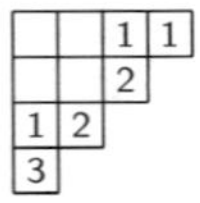

To define a partial map, we have to assign to each box $\boxed{3}$ a corresponding box $\boxed{2}$ and to each box $\boxed{2}$ a corresponding box $\boxed{1}$. Note that we can do this in four different ways. To specify the maps, we distinguish boxes of the same entry by indicating the row on the right; for the two boxes in the first row, we use the letters L and R. Consider the partial map g defined as follows:

$$g : \boxed{3}_4 \mapsto \boxed{2}_3,\ \boxed{2}_3 \mapsto \boxed{1}_L,\ \boxed{2}_2 \mapsto \boxed{1}_R$$

Note that the partial map g has three orbits (one consists only of the box $\boxed{1}_3$). It satisfies (EBP) since there is a column of 2 empty boxes corresponding to the jump $\boxed{2}_3 \mapsto \boxed{1}_1$. However, the partial map

$$g' : \boxed{3}_4 \mapsto \boxed{2}_2,\ \boxed{2}_3 \mapsto \boxed{1}_L,\ \boxed{2}_2 \mapsto \boxed{1}_R$$

does not satisfy (EBP), because there is no column of 3 empty boxes corresponding to the jump at $\boxed{3}_4 \mapsto \boxed{2}_2$.

We collect some properties of tableaux for cyclic embeddings.

PROPOSITION 2.19. *Suppose the embedding $(A \subset B)$ is cyclic with A a module of Loewy length r.*

(1) *In the tableau, each entry $1, \ldots, r$ occurs exactly once.*
(2) *The height sequence (m_i) of the submodule generator a determines the rows in Γ in which the entries occur, and conversely. More precisely, the entry e occurs in row $m_{e-1} + 1$.*
(3) *The LR-tableau Γ of a cyclic embedding is a union of columns with disjoint entries.*
(4) *There is a unique partial map on Γ; it satisfies the (EBP).*

PROOF. (1) For the first statement note that for each $0 < e \le r$, the module $B/p^{e-1}A$ is a factor module of B/p^eA of colength one.

(2) For any embedding, the number of boxes in the first m rows of $\gamma^{(e)}$ is the length of $(B/p^eA)/((p^mB + p^eA)/p^eA) = B/(p^eA + p^mB)$. Here, the modules $B/(p^{e-1}A + p^mB)$ and $B/(p^eA + p^mB)$ have the same length if $m \le m_{e-1}$ and different lengths otherwise since $p^{e-1}a \in p^{m_{e-1}}B \setminus p^{m_{e-1}+1}B$.

(3) By construction of $P((m_i))$, each gap in the height sequence gives rise to a new summand of B and hence to a new column in the tableau. It follows from (2) that Γ is a union of columns.

(4) There is a unique partial map on Γ given by assigning to a box $\boxed{e}$ where $e > 1$ the unique box with entry $e - 1$. Since Γ is a union of columns, the (EBP) is satisfied. $\square$

COROLLARY 2.20. (1) *There is a one-to-one correspondence between cyclic embeddings up to isomorphy and LR-tableaux in which each entry occurs exactly once.*

(2) *Under this correspondence, a pole corresponds to a tableau in which the number of columns equals the number of gaps in the height sequence. An empty embedding E_β corresponds to the empty tableaux on the Young diagram for β.*

PROOF. (1) By the Green-Klein theorem, any LR-tableau Γ can be realized as the tableau of an embedding $(A \subset B)$. If Γ has exactly one box $\boxed{1}$, then A is an indecomposable Λ-module since $\dim A/pA = \dim B/pA - \dim B/A = \#\{\text{boxes } \boxed{1}\}$. The height sequence of the generator of A can be read off using Proposition 2.19, (2). The uniqueness of the isomorphism type follows from Proposition 2.8.

(2) Regarding poles, we have seen in the proof of Proposition 2.3 that the summands of the ambient space of a pole correspond to the gaps in the height sequence of the submodule generator. The result follows from Proposition 2.8. □

3. Direct sums of cyclic embeddings

We can now generalize Kaplansky's results and give a combinatorial description for finite direct sums of cyclic embeddings, up to isomorphy.

DEFINITION 3.1.

- Two partial maps g, g' on an LR-tableau Γ are *equivalent* if $g' = \pi^{-1} \circ g \circ \pi$ for some permutation π on the set of non-empty boxes in Γ which preserves the entry and the row.
- Two pairs (Γ, g), (Δ, h), each consisting of an LR-tableau and a partial map on the tableau, are *equivalent* if $\Gamma = \Delta$ and if the partial maps g, h are equivalent. In this case we write $(\Gamma, g) \sim (\Delta, h)$.

EXAMPLE 3.2. Consider the LR-tableau from the example in Section 2.2.

$$\Gamma : \quad \begin{array}{|c|c|c|c|} \hline \ & \ & 1 & 1 \\ \hline \ & \ & 2 \\ \cline{1-3} 1 & 2 \\ \cline{1-2} 3 \\ \cline{1-1} \end{array}$$

The maps g and g' are not equivalent. But g is equivalent to the partial map

$$h : \boxed{3} \mapsto \boxed{2}\text{-}3,\ \boxed{2}\text{-}3 \mapsto \boxed{1}\text{-}R,\ \boxed{2}\text{-}2 \mapsto \boxed{1}\text{-}L.$$

(There is a fourth partial map on Γ, it is equivalent to g'.)

Clearly, given two equivalent partial maps on an LR-tableau, one satisfies (EBP) if and only if the other does.

We recall from [**9**, Theorem 2.4]:

THEOREM 3.3. *There is a one-to-one correspondence*

$$\{\bigoplus \textit{cyclic embeddings}\}/_{\cong} \quad \overset{1-1}{\longleftrightarrow} \quad \{(\Gamma, g) : g \textit{ partial map on } \Gamma \textit{ with (EBP)}\}/_{\sim}$$

between the isomorphism types of direct sums of cyclic embeddings, and the equivalence classes of pairs (Γ, g) where Γ is an LR-tableau and g a partial map on Γ which satisfies (EBP).

There is a different tool to decide for a given LR-tableau if it is the tableau of a direct sum of cyclic embeddings.

COROLLARY 3.4. *An LR-tableau* Γ *is the tableau of a direct sum of cyclic embeddings if and only if* Γ *is a union of columns.*

For the proof, see [**9**, Lemma 2.5].

We conclude this section with examples from [**9**, pp. 251–252].

EXAMPLE 3.5. In general, the embedding corresponding to a given LR-tableau is not determined uniquely, up to isomorphy, by the tableau alone. In each case, the following embeddings will have the same given LR-tableau.

(1) Two nonisomorphic sums of poles, given by nonequivalent partial maps.

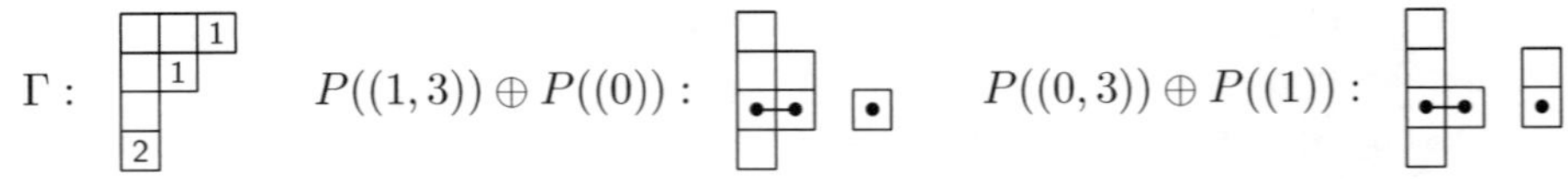

(2) A sum of two poles vs. a sum of two poles and an empty embedding.

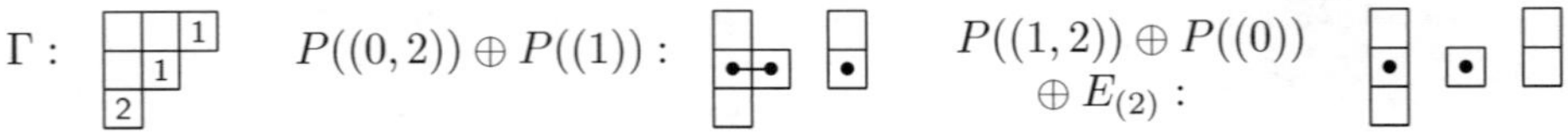

(3) The sum of poles is determined uniquely by the partial map; but there is also an embedding which is not a sum of poles.

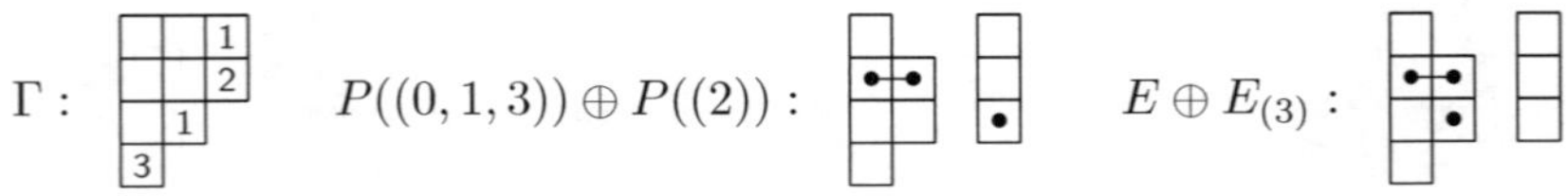

(4) Two partial maps, up to equivalence, only one has the (EBP).

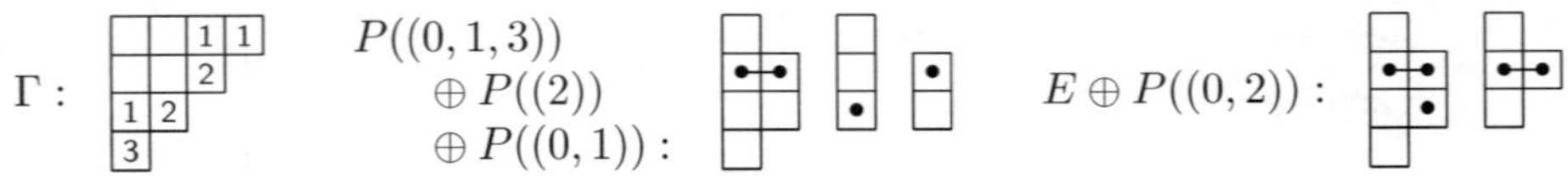

(5) Two partial maps with (EBP), up to equivalence.

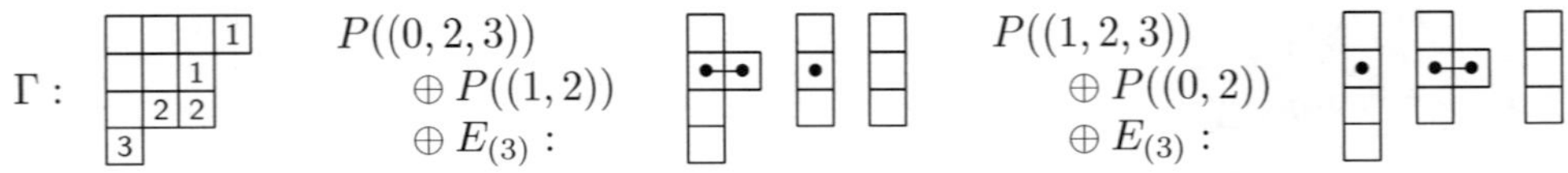

4. Applications of height sequences

We describe four situations in which height sequences or direct sums of cyclic embeddings play a role.

4.1. p^2**-bounded embeddings in infinite length modules.** In this paragraph, B may be any Λ-module, not necessarily finitely generated.

COROLLARY 4.1. *Suppose that* $(A \subset B)$ *is an embedding of* Λ*-modules where* A *is* p^2*-bounded and where* B *is either finitely generated or bounded. Then the embedding is a direct sum of cyclic embeddings.*

PROOF. The case where B is finitely generated is covered in [**13**, Corollary 5.4]. There is exactly one indecomposable embedding where B is not bounded, namely $(0 \subset \Lambda)$.

For the bounded case, note that an embedding $(A \subset B)$ with $p^2A = 0 = p^nB$ can be considered to be a module over the ring

$$R = \begin{pmatrix} \Lambda/(p^2) & \Lambda/(p^2) \\ 0 & \Lambda/(p^n) \end{pmatrix},$$

so [**12**, Theorem 1] yields the claim since R has finite representation type. □

4.2. Direct sums of cyclic embeddings in the Green-Klein Theorem. Let α, β, γ be partitions. Recall from [**6**] that there exists a short exact sequence

$$\mathcal{E} : 0 \to N_\alpha \to N_\beta \to N_\gamma \to 0$$

if and only if there exists an LR-tableau of shape $\beta \setminus \gamma$ and content α. As a consequence, we obtain

COROLLARY 4.2. *Given partitions α, β, γ, there exists an embedding $N_\alpha \to N_\beta$ with cokernel N_γ which is a direct sum of cyclic embeddings if and only if there is an LR-tableau of shape $\beta \setminus \gamma$ and content α which is a union of columns.*

PROOF. Using Corollary 3.4 it is possible to decide whether there exists an embedding $N_\alpha \to N_\beta$ with cokernel N_γ which is a direct sum of cyclic embeddings. □

4.3. Tableaux which are horizontal strips. The skew diagram $\beta \setminus \gamma$ is said to be a *horizontal strip* if $\beta_i \leq \gamma_i + 1$ holds for all i, and a *vertical strip* if $\beta' \setminus \gamma'$ is a horizontal strip. We are particularly interested in the following situation.

COROLLARY 4.3. *Suppose Γ is an LR-tableau of shape (α, β, γ) such that $\beta \setminus \gamma$ is a horizontal strip. Then there exists an embedding $(A \subset B)$ which is a direct sum of cyclic embeddings and which has tableau Γ.*

PROOF. Since Γ is an LR-tableau, there exists a partial map g on Γ. Since $\beta \setminus \gamma$ is a horizontal strip, each column in the tableau Γ has at most one entry, so the (EBP) is trivially satisfied. The embedding corresponding to the partial map g is a direct sum of cyclic embeddings, and it has the tableau Γ. □

REMARK 4.4. In the situation of Corollary 4.3, finite direct sums of cyclic embeddings turn out to be a useful tool for the study of the variety $\mathbb{V}^\beta_{\alpha,\gamma}$ of all embeddings $N_\alpha \to N_\beta$ with cokernel N_γ; here, $\Lambda = k[[T]]$ is the power series ring with coefficients in an algebraically closed field. Namely, each irreducible component of $\mathbb{V}^\beta_{\alpha,\gamma}$ has the form $\overline{\mathbb{V}}_\Gamma$ for some LR-tableau Γ of shape (α, β, γ). The boundary relation for tableaux Γ, Δ is given by the condition that $\overline{\mathbb{V}}_\Gamma \cap \mathbb{V}_\Delta \neq \emptyset$. In [**9**, Theorem 5.1] it is shown that box moves in LR-tableaux imply the boundary relation by constructing one-parameter families of embeddings which are all finite direct sums of cyclic embeddings.

4.4. Endo-submodules of Λ-modules. Given a Λ-module B, we obtain a precise description for the submodules of B when considered as a module over its endomorphism ring.

We recall from [**5**, Theorem 24] that a height sequence (m_i) defines an End(B) submodule consisting of all elements of B with height sequence $\geq (m_i)$ (this is the "fully transitive" property). The partial ordering on height sequences is given by $(m_i) \geq (q_i)$ if $m_i \geq q_i$ holds for all $i \in \mathbb{N}$.

It is shown in [**5**, Theorem 25] that every End(B)-submodule of B is cyclic (as an $\mathrm{End}_\Lambda(B)$-module), and hence is determined uniquely, up to equality, by the height sequence of a generator.

Suppose $a \in B$ has height sequence (m_i) with gaps exactly after $m_{i_1} > m_{i_2} > \cdots > m_{i_s}$. We can specify the corresponding End(B)-submodule X explicitly. Namely, with $a = \sum_j p^{\ell_j} b_{\beta_j}$ as above, also the summands $p^{\ell_j} b_{\beta_j}$ are in X, and so are their images under endomorphisms of B. Put $k_j = \beta_j - \ell_j$ for $1 \leq j \leq s$, then

$$\mathrm{End}(B) \cdot a = \sum_{j=1}^{s} (\mathrm{rad}^{\ell_j} B \cap \mathrm{soc}^{k_j} B).$$

With this notation, the endo-submodule End$(B) \cdot a$ can also be written as an intersection of sums:

$$\mathrm{End}(B) \cdot a = \mathrm{rad}^{\ell_1} B \cap (\mathrm{soc}^{k_1} B + \mathrm{rad}^{\ell_2} B) \cap \cdots \cap (\mathrm{soc}^{k_{s-1}} B + \mathrm{rad}^{\ell_s} B) \cap \mathrm{soc}^{k_s} B$$

To verify this, use induction and the modular law.

The following result leads to a quick proof for the formula for the number of End(B)-submodules of a finite length Λ-module B.

COROLLARY 4.5. *Suppose the Λ-module B is given by a partition β. Any two of the following sets are in one-to-one correspondence.*

(1) *The set of* End(B)*-submodules of B.*
(2) *The set of embeddings of the form $((a) \subset B)$, up to isomorphy.*
(3) *The set of LR-tableaux with outer shape β in which each entry occurs exactly once.*

As a consequence, the Λ-module $B = N_\beta$ has exactly $\prod_i (1 + \beta_i - \beta_{i+1})$ many endo-submodules.

PROOF. In each of the sets, the elements are given by height sequences. The correspondence follows from Corollary 2.20, and Proposition 2.19, 2., together with the isomorphy between the lattices of End(B)-submodules of B and of height sequences of elements in B in [**5**, Theorem 25]. For the formula we count the third set. Note that tableau properties require that each entry be the rightmost entry in its row since there are no multiple entries. Hence in the i-th column, there may be between 0 and $\beta_i - \beta_{i+1}$ many entries. The expression is the product of the number of choices for the number of entries in each column. □

Acknowledgment

The authors wish to dedicate this paper to Professor José Antonio de la Peña on the occasion of his 65th birthday. Throughout their careers, Professor de la Peña has shown an active interest in their work, supported them through invitations to important conferences, and inspired them through his dedication to representation theory and his interest in applications. His exemplary paper [**2**] has been a model for our work on this manuscript and the related papers [**9**]–[**11**]: Combinatorial invariants control families of modules which turn out to play a meaningful role geometrically.

References

[1] Garrett Birkhoff, *Subgroups of Abelian Groups*, Proc. London Math. Soc. (2) **38** (1935), 385–401, DOI 10.1112/plms/s2-38.1.385. MR1576323

[2] Thomas Brüstle, José Antonio de la Peña, and Andrzej Skowroński, *Tame algebras and Tits quadratic forms*, Adv. Math. **226** (2011), no. 1, 887–951, DOI 10.1016/j.aim.2010.07.007. MR2735778

[3] Calin Chindris, Harm Derksen, and Jerzy Weyman, *Counterexamples to Okounkov's log-concavity conjecture*, Compos. Math. **143** (2007), no. 6, 1545–1557, DOI 10.1112/S0010437X07003090. MR2371381

[4] Roger Hunter, Fred Richman, and Elbert Walker, *Finite direct sums of cyclic valuated p-groups*, Pacific J. Math. **69** (1977), no. 1, 97–104. MR435252

[5] Irving Kaplansky, *Infinite abelian groups*, University of Michigan Press, Ann Arbor, 1954. MR0065561

[6] T. Klein, *The multiplication of Schur-functions and extensions of p-modules*, J. London Math. Soc. **43** (1968), 280–284, DOI 10.1112/jlms/s1-43.1.280. MR228481

[7] Justyna Kosakowska and Markus Schmidmeier, *Operations on arc diagrams and degenerations for invariant subspaces of linear operators*, Trans. Amer. Math. Soc. **367** (2015), no. 8, 5475–5505, DOI 10.1090/S0002-9947-2014-06206-5. MR3347180

[8] Justyna Kosakowska and Markus Schmidmeier, *Arc diagram varieties*, Expository lectures on representation theory, Contemp. Math., vol. 607, Amer. Math. Soc., Providence, RI, 2014, pp. 205–224, DOI 10.1090/conm/607/12088. MR3205112

[9] Justyna Kosakowska and Markus Schmidmeier, *Box moves on Littlewood-Richardson tableaux and an application to invariant subspace varieties*, J. Algebra **491** (2017), 241–264, DOI 10.1016/j.jalgebra.2017.07.025. MR3699096

[10] Justyna Kosakowska and Markus Schmidmeier, *The boundary of the irreducible components for invariant subspace varieties*, Math. Z. **290** (2018), no. 3-4, 953–972, DOI 10.1007/s00209-018-2047-8. MR3856839

[11] Justyna Kosakowska, Markus Schmidmeier, and Hugh Thomas, *Two partial orders for standard Young tableaux*, Electron. J. Combin. **26** (2019), no. 3, Paper No. 3.20, 18, DOI 10.37236/5967. MR3992295

[12] Audrey Moore, *The Auslander and Ringel-Tachikawa theorem for submodule embeddings*, Comm. Algebra **38** (2010), no. 10, 3805–3820, DOI 10.1080/00927870903286843. MR2760692

[13] Carla Petroro and Markus Schmidmeier, *Abelian groups with a p^2-bounded subgroup, revisited*, J. Algebra Appl. **10** (2011), no. 3, 377–389, DOI 10.1142/S0219498811004823. MR2812117

[14] Heinz Prüfer, *Untersuchungen über die Zerlegbarkeit der abzählbaren primären Abelschen Gruppen* (German), Math. Z. **17** (1923), no. 1, 35–61, DOI 10.1007/BF01504333. MR1544601

Faculty of Mathematics and Computer Science, Nicolaus Copernicus University, ul. Chopina 12/18, 87-100 Toruń, Poland

Email address: justus@mat.umk.pl

Department of Mathematical Sciences, Florida Atlantic University, 777 Glades Road, Boca Raton, Florida 33431

Email address: markus@math.fau.edu

Contemporary Mathematics
Volume **761**, 2021
https://doi.org/10.1090/conm/761/15315

The algebraic theory of fuchsian singularties

Helmut Lenzing

Abstract. This article focusses on fuchsian singularities of the first kind. It has the following aims:

(1) Extend their theory to algebraically closed base fields of arbitrary characteristic.
(2) Relate them to mathematical objects of different nature.
(3) Provide for them a purely ring-theoretic definition.
(4) Explore their singularity categories and Grothendieck groups.

1. Introduction

A fuchsian singularity, as considered in this article, is classically defined — for a discrete cocompact subgroup G of the automorphism group $PSL(2,\mathbb{R})$ of the upper complex halfplane $\mathbb{H}$ — as the graded $\mathbb{C}$-algebra of G-invariant holomorphic differential forms. These graded rings of automorphic forms relate to many mathematical objects of quite different appearance.

This article has three major aims:

(1) Extend the notion of a fuchsian singularity to algebraically closed base field of arbitrary characteristic.
(2) Discuss their relationship to mathematical objects of a different nature.
(3) Provide for them a purely ring-theoretic definition.
(4) Explore their singularity categories and Grothendieck groups.

In particular, the last task provides an easy, systematic, access to produce members of the rare species of fuchsian singularities that are hypersurfaces or complete intersection. We recall for this, that for underlying genus zero there are just 22 fuchsian hypersurface singularities, with 14 of them forming Arnold's famous strange duality list of exceptional unimodal singularities, compare [**9**] and [**22**].

More specifically, and working over an algebraically closed field of arbitrary characteristic, we are going to show that there exist natural bijective correspondences between isomorphism classes of the following mathematical objects:

(i) Weighted smooth projective curves (WPC's) of negative orbifold Euler characteristic.
(ii) Hereditary noetherian categories with Serre duality (HNC's) of negative orbifold Euler characteristic.
(iii) $\mathbb{Z}$-graded local isolated Gorenstein singularities of Krull dimension two and Gorenstein parameter -1 (IGS's)

(iv) $\mathbb{Z}$-graded orbit algebras, corresponding to the τ-orbit of the structure sheaf on a (WPC) of negative orbifold Euler characteristic.

(v) For $k = \mathbb{C}$

(a) Fuchsian subgroups of $\mathrm{Aut}(\mathbb{H})$, the group of biholomorphic maps of the hyperbolic plane $\mathbb{H}$.

(b) The graded algebras of G-invariant differential (automorphic) forms on the upper complex half-plane, attached to a fuchsian group G.

In Section 2 we provide the definitions of the above concepts and establish natural bijections between their isomorphism classes.

In Section 3 we introduce the triangulated singularity category $\mathcal{T}$ attached to a graded fuchsian algebra R. Let $\mathbb{X}$ be the (WPC), attached to R; we show that $\mathcal{T}$ is a one-point extension of the bounded derived category $\mathrm{D}^b(\mathrm{coh}(\mathbb{X}))$ of coherent sheaves on $\mathbb{X}$. We further discuss various incarnations of the singularity category, describe their Grothendieck groups resp. their reduced or numerical Grothendieck groups and their relationship to the respective Grothendieck groups of $\mathrm{coh}(\mathbb{X})$. In particular, we establish a connection between the Coxeter polynomials for $\mathrm{coh}(\mathbb{X})$ and for $\mathcal{T}$ with the Poincaré series of R, compare [**9**] for a similar result.

Finally, in Appendix A we systematically summarize properties of coherent sheaves on smooth projective curves (resp. compact Riemann surfaces), and more generally on (WPC's) (Appendix B).

This paper is a continuation of previous work with José Antonio de la Peña dealing with the special case of underlying genus zero, that is, of fuchsian singularities associated to weighted projective lines and their relationship to extended canonical algebras [**22**].

2. Basic definitions and description of the correspondences

2.1. Weighted projective curves.

DEFINITION 2.1. A ***weighted smooth projective curve***, weighted projective curve or (WPC) for short, is a pair $\mathbb{X} = (X, w)$ consisting of a smooth projective curve X and a function $w : X \to \mathbb{N}_{\geq 1}$ to the integers ≥ 1 such that $w(x) > 1$ for only finitely many points.

An isomorphism of (WPC's) $f := (X, w) \to (X', w')$ is a biregular map $f : X \to X'$ commuting with the weight functions, that is, satisfying $w = w' \circ f$.

If $\mathbb{X} = (X, w)$ is a (WPC), the points $x_1, x_2, \ldots, x_t$ with $w(x_i) > 1$ are called ***weighted points***, the other ones ***ordinary points*** and $w_1 = w(x_1), \ldots, w_t = w(x_t)$ are called the ***weights*** of $\mathbb{X}$. In contexts, where the position of the weights does not matter, we also use the notation $\mathbb{X} = X\langle w_1, w_2, \ldots, w_t\rangle$.

A (WPC) $\mathbb{X} = (X, w)$ has an ***orbifold Euler characteristic***, see Theorem B.1, given by:

$$\chi_{\mathbb{X}} = \chi_X - \sum_{x \in X} \left(1 - \frac{1}{w(x)}\right).$$

We note that in the above sum only the finitely many points x with $w(x) > 1$ matter.

2.2. The category of coherent sheaves on a (WPC). We now describe a two step procedure associating to a (WPC) $\mathbb{X} = (X, w)$ a category of coherent sheaves $\mathrm{coh}(\mathbb{X})$. We first form the category $\mathrm{coh}(X)$ of (algebraic) coherent sheaves on X. (If $k = \mathbb{C}$ and X is given as a compact Riemann surface then we form instead the category $\mathrm{coh}(X)$ of holomorphic coherent sheaves.) Next, for each weighted point x_i of X having weight w_i we form, following [**16**] or [**19**], iteratively the so-called category of w_i-cycles. The result is an enlargement $\mathrm{coh}(\mathbb{X})$ of $\mathrm{coh}(X)$, containing $\mathrm{coh}(X)$ as a full subcategory, and serving as the category of coherent sheaves on $\mathbb{X}$. In particular, the structure sheaf $\mathcal{O}_X$ will also be chosen as the structure sheaf $\mathcal{O}_{\mathbb{X}}$ for $\mathbb{X}$. Alternatively, one can construct $\mathrm{coh}(\mathbb{X})$ by means of a sheaf of orders, see [**27**] for details.

A visible effect of the iterated cycle construction is that for an ordinary point p there is still a unique simple sheaf, concentrated in p. By contrast, for an 'exceptional point' x_i of weight w_i a full Auslander-Reiten orbit of w_i simple sheaves is concentrated in x_i. For the subcategory of vector bundles the effect of the iterated cycle construction is not so easily seen; it may however be more dramatic than for the simple sheaves.

The category $\mathrm{coh}(\mathbb{X})$ then is an abelian k-category that is connected, noetherian Hom- and Ext-finite with Serre duality which satisfies all axioms of a hereditary noetherian category with Serre duality (HNC).

2.3. Hereditary noetherian categories with Serre duality. Hereditary noetherian categories with Serre duality (HNC's) have an axiomatic description. We follow here the presentation of [**23**] which also contains the proofs of the derived statements.

By a *hereditary noetherian category* we mean in the sequel a small k-linear category satisfying the following assumptions:

(H 1) *$\mathcal{H}$ is a connected abelian k-category, and each object A in $\mathcal{H}$ is noetherian, meaning that ascending chains of subobjects for A become stationary.*

(H 2) *$\mathcal{H}$ is Ext-finite, that is, all morphism and extension spaces in $\mathcal{H}$ are finite dimensional k-vector spaces.*

(H 3) *$\mathcal{H}$ satisfies **Serre duality** $\mathrm{Ext}^1(X,Y) = D\mathrm{Hom}(Y,\tau X)$ where τ is a self-equivalence of $\mathcal{H}$, and D denotes the usual k-dual. (The self-equivalence τ is furtheron called **Auslander-Reiten translation**.*

(H 4) *$\mathcal{H}$ contains an object of infinite length.*

By $\mathcal{H}_0$ we denote the Serre subcategory of $\mathcal{H}$ consisting of all objects having finite length. We then request:

(H 5) *Each object in the quotient category $\mathcal{H}/\mathcal{H}_0$ has finite length.*

By $\mathcal{H}_+$ we denote the full subcategory of all objects of $\mathcal{H}$ not having any simple subobject. The members of $\mathcal{H}_+$ will be called ***vector bundles***, those of $\mathcal{H}_0$ will be called ***torsion***.

PROPOSITION 2.2. *Each indecomposable object from $\mathcal{H}$ is either a vector bundle, that is, belongs $\mathcal{H}_+$, or it has finite length and thus is a member of $\mathcal{H}_0$.*

Moreover, for some index set $C = C(\mathcal{H})$ the category $\mathcal{H}_0$ decomposes into a coproduct $\coprod_{x\in C}\mathcal{U}_x$, of connected uniserial length categories $\mathcal{U}_x$.

The simple objects in $\mathcal{U}_x$ form a single Auslander-Reiten orbit of finite cardinality $w(x)$. Only for finitely many points $x \in C(\mathcal{H})$ the cardinality $w(x)$ is > 1. □

The members of $C = C(\mathcal{H})$ are called ***points of** $\mathcal{H}$*.

(H 6) *$\mathcal{H}$ has infinitely many points*

This completes the requests on $\mathcal{H}$. It can be shown, see [**23**], that (H 5) is implied by the other axioms.

For each object X in $\mathcal{H}$ the ***rank*** $\mathrm{rk}(X)$ of X is defined as the length of X in the Serre quotient $\mathcal{H}/\mathcal{H}_0$. The rank is zero exactly for the members of $\mathcal{H}_0$. Among vector bundles, we choose a member of rank one and call it the ***structure sheaf*** $\mathcal{O}_C$ of $\mathcal{H}$.

PROPOSITION 2.3. *The endomorphism ring of $\mathcal{O}_C$ in the Serre quotient $\mathcal{H}/\mathcal{H}_0$ is an algebraic function field $k(\mathcal{H})/k$ in one variable, that is, $k(\mathcal{H})$ is a finite algebraic extension of the field $k(X)$ of rational functions in one variable.*

Classical valuation theory, see [**6**], associates to such a function field a smooth projective curve C whose points are in one-to-one correspondence with the discrete valuations of $k(\mathcal{H})/k$. It is not difficult to show that these are in one-to-one correspondence with the members of $C(\mathcal{H})$. In this way, $C(\mathcal{H})$ becomes a smooth projective curve, equipped with the weight function w from Proposition 2.2.

An important tool in investigating $\mathcal{H}$ is the ***Euler form***, a bilinear form on the Grothendieck group $\mathrm{K}_0(\mathcal{H})$ modulo short exact sequences. It is defined on classes of objects by the expression

$$\langle [E],[F]\rangle = \dim_k \mathrm{Hom}(E,F) - \dim_k \mathrm{Ext}^1(E,F).$$

We are now going to describe how to get back from $\mathrm{coh}(\mathbb{X})$ to $\mathrm{coh}(X)$, where $x_1, x_2, \ldots, x_t$ are the weighted points and $w_1, w_2, \ldots, w_t$ are their weights. One uses that for each τ-orbit of simple sheaves concentrated in x_i, $i = 1, \ldots, t$, the structure sheaf $\mathcal{O}_{\mathbb{X}}$ admits a non-zero homomorphism to exactly one member of each τ-orbit $(\tau^j S_{x_i})$, $j \in \mathbb{Z}_{w_i}$.

We may assume that $\mathrm{Hom}(\mathcal{O}_{\mathbb{X}}, S_{x_i}) = k$ for $i = 1, \ldots, t$. We then form the finite set $\mathfrak{S}$ of all simple sheaves S with $\mathrm{Hom}(\mathcal{O}_{\mathbb{X}}, S) = 0$.

The full subcategory of $\mathrm{coh}(\mathbb{X})$, right perpendicular to $\mathfrak{S}$ is then (equivalent to) the original category $\mathrm{coh}(X)$ of coherent sheaves on the underlying smooth projective curve, we started with, compare [**12**]. By construction, the simple sheaves S_{x_i} with $\mathrm{Hom}(\mathcal{O}_{\mathbb{X}}, S_{x_i}) = k$, $i = 1, \ldots, t$, belong to $\mathfrak{S}^{\perp} = \mathrm{coh}(X)$. The points $x_1, \ldots, x_t$, together with their weights $w_1, \ldots, w_t$, then define a (WPC) (X, w).

2.4. Graded isolated Gorenstein singularities. By a ***positively $\mathbb{Z}$-graded k-algebra*** $R = \bigoplus_{n\geq 0} R_n$ we understand a commutative algebra R which is affine, hence finitely generated as a k-algebra. We assume that the homogeneous components R_n are finite dimensional over k and satisfy the usual grading condition $R_n \cdot R_m \subseteq R_{n+m}$ for all $n, m \geq 0$. We further request that $R_0 = k$, an assumption implying that R is ***graded local***, that is, has a unique maximal graded ideal $\mathfrak{m} = \bigoplus_{n>0} R_n$ and ***residue class field*** $R/\mathfrak{m} = k$.

A graded algebra R, as above, is said to be ***graded isolated singularity*** if for each homogeneous prime ideal $\mathfrak{p}$, different from the maximal ideal $\mathfrak{m}$, the $\mathbb{Z}$-graded localization $S = R_{\mathfrak{p}}$ has finite global dimension. We note that the homogeneous component S_n consists of all fractions a/b with a, b homogeneous, say of degree r,s respectively, with $b \notin \mathfrak{p}$, and such that $r - s = n$.

DEFINITION 2.4. A positively $\mathbb{Z}$-graded commutative noetherian algebra R is called ***graded Gorenstein*** if its injective dimension, calculated in the category $\mathrm{Mod}^{\mathbb{Z}}(R)$ of all graded R-modules, is finite.

This dimension then agrees with the ***graded Krull dimension*** n of R, which may be defined as the maximal length of a proper chain of graded prime ideals

$$0 \subseteq \mathfrak{p}_0 \subset \mathfrak{p}_1 \subset \cdots \subset \mathfrak{p}_{n-1} \subset \mathfrak{p}_n = \mathfrak{m}$$

in R

An important invariant for a $\mathbb{Z}$-graded Gorenstein algebra R of Krull dimension n is the ***Gorenstein parameter***. To determine this invariant we consider a minimal graded injective resolution in $\mathrm{Mod}^{\mathbb{Z}}(R)$

$$0 \to R \to E^0 \to E^1 \to \cdots \to E^{n-1} \to E^n \to 0$$

by $\mathbb{Z}$-graded injective modules E^i, $i = 0, \dots, n$. If E^n is isomorphic to the a-fold degree shift of the graded injective hull $E(k)$ of the R-module $k = R/\mathfrak{m}$, then a is called the ***Gorenstein parameter*** of R. To have Gorenstein parameter a we thus request that the n-th cosyzygy of R equals $E(k(a)) = (E(k)(a)$.

In case of a graded complete intersection $R = k[x_1, x_2, \dots, x_t]/(f_1, f_2 \dots, f_r)$, where R is positively graded by assigning $x_1, x_2, \dots, x_t$ positive degrees $d_1, d_2, \dots, d_t$ and where the homogeneous polynomials $f_1, f_2 \dots, f_r$ of degrees $h_1, h_2, \dots, h_r$ are assumed to form a homogenous regular sequence, the Gorenstein parameter a of R is determined by the expression

$$a = \sum_{j=1}^{t} d_j - \sum_{i=1}^{r} h_i.$$

In particular, a graded hypersurface $R = k[x_1, x_2, x_3]/(f)$ of Krull dimension 2, is graded Gorenstein of Gorenstein parameter $\sum_{i=1}^{3} \deg(x_i) - \deg(f)$.

2.4.1. *Graded Gorenstein isolated singularities of Krull dimension two and weighted projective curves.* We next describe how graded Gorenstein isolated singularities of *Krull dimension two* relate to weighted projective curves.

PROPOSITION 2.5. *Let R be a graded Gorenstein isolated singularity of Krull dimension two and Gorenstein parameter a. Then R is graded integral, that is, $x \cdot y = 0$ for homogeneous elements implies $x = 0$ or $y = 0$. Moreover, sheafification of R, say by Serre construction, yields a hereditary noetherian k-category (HNC) with Serre duality*

$$\mathcal{H} = \frac{\mathrm{mod}^{\mathbb{Z}}(R)}{\mathrm{mod}_0^{\mathbb{Z}}(R)}, \tag{1}$$

where the AR-translation is induced by degree shift $X \mapsto X(a)$.

PROOF. Following the model of [**11**], see also [**4, 5**], we construct a $\mathbb{Z}$-graded sheaf theory from R on the $\mathbb{Z}$-graded (maximal) projective spectrum X, consisting of the non-zero homogeneous non-maximal prime ideals $\mathfrak{p}$ of R. There results a structure sheaf $\mathcal{O}$ on X whose stalks are the graded localizations $R_{\mathfrak{p}}$, defined by means of homogeneous quotients a/b with $a \in R$ and $b \notin \mathfrak{p}$. The stalks are graded regular local of Krull dimension one and hence are graded discrete valuation rings. Since all stalks have global dimension one, it follows that the abelian category $\mathcal{H} = \mathrm{coh}(X)$ of graded coherent sheaves has global dimension one. Moreover, since R is graded noetherian, the category $\mathcal{H}$ is also noetherian. Finally, since R is graded-Gorenstein, say of Gorenstein parameter a, the category $\mathcal{H}$ has Serre duality in the form $\mathrm{Ext}^1(A, B) = \mathrm{Hom}(B, A(a))$, where $A(a)$ denotes the degree shift of A with the Gorenstein parameter $a \in \mathbb{Z}$, see for instance [**26**].

Alternatively, again following [**11**], $\mathcal{H}$ can be obtained by Serre construction as the abelian quotient category of the category $\mathrm{mod}^{\mathbb{Z}}(R)$ of finitely generated graded R-modules modulo its Serre subcategory $\mathrm{mod}_0^{\mathbb{Z}}(R)$ of graded R-modules of finite length.

It is left to the reader to check that $\mathcal{H}$ fulfills all remaining requirements for a hereditary noetherian category (HNC) as introduced above. It then follows that $X = C(\mathcal{H})$ carries the structure of a (WPC). □

2.4.2. *Gorenstein parameter* −1. Next, we are going to show how, conversely, each (HNC) $\mathcal{H}$ of *negative orbifold Euler characteristic* gives rise to a $\mathbb{Z}$-graded Gorenstein isolated singularity of Krull dimension two and Gorenstein parameter -1. For this, we interpret $\mathcal{H}$ as the category of coherent sheaves $\mathrm{coh}(\mathbb{X})$ on a (WPC) $\mathbb{X} = (X, w)$. Let $\mathcal{O}_{\mathbb{X}}$ be the structure sheaf of $\mathbb{X}$ and let $\tau : \mathcal{H} \to \mathcal{H}$ be the Auslander-Reiten translation of $\mathrm{coh}(\mathbb{X})$. Following the model of [**15**] we form the ***$\mathbb{Z}$-graded orbit-algebra of τ on $\mathcal{O}_{\mathbb{X}}$***:

$$(2) \qquad R = \bigoplus_{n\geq 0} \mathrm{Hom}(\mathcal{O}_{\mathbb{X}}, \tau^n \mathcal{O}_{\mathbb{X}}) \text{ with multiplication } r_n \cdot r_m = [\tau^m(r_n) \circ r_m].$$

From the definition of the multiplication it is easily shown that R is an associative, positively $\mathbb{Z}$-graded algebra. To show that R is also commutative, we need the concept of the ***point shift automorphism*** σ_x with $x \in \mathbb{X}$, compare [**18**, Section 10.3].

For each $x \in \mathbb{X}$ there is an automorphism σ_x of $\mathrm{coh}(\mathbb{X})$ which on vector bundles E is given by the universal extension $E \mapsto E(x)$

$$(3) \qquad 0 \to E \xrightarrow{x_E} E(x) \to E_x \to 0, \text{ where } E_x = \bigoplus_{j \in \mathbb{Z}_{w(x)}} \mathrm{Ext}^1(\tau^j S_x, E) \otimes_k \tau^j S_x.$$

Obviously, each σ_x, $x \in \mathbb{X}$, induces the identity map on $\mathbb{X} = C(\mathcal{H})$. By definition, all the σ_x, $x \in \mathbb{X}$, generate the Picard group $\mathrm{Pic}(\mathbb{X})$ which is a commutative group acting transitively on isomorphism classes of line bundles [**20**].

PROPOSITION 2.6. *The Auslander-Reiten translation τ of* $\mathrm{coh}(\mathbb{X})$ *is a member of the* ***Picard group*** $\mathrm{Pic}(\mathbb{X})$. *Further, the orbit algebra R is commutative.*

PROOF. Since $\mathrm{Pic}(\mathbb{X})$ acts transitively on line bundles, there exists $\sigma \in \mathrm{Pic}(\mathbb{X})$ with $\tau(\mathcal{O}_{\mathbb{X}}) = \sigma(\mathcal{O}_{\mathbb{X}})$. Since $\sigma^{-1}\tau$ fixes $\mathcal{O}_{\mathbb{X}}$, it is an automorphism of $\mathbb{X}$ which additionally fixes each point of $\mathbb{X}$, and hence is (isomorphic to) the identity, and $\tau = \sigma$ follows. This proves the first assertion.

Now, τ is a finite product of shift functors σ_x or their inverses σ_x^{-1}. Passing to the quotient category $\mathcal{H}/\mathcal{H}_0$, each shift functor σ_x is isomorphic to the identity functor as follows from the functorial construction (3). This property extends to the product τ. We may thus assume that $\tau = Id$ on $\mathcal{H}/\mathcal{H}_0$, and correspondingly $\tau^n \mathcal{O}_{\mathbb{X}} = \mathcal{O}_{\mathbb{X}}$ in $\mathcal{H}/\mathcal{H}_0$. Passing with the two products $v_n \circ \tau^m(u_m)$ and $u_m \circ \tau^n(u_n)$ from $\mathrm{Hom}(\mathcal{O}_{\mathbb{X}}, \tau^{n+m}\mathcal{O}_{\mathbb{X}})$ to $\mathcal{H}/\mathcal{H}_0$, they simplify to the expressions $v_n \circ u_m$ and $u_m \circ v_n$ with v_m, u_n lying in the endomorphism ring K of $\mathcal{O}_{\mathbb{X}}$ in the category $\mathcal{H}/\mathcal{H}_0$. As the function field of a smooth projective curve, K is commutative, the identity $v_n \circ u_m = u_m \circ v_n$ in K thus descends to the equality $v_n \circ \tau^m(u_m) = u_m \circ \tau^n(u_n)$ in $\mathrm{Hom}(\mathcal{O}_{\mathbb{X}}, \tau^{n+m}\mathcal{O}_{\mathbb{X}})$, which establishes the claimed commutativity of R. □

In order to establish further properties of R we prove an important property of the pair $(\mathcal{O}_{\mathbb{X}}, \tau)$:

PROPOSITION 2.7. *If $\mathcal{H} = \mathrm{coh}(\mathbb{X})$ has negative Euler characteristic, the pair $(\mathcal{O}_{\mathbb{X}}, \tau)$ is* ***ample*** *in $\mathcal{H}$ in the sense of* [**1**], *meaning that the following two conditions are satisfied:*

(i) *For each H in $\mathcal{H}$ there exists an epimorphism* $\bigoplus_{i=1}^{r} \tau^{-n_i} \mathcal{O}_{\mathbb{X}} \to H$ *with exponents* $n_i > 0$.

(ii) *For every epimorphism* $H \to K$ *in* $\mathcal{H}$ *the induced map* $\mathrm{Hom}(\tau^{-N}\mathcal{O}_{\mathbb{X}}, H) \to \mathrm{Hom}(\tau^{-N}\mathcal{O}_{\mathbb{X}}, K)$ *is surjective for large* N.

PROOF. Before entering the proof of items (i) and (ii), we recall some properties of $\mathcal{H}$.

(a) Each object H of $\mathcal{H}$ splits into a direct sum $H = H_+ \oplus H_0$, where H_0 has finite length and the vector bundle H_+ has a finite filtration by line bundles, compare [**11**].

(b) For each non-zero object $A \in \mathrm{coh}(\mathbb{X})$ with $\mathbb{X} = X\langle a_1, a_2, \dots, a_t\rangle$ we obtain the slope identity $\mu(\tau A) = \mu(A) - \bar{a}\chi_{\mathcal{H}}$, compare [**23**, Prop. 3.5][1]. Here, $\bar{a} = \mathrm{lcm}(a_1, a_2, \dots, a_t)$.

(c) Given H in $\mathcal{H}$, then $\mathrm{Ext}^1(\tau^{-N}\mathcal{O}_{\mathbb{X}}, H) = 0$ holds for large N.

(d) Assume the sequence $0 \to A \xrightarrow{u} B \to C \to 0$ is exact where A and C satisfy property (i) and, moreover, C has finite length. Then also B satisfies property (i).

In order to prove (c) it is sufficient, in view of (a), to assume that H is either a simple sheaf or a line bundle. Invoking Serre duality, the expression $\mathrm{Ext}^1(\tau^{-N}\mathcal{O}_{\mathbb{X}}, H) = D\mathrm{Hom}(H, \tau^{-N+1}\mathcal{O}_{\mathbb{X}})$ is zero if H is simple (this needs no assumption on N). If H is a line bundle, then we may choose N large enough that $\deg(H) > \deg(\tau^{-N+1}\mathcal{O}_{\mathbb{X}})$, implying that $\mathrm{Hom}(H, \tau^{-N+1}\mathcal{O}_{\mathbb{X}})$ is zero for large N.

Concerning (d), let $\bigoplus_{i=1}^{c} \tau^{-n_i}\mathcal{O}_{\mathbb{X}} \xrightarrow{v} C$ be an epimorphism. Since τ is periodic on $\mathcal{H}_0$, we may choose the n_i large enough that $\mathrm{Ext}^1(\tau^{-n_i}\mathcal{O}_{\mathbb{X}}, A) = D\mathrm{Hom}(A, \tau^{-n_i+1}\mathcal{O}_{\mathbb{X}})$ equals zero. Hence the map v lifts to a map $\bar{v} : \bigoplus_{i=1}^{c} \tau^{-n_i}\mathcal{O}_{\mathbb{X}} \to B$ and, clearly, then $(u, \bar{v}) : A \oplus \bigoplus_{i=1}^{c} \tau^{-n_i}\mathcal{O}_{\mathbb{X}} \xrightarrow{(u,\bar{v})} C$ is an epimorphism. We are done, since A satisfies (i).

We next prove property (ii): The epimorphism $u : H \to K$ yields an exact sequence $0 \to N \to H \xrightarrow{u} K \to 0$. Invoking (c), $\mathrm{Ext}^1(\tau^{-N}\mathcal{O}_{\mathbb{X}}, N) = 0$ for large N, implying that $\mathrm{Hom}(\tau^{-N}\mathcal{O}_{\mathbb{X}}, u)$ is surjective.

Finally we prove property (i), where we first deal with the cases where H either has finite length or where H is a vector bundle. Since $\mathcal{O}_{\mathbb{X}}$ maps nontrivially to a member S of each τ-orbit of simples, it is obvious that there exists an epimorphism $\tau^{-n}\mathcal{O}_{\mathbb{X}} \to S$ with $n \geq 0$ which deals with the case of a simple sheaf. If H has finite length, we consider a sequence $0 \to S \to H \to C \to 0$, where S is simple, and, using induction on the length, apply (d).

If $H = L$ is a line bundle, we invoke negative Euler characteristic and the weighted Riemann-Roch theorem B.2 to establish a non-zero morphism $u : \tau^{-n}\mathcal{O}_{\mathbb{X}} \to L$ for some $n > 0$. In the resulting exact sequence $0 \to \tau^{-n}\mathcal{O}_{\mathbb{X}} \xrightarrow{u} L \to C \to 0$, the cokernel-term has finite length. Invoking the preceding argument, property (d) shows that L satisfies property (i).

In view of property (a) this finishes the proof. □

THEOREM 2.8. *Let* $\mathbb{X} = (X, w)$ *be a smooth weighted projective curve (WPC) of negative orbifold Euler characteristic* $\chi_{\mathbb{X}}$ *with category* $\mathcal{H} = \mathrm{coh}(\mathbb{X})$ *of coherent*

[1] Due to a different normalization, the factor $\bar{a}$ is missing in [**23**]

sheaves and structure sheaf $\mathcal{O}_{\mathbb{X}}$*. With respect to the AR-translation of* $\mathcal{H}$ *let*

$$R = \bigoplus_{n\geq 0} \mathrm{Hom}(\mathcal{O}_{\mathbb{X}}, \tau^n \mathcal{O}_{\mathbb{X}})$$

be the positively $\mathbb{Z}$*-graded orbit algebra of* τ *on* $\mathcal{O}_{\mathbb{X}}$*.*

Then R *is a commutative affine* k*-algebra which is a graded isolated singularity of Krull dimension two. Morover* R *is* $\mathbb{Z}$*-graded Gorenstein of Gorenstein parameter* -1.

Sheafification of R *by means of Serre construction or by a* $\mathbb{Z}$*-graded sheaf theory leads back to* $\mathrm{coh}(\mathbb{X})$*. Moreover, the degree shift by* -1 *on* $\mathrm{mod}^{\mathbb{Z}}(R)$ *induces on* $\mathrm{coh}(\mathbb{X})$ *the AR-translation* τ*.*

PROOF. In Proposition 2.6 we have shown that R is commutative. Since the pair $(\mathcal{O}_{\mathbb{X}}, \tau)$ is ample in $\mathrm{coh}(\mathbb{X})$ it follows from [**1**, Theorem 4.5] that R is graded noetherian and sheafification of R leads back to $\mathrm{coh}(\mathbb{X})$. Since $R_0 = k$, R is graded local and a finite system $x_1, x_2, \ldots, x_s$ of generators for the graded maximal ideal $\mathfrak{m}$ of R generates R as a k-algebra, which is thus affine over k. Serre construction shows that the category $\mathrm{mod}^{\mathbb{Z}}(R)$, hence R, has graded Krull dimension two. Using a $\mathbb{Z}$-graded sheaf theory — instead of Serre construction — shows that the graded localizations $R_{\mathfrak{p}}$ for non-maximal primes, that is, just the stalks of $\mathcal{O}_{\mathbb{X}}$ are hereditary, which shows that R is graded isolated singularity.

Before showing the claimed Gorenstein properties, we recall the concept of a graded dual: If $M := \bigoplus_{n\in\mathbb{Z}} M_n$ is a graded module, then $DM = \bigoplus_{n\in\mathbb{Z}} DM_{-n}$ is called the ***graded dual*** of M. Since R is indecomposable graded projective it follows that $DR = \bigoplus_{n\leq 0} DR_{-n}$ is indecomposable graded injective, and since $DR_0 = k$ is contained in DR, it is the graded injective envelope $E(k)$ of k. We next show that R is graded Gorenstein of Gorenstein parameter -1. For this we start with a minimal injective resolution

$$0 \to \mathcal{O}_{\mathbb{X}} \to \mathcal{I}^0 \to \mathcal{I}^1 \to 0 \tag{4}$$

in the Grothendieck category $\mathrm{Qcoh}(\mathbb{X})$ of quasi-coherent sheaves. Such a resolution exists since $\mathrm{Qcoh}(\mathbb{X})$ has global dimension one because $\mathrm{coh}(\mathbb{X})$ has the same property and is additionally noetherian[2]. We apply graded global sections $\Gamma^* = \bigoplus_{n\in\mathbb{Z}} \mathrm{Hom}(\tau^{-n}\mathcal{O}_{\mathbb{X}}, -)$ to (4) and obtain exactness of

$$0 \to R \to \Gamma^*(\mathcal{I}_0) \to \Gamma^*(\mathcal{I}^1) \to \bigoplus_{n\in\mathbb{Z}} \mathrm{Ext}^1(\tau^{-n}\mathcal{O}_{\mathbb{X}}, \mathcal{O}_{\mathbb{X}}) \to 0 \tag{5}$$

in the category $\mathrm{Mod}^{\mathbb{Z}}(R)$ of all $\mathbb{Z}$-graded R-modules. Due to Serre duality the Ext-term evaluates to $\bigoplus_{n\in\mathbb{Z}} D\mathrm{Hom}(\mathcal{O}_{\mathbb{X}}, \tau^{-(n-1)}\mathcal{O}_{\mathbb{X}})$, the graded dual $(DR)(-1)$ of $R(-1)$, thus the graded injective envelope of $k(-1)$. This establishs that R has graded injective dimension two with Gorenstein parameter -1. □

COROLLARY 2.9 (Uniqueness). *Assuming negative orbifold Euler characteristic of* $\mathbb{X}$*, a positively* $\mathbb{Z}$*-graded isolated Gorenstein singularity* S *of Krull dimension two and Gorenstein parameter* -1 *is isomorphic to the orbit algebra* $R = \bigoplus_{n\geq 0} \mathrm{Hom}(\mathcal{O}_{\mathbb{X}}, \tau^n\mathcal{O}_{\mathbb{X}})$ *if and only if its associated category of coherent sheaves is equivalent to* $\mathrm{coh}(\mathbb{X})$*.*

---

[2]Even, if one only knows the noetherian category $\mathcal{H}$ of coherent sheaves, the category of quasicoherent sheaves is easily constructed as the Grothendieck category $\mathbf{Lex}(\mathcal{H}^{op}, \mathbf{Ab})$ of left exact additive functors on $\mathcal{H}^{op}$ with values in the category $\mathbf{Ab}$ of abelian groups, see [**10**].

PROOF. Assume Serre construction for S yields the category $\mathrm{coh}(\mathbb{X})$. Then sheafification $\tilde{S}$ of S yields a line bundle in $\mathrm{coh}(\mathbb{X})$. Applying a suitable automorphism of $\mathrm{coh}(\mathbb{X})$, we may assume that $\tilde{S} = \mathcal{O}_{\mathbb{X}}$. Since S is graded Gorenstein of Gorenstein parameter -1, the degree shift by -1 is equal to τ, and then

$$S_r = \mathrm{Hom}(\tilde{S}, \tilde{S}(r) = \mathrm{Hom}(\mathcal{O}_{\mathbb{X}}, \tau^r \mathcal{O}_{\mathbb{X}}) = R_r \tag{6}$$

yielding an isomorphism between R and S as graded algebras. □

For $\mathbb{X}$ a (WPC) of negative Euler characteristic we call this uniqely determined graded isolated singularity of Gorenstein parameter -1 the ***fuchsian singularity associated to*** $\mathbb{X}$ ***or*** $\mathrm{coh}(\mathbb{X})$.

If R is a fuchsian singularity with attached weighted projective curve $\mathbb{X} = (X, w)$ the ***signature*** of R is defined as the tuple $(g_X; a_1, a_2, \dots, a_t)$, where g_X is the genus of X and $(a_1, a_2, \dots, a_t)$ is the weight sequence of $\mathbb{X}$. We are next showing a couple of polynomials f in three variables defining ***fuchsian hypersurfaces*** $R = k[x_1, x_2, x_3]/(f)$ and display their signatures. It was already mentioned that fuchsian hypersurfaces are a rare species. Indeed for genus 0 there only exists finitely many and it was moreover shown by Wagreich [**31**], see also [**8**], [**28**] and [**22**], that also for higher genus there are only finitely many signatures which can be represented by a fuchsian hypersurface. Ebeling [**9**] extended this also to complete intersections.

	f	$deg(x,y,z\vert f)$	*signature*
1	$x^2+y^3+z^7$	$(21,14,6\vert 42)$	$(0;2,3,7)$
2	$x^2+y^3+yz^5$	$(15,10,4\vert 30)$	$(0;2,4,5)$
3	$x^2+zy^3+yz^4$	$(11,6,4\vert 22)$	$(0;2,4,6)$
4	$x^4+xy^2+yz^2$	$(4,6,5\vert 16)$	$(0;2,5,6)$
5	$x^2+y^6+z^6$	$(3,1,1\vert 6)$	$(2;-)$
6	$x^4+y^4+z^4$	$(1,1,1\vert 4)$	$(3;-)$
7	$xy^3+yz^3+zx^3$	$(1,1,1\vert 4)$	$(3,-)$

Fuchsian singularities R of genus zero with a signature having at most three weights are uniquely determined by their signature, because then the weight sequence determines the weighted projective line attached to R.

In general, however, fuchsian singularities with the same signature may not be isomorphic as show the table etems 6 and 7, the ***Fermat quartic*** and the***Klein's quartic***. Assuming $k = \mathbb{C}$, the two curves have non-isomorphic automorphism groups: in case of the Fermat quartic this is the semi-direct product of the symmetric group S_3 with the action on the product of three cyclic groups of order 4 modulo their diagonal [**30**]. By contrast, the automorphism group of Klein's quartic is the nonabelian simple group G_{168} of order 168, compare the papers in [**24**].

2.5. Fuchsian groups and their automorphic forms. For this subsection, the base field is the field $\mathbb{C}$ of complex numbers. Instead of weighted projective curves, we consider weighted compact Riemann surfaces (WCRS's) equipped with holomorphic coherent sheaves. For Fuchsian groups, in general, we refer to [**2**] and [**13**].

Within this article, a discrete subgroup G of $\mathrm{Aut}(\mathbb{H})$, the group of holomorphic automorphisms of the upper complex halfplane $\mathbb{H}$ is called ***a fuchsian group*** if it is finitely generated and cocompact, meaning that the quotient $\mathcal{H}/G$ is compact. We assume moreover that the stabilizer groups G_x, with $x \in \mathbb{H}$, are all finite and then, automatically, cyclic, see [**13**]. In this case G has a fundamental domain that is a hyperbolic convex polygon with finitely many sides in the hyperbollic plane $\mathbb{H}$.

PROPOSITION 2.10 ([**7**],Theorem 3.2.10). *Two fuchsian subgroups G and G' of* $\mathrm{Aut}(\mathbb{H})$ *are isomorphic if and only if they are conjugate as subgroups of* $\mathrm{Aut}(\mathbb{H})$. *Accordingly, the (WCRS's) $\mathbb{H}/G$ and $\mathbb{H}/G'$ are isomorphic if and only if G and G' are isomorphic.* □

For any fuchsian group G the quotient $\mathbb{H}/G$ is a compact holomorphic orbifold $\mathbb{X}$ of complex dimension one. In other words, $\mathbb{X}$ has an underlying compact Riemann surface with a finite number of cone points. These cone points, and their orders, correspond to the G-orbits with non-trivial stabilizer group, which are necessarily cyclic, see [**13**].

In other words, the quotient $\mathbb{H}/G$ is a weighted compact Riemann surface, or just a weighted projective curve (X, w) over $\mathbb{C}$.

Conversely, we can pass back from a weighted Riemann surface $\mathbb{X} = (X, w)$ of negative orbifold Euler characteristic to the action of a fuchsian group on $\mathcal{H}$:

We refer to [**29**, Chapter 13] and [**19**] for the concepts ***orbifold fundamental group*** and ***universal orbifold cover***. The next theorem combines properties from [**19**, Proposition 6 and Theorem 7]:

THEOREM 2.11. *Let $\mathbb{X} = X\langle a_1, a_2, \dots, a_t\rangle$ be a compact weighted Riemann surface (WCRS) of negative Euler characteristic $\chi_{\mathbb{X}}$, then the following holds:*

(i) *The orbifold fundamental group $\pi_1^{orb}(\mathbb{X})$ has generators $\alpha_1, \alpha_2, \dots, \alpha_g$, $\beta_1, \beta_2, \dots, \beta_g$, $\sigma_1, \sigma_2, \dots, \sigma_t$ and is subject to the relations*

$$\sigma_1^{a_1} = \sigma_2^{a_2} = \cdots = \sigma_t^{a_t} = 1 = \sigma_1\sigma_2\cdots\sigma_t\,[\alpha_1, \beta_1][\alpha_2, \beta_2]\cdots[\alpha_g, \beta_g],$$

where $[a, b]$ denotes the commutator $aba^{-1}b^{-1}$ of a and b.

(ii) *The orbifold fundamental group $\pi_1^{orb}(\mathbb{X})$ acts on the universal orbifold cover $\mathbb{H} = \widetilde{\mathbb{X}}$ as group of deck transformations (the members of* $\mathrm{Aut}\mathbb{H}$ *commuting with the projection $\pi : \mathbb{H} = \widetilde{\mathbb{X}} \to \mathbb{X}$). This action is discrete on $\mathbb{H}$ and represents $\mathbb{X}$ as orbifold quotient*

$$\mathbb{X} = \widetilde{\mathbb{X}}/\pi_1^{orb}(\mathbb{X}).$$

(iii) *In particular, the orbifold fundamental group is fuchsian, and each fuchsian group (in the sense of the paper) arises that way.*

PROOF. For the first two items we refer to [**19**, Proposition 6 and Theorem 7]. The last assertion follows from the other two. □

In the description of automorphic forms we follow Milnor [**25**]. Let G be a fuchsian group and U a G-stable open subset of the upper complex half-plane $\mathbb{H}$. A ***differential form of degree*** n ***on*** U has the form $\Phi = f(z)z^k$. For any $g \in G$ the pull-back of Φ along g is given as $g^*(\Phi) = f(g(z))\,\dot{g}\,(z)^k dz^k$, where $\dot{g}\,(z)$ denotes the derivative $dg(z)/dz$. Differential forms Φ_1 and Φ_2 of degrees m and n may be multiplied; this yields a differential form of degree $m+n$. The G-invariant differential forms of degree $n \geq 0$, or ***automorphic forms*** on G-invariant open subsets of $\mathbb{H}$, descend to the quotient $\mathbb{X} = \mathbb{H}/G$, yielding there holomorphic coherent sheaves of differentials Ω^n for each degree n.

REMARK 2.12. One can show that $\Omega^n = \tau^n \mathcal{O}_{\mathbb{X}}$, for each $n \geq 0$, implying that the graded algebra

$$\bigoplus_{n\geq 0} \Gamma(\mathbb{X}, \Omega^n)$$

of differential forms on $\mathbb{X}$ is isomorphic to the orbit algebra of the Auslander-Reiten translation of $\mathrm{coh}(\mathbb{X})$ on $\mathcal{O}_{\mathbb{X}}$, thus providing a link to the classical theory of fuchsian singularities.

3. The singularity category and its Grothendieck group(s)

There are several incarnations of the singularity category of a fuchsian singularity R, where each one has its own advantages.

Following Buchweitz [**3**], for the ungraded and Orlov [**26**] for the graded case, the ***category of graded singularities*** $\mathrm{Sing}^{\mathbb{Z}}(R)$ is defined as the Verdier quotient

$$\mathrm{Sing}^{\mathbb{Z}}(R) = \frac{\mathrm{D}^b(\mathrm{mod}^{\mathbb{Z}}(R))}{\mathrm{D}^b(\mathrm{proj}^{\mathbb{Z}}(R))}, \tag{7}$$

where $\mathrm{proj}^{\mathbb{Z}}(R)$ denotes the category of finitely generated projetive R-modules.

In the case of a graded Gorenstein algebra, the singularity category $\mathrm{Sing}^{\mathbb{Z}}(R)$ is equivalent to the ***stable category of*** $\mathbb{Z}$***-graded Cohen-Macaulay modules over*** R $\underline{\mathrm{CM}}^{\mathbb{Z}}(R)$. In our case of graded Krull-dimension two, a finitely generated $\mathbb{Z}$-graded R-module M is called (maximal) ***Cohen-Macaulay*** if M satisfies the two conditions $\mathrm{Hom}(k(n), M) = 0 = \mathrm{Ext}^1(k(n), M)$ for each integer n.

The category $\mathrm{CM}^{\mathbb{Z}}(R)$ carries an exact structure in the sense of Quillen consisting of all short exact sequences with members in $\mathrm{CM}^{\mathbb{Z}}(R)$ that are exact in $\mathrm{mod}^{\mathbb{Z}}(R)$. The full subcategory $\mathrm{proj}^{\mathbb{Z}}(R)$ of finitely generated projective modules consists exactly of the relative projectives in $\mathrm{CM}^{\mathbb{Z}}(R)$; moreover, since R is graded Gorenstein, it is known that the relative projectives coincide with the relative injectives. Further, $\mathrm{CM}^{\mathbb{Z}}(R)$ has sufficiently many relative projectives resp. injectives, and thus is a ***Frobenius category***.

Stabilizing $\mathrm{CM}^{\mathbb{Z}}(R)$, that is forming the factor category $\underline{\mathrm{CM}}^{\mathbb{Z}}(R) = \mathrm{CM}^{\mathbb{Z}}(R)/[\mathrm{proj}^{\mathbb{Z}}(R)]$ modulo all moprhisms that factor through projectives, yields a category that is triangulated. It is due to Buchweitz [**3**] that the functor

$$\underline{\mathrm{CM}}^{\mathbb{Z}}(R) \to \mathrm{Sing}^{\mathbb{Z}}(R), \quad M \mapsto (M),$$

sending a graded CM-module M to the stalk complex (M), concentrated in degree zero, induces an equivalence of $\underline{\mathrm{CM}}^{\mathbb{Z}}(R)$ to the singularity category $\mathrm{Sing}^{\mathbb{Z}}(R)$.

In the special case of the orbit algebra $R = \bigoplus_{n\geq 0} \mathrm{Hom}(\mathcal{O}_{\mathbb{X}}, \tau^n \mathcal{O}_{\mathbb{X}})$ associated to a (WPC) of negative orbifold Euler characteristic, the category $\mathrm{CM}^{\mathbb{Z}}(R)$ has an even more accessible incarnation as the ***stable category of vector bundles*** on $\mathbb{X}$.

PROPOSITION 3.1. *There is an equivalence between the category* $\mathrm{CM}^{\mathbb{Z}}(R)$ *of finitely generated graded (maximal) Cohen-Macaulay modules over* R *and the category* $\mathrm{vect}(\mathbb{X})$ *of vector bundles on* $\mathbb{X}$ *induced by restricting the quotient functor* $q : \mathrm{mod}^{\mathbb{Z}}(R) \to \mathrm{coh}(\mathbb{X})$ *to the subcategory* $\mathrm{CM}^{\mathbb{Z}}(R)$. *The inverse functor is given by taking graded global sections* $\Gamma^* = \bigoplus_{n\in\mathbb{Z}} \mathrm{Hom}(\tau^{-n}\mathcal{O}_{\mathbb{X}}, -)$ *of vector bundles. Moreover the equivalence* $q : \mathrm{CM}^{\mathbb{Z}}(R) \to \mathrm{vect}(\mathbb{X})$ *induces an equivalence between indecomposable projective* R*-modules and members of the orbit* $\tau^{\mathbb{Z}}\mathcal{O}_{\mathbb{X}}$.

PROOF. The proof of the corresponding result in [**11**] carries over. □

We define the ***stable category of vector bundles*** on $\mathbb{X}$ as the factor category $\underline{\mathrm{vect}}(\mathbb{X}) = \mathrm{vect}(\mathbb{X})/[\tau^{\mathbb{Z}}\mathcal{O}_{\mathbb{X}}]$ of $\mathrm{vect}(\mathbb{X})$ by the ideal of all morphisms factoring through a member from the additive closure of $\tau^{\mathbb{Z}}\mathcal{O}_{\mathbb{X}}$.

COROLLARY 3.2. *The stable category* $\underline{\mathrm{vect}}(\mathbb{X}) = \mathrm{vect}(\mathbb{X})/[\tau^{\mathbb{Z}}\mathcal{O}_{\mathbb{X}}]$ *is triangle equivalent to the stable category* $\underline{\mathrm{CM}}^{\mathbb{Z}}(R)$, *and hence triangle-equivalent to the singulrity category* $\mathrm{Sing}^{\mathbb{Z}}(R)$ □

We recall that a triangulated category $\mathcal{T}$ is ***homologically finite*** if for all objects X, Y from $\mathcal{T}$ we have $\mathrm{Hom}(X, Y[n]) = 0$ for $|n| \gg 0$.

THEOREM 3.3 (Serre duality). *The singularity category* $\underline{\mathrm{vect}}(\mathbb{X})$ *is triangulated, Hom-finite, Krull-Schmidt and homologically finite. Moreover,* $\underline{\mathrm{vect}}(\mathbb{X})$ *has Serre duality given by functorial isomorphisms* $\mathrm{Hom}(X, Y[1]) = D\mathrm{Hom}(Y, \tau X)$, *where* τ *is an equivalence induced from the AR-translation of* $\mathrm{vect}(\mathbb{X})$. *In particular,* $\underline{\mathrm{vect}}(\mathbb{X})$ *has AR-triangles, and the AR-translation for* $\mathrm{vect}(\mathbb{X})$ *induces the AR-translation for* $\underline{\mathrm{vect}}(\mathbb{X})$.

PROOF. As a factor category of a Hom-finite category, $\underline{\mathrm{vect}}(\mathbb{X})$ inherits Hom-finiteness, and the Krull-Schmidt property from $\mathrm{vect}(\mathbb{X})$. Concerning Serre duality we apply the functors $\mathrm{Hom}(X, -)$ (resp. $\mathrm{Hom}(-, \tau X)$ to the exact sequence $\mu : 0 \to Y \to \mathcal{I}(Y) \to Y[1] \to 0$, where $\mathcal{I}(Y)$ denotes an injective hull of Y in the Frobenius category $\mathrm{vect}(\mathbb{X})$, and obtain exact sequences

$$\mathrm{Hom}(X, \mathcal{I}(Y)) \to \mathrm{Hom}(X, Y[1]) \to \underline{\mathrm{Hom}}(X, Y[1]) \to 0, \tag{8}$$

$$\mathrm{Hom}(\mathcal{I}(Y), \tau X) \to \mathrm{Hom}(Y, \tau X) \to \underline{\mathrm{Hom}}(Y, \tau X) \to 0. \tag{9}$$

By Serre duality in $\mathrm{coh}(\mathbb{X})$, dualization of the lower sequence yields exactness of

$$0 \to D\underline{\mathrm{Hom}}(Y, \tau X) \to \mathrm{Ext}^1(X, Y) \to \mathrm{Ext}^1(X, \mathcal{I}(Y)),$$

where $\underline{\mathrm{Hom}}$ denotes morphisms in $\underline{\mathrm{vect}}(\mathbb{X})$. Invoking the long exact Hom-Ext sequence $\mathrm{Hom}(X, \mu)$ we obtain from the two sequences (8) and (9) a natural isomorphism $\underline{\mathrm{Hom}}(X, Y[1]) = D\underline{\mathrm{Hom}}(Y, \tau X)$.

To show homological finiteness of $\mathrm{vect}(\mathbb{X})$ we use that in the Frobenius category of vector bundles the suspension functor is given by a distinguished exact sequence $0 \to E \to \mathcal{I}(E) \to E[1] \to 0$, where $\mathcal{I}(E)$ denotes the injective hull in the Frobenius category. Because $\chi_{\mathbb{X}} < 0$, iteration of the construction leads to $E[n]$ which obtains an arbitrary large slope for $n \gg 0$. Invoking a line a bundle filtration for F, this implies that $\mathrm{Hom}(E[n], F) = 0$, and then also $\underline{\mathrm{Hom}}(E[n], F) = 0$ for $n >> 0$. Concerning $\underline{\mathrm{Hom}}(F, E[n])$, we invoke Serre duality for $\underline{\mathrm{vect}}(\mathbb{X})$ and obtain $D\underline{\mathrm{Hom}}(E[n-1], \tau F)$ which for $n \gg 0$ is also zero. □

COROLLARY 3.4. *All AR-components of the singularity category have shape* $\mathbb{Z}\mathbb{A}_\infty$.

PROOF. Arguing as in [**21**, Proposition 4.5] one shows first, invoking negative orbifold Euler characteristic, that all AR-components in $\mathrm{vect}(\mathbb{X})$ have shape $\mathbb{Z}\mathbb{A}_\infty$. Next, one observes that line bundles always sit at the border-orbit of their component, which implies that passage to the stable category $\underline{\mathrm{vect}}(\mathbb{X})$ consists in deleting the border, and thus also preserves the $\mathbb{Z}\mathbb{A}_\infty$-shape of the component containing $\tau^{\mathbb{Z}}\mathcal{O}_{\mathbb{X}}$. □

As a fuchsian singularity R has Gorenstein parameter -1, further Serre construction $\mathrm{mod}^{\mathbb{Z}}(R)/\mathrm{mod}_0^{\mathbb{Z}}(R)$ yields the category $\mathrm{coh}(\mathbb{X})$ of coherent sheaves on a weighted projective curve $\mathbb{X}$. It then follows from [**26**] that the object $k = R/\mathfrak{m}$ is exceptional in $\mathrm{Sing}^{\mathbb{Z}}(R)$ and its right perpendicular category, consisting of all objects X in $\mathrm{Sing}^{\mathbb{Z}}(R)$ satisfying $\mathrm{Hom}(k, X[n]) = 0$ for each integer n, is equivalent to $\mathrm{D}^b(\mathrm{coh}(\mathbb{X}))$. More is true, $\mathrm{Sing}^{\mathbb{Z}}(R)$ behaves like a ***one-point extension*** of $\mathrm{D}^b(\mathrm{coh}(\mathbb{X}))$ by the structure sheaf $\mathcal{O}_{\mathbb{X}}$:

THEOREM 3.5 (One-point extension). *Let R be a fuchsian singularity with associated weighted projective curve $\mathbb{X}$. Then the right perpendicular category in $\mathcal{T} = \mathrm{Sing}^{\mathbb{Z}}(R)$ to the exceptional object $k = R/\mathfrak{m}$ is equivalent to $\mathrm{D}^b(\mathrm{coh}(\mathbb{X}))$.*

Moreover, the right adjoint $r : \mathcal{T} \to k^{\perp}$ to inclusion $k^{\perp} \hookrightarrow \mathcal{T}$ sends k to $\mathcal{O}_{\mathbb{X}}$ and, hence induces functorial isomorphisms $\mathrm{Hom}(X, k) = \mathrm{Hom}(X, \mathcal{O}_{\mathbb{X}})$ for each $X \in \mathrm{coh}(\mathbb{X})$.

PROOF. The first assertion is an immediate consequence of Orlov's theorem [**26**] since R is graded Gorenstein of Gorenstein parameter -1.

We now turn to the second assertion. Since the Picard group acts transitively on the line bundles of $\mathrm{coh}(\mathbb{X})$, it suffices to show that the left adjoint $\ell : \mathcal{T} \to {}^{\perp}k$ to the inclusion $j : {}^{\perp}k \hookrightarrow \mathcal{T}$ maps k to a line bundle in $\mathrm{D}^b(\mathrm{coh}(\mathbb{X}))$ up to suspension. For the proof of this claim, we refer to the related proof in [**22**, Section 3.8]. □

An immediate consequence is the structure of the Grothendieck group $\mathrm{K}_0(\mathrm{Sing}^{\mathbb{Z}}(R)$ as well as the structure of the the reduced Grothendieck group.

PROPOSITION 3.6. *The Grothendieck group $\mathrm{K}_0(\mathrm{Sing}^{\mathbb{Z}}(R))$ is the K-theoretic one-point extension of $\mathrm{K}_0(\mathbb{X}) = \mathrm{K}_0(\mathrm{D}^b(\mathrm{coh}(\mathbb{X}))$ by the class $[\mathcal{O}_{\mathbb{X}}]$ of the structure sheaf of $\mathbb{X}$.*

PROOF. In more detail this means that for the class $e = [E]$ of some exceptional object $E = k$ in $\mathcal{T} = \mathrm{Sing}^{\mathbb{Z}}(R)$, we get that $e^{\perp} = \{x \in \mathrm{K}_0(\mathcal{T}) \mid \langle e, x\rangle_{\mathcal{T}} = 0\}$ equals $\mathrm{K}_0(\mathbb{X})$. Further $\mathrm{K}_0(\mathcal{T}) = \mathrm{K}_0(\mathbb{X}) \oplus \mathbb{Z}.e$ where the Euler form $\langle -,-\rangle_{\mathcal{T}}$ satisfies the following conditions

$$\begin{aligned}&\langle e, e\rangle_{\mathcal{T}} = 1\\&\langle x, y\rangle_{\mathcal{T}} = \langle x, y\rangle_{\mathbb{X}} \text{ if } x \text{ and } y \text{ belong to } \mathrm{K}_0(\mathbb{X})\\&\langle e, x\rangle_{\mathcal{T}} = 0 \text{ and } \langle x, e\rangle_{\mathcal{T}} = \langle x, [\mathcal{O}_{\mathbb{X}}]\rangle_{\mathbb{X}} \text{ if } x \text{ belongs to } \mathrm{K}_0(\mathbb{X})\end{aligned}$$

These properties follow immediately from Theorem 3.5. □

Let $\mathbb{X} = (X, w)$ with a smooth projective curve X of genus g. For $g \geq 1$ the Grothendieck group $\mathrm{K}_0(\mathbb{X})$, and hence $\mathrm{Sing}^{\mathbb{Z}}(R)$ are not finitely generated, compare [**19**, **20**]. In particular, for $g \geq 1$ neither $\mathrm{coh}(\mathbb{X})$ nor $\mathrm{Sing}^{\mathbb{Z}}(R)$ has a tilting object. By contrast, for $g = 0$, where we deal with weighted projective lines both categories have tilting objects, where the endomorphism rings are ***canonical algebras*** in case $\mathrm{coh}(\mathbb{X})$, see [**11**], respectively ***extended canonical algebras*** in case $\mathrm{Sing}^{\mathbb{Z}}(R)$, see [**22**]. In both cases the Euler forms are non-degenerate.

For genus $g > 0$, the Euler form for $\mathrm{coh}(\mathbb{X}))$ always has a large kernel

$$K = \{x \in \mathrm{K}_0 \mid \langle x, y\rangle_{\mathbb{X}} = 0 \text{ for each } x \in \mathrm{K}_0(\mathbb{X})\}.$$

Due to Serre duality, the kernel equivalently can be described as the set of all $y \in \mathrm{K}_0(\mathbb{X})$ such that $\langle x, y\rangle_{\mathbb{X}} = 0$ for all X. Similar assertions hold for the kernel of the Euler form on $\mathrm{K}_0(\mathrm{Sing}^{\mathbb{Z}}(R))$. It is further easily seen that K is generated by all differences $[S]-[T]$, where S, T run through all ordinary simple sheaves on $\mathbb{X}$.

If one is mainly interested in the Euler form and related constructions, then both for the category of coherent sheaves and for the singularity category it is advisable to pass to the quotients

$$\mathrm{K}_0^{\mathrm{red}}(\mathbb{X}) = \mathrm{K}_0(\mathbb{X})/K \text{ and } \mathrm{K}_0^{\mathrm{red}}(\mathrm{Sing}^{\mathbb{Z}}(R)) = \mathrm{K}_0(\mathrm{Sing}^{\mathbb{Z}}(R))/K.$$

Both groups, called ***reduced*** or ***numerical*** Grothendieck groups, are finitely generated free abelian groups. Moreover, the Euler form in question $\langle -,-\rangle_{\mathbb{X}}$, respectively $\langle -,-\rangle_{\mathcal{T}}$, descends to non-degenerate bilinear forms on the reduced groups.

EXAMPLE 3.7. To illustrate the concept, we consider a smooth projective curve (resp. a compact Riemann surface) X of genus g. Then group $\mathrm{K}_0^{\mathrm{red}}(\mathbb{X})$ has a basis $a=[\mathcal{O}_{\mathbb{X}}]$, $b=[S]$, where S is any simple sheaf on X. For $g \geq 2$, and R the orbit algebra of the AR-translation τ of $\mathrm{coh}(\mathbb{X})$, we have $\mathrm{K}_0^{\mathrm{red}}(\mathrm{Sing}^{\mathbb{Z}}(R)) = \mathbb{Z}^3$ with basis $a=[\mathcal{O}_{\mathbb{X}}]$, $b=[S]$ with S a simple sheaf, and $e=[E]$ the class of the extending object.

With regard to the above bases, the two Euler forms $\langle -,-\rangle_{\mathbb{X}}$ and $\langle -,-\rangle_{\mathcal{T}}$ can be expressed by their ***Cartan matrices***

$$C_X = \begin{bmatrix} 1-g & -1 \\ -1 & 0 \end{bmatrix} \text{ and } C_{\mathcal{T}} = \begin{bmatrix} 1-g & 1 & 1-g \\ -1 & 0 & -1 \\ 0 & 0 & 1 \end{bmatrix}. \tag{10}$$

Further we can form the ***Coxeter transformations***

$$\Phi_X = -C_X^{-1}C_X^{tr} = \begin{bmatrix} 1 & 0 \\ 2(g-1) & 1 \end{bmatrix} \text{ and } \Phi_{\mathcal{T}} = -C_{\mathcal{T}}^{-1}C_{\mathcal{T}}^{tr} = \begin{bmatrix} 2-g & -1 & 1 \\ 2(g-1) & 1 & 0 \\ g-1 & 1 & -1 \end{bmatrix}. \tag{11}$$

The characteristic polynomial of the Coxeter transformation Φ_X (resp. $\Phi_{\mathcal{T}}$) is called ***Coxeter polynomial*** of X (resp. $\mathcal{T}$). If X is a smooth projective curve of genus g, then the respective Coxeter polynomials are

$$\varphi_X = (x-1)^2 \text{ and } \varphi_{\mathcal{T}} = x^3 + (g-2)x^2 + (g-2)x + 1. \tag{12}$$

Note that φ_X does not depend on g, while $\varphi_{\mathcal{T}}$ does. For $g \leq 5$ $\varphi_{\mathcal{T}}$ factors into cyclotomic polynomials

$$g{=}2 \text{ then } \varphi_{\mathcal{T}}{=}\Phi_2\Phi_6,\ g{=}3 \text{ then } \varphi_{\mathcal{T}}{=}\Phi_2\Phi_4,\ g{=}4 \text{ then } \Phi_{\mathcal{T}}\Phi_2\Phi_3,\ g{=}5 \text{ then } \varphi_{\mathcal{T}}{=}\Phi_2^3. \tag{13}$$

Moreover, the Coxeter transformations $\Phi_{\mathcal{T}}$ for $g=2$, $g=3$, $g=4$ are periodic of order $6, 4$, and 6, respectively, while for $g=5$ it is not periodic. Since for $g \geq 6$ the factorization of $\varphi_{\mathcal{T}}$ in irreducible factors (over $\mathbb{Z}$) is $\varphi_{\mathcal{T}} = \varphi_2(x^2+(g-3)x+1)$ is not a product of cyclotomic polynomials, $\Phi_{\mathcal{T}}$ can also not be periodic for $g \geq 6$.

We now comment on the weighted situation. Let $\mathbb{X} = X\langle a_1, a_2, \ldots, a_t\rangle$ be a weighted projective curve, then the Grothendieck group $\mathrm{K}_0(\mathbb{X}) = \mathrm{K}_0(\mathrm{coh}(\mathbb{X}))$ is obtained from $\mathrm{K}_0(X)$ by attaching for each weight a K-theoretic tube T_i of rank a_i, see [**20**] for details. The same procedure works to obtain the reduced Grothendieck group $\mathrm{K}_0^{\mathrm{red}}(\mathbb{X})$ from $\mathrm{K}_0^{\mathrm{red}}(X)$. It follows that the Coxeter polynomial of $\mathbb{X}$ equals $\varphi_{\mathbb{X}} = (x-1)^2\prod_{i=1}^t v_{a_i}$ which agrees with the Coxeter polynomial of the weighted projective line $\mathbb{P}^1\langle a_1, a_2, \ldots, a_t\rangle$.

Concerning the Grothendieck group $\mathrm{K}_0(\mathcal{T})$ (resp. the reduced Grothendieck group $\mathrm{K}_0^{\mathrm{red}}(\mathcal{T})$) of $\mathcal{T} = \mathrm{Sing}^{\mathbb{Z}}(R)$, where R is the orbit algebra attached to $\mathbb{X}$, the situation is more interesting. In both cases, it is obtained as the one-point extension of $\mathrm{K}_0(\mathbb{X})$, resp. $\mathrm{K}_0^{\mathrm{red}}(\mathbb{X})$, by the respective class of the structure sheaf $\mathcal{O}_{\mathbb{X}}$, yielding, for instance, $\mathrm{K}_0^{\mathrm{red}}(\mathcal{T}) = \mathrm{K}_0^{\mathrm{red}}(\mathbb{X}) \oplus \mathbb{Z}e$, where $\langle e,e\rangle = 1$ and further $\langle e,x\rangle = 0$ and $\langle x,e\rangle = \langle x,[\mathcal{O}_{\mathbb{X}}]\rangle$ for all $x,y \in \mathrm{K}_0^{\mathrm{red}}(\mathbb{X})$.

The ***Poincaré* series** $p_R = \sum_{i=0}^{\infty} \dim_k(R_i)x^i$ of a fuchsian singularity R of signature $(g;a_1,a_2...,a_t)$ was determined in [**31**]. As a rational function of x it is given by the expression

$$p_R = \frac{1+(g-2)x+(g-2)x^2+x^3}{(1-x)^2} + \frac{x^2}{(1-x)^2}\sum_{i=1}^{t}\frac{v_{a_i-1}}{v_{a_i}},$$

where $v_a = \frac{x^n-1}{x-1}$, see [**9**, Section 3].

PROPOSITION 3.8. *Let $\mathbb{X} = X\langle a_1,a_2...,a_t\rangle$ be a (WPC) of negative Euler characteristic. Let $\varphi_{\mathbb{X}} = (1-x)^2\prod_{i=1}^{t} v_{a_i}$ the Coxeter polynomial of* $\mathrm{coh}(\mathbb{X})$. *Then the Coxeter polynomial $\varphi_{\mathcal{T}}$ of the fuchsian singularity R attached to $\mathbb{X}$ is given by*

$$\varphi_{\mathcal{T}} = p_R\cdot\varphi_{\mathbb{X}} = \prod_{i=1}^{t} v_{a_i}\left[1+(g-2)x+(g-1)x^2+x^3+\sum_{i=1}^{t}\frac{v_{a_i-1}}{v_{a_i}}\right].$$

For $t=0$ the expression for $\varphi_{\mathcal{T}}$ specializes to the Coxeter polynomial of X from Example 3.7. For $g=0$ we obtain the Coxeter polynomial of the extended canonical algebra with the weight data $a_1,a_2,...,a_t$, see [**22**].

PROOF. Comparing the Poincaré series p_R with its K-theoretic variant $p_{\mathbb{X}} = \sum_{i=0}^{\infty}\langle\mathcal{O}_{\mathbb{X}},\tau^i\mathcal{O}_{\mathbb{X}}\rangle x^i$, we first see that $p_R = (g+x)+p_{\mathbb{X}}$. Next, applying a general result for one-point extensions [**17**, Theorem 18.1], we obtain $\varphi_{\mathcal{T}} = \varphi_{\mathbb{X}}\cdot p_R$ as claimed. □

PROPOSITION 3.9. *If $R = k[x,y,z]/(f)$ is a fuchsian hypersurface, then $\tau^{\deg(f)} = [2]$ holds in* $\mathrm{Sing}^{\mathbb{Z}}(R)$, *thus* $\mathrm{Sing}^{\mathbb{Z}}(R)$ *is **fractionally Calabi-Yau** of CY-dimension* $\frac{2}{\deg(f)}$. *Accordingly, the Coxeter transformation of R is periodic of period* $\deg(f)$, *and the Coxeter polynomial of R is a product of cyclotomic polynomials.*

PROOF. The first assertion is well known, compare [**14**]; it immediately implies the claimed periodicity of the Coxeter transformation. □

Appendix A. The category of coherent sheaves on a smooth projective curve

A.1. The category of coherent sheaves. Let k be an algebraically closed field, and X a smooth projective curve, or for $k=\mathbb{C}$ a compact Riemann surface. Then X is equipped with a category $\mathcal{H}$ of coherent sheaves (algebraic in case of a curve, holomorphic in case of a Riemann surface). We are giving here an axiomatic description of $\mathcal{H}$.

($H1$) It is a connected k-linear category which is Hom- and Ext-finite over k.
($H2$) $\mathcal{H}$ is hereditary (i.e. $\mathrm{Ext}^2(-,-)=0$) and noetherian (i.e. ascending chains of subobjects become stationary).
($H3$) $\mathcal{H}$ satisfies Serre duality in the form $D\mathrm{Ext}^1(A,B) = \mathrm{Hom}(B,\tau A$, where τ is an auto-equivalence of $\mathcal{H}$.

($H4$) The function field $K = k(X)$ of $\mathcal{H}$ (or X), see Section 2.3, has infinite dimension over k.

COMMENT A.1. (i) Due to ($H2$) the category $\mathcal{H}$ has no non-zero projective or injective objects.

(ii) As a further consequence of Serre duality, the category $\mathcal{H}$ has almost split sequences; moreover the functor τ serves as the Auslander-Reiten translation.

(iii) The sheaf $\mathcal{O}_X$ of regular functions on X is called the *structure sheaf* of X.

(iv) The *dualizing sheaf* $\omega = \tau\mathcal{O}_X$ is the sheaf of differential forms of degree one over X. Accordingly $\tau^n\mathcal{O}_X = \omega^{\otimes n}$, the n-fold tensor power of ω, is the sheaf of differential forms of degree n.

(v) The function field $\mathbb{C}(X)$, classically defined as the global sections of the sheaf of meromorphic functions on X, allows the folowing alternative description: Let $\mathcal{H}_0$ be the Serre subcategory of all objects Y of $\mathcal{H}$ having finite length (meaning that Y has a finite filtration with factors that are simple in the category $\mathcal{H}$). Then the structure sheaf $\mathcal{O}_X$, when viewed as an object in the quotient category $\mathcal{H}/\mathcal{H}_0$, has an endomorphism ring, isomorphic to $K = k(X)$.

(vi) Actually there is an equivalence $\mathcal{H}/\mathcal{H}_0 \to \mathrm{mod}(K)$ with $\mathcal{O}_X$ viewed as a member of of $\mathcal{H}/\mathcal{H}_0$ corresponding to $K = k(X)$. In particular $\mathcal{H}/\mathcal{H}_0$ is a *length category*, that is, an abelian category in which each object has finite length.

(vii) Condition ($H4$) is introduced in order to avoid some degenerate examples like the category $\mathrm{mod}^{\mathbb{Z}\times\mathbb{Z}}(k[x,y])/\mathrm{mod}_0^{\mathbb{Z}\times\mathbb{Z}}(k[x,y])$ which satisfies $(H1),(H2),(H3)$, but not $(H4)$.

DEFINITION A.2. The *genus* g_X of $\mathcal{H}$ (or X) is the k-dimension of $\mathrm{Ext}^1(\mathcal{O}_X,\mathcal{O}_X)$, equivalently the dimension of $\mathrm{Hom}(\mathcal{O}_X,\omega)$.

Often the *Euler characteristic* χ_X is more useful than the genus. It relates to the genus by means of $\chi_X = 2(1-g_X)$.

A.2. Euler form, rank, degree and Riemann-Roch. In this section $\mathcal{H} = \mathrm{coh}(X)$ denotes the category of algebraic coherent sheaves on a smooth projective curve X (resp. of holomorphic coherent sheaves on a compact Riemann surface). By $\mathrm{K}_0(\mathcal{H})$ or just $\mathrm{K}_0(X)$ we denote the *Grothendieck group* of the category $\mathcal{H}$ modulo short exact sequences. We always consider $\mathrm{K}_0(\mathcal{H})$ to be equipped with the *Euler form* which is the non-symmetric bilinear form given on classes of objects by the expression

$$\langle [A],[B]\rangle = \dim_k \mathrm{Hom}(A,B) - \dim_k \mathrm{Ext}^1(A,B).$$

This uses that $\mathcal{H}$ is hereditary. Often, we will write $\langle A,B\rangle$ instead of $\langle [A],[B]\rangle$.

There are two important linear forms $\mathrm{K}_0(\mathcal{H}) \to \mathbb{Z}$, called rank and degree.

DEFINITION A.3. The *rank* of a coherent sheaf A, notation $\mathrm{rk}(A)$, is defined as the length of A, when viewed as an object of $\mathcal{H}/\mathcal{H}_0$.

The *degree* of a coherent sheaf A is defined as

$$\deg(A) = \langle \mathcal{O}_X, A\rangle - \mathrm{rk}(A)\cdot\langle \mathcal{O}_X,\mathcal{O}_X\rangle.$$

As follows from the next proposition, rank and degree satisfy largely dual properties.

PROPOSITION A.4. *The rank of a coherent sheaf A is ≥ 0 with equality if and only if A has finite length. Further, we always have* $\mathrm{rk}(A) = \mathrm{rk}(\tau A)$.

The degree of the structure sheaf is 0. The degree of any simple sheaf is 1. A coherent sheaf of finite length always has degree ≥ 0 with equality if and only if $A = 0$. Further, we have $\deg(A) = \deg(\tau A)$ *if A has finite length.*

A coherent sheaf with rank and degree zero is already 0. □

For each non-zero coherent sheaf A its ***slope*** $\mu(A) = \frac{\deg(A)}{\mathrm{rk}(A}$ is thus well-defined.

In order to deal with Riemann-Roch, it is useful to introduce before the *reduced (or* numerical) Grothendieck group $\mathrm{K}_0^{\mathrm{red}}(X)$ which is the factor group of $\mathrm{K}_0(X)$ modulo the *radical* $\{x \in \mathrm{K}_0(X) \mid \langle x, - \rangle = 0\}$ of the Euler form. Obviously, the Euler form descends from $\mathrm{K}_0(X)$ to $\mathrm{K}_0^{\mathrm{red}}(X)$, that is, denoting classes of coherent sheaves in $\mathrm{K}_0^{\mathrm{red}}(X)$ by double brackets, we have $\langle [\![X]\!], [\![Y]\!] \rangle = \langle X, Y \rangle$ for all $X, Y \in \mathcal{H}$. Denoting the simple sheaf, concentrated in $x \in X$, by $\mathcal{S}_x$, it follows that all classes $[\![\mathcal{S}_x]\!]$ of simple sheaves agree in $\mathrm{K}_0^{\mathrm{red}}(X)$. As a consequence $\mathrm{K}_0^{\mathrm{red}}(X) = \mathbb{Z}[\![\mathcal{O}_X]\!] \oplus \mathbb{Z}[\![\mathcal{S}]\!]$, where $\mathcal{S}$ is any simple sheaf.

THEOREM A.5 (Riemann-Roch). *For any two coherent sheaves A, B on X we have*

$$\langle [\![A]\!], [\![B]\!] \rangle = (1 - g_X)\mathrm{rk}(A)\mathrm{rk}(B) + \begin{vmatrix} \mathrm{rk}(A) & \mathrm{rk}(B) \\ \deg(A) & \deg(B) \end{vmatrix}.$$

PROOF. It suffices to check the validity of the RR-formula for $A, B \in \{\mathcal{O}_X, \mathcal{S}\}$, where it is obviously satisfied. □

COROLLARY A.6 (Existence of morphisms). *Let A and B be non-zero vector bundles on X. Then*

$$\frac{\langle [\![A]\!], [\![B]\!] \rangle}{\mathrm{rk}(A) \cdot \mathrm{rk}(B)} = (1 - g_X) + (\mu(B) - \mu(A)).$$

If $\mu(B) - \mu(A) > g_X - 1 = -1/2\chi_X$, *then* $\mathrm{Hom}(X, Y) \neq 0$. □

We recall that an object E of $\mathcal{H}$ is called *exceptional* if $\mathrm{End}(E) = k$ and $\mathrm{Ext}^1(E, E) = 0$. In particular, an exceptional object is indecomposable.

COROLLARY A.7. *The category* $\mathrm{coh}(X)$ *has an exceptional object E if and only if $g_X = 0$ and, additionally, E has rank one, that is, E is a line bundle. In particular,* $\mathrm{coh}(X)$ *has a tilting object if and only if X is (isomorphic to) the Riemann sphere* $\mathbb{S}$.

PROOF. Assume E exceptional. By Riemann-Roch we obtain $1 = \langle E, E \rangle = (1 - g_X)\mathrm{rk}(E)^2$, hence $g_X = 0$ and $\mathrm{rk}(E) = 1$. Concerning the last assertion, it is well known that $\mathrm{coh}(\mathbb{S})$ has a tilting object $\mathcal{O}_X \oplus \mathcal{O}_X(x)$, where $\mathcal{O}_X(x)$ is the central term of 'the' non-split exact sequence $0 \to \mathcal{O}_X \to \mathcal{O}_X(x) \to \mathcal{S}_x \to 0$. □

Appendix B. The category of coherent sheaves on a weighted projective curve

Many aspects have already been dealt with in Sections 2.2 and 2.3.

Let $\mathbb{X} = (X, w)$ be an (WPC) and $\mathrm{coh}(\mathbb{X})$ its category of coherent sheaves. Let $\mathrm{coh}_0(\mathbb{X})$ denote the full subcategory of coherent sheaves of finite length. It is easily seen that the quotient categories $\mathrm{coh}(\mathbb{X})/\mathrm{coh}_0(\mathbb{X})$ and $\mathrm{coh}(X)/\mathrm{coh}_0(X)$ are equivalent. Hence the ***function fields*** $k(\mathbb{X})$ and $k(X)$ are isomorphic. In particular, the (WPC) $\mathbb{X}$ determines the underlying smooth projective curve X. The most important invariant of $\mathbb{X}$, or $\mathrm{coh}(\mathbb{X})$, next to the function field $k(\mathbb{X})$, is the ***orbifold Euler characteristic***, or just Euler characteristic $\chi_{\mathbb{X}}$ of $\mathbb{X}$. The ***Euler form*** is the bilinear form on the Grothendieck group $\mathrm{K}_0(\mathbb{X})$ of the category $\mathrm{coh}(\mathbb{X})$ of coherent sheaves on $\mathbb{X}$ which is given on (classes of) coherent sheaves by the expression $\langle A, B\rangle = \dim \mathrm{Hom}(A,B) - \dim \mathrm{Ext}^1(A,B)$.

For the convenience of the reader, we include the following treatment of orbifold Euler characteristic and weighted Riemann-Roch theorem from [**20**]. Let $a_1, a_2, \dots, a_t$ be the weight sequence associated to the weighted points of $\mathbb{X}$. We put $\bar{a} = \mathrm{lcm}(a_1, a_2, \dots, a_t)$. By means of the ***averaged Euler form***

$$\langle\langle X, Y\rangle\rangle = \frac{1}{\bar{a}}\sum_{j=0}^{\bar{a}-1}\langle \tau^j X, Y\rangle,$$

we define the ***degree*** as the linear form deg on $\mathrm{K}_0(\mathbb{X})$, given by the expression[3]

$$\deg(A) = \bar{a}\langle\langle \mathcal{O}_{\mathbb{X}}, A\rangle\rangle - \bar{a}\langle\langle \mathcal{O}_{\mathbb{X}}, \mathcal{O}_{\mathbb{X}}\rangle\rangle.$$

Rank and degree then are uniquely determined by the following complementry properties:

(i) For A in $\mathrm{coh}(\mathbb{X})$ we always have $\mathrm{rk}A \geq 0$, further the structure sheaf $\mathcal{O}_{\mathbb{X}}$ has rank one. Moreover, $\mathrm{rk}A = 0$ holds exactly if A has finite length. Also rk is preserved under automorphisms of $\mathrm{coh}(\mathbb{X})$, in particular under the Auslander-Reiten translation.

(ii) A non-zero object A of finite length has degree > 0, and the structure sheaf has degree zero. Further each simple sheaf S_x, concentrated in a point x of $\mathbb{X}$, has degree $\frac{\bar{a}}{w(x)}$.

We next define the ***orbifold Euler characteristic*** of $\mathbb{X}$ as $\chi_{\mathbb{X}} = \frac{2}{\bar{a}}\langle\langle \mathcal{O}_{\mathbb{X}}, \mathcal{O}_{\mathbb{X}}\rangle\rangle$. Clearly, for non-weighted smooth projective curves the orbifold Euler characteristic agrees with the ordinary Euler characteristic.

THEOREM B.1. *Let $\mathbb{X} = X\langle a_1, a_2, \dots, a_t\rangle$ be a weighted projective curve with underlying smooth projective curve X. Then the orbifold Euler characteristic of $\mathbb{X}$ is given by the expression*

$$\chi_{\mathbb{X}} = \frac{2}{\bar{a}}\langle\langle \mathcal{O}_{\mathbb{X}}, \mathcal{O}_{\mathbb{X}}\rangle\rangle = \chi_X - \sum_{i=1}^{t}\left(1 - \frac{1}{a_i}\right). \tag{14}$$

PROOF. We work in the reduced Grothendieck group $\mathrm{K}_0^{\mathrm{red}}(\mathbb{X})$. We put $a = [\![\mathcal{O}_{\mathbb{X}}]\!]$, $s_0 = [\![S_x]\!]$ where S_x is the simple sheaf concentrated in a non-weighted point, $s_i = [\![S_i]\!]$, where S_i is the simple sheaf concentrated in x_i with $\mathrm{Hom}(\mathcal{O}_{\mathbb{X}}, S_i) = k$ $(i = 1, 2, \dots, t)$. Then we have

$$\tau a - a = -\sum_{i=1}^{t} s_i + (t - \chi_X)s_0, \tag{15}$$

where $\chi_X = 2\langle a, a\rangle = 2(1-g)$ is the Euler characteristic of the underlying smooth projective curve X. The validity of formula (15) is easily checked by forming for both sides the Euler product with each member of the generating system $a, s_0, \tau^j s_i$,

[3] The factor $\bar{a}$ is introduced to yield integer values for the degree.

$i = 1, 2, \ldots, t$ and using non-degeneracy of the Euler form on $\mathrm{K}_0^{\mathrm{red}}(\mathbb{X})$. Then, applying $\sum_{j=0}^{\bar{a}-1} \tau^j$ to (15) implies

$$\tau^{\bar{a}} a - a = \bar{a}\,\delta\, s_0, \text{ where } \delta = \sum_{j=1}^{t}(1 - \frac{1}{a_i}) - \chi_X. \tag{16}$$

Finally, we obtain

$$2\langle\langle a, a\rangle\rangle + \bar{a}\delta = \langle\langle a, a\rangle\rangle + \langle\langle a, \tau^{\bar{a}} a\rangle\rangle = \frac{1}{\bar{a}}\left(\sum_{j=0}^{\bar{a}-1}\langle \tau^j a, a\rangle - \sum_{j=0}^{\bar{a}-1}\langle \tau^{(\bar{a}-1)-j} a, a\rangle\right) = 0, \tag{17}$$

proving the claim. □

We note that the rank allows a purely K-theoretic definition as $\mathrm{rk}(A) = \langle A, S\rangle$ where S is a simple sheaf concentrated in an ordinary point. This K-theoretic rank agrees with the sheaf-theoretic rank, defined for a coherent sheaf A as its length in the Serre quotient $\mathrm{coh}(\mathbb{X})/\mathrm{coh}_0(\mathbb{X})$.

THEOREM B.2 (Riemann-Roch). *Assume b, c are members of $\mathrm{K}_0(\mathbb{X})$, where $\mathbb{X} = X\langle a_1, a_2, \ldots, a_t\rangle$ then*

$$\langle\langle b, c\rangle\rangle = \langle\langle \mathcal{O}_{\mathbb{X}}, \mathcal{O}_{\mathbb{X}}\rangle\rangle \cdot \mathrm{rk} b \cdot \mathrm{rk} c + \frac{1}{\bar{a}} \cdot \begin{vmatrix} \mathrm{rk}(b) & \mathrm{rk}(c) \\ \deg(b) & \deg(c) \end{vmatrix}. \tag{18}$$

PROOF. Passing to the reduced Grothendieck group $\mathrm{K}_0^{\mathrm{red}}(\mathbb{X})$ one checks the equality of both sides of (18) on the generating system $a, s_0, \tau^j s_i$, $i = 1, 2, \ldots, t$, $j \in \mathbb{Z}_{a_i}$. □

References

[1] M. Artin and J. J. Zhang, *Noncommutative projective schemes*, Adv. Math. **109** (1994), no. 2, 228–287, DOI 10.1006/aima.1994.1087. MR1304753

[2] Alan F. Beardon, *The geometry of discrete groups*, Graduate Texts in Mathematics, vol. 91, Springer-Verlag, New York, 1995. Corrected reprint of the 1983 original. MR1393195

[3] Ragnar-Olaf Buchweitz, *Maximal Cohen-Macaulay Modules and Tate-Cohomology over Gorenstein Rings*, preprint, 1986.

[4] Alberto Canonaco, *A Beilinson-type theorem for coherent sheaves on weighted projective spaces*, J. Algebra **225** (2000), no. 1, 28–46, DOI 10.1006/jabr.1999.8090. MR1743649

[5] Alberto Canonaco, *Beilinson resolutions on weighted projective spaces* (English, with English and French summaries), C. R. Math. Acad. Sci. Paris **336** (2003), no. 1, 35–40, DOI 10.1016/S1631-073X(02)00004-3. MR1968899

[6] Claude Chevalley, *Introduction to the Theory of Algebraic Functions of One Variable*, Mathematical Surveys, No. VI, American Mathematical Society, New York, N. Y., 1951. MR0042164

[7] D. J. Collins, R. I. Grigorchuk, P. F. Kurchanov, and H. Zieschang, *Combinatorial group theory and applications to geometry*, Springer-Verlag, Berlin, 1998. Translated from the 1990 Russian original by P. M. Cohn; Reprint of the original English edition from the series Encyclopaedia of Mathematical Sciences [*Algebra. VII*, Encyclopaedia Math. Sci., 58, Springer, Berlin, 1993; MR1265269 (95g:57004)]. MR1658468

[8] I. V. Dolgačev, *Automorphic forms, and quasihomogeneous singularities* (Russian), Funkcional. Anal. i Priložen. **9** (1975), no. 2, 67–68. MR0568895

[9] Wolfgang Ebeling, *The Poincaré series of some special quasihomogeneous surface singularities*, Publ. Res. Inst. Math. Sci. **39** (2003), no. 2, 393–413. MR1987870

[10] Pierre Gabriel, *Des catégories abéliennes* (French), Bull. Soc. Math. France **90** (1962), 323–448. MR232821

[11] Werner Geigle and Helmut Lenzing, *A class of weighted projective curves arising in representation theory of finite-dimensional algebras*, Singularities, representation of algebras, and vector bundles (Lambrecht, 1985), Lecture Notes in Math., vol. 1273, Springer, Berlin, 1987, pp. 265–297, DOI 10.1007/BFb0078849. MR915180

[12] Werner Geigle and Helmut Lenzing, *Perpendicular categories with applications to representations and sheaves*, J. Algebra **144** (1991), no. 2, 273–343, DOI 10.1016/0021-8693(91)90107-J. MR1140607

[13] Svetlana Katok, *Fuchsian groups*, Chicago Lectures in Mathematics, University of Chicago Press, Chicago, IL, 1992. MR1177168

[14] Dirk Kussin, Helmut Lenzing, and Hagen Meltzer, *Triangle singularities, ADE-chains, and weighted projective lines*, Adv. Math. **237** (2013), 194–251, DOI 10.1016/j.aim.2013.01.006. MR3028577

[15] H. Lenzing, *Wild canonical algebras and rings of automorphic forms*, Finite-dimensional algebras and related topics (Ottawa, ON, 1992), NATO Adv. Sci. Inst. Ser. C Math. Phys. Sci., vol. 424, Kluwer Acad. Publ., Dordrecht, 1994, pp. 191–212. MR1308987

[16] Helmut Lenzing, *Representations of finite-dimensional algebras and singularity theory*, Trends in ring theory (Miskolc, 1996), CMS Conf. Proc., vol. 22, Amer. Math. Soc., Providence, RI, 1998, pp. 71–97. MR1491919

[17] Helmut Lenzing, *Coxeter transformations associated with finite-dimensional algebras*, Computational methods for representations of groups and algebras (Essen, 1997), Progr. Math., vol. 173, Birkhäuser, Basel, 1999, pp. 287–308. MR1714618

[18] Helmut Lenzing, *Hereditary categories*, Handbook of tilting theory, London Math. Soc. Lecture Note Ser., vol. 332, Cambridge Univ. Press, Cambridge, 2007, pp. 105–146, DOI 10.1017/CBO9780511735134.006. MR2384609

[19] Helmut Lenzing, *Weighted projective lines and Riemann surfaces*, Proceedings of the 49th Symposium on Ring Theory and Representation Theory, Symp. Ring Theory Represent. Theory Organ. Comm., Shimane, 2017, pp. 67–79. MR3644095

[20] Helmut Lenzing, *On the K-theory of weighted projective curves*, Representations of algebras, Contemp. Math., vol. 705, Amer. Math. Soc., Providence, RI, 2018, pp. 131–153, DOI 10.1090/conm/705/14203. MR3799493

[21] Helmut Lenzing and José Antonio de la Peña, *Wild canonical algebras*, Math. Z. **224** (1997), no. 3, 403–425, DOI 10.1007/PL00004589. MR1439198

[22] Helmut Lenzing and José A. de la Peña, *Extended canonical algebras and Fuchsian singularities*, Math. Z. **268** (2011), no. 1-2, 143–167, DOI 10.1007/s00209-010-0663-z. MR2805427

[23] Helmut Lenzing and Idun Reiten, *Hereditary Noetherian categories of positive Euler characteristic*, Math. Z. **254** (2006), no. 1, 133–171, DOI 10.1007/s00209-006-0938-6. MR2232010

[24] Silvio Levy (ed.), *The eightfold way*, Mathematical Sciences Research Institute Publications, vol. 35, Cambridge University Press, Cambridge, 1999. The beauty of Klein's quartic curve. MR1722410

[25] John Milnor, *On the 3-dimensional Brieskorn manifolds* $M(p,q,r)$, Knots, groups, and 3-manifolds (Papers dedicated to the memory of R. H. Fox), Princeton Univ. Press, Princeton, N. J., 1975, pp. 175–225. Ann. of Math. Studies, No. 84. MR0418127

[26] Dmitri Orlov, *Derived categories of coherent sheaves and triangulated categories of singularities*, Algebra, arithmetic, and geometry: in honor of Yu. I. Manin. Vol. II, Progr. Math., vol. 270, Birkhäuser Boston, Boston, MA, 2009, pp. 503–531, DOI 10.1007/978-0-8176-4747-6_16. MR2641200

[27] I. Reiten and M. Van den Bergh, *Noetherian hereditary abelian categories satisfying Serre duality*, J. Amer. Math. Soc. **15** (2002), no. 2, 295–366, DOI 10.1090/S0894-0347-02-00387-9. MR1887637

[28] I. G. Ščerbak, *Algebras of automorphic forms with three generators* (Russian), Funkcional. Anal. i Priložen. **12** (1978), no. 2, 93–94. MR0568900

[29] William P. Thurston, *The Geometry and Topology of Three-Manifolds*, Electronic version 1.1 — March 2002, available at `http://www.msri.org/publications/books/gt3m/PDF/13.pdf`.

[30] Pavlos Tzermias, *The group of automorphisms of the Fermat curve*, J. Number Theory **53** (1995), no. 1, 173–178, DOI 10.1006/jnth.1995.1085. MR1344839

[31] Philip Wagreich, *Algebras of automorphic forms with few generators*, Trans. Amer. Math. Soc. **262** (1980), no. 2, 367–389, DOI 10.2307/1999833. MR586722

Institut für Mathematik, Universität Paderborn, 33095 Paderborn, Germany
Email address: `helmut@math.uni-paderborn.de`

Contemporary Mathematics
Volume **761**, 2021
https://doi.org/10.1090/conm/761/15316

Cycle-finite modules over artin algebras

Piotr Malicki and Andrzej Skowroński

Dedicated to José Antonio de la Peña on the occasion of his 60th birthday

Abstract. We describe the representation theory of finitely generated indecomposable modules over artin algebras which do not lie on cycles of indecomposable modules involving homomorphisms from the infinite Jacobson radical of the module category.

1. Introduction

Throughout the paper, by an algebra is meant an artin algebra over a fixed commutative artin ring K, which we shall assume (without loss of generality) to be basic and indecomposable. For an algebra A, we denote by $\operatorname{mod} A$ the category of finitely generated right A-modules and by $\operatorname{ind} A$ the full subcategory of $\operatorname{mod} A$ formed by the indecomposable modules. The Jacobson radical rad_A of $\operatorname{mod} A$ is the ideal generated by all nonisomorphisms between modules in $\operatorname{ind} A$, and the infinite radical $\operatorname{rad}_A^\infty$ of $\operatorname{mod} A$ is the intersection of all powers rad_A^i, $i \geq 1$, of rad_A. By a result of M. Auslander [**6**], $\operatorname{rad}_A^\infty = 0$ if and only if A is of finite representation type, that is, $\operatorname{ind} A$ admits only a finite number of pairwise nonisomorphic modules (see also [**25**] for an alternative proof of this result). On the other hand, if A is of infinite representation type then $(\operatorname{rad}_A^\infty)^2 \neq 0$, by a result proved in [**12**].

An important combinatorial and homological invariant of the module category $\operatorname{mod} A$ of an algebra A is its Auslander-Reiten quiver Γ_A. Recall that Γ_A is a valued translation quiver whose vertices are the isomorphism classes $\{X\}$ of modules X in $\operatorname{ind} A$, the arrows correspond to irreducible homomorphisms between modules in $\operatorname{ind} A$, and the translation is the Auslander-Reiten translation $\tau_A = D\operatorname{Tr}$. We shall not distinguish between a module in X in $\operatorname{ind} A$ and the corresponding vertex $\{X\}$ of Γ_A. If A is an algebra of finite representation type, then every nonzero nonisomorphism in $\operatorname{ind} A$ is a finite sum of composition of irreducible homomorphisms between modules in $\operatorname{ind} A$, and hence in most of the cases we may recover $\operatorname{mod} A$ from the translation quiver Γ_A. In general, Γ_A describes only the quotient category $\operatorname{mod} A/\operatorname{rad}_A^\infty$.

2010 *Mathematics Subject Classification.* Primary 16G10, 16G60, 16G70; Secondary 16E10, 16E30.

Key words and phrases. Artin algebra, Auslander-Reiten quiver, Cycle-finite algebra, Cycle-finite module.

The second author passed away on October 22, 2020.

A prominent role in the representation theory of algebras is played by cycles of indecomposable modules (see [**33**], [**40**], [**49**], [**55**]). Recall that a *cycle* in $\operatorname{ind} A$ is a sequence

$$M_0 \xrightarrow{f_1} M_1 \to \cdots \to M_{r-1} \xrightarrow{f_r} M_r = M_0$$

of nonzero nonisomorphisms in $\operatorname{ind} A$ [**49**], and such a cycle is said to be *finite* if the homomorphisms $f_1, \dots, f_r$ do not belong to $\operatorname{rad}_A^\infty$ (see [**3**], [**4**]). A module M in $\operatorname{ind} A$ is said to be *cycle-finite* if every cycle in $\operatorname{ind} A$ passing through M is finite. We note that this definition is more general then the one presented in [**35**]. Namely, every module M in $\operatorname{ind} A$ which does not lie on a cycle (*directing module* in the sense of [**49**]) is cycle-finite.

If A is an algebra of finite representation type, then all cycles in $\operatorname{ind} A$ are finite, and much of their properties is visible in the combinatorial structure of the finite Auslander-Reiten quiver Γ_A of A. On the other hand, if A is an indecomposable algebra of infinite representation type, then cycle-finite modules in $\operatorname{ind} A$ lie in infinite components of the Auslander-Reiten quiver Γ_A and usually belong to cycles containing arbitrary large number of pairwise nonisomorphic indecomposable modules. Hence the study of cycle-finite modules over algebras of infinite representation type is a rather complicated problem.

The aim of this survey article is to present a rather complete representation theory of cycle-finite indecomposable modules over artin algebras.

For general results on the relevant representation theory we refer to the books [**2**], [**8**], [**49**], [**52**], [**53**], [**61**], [**62**] and the survey articles [**15**], [**33**], [**40**], [**48**], [**50**], [**55**].

2. Preliminaries

Let A be an algebra and M a module in $\operatorname{ind} A$. An important information concerning the structure of M is coded in the structure and properties of its support algebra $\operatorname{Supp}(M)$ defined as follows. Consider a decomposition $A = P_M \oplus Q_M$ of A in $\operatorname{mod} A$ such that the simple summands of the semisimple module $P_M/\operatorname{rad} P_M$ are exactly the simple composition factors of M. Then $\operatorname{Supp}(M) = A/t_A(M)$, where $t_A(M)$ is the ideal in A generated by the images of all homomorphisms from Q_M to A in $\operatorname{mod} A$. We note that M is an indecomposable module over $\operatorname{Supp}(M)$. Clearly, we may realistically hope to describe the structure of $\operatorname{Supp}(M)$ only for modules M having some distinguished properties. For example, if M is a directing module in $\operatorname{ind} A$, then the support algebra $\operatorname{Supp}(M)$ of M over an algebra A is a tilted algebra $\operatorname{End}_H(T)$, for a hereditary algebra H and a tilting module T in $\operatorname{mod} H$, and M is isomorphic to the image $\operatorname{Hom}_H(T, I)$ of an indecomposable injective module I in $\operatorname{mod} H$ via the functor $\operatorname{Hom}_H(T, -) : \operatorname{mod} H \to \operatorname{mod} \operatorname{End}_H(T)$ (see [**49**] and [**23**], [**24**] for the corresponding result over arbitrary artin algebra).

Let A be an algebra and Γ_A be the Auslander-Reiten quiver of A. By a component of Γ_A we mean a connected component of the quiver Γ_A. For a component $\mathscr{C}$ of Γ_A, we denote by $\operatorname{ann}_A(\mathscr{C})$ the annihilator of $\mathscr{C}$ in A, that is, the intersection of the annihilators $\{a \in A \mid Ma = 0\}$ of all modules M in $\mathscr{C}$, and by $B(\mathscr{C})$ the quotient algebra $A/\operatorname{ann}_A(\mathscr{C})$, called the *faithful algebra* of $\mathscr{C}$. We note that $\mathscr{C}$ is a faithful component of $\Gamma_{B(\mathscr{C})}$. For a module M in Γ_A, the τ_A-*orbit* of M is the set $\mathcal{O}(M)$ of all possible vertices in Γ_A of the form $\tau_A^n M$, $n \in \mathbb{Z}$. The τ_A-orbit $\mathcal{O}(M)$ of M (respectively, the module M) is said to be *periodic* if $M \cong \tau_A^n M$ for some $n \geq 1$. A module M in Γ_A is said to be an *acyclic module* if M does not lie on an oriented cycle in Γ_A, and otherwise a *cyclic module*. A component $\mathscr{C}$ of Γ_A without oriented cycles is said to be *acyclic*. Dually, a component $\mathscr{C}$ of Γ_A is said to be *cyclic* if any module in $\mathscr{C}$ lies on an oriented cycle of $\mathscr{C}$. Following [**37**], we denote by ${}_c\mathscr{C}$ the full translation subquiver of $\mathscr{C}$ obtained by removing all acyclic modules and the arrows attached to them, and call it the *cyclic part* of $\mathscr{C}$. The connected translation subquivers of ${}_c\mathscr{C}$ are said to be *cyclic components* of $\mathscr{C}$. It was shown in [**37**, Proposition 5.1] that two modules M and N in ${}_c\mathscr{C}$ belong to the same cyclic component of $\mathscr{C}$ if there is an oriented cycle in $\mathscr{C}$ passing through M and N. All the modules in a cyclic component $\mathscr{C}$ of Γ_A containing a cyclic-finite module are cyclic-finite, and such an cyclic component $\mathscr{C}$ is said to be a *cycle-finite cyclic component*. For a subquiver $\mathscr{C}$ of Γ_A, we consider a decomposition $A = P_{\mathscr{C}} \oplus Q_{\mathscr{C}}$ of A in $\operatorname{mod} A$ such that the simple summands of the semisimple module $P_{\mathscr{C}}/\operatorname{rad} P_{\mathscr{C}}$ are exactly the simple composition factors of indecomposable modules in $\mathscr{C}$, the ideal $t_A(\mathscr{C})$ in A generated by the images of all homomorphisms from $Q_{\mathscr{C}}$ to A in $\operatorname{mod} A$, and call the quotient algebra $\operatorname{Supp}(\mathscr{C}) = A/t_A(\mathscr{C})$ the *support algebra* of $\mathscr{C}$. Let M be a nondirecting cycle-finite module in $\operatorname{ind} A$. Observe that M belongs to a unique cyclic component $\mathscr{C}(M)$ of Γ_A consisting entirely of cycle-finite indecomposable modules, and the support algebra $\operatorname{Supp}(M)$ of M is a quotient algebra of the support algebra $\operatorname{Supp}(\mathscr{C}(M))$ of $\mathscr{C}(M)$. Moreover, by a result stated in [**35**, Corollary 1.3], the support algebra $\operatorname{Supp}(\mathscr{C})$ of a cycle-finite cyclic component $\mathscr{C}$ of Γ_A is isomorphic to an algebra of the form $e_{\mathscr{C}}Ae_{\mathscr{C}}$ for an idempotent $e_{\mathscr{C}}$ of A whose primitive summands correspond to the vertices of a convex subquiver of the valued quiver Q_A of A. On the other hand, the support algebra $\operatorname{Supp}(M)$ of a cycle-finite module M in $\operatorname{ind} A$ is not necessarily an algebra of the form eAe for an idempotent e of A (see [**35**, Section 6]). A component $\mathscr{C}$ of Γ_A is called *regular* if $\mathscr{C}$ contains neither a projective module nor an injective module, and *semiregular* if $\mathscr{C}$ does not contain both a projective and an injective module. It has been shown in [**28**] and [**63**] that a regular component $\mathscr{C}$ of Γ_A contains an oriented cycle if and only if $\mathscr{C}$ is a *stable tube* (is of the form $\mathbb{Z}\mathbb{A}_\infty/(\tau^r)$, for a positive integer r). Moreover, S. Liu proved in [**29**] that a semiregular component $\mathscr{C}$ of Γ_A contains an oriented cycle if and only if $\mathscr{C}$ is a *ray tube* (obtained from a stable tube by a finite number (possibly zero) of ray insertions) or a *coray tube* (obtained from a stable tube by a finite number (possibly zero) of coray insertions). A component $\mathscr{C}$ of Γ_A is called *postprojective* if $\mathscr{C}$ is acyclic and every module in $\mathscr{C}$ belongs to the τ_A-orbit of a projective module. Dually, a component $\mathscr{C}$ of Γ_A is called *preinjective* if $\mathscr{C}$ is acyclic and every module in $\mathscr{C}$ belongs to the τ_A-orbit of an injective module. An indecomposable module X in a component $\mathscr{C}$ of Γ_A is said to be *right coherent* if there is in $\mathscr{C}$ an infinite sectional path

$$X = X_1 \longrightarrow X_2 \longrightarrow \cdots \longrightarrow X_i \longrightarrow X_{i+1} \longrightarrow X_{i+2} \longrightarrow \cdots$$

Dually, an indecomposable module Y in $\mathscr{C}$ is said to be *left coherent* if there is in $\mathscr{C}$ an infinite sectional path

$$\cdots \longrightarrow Y_{j+2} \longrightarrow Y_{j+1} \longrightarrow Y_j \longrightarrow \cdots \longrightarrow Y_2 \longrightarrow Y_1 = Y.$$

A module Z in $\mathscr{C}$ is said to be *coherent* if Z is left and right coherent. A component $\mathscr{C}$ of Γ_A is said to be *coherent* [**37**] (see also [**16**]) if every projective module in $\mathscr{C}$ is right coherent and every injective module in $\mathscr{C}$ is left coherent. Further, a component $\mathscr{C}$ of Γ_A is said to be *almost cyclic* if its cyclic part ${}_c\mathscr{C}$ is a cofinite subquiver of $\mathscr{C}$. We note that the stable tubes, ray tubes and coray tubes of Γ_A

are special types of almost cyclic coherent components. In general, it has been proved in [**37**] that a component $\mathscr{C}$ of Γ_A is almost cyclic and coherent if and only if $\mathscr{C}$ is a *generalized multicoil*, obtained from a finite family of stable tubes by a sequence of admissible operations (ad 1)-(ad 5) and their duals (ad 1^*)-(ad 5^*). We refer to [**37**, Section 2] for a detailed description of these admissible operations and generalized multicoils. In particular, one knows that all arrows of a generalized multicoil have trivial valuation. On the other hand, a component $\mathscr{C}$ of Γ_A is said to be *almost acyclic* if all but finitely many modules of $\mathscr{C}$ are acyclic. It has been proved by I. Reiten and the second named author in [**46**] that a component $\mathscr{C}$ of Γ_A is almost acyclic if and only if $\mathscr{C}$ admits a multisection Δ. Moreover, for an almost acyclic component $\mathscr{C}$ of Γ_A, there exists a finite convex subquiver $c(\mathscr{C})$ of $\mathscr{C}$ (possibly empty), called the *core* of $\mathscr{C}$, containing all modules lying on oriented cycles in $\mathscr{C}$ (see [**46**] for details).

Let A be an algebra. A family $\mathscr{C} = (\mathscr{C}_i)_{i\in I}$ of components of Γ_A is said to be *generalized standard* if $\operatorname{rad}_A^\infty(X,Y) = 0$ for all modules X and Y in $\mathscr{C}$ [**54**], and *sincere* if every simple module in $\operatorname{mod} A$ occurs as a composition factor of a module in $\mathscr{C}$. Two components $\mathscr{C}$ and $\mathscr{D}$ of an Auslander-Reiten quiver Γ_A are said to be *orthogonal* if $\operatorname{Hom}_A(X,Y) = 0$ and $\operatorname{Hom}_A(Y,X) = 0$ for all modules $X \in \mathscr{C}$ and $Y \in \mathscr{D}$. We also note that if $\mathscr{C}$ and $\mathscr{D}$ are distinct components of Γ_A then $\operatorname{Hom}_A(X,Y) = \operatorname{rad}_A^\infty(X,Y)$ for all modules $X \in \mathscr{C}$ and $Y \in \mathscr{D}$. Observe that a family $\mathscr{C} = (\mathscr{C}_i)_{i\in I}$ of components of Γ_A is generalized standard if and only if the components $\mathscr{C}_i$, $i \in I$, are generalized standard and pairwise orthogonal. A prominent role in the representation theory of algebras is played by the algebras with separating families of Auslander-Reiten components. A concept of a separating family of tubes has been introduced by C. M. Ringel in [**48**], [**49**] who proved that they occur in the Auslander-Reiten quivers of hereditary algebras of Euclidean type, tubular algebras, and canonical algebras. More generally, following I. Assem, A. Skowroński and B. Tomé [**5**], a family $\mathscr{C} = (\mathscr{C}_i)_{i\in I}$ in Γ_A is said to be *separating* in $\operatorname{mod} A$ if the components of Γ_A split into three disjoint families $\mathscr{P}^A$, $\mathscr{C}^A = \mathscr{C}$ and $\mathscr{Q}^A$ such that the following conditions are satisfied:

(S1) $\mathscr{C}^A$ is a sincere generalized standard family of components;
(S2) $\operatorname{Hom}_A(\mathscr{Q}^A, \mathscr{P}^A) = 0$, $\operatorname{Hom}_A(\mathscr{Q}^A, \mathscr{C}^A) = 0$, $\operatorname{Hom}_A(\mathscr{C}^A, \mathscr{P}^A) = 0$;
(S3) any homomorphism from $\mathscr{P}^A$ to $\mathscr{Q}^A$ in $\operatorname{mod} A$ factors through the additive category $\operatorname{add}(\mathscr{C}^A)$ of $\mathscr{C}^A$.

Then we say that $\mathscr{C}^A$ separates $\mathscr{P}^A$ from $\mathscr{Q}^A$ and write

$$\Gamma_A = \mathscr{P}^A \cup \mathscr{C}^A \cup \mathscr{Q}^A.$$

We note that then $\mathscr{P}^A$ and $\mathscr{Q}^A$ are uniquely determined by $\mathscr{C}^A$ (see [**5**, (2.1)] or [**49**, (3.1)]). Moreover, we have $\operatorname{ann}_A(\mathscr{C}^A) = 0$, so $\mathscr{C}^A$ is a faithful family of components of Γ_A. We note that if A is an algebra of finite representation type that $\mathscr{C}^A = \Gamma_A$ is trivially a unique separating component of Γ_A, with $\mathscr{P}^A$ and $\mathscr{Q}^A$ being empty.

3. Quasitilted algebras

In the representation theory of algebras an important role is played by the canonical algebras introduced by C. M. Ringel in [**49**] and [**51**]. We refer to [**51**, Appendix] for an elementary definition of the canonical algebras proposed by W. Crawley-Boevey. Every canonical algebra Λ is of global dimension at most two.

Moreover, the following theorem due to C. M. Ringel [**51**] describes the general shape of the Auslander-Reiten quiver of a canonical algebra.

THEOREM 3.1. *Let Λ be a canonical algebra. Then the general shape of the Auslander-Reiten quiver Γ_Λ of Λ is as follows*

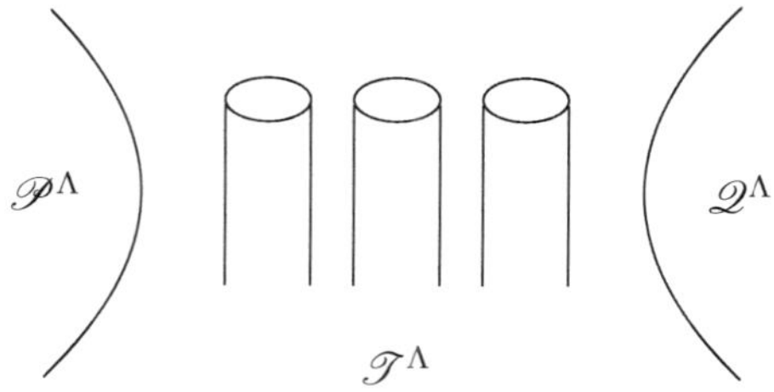

where $\mathscr{P}^\Lambda$ is a family of components containing a unique postprojective component $\mathscr{P}(\Lambda)$ and all indecomposable projective Λ-modules, $\mathscr{Q}^\Lambda$ is a family of components containing a unique preinjective component $\mathscr{Q}(\Lambda)$ and all indecomposable injective Λ-modules, and $\mathscr{T}^\Lambda$ is an infinite family of faithful generalized standard stable tubes separating $\mathscr{P}^\Lambda$ from $\mathscr{Q}^\Lambda$. In particular, we have $\operatorname{pd}_\Lambda X \leq 1$ for all modules X in $\mathscr{P}^\Lambda \cup \mathscr{T}^\Lambda$ and $\operatorname{id}_\Lambda Y \leq 1$ for all modules Y in $\mathscr{T}^\Lambda \cup \mathscr{Q}^\Lambda$.

Let Λ be a canonical algebra. An algebra C of the form $\operatorname{End}_\Lambda(T)$, where T is a tilting Λ-module from the additive category $\operatorname{add}(\mathscr{P}^\Lambda)$ of $\mathscr{P}^\Lambda$ is called a *concealed canonical algebra of type* Λ. Then we have the following theorem describing the general shape of the Auslander-Reiten quiver of a concealed canonical algebra, which is a consequence of Theorem 3.1 and the tilting theory.

THEOREM 3.2. *Let Λ be a canonical algebra, T a tilting module in $\operatorname{add}(\mathscr{P}^\Lambda)$, and $C = \operatorname{End}_\Lambda(T)$ the associated concealed canonical algebra. Then the general shape of the Auslander-Reiten quiver Γ_C of C is as follows*

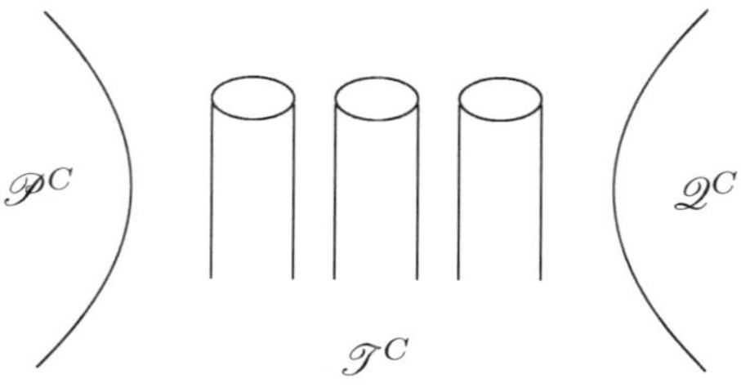

where $\mathscr{P}^C$ is a family of components containing a unique postprojective component $\mathscr{P}(C)$ and all indecomposable projective C-modules, $\mathscr{Q}^C$ is a family of components containing a unique preinjective component $\mathscr{Q}(C)$ and all indecomposable injective C -modules, $\mathscr{T}^C = \operatorname{Hom}_\Lambda(T, \mathscr{T}^\Lambda)$ is an infinite family of faithful pairwise orthogonal generalized standard stable tubes separating $\mathscr{P}^C$ from $\mathscr{Q}^C$. In particular, we have $\operatorname{pd}_C X \leq 1$ for all modules X in $\mathscr{P}^C \cup \mathscr{T}^C$, $\operatorname{id}_C Y \leq 1$ for all modules Y in $\mathscr{T}^C \cup \mathscr{Q}^C$, and $\operatorname{gl.dim} C \leq 2$.

The following characterization of concealed canonical algebras has been established by J. A. de la Peña and H. Lenzing in [**26**].

THEOREM 3.3. *Let A be an algebra. The following statements are equivalent.*

(1) *A is a concealed canonical algebra.*
(2) *Γ_A admits a separating family $\mathscr{T}^A$ of stable tubes.*

The concealed canonical algebras form a distinguished class of *quasitilted algebras*, which are the endomorphism algebras $\mathrm{End}_{\mathscr{H}}(T)$ of tilting objects T in abelian hereditary K-categories $\mathscr{H}$ [**22**]. The following characterization of quasitilted algebras has been established by D. Happel, I. Reiten and S. O. Smalø in [**22**].

THEOREM 3.4. *Let A be an algebra. The following statements are equivalent.*

(1) *A is a quasitilted algebra.*
(2) $\mathrm{gl.\,dim}\, A \leq 2$ *and every module X in* $\mathrm{ind}\, A$ *satisfies* $\mathrm{pd}_A X \leq 1$ *or* $\mathrm{id}_A X \leq 1$.

Recall briefly that the *tilted algebras* are the endomorphism algebras $\mathrm{End}_H(T)$ of tilting modules T over hereditary algebras H and the *quasitilted algebras of canonical type* are the endomorphism algebras $\mathrm{End}_{\mathscr{H}}(T)$ of tilting objects T in abelian hereditary categories $\mathscr{H}$ whose derived category $D^b(\mathscr{H})$ is equivalent to the derived category $D^b(\mathrm{mod}\,\Lambda)$ of the module category $\mathrm{mod}\,\Lambda$ of a canonical algebra Λ. The next result proved in [**21**] shows that there are only two classes of quasitilted algebras.

THEOREM 3.5. *Let A be a quasitilted algebra. Then A is either a tilted algebra or a quasitilted algebra of canonical type.*

The structure of representation-infinite quasitilted algebras of canonical type has been described by H. Lenzing and A. Skowroński in [**27**] (see also [**57**]). In [**27**] the concept of a semiregular branch enlargement of a concealed canonical algebra has been introduced and the following characterization of quasitilted algebras of canonical type established, which extends characterizations of concealed canonical algebras proved in [**26**], [**44**] and [**56**].

THEOREM 3.6. *Let A be an algebra. The following statements are equivalent.*

(1) *A is a representation-infinite quasitilted algebra of canonical type.*
(2) *A is a semiregular branch enlargement of a concealed canonical algebra.*
(3) *Γ_A admits a separating family $\mathscr{T}^A$ of semiregular tubes.*

Here, by a semiregular tube we mean a ray tube or a coray tube.

We may visualise the shape of the Auslander-Reiten quiver Γ_A of a representation-infinite quasitilted algebra A of canonical type as follows

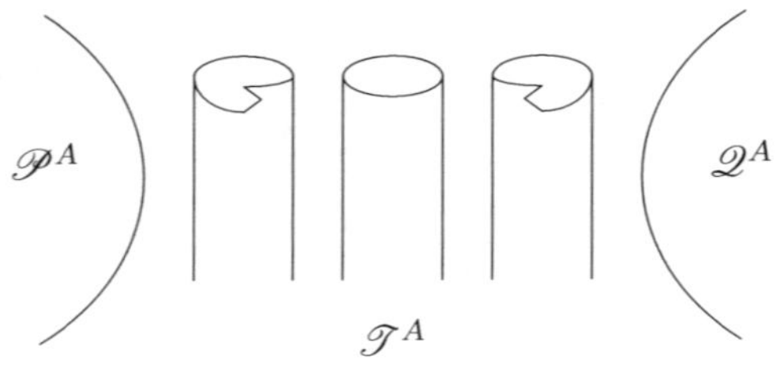

where $\mathscr{T}^A$ is a faithful infinite family of pairwise orthogonal, generalized standard ray or coray tubes, separating $\mathscr{P}^A$ from $\mathscr{Q}^A$. Then every indecomposable projective A-module lies in $\mathscr{P}^A \cup \mathscr{T}^A$ while every indecomposable injective A -module lies in $\mathscr{T}^A \cup \mathscr{Q}^A$.

The following example from [**40**] illustrates the above theorem.

EXAMPLE 3.7. Let K be an algebraically closed field, Q the quiver

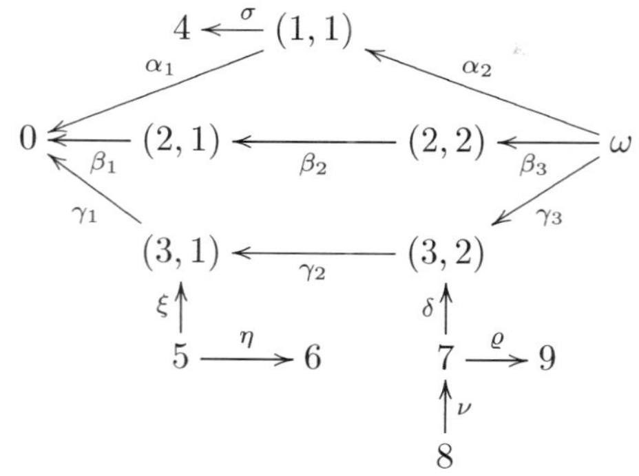

I the ideal of KQ generated by the elements $\alpha_2\alpha_1 + \beta_3\beta_2\beta_1 + \gamma_3\gamma_2\gamma_1$, $\alpha_2\sigma$, $\xi\gamma_1$, $\delta\gamma_2$, $\nu\varrho$, and $A = KQ/I$ the associated bound quiver algebra. Then A is a representation-infinite quasitilted algebra of canonical type. Moreover, A is a semiregular branch enlargement of the canonical algebra $C = K\Delta/J$, where Δ is the full subquiver of Q given by the vertices 0, ω, $(1,1)$, $(2,1)$, $(2,2)$, $(3,1)$, $(3,2)$, and J is the ideal of $K\Delta$ generated by $\alpha_2\alpha_1 + \beta_3\beta_2\beta_1 + \gamma_3\gamma_2\gamma_1$. The Auslander-Reiten quiver Γ_A of A has a disjoint union form

$$\Gamma_A = \mathscr{P}^A \cup \mathscr{T}^A \cup \mathscr{Q}^A$$

where $\mathscr{T}^A$ a family of semiregular tubes separating $\mathscr{P}^A$ from $\mathscr{Q}^A$. Moreover, $\mathscr{T}^A$ is a $\mathbb{P}_1(K)$ -family $(\mathscr{T}^A_\lambda)_{\lambda\in\mathbb{P}_1(K)}$ of semiregular tubes consisting of:

- a coray tube $\mathscr{T}^A_\infty$ obtained from the stable tube $\mathscr{T}^C_\infty$ of Γ_C of rank 2 by one coray insertion,
- a stable tube $\mathscr{T}^A_0 = \mathscr{T}^C_0$ of rank 3 formed by indecomposable C-modules,
- a ray tube $\mathscr{T}^A_1$ obtained from the stable tube $\mathscr{T}^C_1$ of Γ_C of rank 3 by 5 ray insertions,
- the infinite family $\mathscr{T}^A_\lambda, \lambda \in \mathbb{P}_1(K) \setminus \{\infty, 0, 1\}$, of stable tubes of rank 1, consisting of indecomposable C-modules.

We refer to [**40**, Example 6.19] for a detailed description of the families $\mathscr{P}^A$, $\mathscr{T}^A$ and $\mathscr{Q}^A$.

4. Cycle-finite cyclic Auslander-Reiten components

In this section we describe the support algebras of cycle-finite cyclic components of the Auslander-Reiten quivers of artin algebras. The description splits into two cases. In the case when a cycle-finite cyclic component $\mathscr{C}$ of Γ_A is infinite, the support algebra $\operatorname{Supp}(\mathscr{C})$ is a suitable gluing of finitely many generalized multicoil algebras (introduced by the authors in [**38**]) and algebras of finite representation type, and $\mathscr{C}$ is the corresponding gluing of the associated cyclic generalized multicoils via finite translation quivers (Theorem 4.3). In the second case when a cycle-finite cyclic component $\mathscr{C}$ is finite, the support algebra $\operatorname{Supp}(\mathscr{C})$ is a generalized double tilted algebra (in the sense of I. Reiten and A. Skowroński [**46**]) and $\mathscr{C}$ is the core of the connecting component of this algebra (Theorem 4.7).

In order to present the first case, we need the class of generalized multicoil algebras. Recall that following [**38**], an algebra A is called a *generalized multicoil algebra*, if A is a generalized multicoil enlargement of a product $C = C_1 \times \ldots \times$

C_m of concealed canonical algebras $C_1, \dots, C_m$ using modules from the separating families $\mathscr{T}^{C_1}, \dots, \mathscr{T}^{C_m}$ of stable tubes of $\Gamma_{C_1}, \dots, \Gamma_{C_m}$ and a sequence of admissible operations of types (ad 1)-(ad 5) and their duals (ad 1*)-(ad 5 *). The following result has been established in [**38**, Theorem A].

THEOREM 4.1. *Let A be an algebra. The following statements are equivalent.*

(1) *A is a generalized multicoil algebra.*
(2) *Γ_A admits a separating family of almost cyclic coherent components.*

The following consequence of [**38**, Theorem C] describes the structure of the Auslander-Reiten quivers of generalized multicoil algebras.

THEOREM 4.2. *Let A be a generalized multicoil algebra obtained from a family $C_1, \dots, C_m$ of concealed canonical algebras. Then there are unique quotient algebras $A^{(l)}$ and $A^{(r)}$ of A such that the following statements hold:*

(1) *$A^{(l)}$ is a product of quasitilted algebras of canonical type having separating families of coray tubes.*
(2) *$A^{(r)}$ is a product of quasitilted algebras of canonical type having separating families of ray tubes.*
(3) *The Auslander-Reiten quiver Γ_A has a disjoint union decomposition*

$$\Gamma_A = \mathscr{P}^A \cup \mathscr{C}^A \cup \mathscr{Q}^A,$$

where

(a) *$\mathscr{P}^A$ is the left part $\mathscr{P}^{A^{(l)}}$ in a decomposition $\Gamma_{A^{(l)}} = \mathscr{P}^{A^{(l)}} \cup \mathscr{T}^{A^{(l)}} \cup \mathscr{Q}^{A^{(l)}}$ of the Auslander-Reiten quiver $\Gamma_{A^{(l)}}$ of the algebra $A^{(l)}$, with $\mathscr{T}^{A^{(l)}}$ a family of coray tubes separating $\mathscr{P}^{A^{(l)}}$ from $\mathscr{Q}^{A^{(l)}}$;*
(b) *$\mathscr{Q}^A$ is the right part $\mathscr{Q}^{A^{(r)}}$ in a decomposition $\Gamma_{A^{(r)}} = \mathscr{P}^{A^{(r)}} \cup \mathscr{T}^{A^{(r)}} \cup \mathscr{Q}^{A^{(r)}}$ of the Auslander-Reiten quiver $\Gamma_{A^{(r)}}$ of the algebra $A^{(r)}$, with $\mathscr{T}^{A^{(r)}}$ a family of ray tubes separating $\mathscr{P}^{A^{(r)}}$ from $\mathscr{Q}^{A^{(r)}}$;*
(c) *$\mathscr{C}^A$ is a family of generalized multicoils separating $\mathscr{P}^A$ from $\mathscr{Q}^A$, obtained from stable tubes in the separating families $\mathscr{T}^{C_1}, \dots, \mathscr{T}^{C_m}$ of stable tubes of the Auslander-Reiten quivers $\Gamma_{C_1}, \dots, \Gamma_{C_m}$ of the concealed canonical algebras $C_1, \dots, C_m$ by a sequence of admissible operations of types (ad 1)-(ad 5) and their duals (ad 1*)-(ad 5*), corresponding to the admissible operations leading from $C = C_1 \times \dots \times C_m$ to A;*
(d) *$\mathscr{C}^A$ consists of cycle-finite modules and contains all indecomposable modules of $\mathscr{T}^{A^{(l)}}$ and $\mathscr{T}^{A^{(r)}}$;*
(e) *$\mathscr{P}^A$ contains all indecomposable modules of $\mathscr{P}^{A^{(r)}}$;*
(f) *$\mathscr{Q}^A$ contains all indecomposable modules of $\mathscr{Q}^{A^{(l)}}$.*

Moreover, in the above notation, we have the following consequences of [**38**, Theorem E] describing the homological properties of modules over generalized multicoil algebras

- $\operatorname{gl.dim} A \leq 3$;
- $\operatorname{pd}_A X \leq 1$ for any indecomposable module X in $\mathscr{P}^A$;
- $\operatorname{id}_A Y \leq 1$ for any indecomposable module Y in $\mathscr{Q}^A$;
- $\operatorname{pd}_A M \leq 2$ and $\operatorname{id}_A M \leq 2$ for any indecomposable module M in $\mathscr{C}^A$.

The algebra $A^{(l)}$ is said to be the *left quasitilted algebra* of A and the algebra $A^{(r)}$ is said to be the *right quasitilted algebra* of A.

The following theorem from [**35**, Theorem 1.1] describes the support algebras of infinite cycle-finite cyclic components of the Auslander-Reiten quivers of artin algebras.

THEOREM 4.3. *Let A be an algebra and $\mathscr{C}$ be an infinite cycle-finite component of ${}_c\Gamma_A$. Then there exist infinite full translation subquivers $\mathscr{C}_1, \dots, \mathscr{C}_r$ of $\mathscr{C}$ such that the following statements hold.*

(1) *For each $i \in \{1, \dots, r\}$, $\mathscr{C}_i$ is a cyclic coherent full translation subquiver of Γ_A.*
(2) *For each $i \in \{1, \dots, r\}$,* $\mathrm{Supp}(\mathscr{C}_i) = B(\mathscr{C}_i)$ *and is a generalized multicoil algebra.*
(3) *$\mathscr{C}_1, \dots, \mathscr{C}_r$ are pairwise disjoint full translation subquivers of $\mathscr{C}$ and $\mathscr{C}^{cc} = \mathscr{C}_1 \cup \dots \cup \mathscr{C}_r$ is a maximal cyclic coherent and cofinite full translation subquiver of $\mathscr{C}$.*
(4) *$B(\mathscr{C} \setminus \mathscr{C}^{cc})$ is of finite representation type.*
(5) $\mathrm{Supp}(\mathscr{C}) = B(\mathscr{C})$.

REMARK 4.4. It follows from the above theorem that all but finitely many modules lying in an infinite cycle-finite component $\mathscr{C}$ of ${}_c\Gamma_A$ can be obtained from indecomposable modules in stable tubes of concealed canonical algebras by a finite sequence of admissible operations of types (ad 1)-(ad 5) and their duals (ad 1*)-(ad 5*) (see [**38**, Section 3] for details).

We would like to stress that the cycle-finiteness assumption imposed on the infinite component $\mathscr{C}$ of ${}_c\Gamma_A$ is essential for the validity of the above theorem. Namely, it has been proved in [**58**], [**59**] that, for an arbitrary finite dimensional algebra B over a field K, a module M in $\mathrm{mod}\, B$, and a positive integer r, there exists a finite dimensional algebra A over K such that B is a quotient algebra of A, Γ_A admits a faithful generalized standard stable tube $\mathscr{T}$ of rank r, $\mathscr{T}$ is not cycle-finite, and M is a subfactor of all but finitely many indecomposable modules in $\mathscr{T}$. This shows that in general the problem of describing the support algebras of infinite cyclic components (even stable tubes) of Auslander-Reiten quivers is difficult.

In order to present the second case (when a cycle-finite cyclic component $\mathscr{C}$ is finite), we need the class of generalized double tilted algebras introduced by I. Reiten and A. Skowroński in [**46**] (see also [**1**], [**11**] and [**45**]). A *generalized double tilted algebra* is an algebra B for which Γ_B admits a separating almost acyclic component $\mathscr{C}$.

The following consequence of [**46**, Section 3] describes the structure of the Auslander-Reiten quivers of generalized double tilted algebras.

THEOREM 4.5. *Let B be a generalized double tilted algebra. Then the Auslander-Reiten quiver Γ_B has a disjoint union decomposition $\Gamma_B = \mathscr{P}^B \cup \mathscr{C}^B \cup \mathscr{Q}^B$, where*

(1) *$\mathscr{C}^B$ is an almost acyclic component separating $\mathscr{P}^B$ from $\mathscr{Q}^B$;*
(2) *For each $i \in \{1, \dots, m\}$, there exists hereditary algebra $H_i^{(l)}$ and tilting module $T_i^{(l)} \in \mathrm{mod}\, H_i^{(l)}$ such that the tilted algebra $B_i^{(l)} = \mathrm{End}_{H_i^{(l)}}(T_i^{(l)})$ is a quotient algebra of B and $\mathscr{P}^B$ is the disjoint union of all components of $\Gamma_{B_i^{(l)}}$ contained entirely in the torsion-free part $\mathscr{Y}(T_i^{(l)})$ of $\mathrm{mod}\, B_i^{(l)}$ determined by $T_i^{(l)}$;*

(3) *For each* $j \in \{1, \dots, n\}$, *there exists hereditary algebra* $H_j^{(r)}$ *and tilting module* $T_j^{(r)} \in \operatorname{mod} H_j^{(r)}$ *such that the tilted algebra* $B_j^{(r)} = \operatorname{End}_{H_j^{(r)}}(T_j^{(r)})$ *is a quotient algebra of* B *and* $\mathscr{Q}^B$ *is the disjoint union of all components of* $\Gamma_{B_j^{(r)}}$ *contained entirely in the torsion part* $\mathscr{X}(T_j^{(r)})$ *of* $\operatorname{mod} B_j^{(r)}$ *determined by* $T_j^{(r)}$;

(4) *Every indecomposable module in* $\mathscr{C}^B$ *not lying in the core* $c(\mathscr{C}^B)$ *of* $\mathscr{C}^B$ *is an indecomposable module over one of the tilted algebras* $B_1^{(l)}, \dots, B_m^{(l)}$, $B_1^{(r)}, \dots, B_n^{(r)}$;

(5) *Every nondirecting indecomposable module in* $\mathscr{C}^B$ *is cycle-finite and lies in* $c(\mathscr{C}^B)$;

(6) $\operatorname{pd}_B X \leq 1$ *for all indecomposable modules* X *in* $\mathscr{P}^B$;

(7) $\operatorname{id}_B Y \leq 1$ *for all indecomposable modules* Y *in* $\mathscr{Q}^B$;

(8) *For all but finitely many indecomposable modules* M *in* $\mathscr{C}^B$, *we have* $\operatorname{pd}_B M \leq 1$ *or* $\operatorname{id}_B M \leq 1$.

Then $\mathscr{C}^B$ is called a *connecting component* of Γ_B, $B^{(l)} = B_1^{(l)} \times \dots \times B_m^{(l)}$ is called the *left tilted algebra* of B, and $B^{(r)} = B_1^{(r)} \times \dots \times B_n^{(r)}$ is called the *right tilted algebra* of B. Further, the almost acyclic component $\mathscr{C}^B$ of B admits a multisection. Recall that, following [**46**, Section 2], a full connected subquiver Δ of $\mathscr{C}$ is called a *multisection* if the following conditions are satisfied:

(1) Δ is almost acyclic.
(2) Δ is convex in $\mathscr{C}$.
(3) For each τ_B-orbit $\mathcal{O}$ in $\mathscr{C}$, we have $1 \leq |\Delta \cap \mathcal{O}| < \infty$.
(4) $|\Delta \cap \mathcal{O}| = 1$ for all but finitely many τ_B-orbits $\mathcal{O}$ in $\mathscr{C}$.
(5) No proper full convex subquiver of Δ satisfies (1)–(4).

Moreover, for a multisection Δ of a component $\mathscr{C}$, the following full subquivers of $\mathscr{C}$ were defined in [**46**]:

$$\Delta'_l = \{X \in \Delta; \text{there is a nonsectional path in } \mathscr{C} \text{ from } X \text{ to a projective module } P\},$$

$$\Delta'_r = \{X \in \Delta; \text{there is a nonsectional path in } \mathscr{C} \text{ from an injective module } I \text{ to } X\},$$

$$\Delta''_l = \{X \in \Delta'_l; \tau_A^{-1} X \notin \Delta'_l\}, \qquad \Delta''_r = \{X \in \Delta'_r; \tau_A X \notin \Delta'_r\},$$

$$\Delta_l = (\Delta \setminus \Delta'_r) \cup \tau_A \Delta''_r, \quad \Delta_c = \Delta'_l \cap \Delta'_r, \quad \Delta_r = (\Delta \setminus \Delta'_l) \cup \tau_A^{-1} \Delta''_l.$$

Then Δ_l is called the *left part* of Δ, Δ_r the *right part* of Δ, and Δ_c the *core* of Δ. The following basic properties of Δ have been established in [**46**, Proposition 2.4]:

- Every cycle of $\mathscr{C}$ lies in Δ_c.
- Δ_c is finite.
- Every indecomposable module X in $\mathscr{C}$ is in Δ_c, or a predecessor of Δ_l or a successor of Δ_r in $\mathscr{C}$.

REMARK 4.6. The class of algebras of finite representation type coincides with the class of generalized double tilted algebras B with Γ_B being the connecting component $\mathscr{C}^B$ (equivalently, with the tilted algebras $B^{(l)}$ and $B^{(r)}$ being of finite representation type (possibly empty)).

The following theorem from [**35**, Theorem 1.2] describes the support algebras of finite cycle-finite cyclic components of the Auslander-Reiten quivers of artin algebras.

THEOREM 4.7. *Let A be an algebra and $\mathscr{C}$ be a finite cycle-finite component of ${}_c\Gamma_A$. Then the following statements hold.*

(1) $\operatorname{Supp}(\mathscr{C})$ *is a generalized double tilted algebra.*
(2) $\mathscr{C}$ *is the core $c(\mathscr{C}^{B(\mathscr{C})})$ of the unique almost acyclic connecting component $\mathscr{C}^{B(\mathscr{C})}$ of $\Gamma_{B(\mathscr{C})}$.*
(3) $\operatorname{Supp}(\mathscr{C}) = B(\mathscr{C})$.

REMARK 4.8. It follows from [**35**, Corollary 2.7] that every finite cyclic component $\mathscr{C}$ of an Auslander-Reiten quiver Γ_A contains both a projective module and an injective module, and hence Γ_A admits at most finitely many finite cyclic components.

We may summarize the description of the support algebras of cycle-finite cyclic components of the Auslander-Reiten quivers of artin algebras as follows:

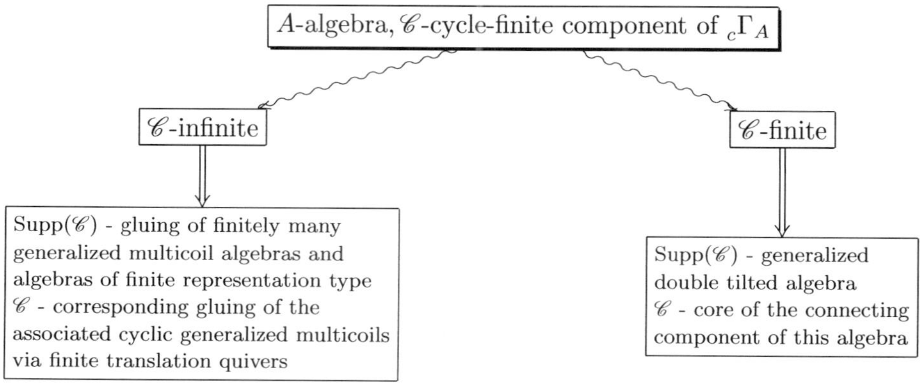

Recall that an idempotent e of an algebra A is called *convex* provided e is a sum of pairwise orthogonal primitive idempotents of A corresponding to the vertices of a convex valued subquiver of the quiver Q_A of A. The following direct consequence of Theorems 4.3, 4.7 and [**35**, Propositions 2.3, 2.4] provides a handy description of the faithful algebra of a cycle-finite component of ${}_c\Gamma_A$.

COROLLARY 4.9. *Let A be an algebra and $\mathscr{C}$ be a cycle-finite component of ${}_c\Gamma_A$. Then there exists a convex idempotent $e_{\mathscr{C}}$ of A such that* $\operatorname{Supp}(\mathscr{C})$ *is isomorphic to the algebra $e_{\mathscr{C}}Ae_{\mathscr{C}}$.*

5. Cycle-finite indecomposable modules

The aim of this section is to present some results describing homological properties of indecomposable cycle-finite modules over artin algebras.

Let A be an algebra and M a module in $\operatorname{mod} A$. We denote by $|M|$ the length of M over the commutative artin ring K. The following theorem is a consequence of Theorems 4.3, 4.7, the results established in [**39**, Theorem 1.3] and the properties of directing modules described in [**49**, 2.4(8)].

THEOREM 5.1. *Let A be an algebra. Then, for all but finitely many isomorphism classes of cycle-finite modules M in* $\operatorname{ind} A$, *the following statements hold.*

(1) $|\operatorname{Ext}_A^1(M,M)| \le |\operatorname{End}_A(M)|$ *and* $\operatorname{Ext}_A^r(M,M) = 0$ *for* $r \ge 2$.
(2) $|\operatorname{Ext}_A^1(M,M)| = |\operatorname{End}_A(M)|$ *if and only if there is a quotient concealed canonical algebra* C *of* A *and a stable tube* $\mathscr{T}$ *of* Γ_C *such that* M *is an indecomposable* C*-module in* $\mathscr{T}$ *of quasi-length divisible by the rank of* $\mathscr{T}$.

In particular, the above theorem shows that, for all but finitely many isomorphism classes of cycle-finite modules M in $\operatorname{ind} A$, the Euler characteristic

$$\chi_A(M) = \sum_{i=0}^{\infty}(-1)^i|\operatorname{Ext}_A^i(M,M)|$$

of M is well defined and nonnegative.

Let A be an algebra and $K_0(A)$ the Grothendieck group of A. For a module M in $\operatorname{mod} A$, we denote by $[M]$ the image of M in $K_0(A)$. Then $K_0(A)$ is a free abelian group with a $\mathbb{Z}$-basis given by $[S_1], \dots, [S_n]$ for a complete family $S_1, \dots, S_n$ of pairwise nonisomorphic simple modules in $\operatorname{mod} A$. Thus, for modules M and N in $\operatorname{mod} A$, we have $[M] = [N]$ if and only if the modules M and N have the same composition factors including the multiplicities. In particular, it would be interesting to find sufficient conditions for a module M in $\operatorname{ind} A$ to be uniquely determined (up to isomorphism) by its composition factors (see [**47**] for a general result in this direction).

The next theorem provides information on the composition factors of cycle-finite modules, and is a direct consequence of Theorems 4.3, 4.7, 5.1 and the results established in [**31**, Theorems A and B].

THEOREM 5.2. *Let* A *be an algebra. The following statements hold.*

(1) *There is a positive integer* m *such that, for any cycle-finite module* M *in* $\operatorname{ind} A$ *with* $|\operatorname{End}_A(M)| \neq |\operatorname{Ext}_A^1(M,M)|$, *the number of isomorphism classes of modules* X *in* $\operatorname{ind} A$ *with* $[X] = [M]$ *is bounded by* m.
(2) *For all but finitely many isomorphism classes of cycle-finite modules* M *in* $\operatorname{ind} A$ *with* $|\operatorname{End}_A(M)| = |\operatorname{Ext}_A^1(M,M)|$, *there are infinitely many pairwise nonisomorphic modules* X *in* $\operatorname{ind} A$ *with* $[X] = [M]$.

Following M. Auslander and I. Reiten [**7**], one associates with each nonprojective module X in $\operatorname{ind} A$ the number $\alpha(X)$ of indecomposable direct summands in the middle term

$$0 \to \tau_A X \to Y \to X \to 0$$

of an almost split sequence with the right term X. It has been proved by R. Bautista and S. Brenner [**9**] that, if A is an algebra of finite representation type and X is a nonprojective module in $\operatorname{ind} A$, then $\alpha(X) \le 4$, and if $\alpha(X) = 4$ then Y admits a projective-injective indecomposable direct summand P, and hence $X = P/\operatorname{soc}(P)$. In [**30**] S. Liu proved that the same is true for any indecomposable nonprojective module X lying on an oriented cycle of the Auslander-Reiten quiver Γ_A of any algebra A, and consequently for any nonprojective and nondirecting cycle-finite module in $\operatorname{ind} A$.

The following theorem is a direct consequence of Theorems 4.3, 4.7, and [**37**, Corollary B], and provides more information on almost split sequences of nondirecting cycle-finite modules.

THEOREM 5.3. *Let* A *be an algebra. Then, for all but finitely many isomorphism classes of nonprojective and nondirecting cycle-finite modules* M *in* $\operatorname{ind} A$, *we have* $\alpha(M) \le 2$.

6. Cycle-finite artin algebras

Following I. Assem and A. Skowroński [**3**], [**4**], an algebra A is said to be *cycle-finite* if all cycles in $\operatorname{ind} A$ are finite. A generalized multicoil algebra B is called *tame* if the quasitilted algebras $B^{(l)}$ and $B^{(r)}$ (see Section 4 for definitions) are products of tilted algebras of Euclidean types or tubular algebras. Moreover, a generalized double tilted algebra C is called *tame* if the tilted algebras $C^{(l)}$ and $C^{(r)}$ (see Section 4 for definitions) are generically tame (see Section 7 for definition) in the sense of W. Crawley-Boevey [**14**], [**15**]. We note that every tame generalized multicoil algebra and every tame generalized double tilted algebra is a cycle-finite algebra.

The following theorem describe the structure of the category $\operatorname{ind} A$ of an arbitrary cycle-finite algebra A, and is a direct consequence of Theorems 4.3 and 4.7 (see also [**33**, Theorems 7.1, 7.2 and 7.3]).

THEOREM 6.1. *Let A be a cycle-finite algebra. Then there exist tame generalized multicoil algebras $B_1, \dots, B_p$ and tame generalized double tilted algebras $B_{p+1}, \dots, B_q$ which are quotient algebras of A and the following statements hold.*

(1) $\operatorname{ind} A = \bigcup_{i=1}^{q} \operatorname{ind} B_i$.
(2) *All but finitely many isomorphism classes of modules in* $\operatorname{ind} A$ *belong to* $\bigcup_{i=1}^{p} \operatorname{ind} B_i$.
(3) *All but finitely many isomorphism classes of nondirecting modules in* $\operatorname{ind} A$ *belong to generalized multicoils of* $\Gamma_{B_1}, \dots, \Gamma_{B_p}$.

The next theorem extends the homological characterization of strongly simply connected algebras of polynomial growth established in [**42**] to arbitrary cycle-finite algebras, and is a direct consequence of Theorem 5.1 and the properties of directing modules described in [**49**, 2.4(8)].

THEOREM 6.2. *Let A be a cycle-finite algebra. Then, for all but finitely many isomorphism classes of modules M in* $\operatorname{ind} A$, *we have* $|\operatorname{Ext}_A^1(M,M)| \leq |\operatorname{End}_A(M)|$ *and* $\operatorname{Ext}_A^r(M,M) = 0$ *for* $r \geq 2$.

In connection to Theorem 5.3 we present the following theorem proved by J. A. de la Peña and the authors in [**34**, Main Theorem].

THEOREM 6.3. *Let A be a cycle-finite algebra and M be a nonprojective module in* $\operatorname{ind} A$, *and*

$$0 \to \tau_A M \to N \to M \to 0$$

be the associated almost split sequence in $\operatorname{mod} A$. *The following statements hold.*

(1) $\alpha(M) \leq 5$.
(2) *If $\alpha(M) = 5$ then N admits an indecomposable projective-injective direct summand P, and hence $M \simeq P/\operatorname{soc}(P)$.*

The following example of a cycle-finite algebra A from [**32**, Example 5.17], illustrates the above theorem.

EXAMPLE 6.4. Let K be a field and $A = KQ/I$ the bound quiver algebra over K, where Q is the quiver of the form

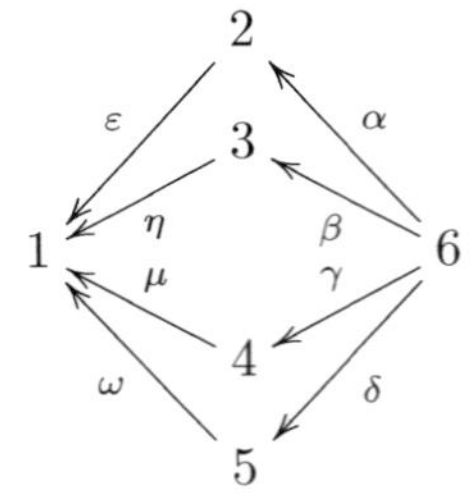

and I the ideal of the path algebra KQ of Q over K generated by the paths $\beta\eta - \alpha\varepsilon$, $\gamma\mu - \alpha\varepsilon$, $\delta\omega - \alpha\varepsilon$. Denote by B the hereditary algebra given by the full subquiver of Q given by the vertices $1, 2, 3, 4, 5$ and by C the hereditary algebra given by the full subquiver of Q given by the vertices $2, 3, 4, 5, 6$. Note that $P_6 = I_1$ is projective-injective A-module. Therefore, applying [**2**, Proposition IV.3.11], we conclude that there is in $\operatorname{mod} A$ an almost split sequence of the form

$$0 \longrightarrow \operatorname{rad} P_6 \longrightarrow S_2 \oplus S_3 \oplus S_4 \oplus S_5 \oplus P_6 \longrightarrow P_6/S_1 \longrightarrow 0,$$

where $S_2 \oplus S_3 \oplus S_4 \oplus S_5 \cong \operatorname{rad} P_6/S_1$. Moreover, $\operatorname{rad} P_6$ is the indecomposable injective B-module I_1^B, whereas P_6/S_1 is the indecomposable projective C-module P_6^C. The component of Γ_A containing $P_6 = I_1$ is the following gluing of the preinjective component of Γ_B with the postprojective component of Γ_C (see details in [**2**, Example VIII.5.7(e)])

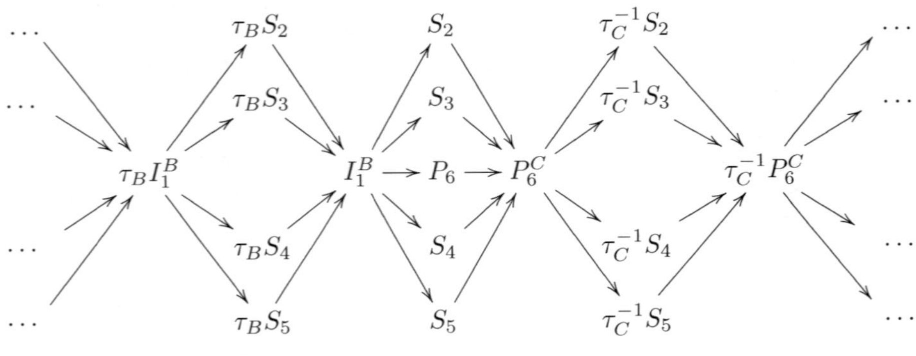

where τ_B and τ_C denote the Auslander-Reiten translations in $\operatorname{mod} B$ and $\operatorname{mod} C$, respectively.

Let A be an algebra. Recall that, following C. M. Ringel [**49**], a module M in $\operatorname{ind} A$ which does not lie on a cycle in $\operatorname{ind} A$ is called *directing*. We note that if all modules in $\operatorname{ind} A$ are directing, then A is of finite representation type [**49**] (see also [**20**] for the corresponding result over arbitrary artin algebra). If A is a cycle-finite algebra and M is a module in $\operatorname{ind} A$, then M is a directing module if and only if M is an acyclic vertex of Γ_A. The following result from [**10**] provides solution of the open problem concerning infinity of directing modules for cycle-finite algebras of infinite representation type.

THEOREM 6.5. *Let A be a cycle-finite algebra of infinite representation type. Then* $\operatorname{ind} A$ *contains infinitely many directing modules.*

In connection to the above theorem the following question occurred naturally: does every cycle-finite algebra of infinite representation type contains at least one directing projective module or a directing injective module?

In general, the answer is negative. Namely, in the joint work with J. A. de la Peña [**36**] we constructed a family of cycle-finite algebras of infinite representation type with all indecomposable projective modules and indecomposable injective modules nondirecting. Moreover, it has been shown (see [**36**, Theorem]) that there are such algebras with an arbitrary large number of almost acyclic Auslander-Reiten components having finite cyclic multisections.

The following example from [**36**, Section 9], illustrates the situation. We refer to [**36**] for the general construction of algebras of this type and more examples.

EXAMPLE 6.6. Let K be an algebraically closed field and $A = K\Sigma/J$ the bound quiver algebra over K, where Σ is the quiver of the form

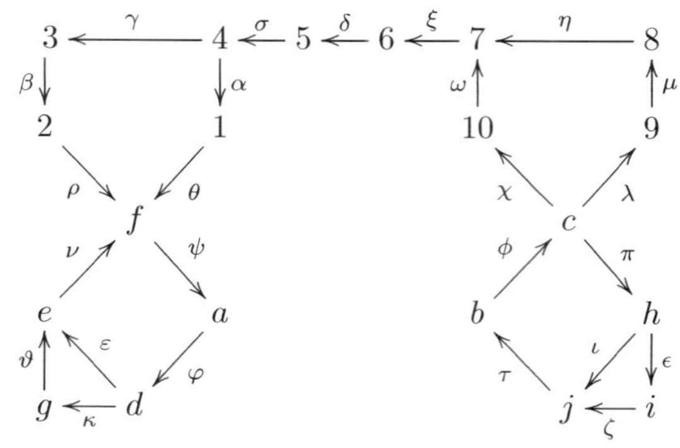

and J is the ideal in $K\Sigma$ generated by the elements $\omega\xi\delta\sigma\alpha$, $\alpha\theta-\gamma\beta\rho$, $\chi\omega-\lambda\mu\eta$, $\rho\psi$, $\theta\psi$, $\nu\psi$, $\varphi\varepsilon$, $\varphi\kappa$, $\kappa\vartheta$, $\vartheta\nu$, $\phi\chi$, $\phi\lambda$, $\phi\pi$, $\pi\epsilon$, $\epsilon\zeta$, $\zeta\tau$, $\iota\tau$. Denote by $B = KQ/I$ the bound quiver algebra given by the subquiver Q of Σ given by the vertices $1, 2, \dots, 10$ and the ideal I in KQ generated by the element $\omega\xi\delta\sigma\alpha$. Then B is a tubular algebra of type $(2, 3, 6)$ in the sense of Ringel [**49**] and following [**36**, Section 7] the algebra B is called *exceptional tubular algebra of type* $(2, 3, 6)$. Moreover, we denote by $H_0 = KQ^{(0)}$ the hereditary algebra of Euclidean type $\widetilde{\mathbb{E}}_8$ given by the full subquiver $Q^{(0)}$ of Σ with the vertices $1, 2, 3, 4, 5, 6, 7, 8, 9$, and by $H_1 = KQ^{(1)}$ the hereditary algebra of Euclidean type $\widetilde{\mathbb{E}}_8$ given by the full subquiver $Q^{(1)}$ of Σ with the vertices $2, 3, 4, 5, 6, 7, 8, 9, 10$.

Then it follows from [**36**, Theorems 5.1, 5.2 and 7.1] that the Auslander-Reiten quiver Γ_A of A has a decomposition

$$\Gamma_A = \mathscr{C}_0 \cup \mathscr{T}_0^B \cup \Big(\bigcup_{q \in \mathbb{Q}_1^0} \mathscr{T}_q^B \Big) \cup \mathscr{T}_1^B \cup \mathscr{C}_1,$$

where

- $\mathscr{C}_0$ is an almost acyclic component of the form $\Delta^{(0)} \cup \mathscr{P}^B$ with a cyclic multisection $\Delta^{(0)}$ such that the left part $\Delta_l^{(0)}$ of $\Delta^{(0)}$ is empty, the right part $\Delta_r^{(0)}$ of $\Delta^{(0)}$ is given by the indecomposable projective modules of the postprojective component $\mathscr{P}^B$ of Γ_B, being the postprojective component $\mathscr{P}^{H_0}$,

- $\mathscr{T}_0^B$ is the $\mathbb{P}_1(K)$-family $(\mathscr{T}_{0,\lambda}^B)_{\lambda\in\mathbb{P}_1(K)}$ of ray tubes obtained from the $\mathbb{P}_1(K)$-family $\mathscr{T}^{H_0}$ of stable tubes of tubular type $(2,3,5)$ by inserting in the unique stable tube of rank 5, say $\mathscr{T}_{0,0}^{H_0}$, one ray as follows

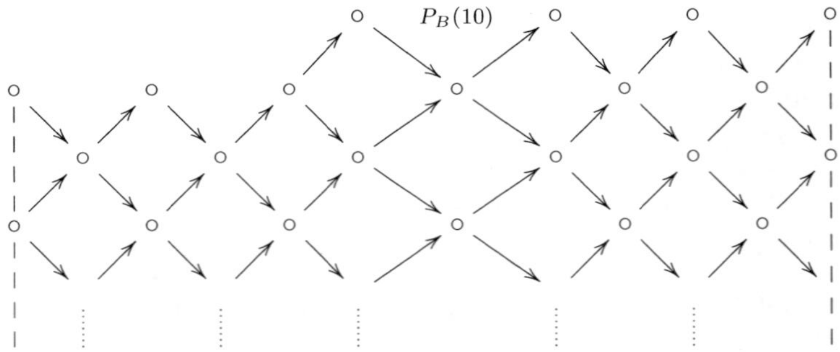

where the vertices along to the dashed vertical lines have to be identified,
- For each $q \in \mathbb{Q}_1^0$, $\mathscr{T}_q^B$ is a $\mathbb{P}_1(K)$-family $(\mathscr{T}_{q,\lambda}^B)_{\lambda\in\mathbb{P}_1(K)}$ of stable tubes of tubular type $(2,3,6)$, where $\mathbb{Q}_1^0 = \mathbb{Q}\cap(0,1)$,
- $\mathscr{T}_1^B$ is the $\mathbb{P}_1(K)$-family $(\mathscr{T}_{1,\lambda}^B)_{\lambda\in\mathbb{P}_1(K)}$ of coray tubes obtained from the $\mathbb{P}_1(K)$-family $\mathscr{T}^{H_1}$ of stable tubes of tubular type $(2,3,5)$ by inserting in the unique stable tube of rank 5, say $\mathscr{T}_{1,0}^{H_1}$, one coray as follows

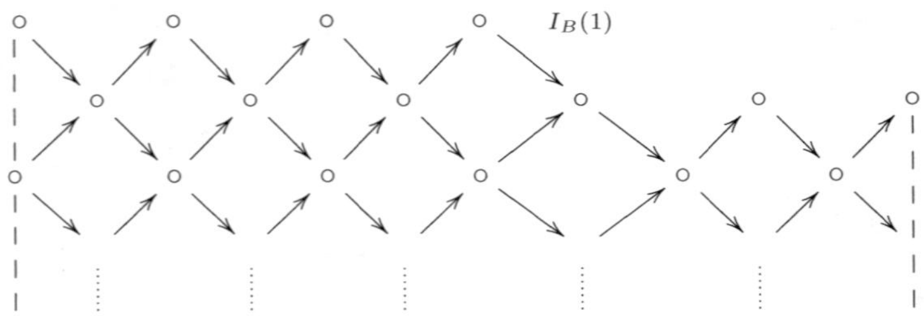

where the vertices along to the dashed vertical lines have to be identified,
- $\mathscr{C}_1$ is an almost acyclic component of the form $\mathscr{Q}^B \cup \Delta^{(1)}$ with a cyclic multisection $\Delta^{(1)}$ such that the right part $\Delta_r^{(1)}$ of $\Delta^{(1)}$ is empty, the left part $\Delta_l^{(1)}$ of $\Delta^{(1)}$ is given by the indecomposable injective modules of the preinjective component $\mathscr{Q}^B$ of Γ_B, being the preinjective component $\mathscr{Q}^{H_1}$.

Therefore, all indecomposable projective modules and indecomposable injective modules in $\operatorname{mod} A$ are nondirecting.

Let A be an algebra. We denote by $\mathcal{F}(A)$ the category of all finitely presented contravariant functors from $\operatorname{mod} A$ to the category $\mathcal{A}b$ of abelian groups. The category $\mathcal{F}(A)$ was intensively studied over the last 40 years, and is considered to be one of the important topics of the modern representation theory of algebras. It is a hard problem to describe the category $\mathcal{F}(A)$ even if the category $\operatorname{mod} A$ is well understood. A natural approach to study the structure of $\mathcal{F}(A)$ is via the associated Krull-Gabriel filtration

$$0 = \mathcal{F}(A)_{-1} \subseteq \mathcal{F}(A)_0 \subseteq \mathcal{F}(A)_1 \subseteq \cdots \subseteq \mathcal{F}(A)_{n-1} \subseteq \mathcal{F}(A)_n \subseteq \cdots$$

of $\mathcal{F}(A)$ by Serre subcategories, where, for each $n \in \mathbb{N}$, $\mathcal{F}(A)_n$ is the subcategory of all functors F in $\mathcal{F}(A)$ which become of finite length in the quotient category

$\mathcal{F}(A)/\mathcal{F}(A)_{n-1}$ [**18**], [**43**]. Following W. Geigle [**19**], we define $KG(A) = \min\{n \in \mathbb{N} \mid \mathcal{F}(A)_n = \mathcal{F}(A)\}$ if such a minimum exists, and set $KG(A) = \infty$ if it is not the case. Then $KG(A)$ is called the *Krull-Gabriel dimension* of A. The interest in the Krull-Gabriel dimension $KG(A)$ is motivated by the fact that the above filtration of $\mathcal{F}(A)$ leads to a hierarchy of exact sequences in $\operatorname{mod} A$, where the almost split sequences form the lowest level (see [**19**]).

The following characterization of cycle-finite algebras with finite Krull-Gabriel dimension, has been established by the second named author in [**60**, Theorem 1.2].

THEOREM 6.7. *Let A be a cycle-finite algebra of infinite representation type. The following statements are equivalent.*

(1) $KG(A) < \infty$.
(2) $KG(A) = 2$.
(3) $\bigcap_{m\geq 1}(\operatorname{rad}_A^{\infty})^m = 0$.
(4) $\operatorname{rad}_A^{\infty}$ *is nilpotent.*
(5) *All but finitely many components of Γ_A are stable tubes of rank one.*
(6) *A does not admit a tubular quotient algebra.*

We end this section with the related open problem (see [**36**, Question 1]).

PROBLEM. *Let A be a cycle-finite algebra of infinite representation type and finite Krull-Gabriel dimension. Is it true that* $\operatorname{ind} A$ *admits a directing projective module or a directing injective module?*

7. Artin algebras with separating Auslander-Reiten components

In this section we discuss the structure of artin algebra A having separating families of components in Γ_A.

Let A be an algebra. A component $\mathscr{C}$ of Γ_A is called *cycle-finite* if all modules in $\mathscr{C}$ are cycle-finite.

We have the following proposition.

PROPOSITION 7.1. *Let A be an algebra with a separating family $\mathscr{C}^A$ of components in Γ_A, and $\Gamma_A = \mathscr{P}^A \cup \mathscr{C}^A \cup \mathscr{Q}^A$ the associated decomposition of Γ_A. Then $\mathscr{C}^A$ is a family of cycle-finite components.*

The next theorem from [**41**, Theorem 1.5] provides solution of the problem concerning the structure of artin algebras admitting a separating family of Auslander-Reiten components, initiated by Ringel [**48**], [**49**], [**50**].

THEOREM 7.2. *Let A be an algebra with a separating family $\mathscr{C}^A$ of components in Γ_A, and $\Gamma_A = \mathscr{P}^A \cup \mathscr{C}^A \cup \mathscr{Q}^A$ the associated decomposition of Γ_A. Then there exist quotient algebras $A^{(l)}$ and $A^{(r)}$ of A such that the following statements hold.*

(1) $A^{(l)} = A_1^{(l)} \times \cdots \times A_m^{(l)} \times A_{m+1}^{(l)} \times \cdots \times A_{m+p}^{(l)}$, *where*
  (a) *For each $i \in \{1, \ldots, m\}$, $A_i^{(l)}$ is a tilted algebra of the form* $\operatorname{End}_{H_i^{(l)}}(T_i^{(l)})$ *for a hereditary algebra $H_i^{(l)}$ and a tilting module $T_i^{(l)}$ in* $\operatorname{mod} H_i^{(l)}$ *without indecomposable preinjective direct summands.*
  (b) *For each $i \in \{m+1, \ldots, m+p\}$, $A_i^{(l)}$ is a quasitilted algebra of canonical type with a separating family of coray tubes in $\Gamma_{A_i^{(l)}}$.*
(2) $A^{(r)} = A_1^{(r)} \times \cdots \times A_n^{(r)} \times A_{n+1}^{(r)} \times \cdots \times A_{n+q}^{(r)}$, *where*

(a) *For each* $j \in \{1, \dots, n\}$, $A_j^{(r)}$ *is a tilted algebra of the form* $\operatorname{End}_{H_j^{(r)}}(T_j^{(r)})$ *for a hereditary algebra* $H_j^{(r)}$ *and a tilting module* $T_j^{(r)}$ *in* $\operatorname{mod} H_j^{(r)}$ *without indecomposable postprojective direct summands.*

(b) *For each* $j \in \{n+1, \dots, n+q\}$, $A_j^{(r)}$ *is a quasitilted algebra of canonical type with a separating family of ray tubes in* $\Gamma_{A_j^{(r)}}$.

(3) $\mathscr{P}^A = \bigcup_{i=1}^{m+p} \mathscr{P}^{A_i^{(l)}}$ *and every component in* $\mathscr{P}^A$ *is either a postprojective component, a ray tube, or obtained from a component of the form* $\mathbb{Z}\mathbb{A}_\infty$ *by a finite number (possibly zero) of ray insertions.*

(4) $\mathscr{Q}^A = \bigcup_{j=1}^{n+q} \mathscr{Q}^{A_j^{(r)}}$ *and every component in* $\mathscr{Q}^A$ *is either a preinjective component, a coray tube, or obtained from a component of the form* $\mathbb{Z}\mathbb{A}_\infty$ *by a finite number (possibly zero) of coray insertions.*

In [**14**], [**15**] Crawley-Boevey introduced the concept of a generically tame algebra. An indecomposable right A-module M over an algebra A is called a *generic module* if M is of infinite length over A but of finite length over $\operatorname{End}_A(M)$, called the *endolength* of M. Then an algebra A is called *generically tame* if, for any positive integer d, there are only finitely many isomorphism classes of generic right A-modules of endolength d. An algebra A is called *generically finite* if there are at most finitely many pairwise non-isomorphic generic right A-modules. Further, A is called *generically of polynomial growth* if there is a positive integer m such that for any positive integer d the number of isomorphism classes of generic right A-modules of endolength d is at most d^m. We note that every algebra A of finite representation type is generically trivial, that is, there is no generic right A-module. We also stress that by a theorem of Crawley-Boevey [**14**, Theorem 4.4], if A is an algebra over an algebraically closed field K, then A is generically tame if and only if A is tame in the sense of Drozd [**17**] (see also [**13**], [**53**]).

Recall also that following [**55**] the *component quiver* Σ_A of an algebra A has the components of Γ_A as vertices and there is an arrow $\mathscr{C} \to \mathscr{D}$ in Σ_A if $\operatorname{rad}_A^\infty(X, Y) \neq 0$, for some modules X in $\mathscr{C}$ and Y in $\mathscr{D}$. In particular, a component $\mathscr{C}$ of Γ_A is generalized standard if and only if there is no loop at $\mathscr{C}$ in Σ_A.

The final theorem is a consequence of [**41**, Theorem 1.10], and characterizes the cycle-finite algebras with separating families of Auslander-Reiten components.

THEOREM 7.3. *Let A be an algebra with a separating family of components in* Γ_A. *The following statements are equivalent:*

(1) *A is cycle-finite.*
(2) *A is generically tame.*
(3) *A is generically of polynomial growth.*
(4) $A^{(l)}$ *and* $A^{(r)}$ *are products of tilted algebras of Euclidean type or tubular algebras.*
(5) Γ_A *is almost periodic.*
(6) Σ_A *is acyclic.*

References

[1] Ibrahim Assem and Flávio U. Coelho, *Two-sided gluings of tilted algebras*, J. Algebra **269** (2003), no. 2, 456–479, DOI 10.1016/S0021-8693(03)00436-8. MR2015287

[2] Ibrahim Assem, Daniel Simson, and Andrzej Skowroński, *Elements of the representation theory of associative algebras. Vol. 1*, London Mathematical Society Student Texts, vol. 65, Cambridge University Press, Cambridge, 2006. Techniques of representation theory. MR2197389

[3] Ibrahim Assem and Andrzej Skowroński, *Algebras with cycle-finite derived categories*, Math. Ann. **280** (1988), no. 3, 441–463, DOI 10.1007/BF01456336. MR936322

[4] Ibrahim Assem and Andrzej Skowroński, *Minimal representation-infinite coil algebras*, Manuscripta Math. **67** (1990), no. 3, 305–331, DOI 10.1007/BF02568435. MR1046991

[5] Ibrahim Assem, Andrzej Skowroński, and Bertha Tomé, *Coil enlargements of algebras*, Tsukuba J. Math. **19** (1995), no. 2, 453–479, DOI 10.21099/tkbjm/1496162881. MR1366647

[6] Maurice Auslander, *Representation theory of Artin algebras. I, II*, Comm. Algebra **1** (1974), 177–268; ibid. 1 (1974), 269–310, DOI 10.1080/00927877408548230. MR349747

[7] Maurice Auslander and Idun Reiten, *Uniserial functors*, Representation theory, II (Proc. Second Internat. Conf., Carleton Univ., Ottawa, Ont., 1979), Lecture Notes in Math., vol. 832, Springer, Berlin, 1980, pp. 1–47. MR607147

[8] Maurice Auslander, Idun Reiten, and Sverre O. Smalø, *Representation theory of Artin algebras*, Cambridge Studies in Advanced Mathematics, vol. 36, Cambridge University Press, Cambridge, 1997. Corrected reprint of the 1995 original. MR1476671

[9] Raymundo Bautista and Sheila Brenner, *On the number of terms in the middle of an almost split sequence*, Representations of algebras (Puebla, 1980), Lecture Notes in Math., vol. 903, Springer, Berlin-New York, 1981, pp. 1–8. MR654699

[10] Jerzy Białkowski and Andrzej Skowroński, *Cycles of modules and finite representation type*, Bull. Lond. Math. Soc. **48** (2016), no. 4, 589–600, DOI 10.1112/blms/bdw030. MR3532135

[11] Flávio Ulhoa Coelho and Marcelo Américo Lanzilotta, *Algebras with small homological dimensions*, Manuscripta Math. **100** (1999), no. 1, 1–11, DOI 10.1007/s002290050191. MR1714397

[12] Flávio U. Coelho, Eduardo do N. Marcos, Héctor A. Merklen, and Andrzej Skowroński, *Module categories with infinite radical square zero are of finite type*, Comm. Algebra **22** (1994), no. 11, 4511–4517, DOI 10.1080/00927879408825084. MR1284342

[13] W. W. Crawley-Boevey, *On tame algebras and bocses*, Proc. London Math. Soc. (3) **56** (1988), no. 3, 451–483, DOI 10.1112/plms/s3-56.3.451. MR931510

[14] William Crawley-Boevey, *Tame algebras and generic modules*, Proc. London Math. Soc. (3) **63** (1991), no. 2, 241–265, DOI 10.1112/plms/s3-63.2.241. MR1114510

[15] William Crawley-Boevey, *Modules of finite length over their endomorphism rings*, Representations of algebras and related topics (Kyoto, 1990), London Math. Soc. Lecture Note Ser., vol. 168, Cambridge Univ. Press, Cambridge, 1992, pp. 127–184. MR1211479

[16] Gabriella D'Este and Claus Michael Ringel, *Coherent tubes*, J. Algebra **87** (1984), no. 1, 150–201, DOI 10.1016/0021-8693(84)90165-0. MR736774

[17] Ju. A. Drozd, *Tame and wild matrix problems*, Representation theory, II (Proc. Second Internat. Conf., Carleton Univ., Ottawa, Ont., 1979), Lecture Notes in Math., vol. 832, Springer, Berlin-New York, 1980, pp. 242–258. MR607157

[18] Carl Faith, *Algebra: rings, modules and categories. I*, Springer-Verlag, New York-Heidelberg, 1973. Die Grundlehren der mathematischen Wissenschaften, Band 190. MR0366960

[19] Werner Geigle, *The Krull-Gabriel dimension of the representation theory of a tame hereditary Artin algebra and applications to the structure of exact sequences*, Manuscripta Math. **54** (1985), no. 1-2, 83–106, DOI 10.1007/BF01171701. MR808682

[20] Dieter Happel and Shi Ping Liu, *Module categories without short cycles are of finite type*, Proc. Amer. Math. Soc. **120** (1994), no. 2, 371–375, DOI 10.2307/2159871. MR1164144

[21] Dieter Happel and Idun Reiten, *Hereditary abelian categories with tilting object over arbitrary base fields*, J. Algebra **256** (2002), no. 2, 414–432, DOI 10.1016/S0021-8693(02)00088-1. MR1939113

[22] Dieter Happel, Idun Reiten, and Sverre O. Smalø, *Tilting in abelian categories and quasitilted algebras*, Mem. Amer. Math. Soc. **120** (1996), no. 575, viii+ 88, DOI 10.1090/memo/0575. MR1327209

[23] Alicja Jaworska, Piotr Malicki, and Andrzej Skowroński, *Tilted algebras and short chains of modules*, Math. Z. **273** (2013), no. 1-2, 19–27, DOI 10.1007/s00209-012-0993-0. MR3010149

[24] Alicja Jaworska, Piotr Malicki, and Andrzej Skowroński, *Modules not being the middle of short chains*, Q. J. Math. **64** (2013), no. 4, 1141–1160, DOI 10.1093/qmath/has034. MR3151608

[25] Otto Kerner and Andrzej Skowroński, *On module categories with nilpotent infinite radical*, Compositio Math. **77** (1991), no. 3, 313–333. MR1092772

[26] Helmut Lenzing and José Antonio de la Peña, *Concealed-canonical algebras and separating tubular families*, Proc. London Math. Soc. (3) **78** (1999), no. 3, 513–540, DOI 10.1112/S0024611599001872. MR1674837

[27] Helmut Lenzing and Andrzej Skowroński, *Quasi-tilted algebras of canonical type*, Colloq. Math. **71** (1996), no. 2, 161–181, DOI 10.4064/cm-71-2-161-181. MR1414820

[28] Shi Ping Liu, *Degrees of irreducible maps and the shapes of Auslander-Reiten quivers*, J. London Math. Soc. (2) **45** (1992), no. 1, 32–54, DOI 10.1112/jlms/s2-45.1.32. MR1157550

[29] Shi Ping Liu, *Semi-stable components of an Auslander-Reiten quiver*, J. London Math. Soc. (2) **47** (1993), no. 3, 405–416, DOI 10.1112/jlms/s2-47.3.405. MR1214905

[30] Shi Ping Liu, *Almost split sequences for nonregular modules*, Fund. Math. **143** (1993), no. 2, 183–190, DOI 10.4064/fm-143-2-183-190. MR1240634

[31] Piotr Malicki, *On the composition factors of indecomposable modules in almost cyclic coherent Auslander-Reiten components*, J. Pure Appl. Algebra **207** (2006), no. 3, 469–490, DOI 10.1016/j.jpaa.2005.10.009. MR2265536

[32] Piotr Malicki, *Auslander-Reiten theory for finite-dimensional algebras*, Homological methods, representation theory, and cluster algebras, CRM Short Courses, Springer, Cham, 2018, pp. 21–63. MR3823387

[33] Piotr Malicki, José A. de la Peña, and Andrzej Skowroński, *Cycle-finite module categories*, Algebras, quivers and representations, Abel Symp., vol. 8, Springer, Heidelberg, 2013, pp. 209–252, DOI 10.1007/978-3-642-39485-0_10. MR3183887

[34] Piotr Malicki, José Antonio de la Peña, and Andrzej Skowroński, *On the number of terms in the middle of almost split sequences over cycle-finite Artin algebras*, Cent. Eur. J. Math. **12** (2014), no. 1, 39–45, DOI 10.2478/s11533-013-0328-3. MR3121820

[35] Piotr Malicki, José Antonio de la Peña, and Andrzej Skowroński, *Finite cycles of indecomposable modules*, J. Pure Appl. Algebra **219** (2015), no. 5, 1761–1799, DOI 10.1016/j.jpaa.2014.07.011. MR3299706

[36] Piotr Malicki, José Antonio de la Peña, and Andrzej Skowroński, *Existence of cycle-finite algebras of infinite representation type without directing projective or injective modules*, Colloq. Math. **148** (2017), no. 2, 165–190, DOI 10.4064/cm7190-2-2017. MR3660285

[37] Piotr Malicki and Andrzej Skowroński, *Almost cyclic coherent components of an Auslander-Reiten quiver*, J. Algebra **229** (2000), no. 2, 695–749, DOI 10.1006/jabr.2000.8314. MR1769296

[38] Piotr Malicki and Andrzej Skowroński, *Algebras with separating almost cyclic coherent Auslander-Reiten components*, J. Algebra **291** (2005), no. 1, 208–237, DOI 10.1016/j.jalgebra.2005.03.021. MR2158519

[39] Piotr Malicki and Andrzej Skowroński, *On the indecomposable modules in almost cyclic coherent Auslander-Reiten components*, J. Math. Soc. Japan **63** (2011), no. 4, 1121–1154. MR2855809

[40] Piotr Malicki and Andrzej Skowroński, *Algebras with separating Auslander-Reiten components*, Representations of algebras and related topics, EMS Ser. Congr. Rep., Eur. Math. Soc., Zürich, 2011, pp. 251–353, DOI 10.4171/101-1/7. MR2931900

[41] Piotr Malicki and Andrzej Skowroński, *The structure and homological properties of generalized standard Auslander-Reiten components*, J. Algebra **518** (2019), 1–39, DOI 10.1016/j.jalgebra.2018.10.015. MR3872856

[42] J. A. de la Peña and A. Skowroński, *Geometric and homological characterizations of polynomial growth strongly simply connected algebras*, Invent. Math. **126** (1996), no. 2, 287–296, DOI 10.1007/s002220050100. MR1411134

[43] N. Popescu, *Abelian categories with applications to rings and modules*, Academic Press, London-New York, 1973. London Mathematical Society Monographs, No. 3. MR0340375

[44] I. Reiten and A. Skowroński, *Sincere stable tubes*, J. Algebra **232** (2000), no. 1, 64–75, DOI 10.1006/jabr.1999.8386. MR1783913

[45] Idun Reiten and Andrzej Skowroński, *Characterizations of algebras with small homological dimensions*, Adv. Math. **179** (2003), no. 1, 122–154, DOI 10.1016/S0001-8708(02)00029-4. MR2004730

[46] Idun Reiten and Andrzej Skowroński, *Generalized double tilted algebras*, J. Math. Soc. Japan **56** (2004), no. 1, 269–288, DOI 10.2969/jmsj/1191418706. MR2027626

[47] I. Reiten, A. Skowroński, and S. O. Smalø, *Short chains and short cycles of modules*, Proc. Amer. Math. Soc. **117** (1993), no. 2, 343–354, DOI 10.2307/2159166. MR1136238

[48] Claus Michael Ringel, *Separating tubular series*, Paul Dubreil and Marie-Paule Malliavin algebra seminar, 35th year (Paris, 1982), Lecture Notes in Math., vol. 1029, Springer, Berlin, 1983, pp. 134–158, DOI 10.1007/BFb0098930. MR732474

[49] Claus Michael Ringel, *Tame algebras and integral quadratic forms*, Lecture Notes in Mathematics, vol. 1099, Springer-Verlag, Berlin, 1984. MR774589

[50] Claus Michael Ringel, *Representation theory of finite-dimensional algebras*, Representations of algebras (Durham, 1985), London Math. Soc. Lecture Note Ser., vol. 116, Cambridge Univ. Press, Cambridge, 1986, pp. 7–79. MR897319

[51] Claus Michael Ringel, *The canonical algebras*, Topics in algebra, Part 1 (Warsaw, 1988), Banach Center Publ., vol. 26, PWN, Warsaw, 1990, pp. 407–432. With an appendix by William Crawley-Boevey. MR1171247

[52] Daniel Simson and Andrzej Skowroński, *Elements of the representation theory of associative algebras. Vol. 2*, London Mathematical Society Student Texts, vol. 71, Cambridge University Press, Cambridge, 2007. Tubes and concealed algebras of Euclidean type. MR2360503

[53] Daniel Simson and Andrzej Skowroński, *Elements of the representation theory of associative algebras. Vol. 3*, London Mathematical Society Student Texts, vol. 72, Cambridge University Press, Cambridge, 2007. Representation-infinite tilted algebras. MR2382332

[54] Andrzej Skowroński, *Generalized standard Auslander-Reiten components*, J. Math. Soc. Japan **46** (1994), no. 3, 517–543, DOI 10.2969/jmsj/04630517. MR1276836

[55] Andrzej Skowroński, *Cycles in module categories*, Finite-dimensional algebras and related topics (Ottawa, ON, 1992), NATO Adv. Sci. Inst. Ser. C Math. Phys. Sci., vol. 424, Kluwer Acad. Publ., Dordrecht, 1994, pp. 309–345. MR1308994

[56] Andrzej Skowroński, *On omnipresent tubular families of modules*, Representation theory of algebras (Cocoyoc, 1994), CMS Conf. Proc., vol. 18, Amer. Math. Soc., Providence, RI, 1996, pp. 641–657. MR1388078

[57] Andrzej Skowroński, *Tame quasi-tilted algebras*, J. Algebra **203** (1998), no. 2, 470–490, DOI 10.1006/jabr.1997.7328. MR1622799

[58] Andrzej Skowroński, *Generalized canonical algebras and standard stable tubes*, Colloq. Math. **90** (2001), no. 1, 77–93, DOI 10.4064/cm90-1-7. MR1874366

[59] Andrzej Skowroński, *A construction of complex syzygy periodic modules over symmetric algebras*, Colloq. Math. **103** (2005), no. 1, 61–69, DOI 10.4064/cm103-1-8. MR2148950

[60] Andrzej Skowroński, *The Krull-Gabriel dimension of cycle-finite Artin algebras*, Algebr. Represent. Theory **19** (2016), no. 1, 215–233, DOI 10.1007/s10468-015-9571-6. MR3465899

[61] Andrzej Skowroński and Kunio Yamagata, *Frobenius algebras. I*, EMS Textbooks in Mathematics, European Mathematical Society (EMS), Zürich, 2011. Basic representation theory. MR2894798

[62] Andrzej Skowroński and Kunio Yamagata, *Frobenius algebras. II*, EMS Textbooks in Mathematics, European Mathematical Society (EMS), Zürich, 2017. Tilted and Hochschild extension algebras. MR3585514

[63] Ying Bo Zhang, *The structure of stable components*, Canad. J. Math. **43** (1991), no. 3, 652–672, DOI 10.4153/CJM-1991-038-1. MR1118014

Faculty of Mathematics and Computer Science, Nicolaus Copernicus University, Chopina 12/18, 87-100 Toruń, Poland
Email address: pmalicki@mat.umk.pl

Faculty of Mathematics and Computer Science, Nicolaus Copernicus University, Chopina 12/18, 87-100 Toruń, Poland
Email address: skowron@mat.umk.pl

Contemporary Mathematics
Volume **761**, 2021
https://doi.org/10.1090/conm/761/15317

Piecewise hereditary incidence algebras

Eduardo do N. Marcos and Marcelo Moreira

Abstract. Let $K\Delta$ be the incidence algebra associated with a finite poset $(\Delta, \preceq)$ over the algebraically closed field K. We study some aspects of incidence algebras $K\Delta$ that are piecewise hereditary, which we denominate PHI algebras. We show that if $K\Delta$ is a PHI algebra, which is not of wild quiver type, then $HH^1(K\Delta)$ is trivial if and only if $K\Delta$ is a simply connected algebra. We show the strong global dimension of a sincere piecewise hereditary algebra is less or equal than three.

1. Introduction

Throughout the paper all algebras are basic K-algebras over an algebraically closed field K. Using Gabriel's theorem, we may assume all algebras to be of the form KQ/I, thus, Q is a finite quiver and I is an admissible ideal. Modules are dimensional right modules.

Incidence algebras were introduced in the mid-1960s as a natural way of studying some combinatorial problems. We will focus our study on incidence algebras $K\Delta$ associated with a finite poset Δ over K. Observe that $K\Delta$ is isomorphic to the algebra KQ/I, where the quiver Q is the Hasse diagram and I is the ideal generated by all commutativity relations. Our purpose is to study incidence algebras which are piecewise hereditary, we call them PHI algebras, for *piecewise hereditary incidence algebras.* We investigate the *strong global dimension*, the *simple conectedeness* and the *one-point extension algebras* of PHI algebras.

The paper is organized as follows. Section 2 is devoted to fixing the notation. Section 3 is dedicated to study the simple conectedeness of PHI algebras.

Section 4 is dedicated to the study of the global dimension of PHI algebras. In particular, we show that the representation-finite PHI algebras have global dimension less or equal to two.

Ringel introduced the notion of strong global dimension, in a private communication to Skowronsky; [**25**], Ringel's definition is the following:

2010 *Mathematics Subject Classification.* Primary 13D05 13D09 14F05 16D10 16E10 16E35 16G20.

The first named author was supported by the tematic project of Fapesp 2014/09310-5 and also a research grant from CNPq 302003/2018-5.

The second named author acknowledges support from CAPES, in the form of a PhD Scholarship, PhD made at programa de Matemática, IME-USP, (Brazil).

DEFINITION. Let A be an algebra. The *strong global dimension* of A, denoted as $\operatorname{sgldim} A$, is defined to be the maximum of the width of indecomposable complexes in $C^b(\mathcal{P}_A)$.

Happel and Zacharia showed in [**16**] that an algebra has finite strong global dimension if and only if it is piecewise hereditary.

In Section 5 we determine an upper bound for the strong global dimension for sincere piecewise hereditary algebras.

Let $p_1, \ldots, p_n$ be a set of positive integers and let $\mathbb{X} = \mathbb{X}(p_1, \ldots, p_n)$ be a weighted projective line of type $p_1, \ldots, p_n$, in the sense of [**11**]. Let $\operatorname{Coh} \mathbb{X}$ be the category of coherent sheaves on $\mathbb{X}(p_1, \ldots, p_n)$. For the PHI algebras $K\Delta$ which are derived equivalent to $\operatorname{Coh} \mathbb{X}$, in Section 6 we study the canonical sincere $K\Delta$ -module M and the one-point extension algebra $K\Delta[M]$. Let A be a representation-infinite quasi-tilted algebra of domestic-sheaf type. We show that the canonical sincere A-module M is exceptional. We conjecture that this is always the case for PHI algebras. Using this we see that the one-point extension algebra $K\Delta[M]$ is also a PHI algebra.

2. Preliminaries

In this Section, for the sake of completeness, we will recall some definitions. We begin with the definition of incidence algebras.

DEFINITION 2.1 (incidence algebra). Let $(\Delta, \preceq)$ be a finite poset. The incidence algebra $K\Delta$ is a quotient of the path algebra of the following quiver Q. The set of vertices, Q_0, is in bijection with the elements of the poset Δ and the set of arrows Q_1 is defined by putting an arrow α from a vertex a to a vertex b, whenever $a \preceq b$ and there is no $a \preceq c \preceq b$, with $c \neq a$ and $c \neq b$. Let I be the ideal generated by all commutativity relations $\gamma - \gamma'$, with γ and γ' parallel paths. The incidence algebra $K\Delta$ is KQ/I.

The quiver Q of the incidence algebra, in the former definition, is also called the Hasse quiver of the poset. We assume always that our incidence algebras are connected, that is the Hasse quiver is connected. Details can be found for instance in [**22**] and [**6**].

REMARK 2.2. All hereditary categories considered in this paper have splitting idempotents, finite dimensional Hom spaces, and tilting objects.

DEFINITION 2.3 (piecewise hereditary algebra). We say that A is a piecewise hereditary algebra of type $\mathcal{H}$ if there exists a hereditary abelian category $\mathcal{H}$ such that $D^b(\mathcal{A})$ is triangle-equivalent to the bounded derived category $D^b(\mathcal{H})$.

Let A be a piecewise hereditary algebra of type $\mathcal{H}$. It follows from Rickard's theorem [**23**], that there exists of a tilting object T in $D^b(\mathcal{H})$ such that $A = \operatorname{End} T$.

We state one of the most important theorems about piecewise hereditary algebras.

THEOREM 2.4 (Happel [**13**]). *Let $\mathcal{H}$ be an abelian hereditary connected K-category with tilting object. Then $\mathcal{H}$ is derived equivalent to* $\operatorname{mod} KQ$ *or derived equivalent to* $\operatorname{Coh} \mathbb{X}$ *for some weighted projective line* $\mathbb{X}$.

An algebra A is called piecewise hereditary of type Q, for some quiver Q, if A and KQ are derived equivalent. It is of sheaf type $\mathbb{X}$, for some weighted projective line $\mathbb{X}$, if A and $\mathrm{Coh}\,\mathbb{X}$ are derived equivalent.

Observe that an algebra can be, at the same time, of quiver and sheaf type.

3. Simple connectedness

In 1993, in [**26**], Skowroński proposed the following problem:

> Describe classes of algebras A such that is simply connected if and only if $HH^1(A) = 0$?

The next statement uses the notions of graph of the tilting objects, see for instance [**15**] and [**10**].

THEOREM 3.1. *Let $K\Delta$ be a PHI algebra which is not derived equivalent to a wild hereditary algebra, then $HH^1(K\Delta) = 0$ if and only if $K\Delta$ is simply connected.*

PROOF. We thank the referee for improving the way this proof was written.

One direction follows using the fact that the first Hochschild cohomology group $HH^1(K\Delta)$ can be expressed using the fundamental group $\pi_1(Q, I)$ (see De la Peña and Saorin [**9**]), then the simple connectedness of the algebra follows. For the other direction, assume that the PHI algebra $K\Delta$ is not of wild-quiver type, and $HH^1(K\Delta) = 0$. Since the algebra is piecewise hereditary, it must be derived equivalent to a hereditary abelian category $\mathcal{H}$. Moreover from our assumption on the quiver type, we get using results of Happel-Unger ([**15**]) and Fu-Geng ([**10**]), that the graph of the tilting object $\mathcal{K}_{\mathcal{H}}$ (see [**15**]) is connected. Now, we are ready to use a theorem due to Le Meur [**19**] that implies the simple connectedness of the algebra. □

It should be noted that for the implication "If $HH^1(K\Delta) = 0$ then $K\Delta$ is simply connected" we need the hypothesis that $K\Delta$ is a PHI algebra. The following is a counterexample, showing that this is not valid, in general. The simplicial complex of the projective plane has the fundamental group isomorphic to $\mathbb{Z}_2$. Knowing that the fundamental group of the simplicial complex is isomorphic to the fundamental group of poset Δ associated with this complex, applying the theorem of De la Peña-Saorin [**9**] yields:

$$\mathrm{Hom}(\mathbb{Z}_2, K^+) \cong HH^1(K\Delta).$$

Then $K\Delta$ is not simply connected but $HH^1(K\Delta) = 0$, if the characteristic of K is not 2.

4. Global dimension

We mention the following result:

THEOREM 4.1 (Happel-Reiten-Smalø [**14**]). *Let $\mathcal{A}$ be a piecewise hereditary, abelian category with finite length and n non-isomorphic simple objects. Then the category $\mathcal{A}$ has global dimension less than or equal to n.*

A natural question is whether there is an upper bound for the global dimension of PHI algebras.

This question was answered by Ladkani [**17**]. To state the result of Ladkani we recall that an algebra is called *sincere* if it admits a sincere indecomposable module in its category of modules, that is, an indecomposable module such that every simple module is a composition factor of it. The following statement is a particular case of Ladkani's statement:

THEOREM 4.2 (Ladkani [**17**]). *A sincere piecewise hereditary algebra has global dimension less than or equal to* 3.

Connected incidence algebras are sincere, as we see next.

PROPOSITION 4.3. *Let* $K\Delta = KQ/I$ *be an incidence algebra. Then* $K\Delta$ *is a sincere algebra.*

PROOF. We need to show the existence of an indecomposable, sincere, module M over $K\Delta$. The candidate M is the module associated with the following representation:

(1) for each vertex a in Q we associate K;
(2) for each arrow $\alpha\colon a \to b$ in Q we associate the identity $1\colon K \to K$.

First, we will show that M is indecomposable. For this, we will study the its endomorphism algebra. Let $f = (f_a)_{a\in Q_0}$ be a non-zero morphism in $\operatorname{End} M$. Thus, there exists $f_a\colon K \to K$ not zero for some $a \in Q_0$, implying that f_a is an isomorphism. Given an arrow $\alpha\colon a \to b$, we have that $f_a 1 = 1 f_b$. If the arrow is in the other direction, we get the same result. Thus, no matter what is the direction of the arrow, we conclude that $f_a = f_b$. So, since the graph is connected, we always have a walk connecting the vertex a to any vertex c :

$$a \text{———} \bullet \text{———} \cdots \text{———} c$$

Inductively, we get that $f_a = f_c$ for any vertex c of Q and consequently $\operatorname{End} M \cong K$, therefore M is an indecomposable module.

It is clear that M is sincere.

Therefore, M is a sincere indecomposable module over an incidence algebra $K\Delta$. □

As a consequence we get the following corollary:

COROLLARY 4.4. *The global dimension of a PHI algebra is less than or equal to* 3.

DEFINITION 4.5. We call the module M in the proposition above the *canonical sincere module.*

The representation-finite piecewise hereditary sincere algebras have global dimension less or equal to two. Before proving this affirmation, we recall the definition of a directed module.

DEFINITION 4.6 (directed module). A *cycle* in the module category is a sequence

$$M_0 \xrightarrow{f_1} M_1 \xrightarrow{f_2} M_2 \xrightarrow{f_3} \dots \to M_{t-1} \xrightarrow{f_t} M_t \cong M_0,$$

where $t \geq 1$, each M_i is indecomposable and each morphism f_i is non-zero, and not an isomorphism.

Let M be an indecomposable module, M is called *directed* if it does not belong to any cycle.

We show now the following statement.

PROPOSITION 4.7. *Let A a representation-finite piecewise hereditary sincere algebra. Then A is a tilted algebra, consequently* $\operatorname{gldim} A \leq 2$.

PROOF. Let A be a piecewise hereditary sincere algebra of finite representation type. Since it is sincere, it must have a sincere and directed indecomposable module (Happel [**12**]), so it is a tilted algebra (Ringel [**24**]), hence its global dimension is at most 2. □

COROLLARY 4.8. *If A is a PHI algebra of finite representation type then its global dimension is less than or equal to* 2.

We now see that the global dimension of an incidence algebra is related to its strongly simple connectedness.

DEFINITION 4.9 (critical algebra [**7**]). Let B be an algebra. We say that B is *critical* if it satisfies the following properties:

(1) Its quiver has no oriented cycle.
(2) B has a unique source a and a unique sink b,
(3) Let S_a be a simple module associated with source a and let S_b be a simple module associated with sink b, then $\operatorname{pd} S_a = 3$ and $\operatorname{id} S_b = 3$. If S is a simple module associated with any other vertex then $\operatorname{pd} S \leq 2$ and $\operatorname{id} S \leq 2$.
(4) Consider the minimal projective resolution of the simple module S_a:
$$0 \longrightarrow P_3 \longrightarrow P_2 \longrightarrow P_1 \longrightarrow P_0 \longrightarrow S_a \longrightarrow 0.$$
Let P be the following projective module, $P = \bigoplus_{k=0}^{3} P_k$. Then all indecomposable projectives are in $\operatorname{add} P$, and each indecomposable projectives is a direct summand of exactly one P_k, for $k \in \{0, \dots, 3\}$.
(5) Consider the minimal injective resolution of the S_b:
$$0 \longrightarrow S_b \longrightarrow I_0 \longrightarrow I_1 \longrightarrow I_2 \longrightarrow I_3 \longrightarrow 0.$$
Consider the B-module $I = \bigoplus_{k=0}^{3} I_k$. Then all indecomposable injectives are in $\operatorname{add} I$, and each indecomposable injective is a direct summand of exactly one I_k, for $k \in \{0, \dots, 3\}$.
(6) B does not contain any proper full subcategory that verifies 1), 2), 3), 4) and 5).

Bordino, Fernández and Trepode obtained the following classification of critical algebras in [**7**].

PROPOSITION 4.10 (Bordino-Fernández-Trepode [**7**]). *Let B be a critical algebra. Then the algebra has one of the following presentations.*

A_1 : 

B_1 : 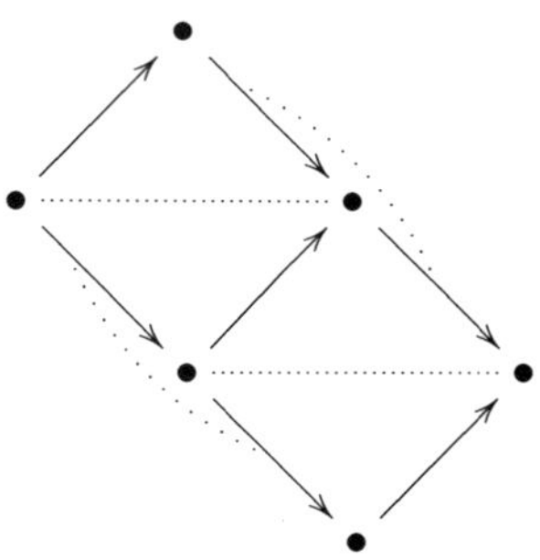

A_l $(l \geq 2)$:

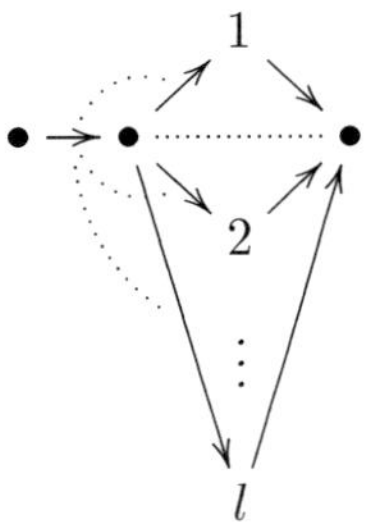

Q_n $(n \geq 2)$:

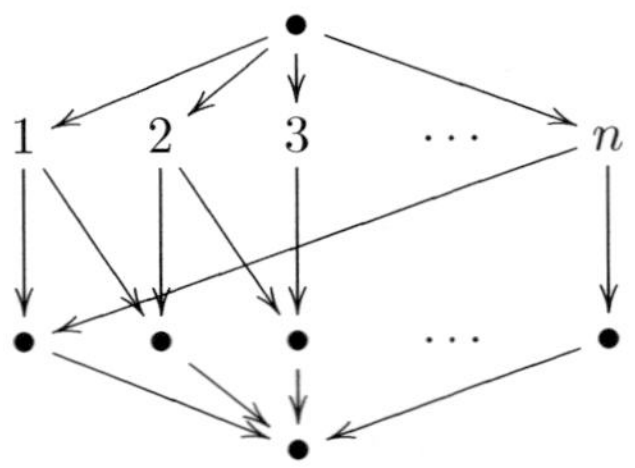

B_m $(m \geq 3)$:

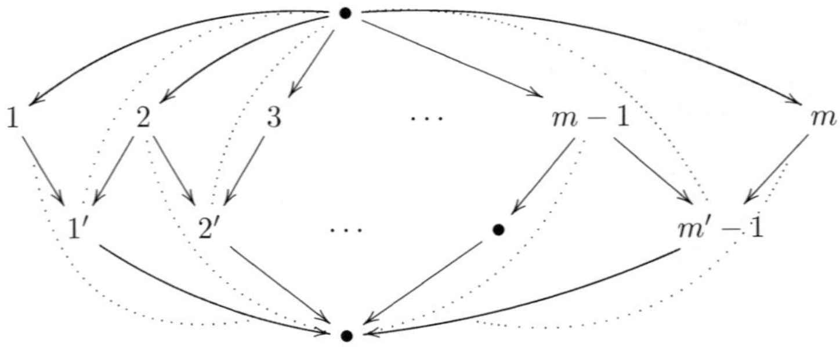

In the case Q_n, above, the relations are given by declaring that two parallel paths are equal.

Using this proposition, we see that the only critical incidence algebras are the ones whose quivers are the Q_n, since the presentations of algebras A_1, A_l, B_1 and B_l have non-commutative relations.

Now we can state a theorem of Bordino, Fernández and Trepode [**7**] which we will use in our proposition 4.13.

THEOREM 4.11 (Bordino-Fernández-Trepode [**7**]). *Let A be a strongly simply connected schurian algebra with global dimension greater than or equal to three. Then there exists a full subcategory B of A such that B is critical.*

In order to state our next theorem we need the definition of crown, which the reader can find in [**2**].

We will use the following result

THEOREM 4.12 (Assem-Castonguay-Marcos-Trepode [**2**]). *Let $K\Delta$ be an incidence algebra, $K\Delta$ is strongly simply connected if and only if $K\Delta$ does not contain a crown.*

Now we have the result on incidence algebras.

PROPOSITION 4.13. *Let $K\Delta$ be a strongly simply connected incidence algebra. Then the global dimension of $K\Delta$ is less than or equal to two.*

PROOF. An incidence algebra is a schurian algebra, so, if its global dimension is greater than or equal to three then it will contain a full critical subcategory B. Then B is of the form Q_n hence the incidence algebra would contain a crown (see [**2**]). This contradicts the result above. $\square$

5. Strong global dimension

In this section A denotes an algebra of the type $A = KQ/I$, where Q is a finite quiver and I an admissible ideal.

Let $\mathcal{P}_A$ be the full subcategory of mod A consisting of the projective modules. We denote $C^b(\mathcal{P}_A)$ the category of bounded complexes with entries in $\mathcal{P}_A$ and $K^b(\mathcal{P}_A)$ the corresponding homotopy category.

DEFINITION 5.1. A complex $P = (P^i, d^i)$ in $C^b(\mathcal{P}_A)$ is called *radical* if the image of each differential d^i is contained in the radical of P^{i-1} .

The following proposition is well known.

PROPOSITION 5.2. *([**5**]) Every complex $P = (P^i, d^i)$ in $C^b(\mathcal{P}_A)$ is homotopic to a unique radical complex, up to isomorphism.*

Due to the former proposition when we consider a complex in the category $K^b(\mathcal{P}_A)$ we will assume that it is radical, since it is isomorphic in $K^b(\mathcal{P}_A)$ to a unique radical complex.

DEFINITION 5.3. If $P = (P^i, d^i) \in K^b(\mathcal{P}_A)$ be a nonzero complex, then there are integers $r \leq s$ such that $P^i = 0$ for all $i < r$ and $P^i = 0$ for all $s < i$, with P_r and P_s not zero. The *length* of P is defined as $\mathrm{l}(P) = s - r$.

We can now define the strong global dimension.

DEFINITION 5.4 (strong global dimension [**25**]). Let A a finite-dimensional algebra. The *strong global dimension* of A is defined in the following way.

$$\operatorname{sgldim} A = \sup\{\mathrm{l}(P) | P \in K^b(\mathcal{P}_A) \ \text{ indecomposable}\}.$$

The next theorem is important in the study of the strong global dimension for PHI algebras.

THEOREM 5.5 (Happel-Zacharia [**16**]). *Let A a finite-dimensional algebra. The algebra is piecewise hereditary if and only if* sgldim A *is finite.*

In particular PHI algebras have finite strong global dimension. A tool to compute the strong global dimension of an algebra is the following theorem.

THEOREM 5.6 (Alvares, Le Meur, Marcos [**1**]). *Let $\mathcal{T}$ be a triangulated category which is triangle equivalent to the bounded derived category of a hereditary abelian category. Let $T \in \mathcal{T}$ be a tilting object.*

There exists a full and additive subcategory $\mathcal{H} \subset \mathcal{T}$ which is hereditary and abelian, such that the embedding $\mathcal{H} \hookrightarrow \mathcal{T}$ extends to a triangle equivalence $D^b(\mathcal{H}) \cong \mathcal{T}$, and

$$T \in \bigvee_{m=0}^{l} \mathcal{H}[m]$$

for some integer $l \geq 0$. Moreover, if $(\operatorname{End}_{\mathcal{T}} T)^{op}$ is not hereditary then there exists such a pair $(\mathcal{H}, l)$ verifying $\operatorname{sgldim}(\operatorname{End}_{\mathcal{T}} T)^{op} = l + 2$

Alvares, Le Meur and Marcos have used an alternative definition of the strong global dimension [**1**].

LEMMA 5.7 (Alvares, Le Meur, Marcos [**1**]). *Let $T \in D^b(\mathcal{H})$ be a tilting object such that $A = (\operatorname{End} T)^{op}$. Given an object Y in $D^b(\mathcal{H})$, we define*

$$\ell_T^+(Y) = \sup\{n \in \mathbb{Z} |\operatorname{Hom}_{D^b(\mathcal{H})}(Y, T[n]) \neq 0\}$$
$$\ell_T^-(Y) = \inf\{n \in \mathbb{Z} |\operatorname{Hom}_{D^b(\mathcal{H})}(T[n], Y) \neq 0\}.$$

Then the strong global dimension of A is $\operatorname{sgldim} A = \sup\{\ell_T^+(Y) - \ell_T^-(Y) | Y \in D^b(\mathcal{H})\}$.

We want to give an upper bound for the strong global dimension of PHI algebras. We have actually a more general result.

THEOREM 5.8. *Let A be a sincere, piecewise hereditary algebra. Then*

$$\operatorname{sgldim} A \leq 3 .$$

PROOF. We consider M a sincere module, which exists since A is sincere, and $A = P_1 \oplus \ldots \oplus P_n$ the decomposition of A in indecomposable modules.

By hypothesis, there exist a hereditary category $\mathcal{H}$ and a triangle equivalence $F \colon D^b(A) \to D^b(\mathcal{H})$.

Since M is sincere, we have that

$$0 \neq \operatorname{Hom}_A(P_i, M) = \operatorname{Hom}_{D^b(A)}(P_i, M) \cong \operatorname{Hom}_{D^b(\mathcal{H})}(FP_i, FM).$$

Let $FA = T$ be a tilting object in $D^b(\mathcal{H})$ such that $(\operatorname{End} T)^{op} \cong A$. For each $i \in \{1, \ldots, n\}$, we denote by $FP_i = T_i[m_i]$ the indecomposable direct summand of T where $T_i \in \mathcal{H}[0]$ such that $D^b(\mathcal{H}) = \bigvee_{m \in \mathbb{Z}} \mathcal{H}[m]$. Also we set $FM = M'[m]$ where $M' \in \mathcal{H}[0]$.

Now, for each $i \in \{1, \ldots, n\}$, we have the following:

$$0 \neq \operatorname{Hom}_{D^b(\mathcal{H})}(FP_i, FM) = \operatorname{Hom}_{D^b(\mathcal{H})}(T_i[m_i], M'[m]).$$

Therefore $m_i = m$ or $m_i = m - 1$, for each $i \in \{1, \ldots, n\}$, implying that the indecomposables direct summands of T are in $\mathcal{H}[m-1]$ or $\mathcal{H}[m]$ which shows that $\operatorname{sgldim} A \leq 3$. □

REMARK 5.9. Ladkani ([**17**]).

Let $K\Delta$ be a PHI algebra . As an immediate consequence of previous theorem, we get that $\operatorname{sgldim} K\Delta \leq 3$.

EXAMPLE 5.10. The incidence algebra, whose quiver is given below, is a PHI algebra which has global dimension equal to three, and therefore it also has strong global dimension three.

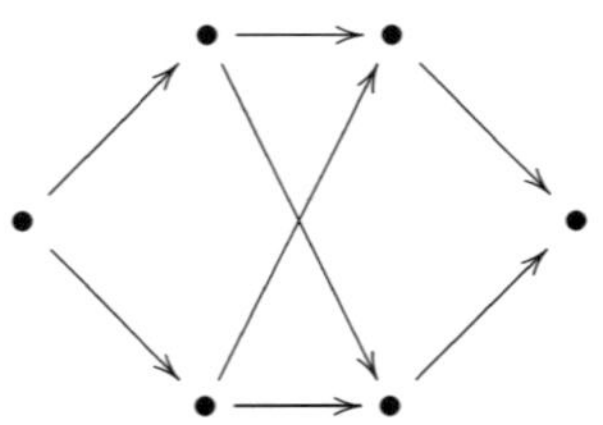

Our result 4.13 implies that the family of PHI algebras that have global dimension and strong global dimension equal to three are not strongly simply connected.

It would be interesting to give a characterization of this family of algebras. Observe that the PHI algebra above is in this class.

6. PHI algebras of sheaf type

The PHI algebras $K\Delta$ which are derived equivalent to Coh $\mathbb{X}$ were studied in [**18**] which in part, relates to our work.

THEOREM 6.1 (Ladkani [**18**]). *Let $K\Delta$ be an incidence algebra. If $K\Delta$ is derived equivalent to $C(p,\lambda)$ then $p = (p_1, p_2, p_3)$ or $p = (p_1, p_2)$, where $p_i \geq 2$. For each weighted projective line with $p = (p_1, p_2, p_3)$ or $p = (p_1, p_2)$, where $p_i \geq 2$, there exist at least one incidence algebra $K\Delta$ such that $K\Delta$ is derived equivalent to $C(p,\lambda)$.*

DEFINITION 6.2.

(1) The *Euler characteristic* $\mathcal{X}(\mathbb{X})$ of a weighted projective line is defined by the formula:
$$\mathcal{X}(\mathbb{X}) = 2 - \sum_{i=1}^{n}(1 - \frac{1}{p_i}).$$
(2) The category Coh($\mathbb{X}$) is called *domestic*, *tubular* or *wild*, if its Euler characteristic is respectively bigger than zero, equal zero or smaller than zero.

An algebra A is called a *piecewise hereditary algebra of domestic-sheaf type* or *of tubular-sheaf type* or *of wild-sheaf type* if A is derived equivalent to Coh($\mathbb{X}$) where Coh($\mathbb{X}$) is of domestic, tubular or wild type, respectively.

For the purpose of exhibiting some families of PHI algebras of sheaf type, we will study one-point extension. The papers of Barot, De la Peña and Lenzing ([**3**], [**4**] and [**8**]) provide conditions on the piecewise hereditary algebras A and the A-modules M in order that $A[M]$ is also piecewise hereditary.

The next example is an example of a one point extension of a PHI algebra by its canonical sincere module, in which the one point extension is also PHI. It is clear that the one point extension, in the example is an incidence algebra. We postpone the proof that it is also piecewise hereditary to 6.7.

EXAMPLE 6.3. Let M be the canonical sincere module (see definition 4.3) over the algebra described below, by quiver and relation, and consider the one-point extension $K\Delta[M]$.

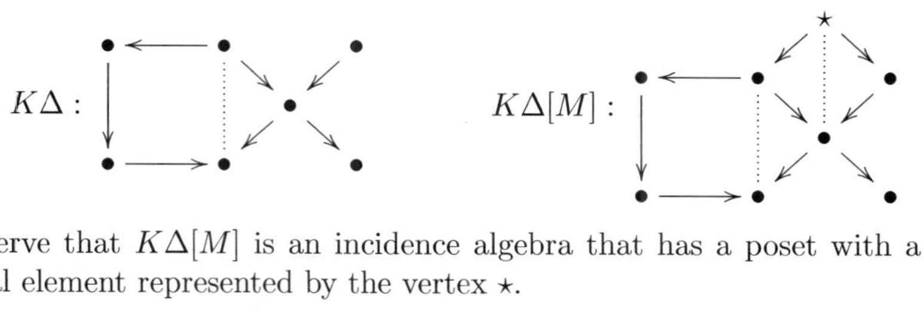

We observe that $K\Delta[M]$ is an incidence algebra that has a poset with a unique maximal element represented by the vertex $\star$.

We denote Coh$(\mathbb{X})_+$ (resp. Coh$(\mathbb{X})_0$) the full subcategory of all vector bundles (resp. sheaves of finite length) on $\mathbb{X}$.

Assume that we have a triangle equivalence:
$$F\colon D^b(K\Delta) \to D^b(\mathrm{Coh}(\mathbb{X})).$$

Inspired by the work of Lenzing and Skowroński, [**21**], we will show in proposition below, that the canonical sincere module M is associated, to a vector bundle, via the canonical equivalence on the derived categories.

PROPOSITION 6.4. *Let A is a representation-infinite quasi-tilted algebra of sheaf type with a sincere indecomposable module M. Assume $F\colon D^b(A) \to D^b(\mathrm{Coh}(\mathbb{X}))$ a triangle equivalence. Then FM is a vector bundle.*

PROOF. We consider $A = P_1 \oplus \ldots \oplus P_n$ the decomposition of A in indecomposable modules. By hypothesis, $D^b(A)$ is equivalent as a triangulated category to $D^b(\mathrm{Coh}(\mathbb{X}))$, so there is a tilting object T in $D^b(\mathrm{Coh}(\mathbb{X}))$ such that

$$(\mathrm{End}_{D^b(\mathrm{Coh}(\mathbb{X}))}(T))^{op} = A.$$

Since M is sincere, for each $i \in \{1, \ldots, n\}$ it follows that

$$0 \neq \mathrm{Hom}_A(P_i, M) = \mathrm{Hom}_{D^b(A)}(P_i, M) \cong \mathrm{Hom}_{D^b(\mathcal{H})}(FP_i, FM).$$

Let $FP_i = Q_i[p]$ and $FM = Y[m]$, where Q_i, Y belongs to $\mathrm{Coh}(\mathbb{X})[0]$, for each $i \in \{1, \ldots, n\}$. Therefore,

$$\mathrm{Hom}_{D^b(\mathrm{Coh}(\mathbb{X}))}(FP_i, FM) = \mathrm{Hom}_{D^b(\mathrm{Coh}(\mathbb{X}))}(Q_i[p], Y[m]) \neq 0.$$

For each $i \in \{1, \ldots, n\}$, $Q_i[p]$ is an indecomposable direct summand of T, where T decomposes into a direct sum $T_0'[-1] \oplus T_+ \oplus T_0''$, see [**21**]. That is, $Q_i[p]$ is isomorphic to an indecomposable direct summand of T consequently $p \in \{-1, 0\}$, and moreover $m = 0$.

We use a proposition of Lenzing-Skowroński [**21**]. If $FM \in \mathrm{Coh}(\mathbb{X})_0'$, we have that

$$\mathrm{Hom}_{D^b(\mathrm{Coh}(\mathbb{X}))}(T_+, FM) = \mathrm{Hom}_{\mathrm{Coh}(\mathbb{X})}(T_+, FM) \neq 0$$

Then $FM[-1] \notin \mathrm{mod}_0^l(B)$ eliminating the case a). Similarly, we eliminate the cases c) and e). This shows that FM is a bundle.

If A is an almost concealed-canonical algebra, see [**21**], then $T = T_+ \oplus T_0''$. This implies that $p = 0$ and $m \in \{0, 1\}$, where $\mathrm{Hom}_{D^b(\mathrm{Coh}(\mathbb{X}))}(Q_i[p], Y[m]) \neq 0$. The case where $FM = Y[0]$ is similar to the preceding one. Let $FM = Y[1]$. If $FM \in \mathrm{Coh}(\mathbb{X})_0''$, we have

$$\mathrm{Hom}_{D^b(\mathrm{Coh}(\mathbb{X}))}(T_+, Y[1]) \cong \mathrm{Ext}_{\mathrm{Coh}(\mathbb{X})}(T_+, Y) \neq 0$$

This contradicts the facts that T_+ is a bundle and Y is a finite length sheaf. Therefore FM is a vector bundle. □

Again, the next proposition is not specific to incidence algebras.

PROPOSITION 6.5. *If A is a representation-infinite quasi-tilted algebra of domestic-sheaf type with a sincere indecomposable module M, then M is exceptional.*

PROOF. We have a triangle equivalence:

$$F\colon D^b(A) \to D^b(\mathrm{Coh}(\mathbb{X})).$$

We now show that $\mathrm{Ext}^1_{\mathrm{Coh}(\mathbb{X})}(FM, FM) = 0$. We have that

$$\mathrm{Ext}^1_{\mathrm{Coh}(\mathbb{X})}(FM, FM) = \mathrm{Hom}_{D^b(\mathrm{Coh}(\mathbb{X}))}(FM, FM[1]) \cong \mathrm{Ext}^1_A(M, M).$$

By proposition 6.4, FM in $\mathrm{Coh}(\mathbb{X})_+$. Therefore, FM is a vector bundle. Then we use the theorem of Lenzing-Reiten [**20**] which states that if $\mathrm{Coh}(\mathbb{X})$ of domestic type, then each indecomposable bundle is exceptional.

Hence, the sincere indecomposable module M is exceptional. □

We now state a particular case of a result of Lenzing- De la Peña, which we will use in 6.7.

THEOREM 6.6 (Lenzing-De la Peña [**8**]). *Let A be an algebra derived equivalent to a canonical algebra Λ of weight type $(p_1, \dots, p_n)$. We fix a triangle-equivalence $D^b(A) \cong D^b(\mathrm{Coh}(\mathbb{X}))$, where $\mathrm{Coh}(\mathbb{X})$ is derived equivalent to Λ. Let M be an exceptional A-module. If the module M corresponds to a shift of an exceptional vector bundle over $\mathbb{X}$, and $\mathcal{X}(\mathbb{X}) > 0$, then $A[M]$ is derived equivalent to the path algebra of a connected wild quiver.*

LEMMA 6.7. *Let A be a representation-infinite quasi-tilted algebra of domestic-sheaf type with a sincere indecomposable module M. Then $A[M]$ is derived equivalent to the path algebra of a connected wild quiver.*

PROOF. We have a triangle equivalence:

$$F\colon D^b(A) \to D^b(\mathrm{Coh}(\mathbb{X})).$$

By proposition 6.5, M is exceptional. Moreover, by proposition 6.4, M associated, by F, with an exceptional vector bundle over $\mathbb{X}$.

We use the theorem of Lenzing-De la Peña [**8**] above. We conclude that $A[M]$ is derived equivalent to the path algebra of a connected wild quiver. □

We can produce several examples of PHI algebras where $K\Delta[M]$ is of wild type.

EXAMPLE 6.8. We give now an example of a PHI algebra $K\Delta$ of type $\widetilde{\mathbb{E}}_6$. Using our lemma, 6.7, we see that the one point extension of $K\Delta$ by the canonical sincere module M is a PHI algebra of wild type.

Acknowledgments

We thank the referee for nice and friendly suggestions. We also thank Ibrahim Assem, for his suggestions.

References

[1] Edson Ribeiro Alvares, Patrick Le Meur, and Eduardo N. Marcos, *The strong global dimension of piecewise hereditary algebras*, J. Algebra **481** (2017), 36–67, DOI 10.1016/j.jalgebra.2017.02.012. MR3639467

[2] I. Assem, D. Castonguay, E. N. Marcos, and S. Trepode, *Strongly simply connected Schurian algebras and multiplicative bases*, J. Algebra **283** (2005), no. 1, 161–189, DOI 10.1016/j.jalgebra.2004.08.020. MR2102077

[3] Michael Barot and Helmut Lenzing, *Derived canonical algebras as one-point extensions*, Trends in the representation theory of finite-dimensional algebras (Seattle, WA, 1997), Contemp. Math., vol. 229, Amer. Math. Soc., Providence, RI, 1998, pp. 7–15, DOI 10.1090/conm/229/03307. MR1676208

[4] Michael Barot and Helmut Lenzing, *One-point extensions and derived equivalence*, J. Algebra **264** (2003), no. 1, 1–5, DOI 10.1016/S0021-8693(03)00124-8. MR1980681

[5] Raymundo Bautista and Shiping Liu, *The bounded derived categories of an algebra with radical squared zero*, J. Algebra **482** (2017), 303–345, DOI 10.1016/j.jalgebra.2017.04.006. MR3646293

[6] Raymundo Bautista R. and R. Martinez-Villa, *Representations of partially ordered sets and 1-Gorenstein Artin algebras*, Ring theory (Proc. Antwerp Conf. (NATO Adv. Study Inst.), Univ. Antwerp, Antwerp, 1978), Lecture Notes in Pure and Appl. Math., vol. 51, Dekker, New York, 1979, pp. 385–433. MR563305

[7] N. Bordino, E. Fernández, and S. Trepode, *A criterion for global dimension two for strongly simply connected schurian algebras*, `arXiv:1110.6160v1` [math.RA] (2011).

[8] Helmut Lenzing and José Antonio de la Peña, *Spectral analysis of finite dimensional algebras and singularities*, Trends in representation theory of algebras and related topics, EMS Ser. Congr. Rep., Eur. Math. Soc., Zürich, 2008, pp. 541–588, DOI 10.4171/062-1/13. MR2484735

[9] José Antonio de la Peña and Manuel Saorín, *On the first Hochschild cohomology group of an algebra*, Manuscripta Math. **104** (2001), no. 4, 431–442, DOI 10.1007/s002290170017. MR1836104

[10] C. Fu and S. Geng, *On cluster-tilting graphs for hereditary categories*, `arXiv:1811.04735v1` [math.RT] (2018), 21.

[11] Werner Geigle and Helmut Lenzing, *A class of weighted projective curves arising in representation theory of finite-dimensional algebras*, Singularities, representation of algebras, and vector bundles (Lambrecht, 1985), Lecture Notes in Math., vol. 1273, Springer, Berlin, 1987, pp. 265–297, DOI 10.1007/BFb0078849. MR915180

[12] Dieter Happel, *On the derived category of a finite-dimensional algebra*, Comment. Math. Helv. **62** (1987), no. 3, 339–389, DOI 10.1007/BF02564452. MR910167

[13] Dieter Happel, *A characterization of hereditary categories with tilting object*, Invent. Math. **144** (2001), no. 2, 381–398, DOI 10.1007/s002220100135. MR1827736

[14] Dieter Happel, Idun Reiten, and Sverre Smalø, *Piecewise hereditary algebras*, Arch. Math. (Basel) **66** (1996), no. 3, 182–186, DOI 10.1007/BF01195702. MR1373243

[15] Dieter Happel and Luise Unger, *On the set of tilting objects in hereditary categories*, Representations of algebras and related topics, Fields Inst. Commun., vol. 45, Amer. Math. Soc., Providence, RI, 2005, pp. 141–159. MR2146246

[16] Dieter Happel and Dan Zacharia, *A homological characterization of piecewise hereditary algebras*, Math. Z. **260** (2008), no. 1, 177–185, DOI 10.1007/s00209-007-0268-3. MR2413349

[17] Sefi Ladkani, *Bounds on the global dimension of certain piecewise hereditary categories*, J. Pure Appl. Algebra **212** (2008), no. 9, 2140–2145, DOI 10.1016/j.jpaa.2007.12.007. MR2422197

[18] Sefi Ladkani, *Which canonical algebras are derived equivalent to incidence algebras of posets?*, Comm. Algebra **36** (2008), no. 12, 4599–4606, DOI 10.1080/00927870802185989. MR2473349

[19] Patrick Le Meur, *Topological invariants of piecewise hereditary algebras*, Trans. Amer. Math. Soc. **363** (2011), no. 4, 2143–2170, DOI 10.1090/S0002-9947-2010-05185-2. MR2746678

[20] Helmut Lenzing and Idun Reiten, *Hereditary Noetherian categories of positive Euler characteristic*, Math. Z. **254** (2006), no. 1, 133–171, DOI 10.1007/s00209-006-0938-6. MR2232010

[21] Helmut Lenzing and Andrzej Skowroński, *Quasi-tilted algebras of canonical type*, Colloq. Math. **71** (1996), no. 2, 161–181, DOI 10.4064/cm-71-2-161-181. MR1414820

[22] Michèle Loupias, *Indecomposable representations of finite ordered sets*, Representations of algebras (Proc. Internat. Conf., Carleton Univ., Ottawa, Ont., 1974), Springer, Berlin, 1975, pp. 201–209. Lecture Notes in Math., Vol. 488. MR0412210

[23] Jeremy Rickard, *Morita theory for derived categories*, J. London Math. Soc. (2) **39** (1989), no. 3, 436–456, DOI 10.1112/jlms/s2-39.3.436. MR1002456
, Lectures Notes in Mathematics, vol. 1099, Springer, 1984.

[24] Claus Michael Ringel, *Tame algebras and integral quadratic forms*, Lecture Notes in Mathematics, vol. 1099, Springer-Verlag, Berlin, 1984. MR774589

[25] Andrzej Skowroński, *On algebras with finite strong global dimension* (English, with Russian summary), Bull. Polish Acad. Sci. Math. **35** (1987), no. 9-10, 539–547. MR937729

[26] Andrzej Skowroński, *Simply connected algebras and Hochschild cohomologies [MR1206961 (94e:16016)]*, Representations of algebras (Ottawa, ON, 1992), CMS Conf. Proc., vol. 14, Amer. Math. Soc., Providence, RI, 1993, pp. 431–447. MR1265301

Dto de Matemática, Instituto de Matemática e Estatística, Universidade São Paulo, Rua do Matão 1010, Cidade Universitária, CEP 05508-090, São Paulo-SP, Brasil
Email address: enmarcos@ime.usp.br

Dto de Matemática, Instituto de Ciências Exatas, Universidade Federal de Alfenas, Campus Sede, Rua Gabriel Monteiro da Silva 700, Centro, CEP 37130-001, Alfenas-MG, Brasil
Email address: marcelo.moreira@unifal-mg.edu.br

Contemporary Mathematics
Volume **761**, 2021
https://doi.org/10.1090/conm/761/15318

Derived categories of gentle algebras and surface dissections

Pierre-Guy Plamondon

ABSTRACT. Gentle algebras form a class of associative algebras whose representation theory is well-understood. The data of a gentle algebra is equivalent to the data of an oriented surface with boundary, marked points, punctures, and a dissection. In this lecture, we explain how the derived category of a gentle algebra can be read off of the associated dissected surface, and how this allows for a complete geometric description of silting and tilting objects.

CONTENTS

1. Surfaces and dissections
2. The gentle algebra of a dissection
3. Objects in the derived category as graded curves on the surface
4. Silting objects as dissections
5. Bibliographical notes
Acknowledgments
References

1. Surfaces and dissections

Let S be an oriented *surface with boundary*, that is, a smooth compact oriented surface with a finite number of open discs removed. Such a surface is determined (up to diffeomorphism) by its genus and by the number of connected components of its boundary.

Let $M = M_\circ \sqcup M_\bullet$ be a finite set of *marked points* on the boundary of S. We require that each boundary component of S contains at least one marked point, and that points $\circ$ in $M_\circ$ and $\bullet$ in $M_\bullet$ alternate on each boundary component. Moreover, let $P = P_\circ \sqcup P_\bullet$ be a finite set of *punctures* in the interior of S.

A $\circ$-*arc* (or $\bullet$-*arc*) is a smooth map from $[0,1]$ to the surface S whose endpoints are in $M_\circ \cup P_\circ$ (or $M_\bullet \cup P_\bullet$, respectively) and whose interior is in the interior of S. We consider arcs up to (smooth) homotopy.

From now on, let (S, M, P) be a surface with marked points and punctures.

DEFINITION 1.1. A finite collection $\Delta = \{\gamma_1, \dots, \gamma_r\}$ of pairwise distinct $\circ$-arcs (or $\bullet$-arcs) on (S, M, P) is *admissible* if

2010 *Mathematics Subject Classification.* Primary 16E35, 55M25.
The author was supported by the French ANR grant SC3A (ANR-15-CE40-0004-01).

(1) the arcs of Δ are pairwise non-intersecting in their interior, and
(2) the arcs of Δ do not enclose a subsurface containing no $\bullet$ (or no $\circ$, respectively).

An *admissible* $\circ$*-dissection* (or *admissible* $\bullet$*-dissection*) is a maximal admissible collection of $\circ$-arcs (or $\bullet$-arcs, respectively).

EXAMPLE 1.2. The following pictures are admissible $\circ$-dissections of a disc with one $\circ$-puncture, two $\bullet$-punctures and six marked points of each kind on its boundary (left), and of an annulus with one marked point of each kind on each boundary component (right).

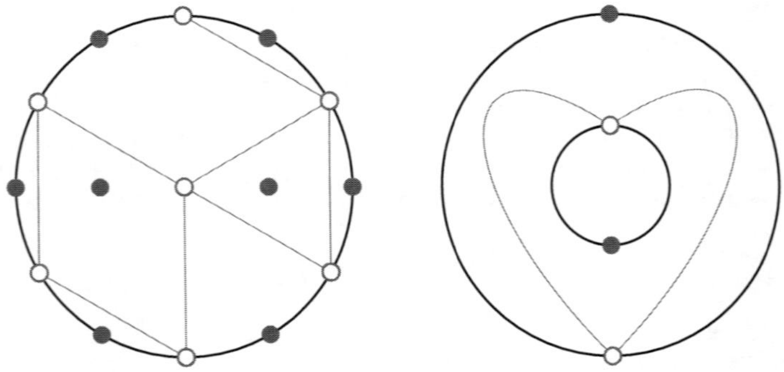

The next picture is an admissible $\circ$-dissection of a torus with one boundary component and two marked points of each kind on its boundary component.

EXAMPLE 1.3. Examples of non-admissible collections of $\circ$-arcs are depicted below; the first one is on a disc, and the second on a torus with one boundary component.

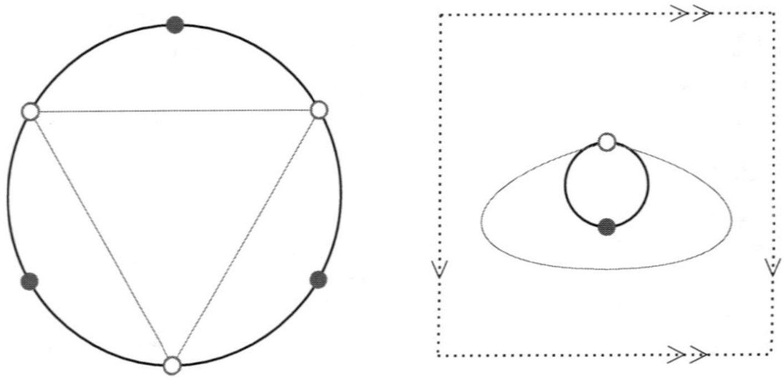

In these examples, the ○-arcs enclose a triangle and a torus with no •, respectively.

While Definition 1.1 introduces admissibility in the form of a compatibility relations for collections of arcs, there exists a more global definition for admissible dissections.

PROPOSITION 1.4 ([**APS19**]). *A collection Δ of pairwise distinct ○-arcs on (S, M, P) is an admissible ○-dissection if and only if the complement of the interiors of the arcs of Δ is a disjoint union of pieces of the following two forms:*

(1) *an unpunctured disc with exactly one • on its boundary, or*
(2) *a disc containing exactly one • puncture and no • on its boundary.*

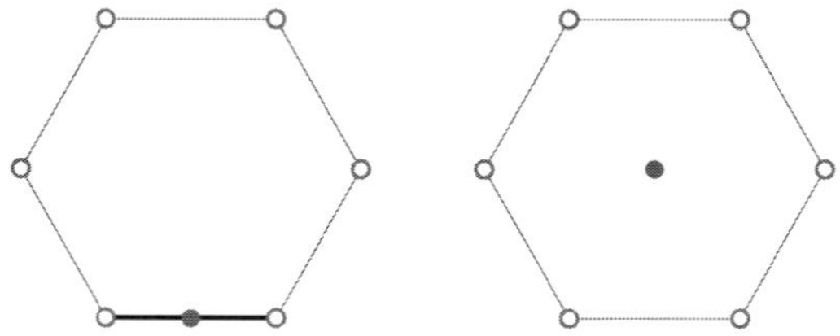

The notions of admissible ○- and • -dissection are dual to each other. Geometrically, this can be expressed as follows:

PROPOSITION 1.5. *Let Δ be an admissible ○-dissection of (S, M, P). Then there exists a unique admissible •-dissection Δ^* of (S, M, P) such that each arc of Δ intersects (in its interior) exactly one arc of Δ^*, and vice versa.*

The dissection Δ^* of Proposition 1.5 is called the *dual dissection* of Δ.

EXAMPLE 1.6. Below are the admissible ○-dissections of Example 1.2, together with their dual dissections.

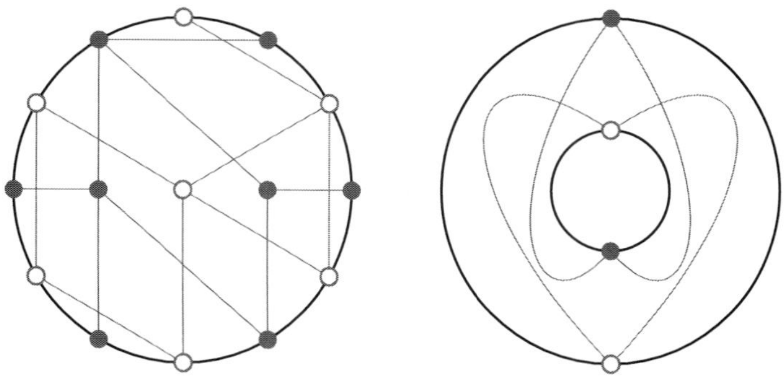

2. The gentle algebra of a dissection

Fix a surface with boundary and marked points (S, M, P), and let $\mathbf{k}$ be a field. To any admissible $\circ$-dissection of (S, M, P), we now associate an associative algebra.

DEFINITION 2.1. Let Δ be an admissible $\circ$-dissection of (S, M, P). Define the algebra $A(\Delta) = \mathbf{k}Q(\Delta)/I(\Delta)$, where

(1) $Q(\Delta)$ is a quiver whose vertices are in bijection with the arcs of Δ, and where there is an arrow from i to j for each $\circ$ shared by i and j and such that j appears immediately after i around that $\circ$ in the counter-clockwise order;

(2) $I(\Delta)$ is the ideal of the path algebra $\mathbf{k}Q(\Delta)$ generated by the paths of length 2 defined as follows: whenever i, j and k are adjacent arcs in a polygon of Δ in the clockwise order, and if $\alpha : i \to j$ and $\beta : j \to k$ are the corresponding arrows, then $\beta\alpha \in I(\Delta)$.

EXAMPLE 2.2. Below is the quiver with relations associated to the first dissection of Example 1.2. Dotted lines represent relations.

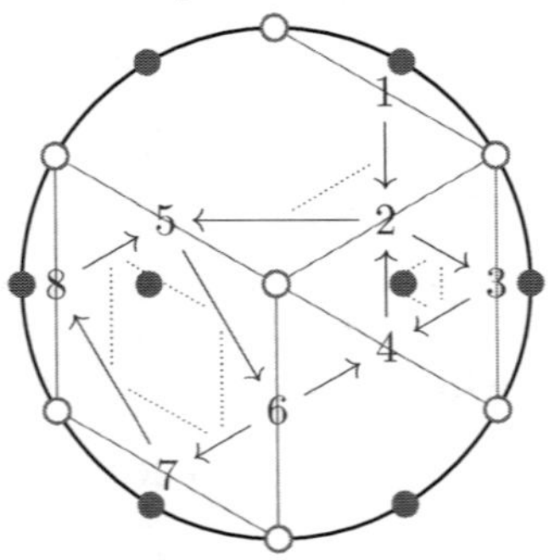

The algebras appearing as $A(\Delta)$ have a nice combinatorial description.

PROPOSITION 2.3. *Let Δ be an admissible $\circ$-dissection of (S, M, P). Then the bound quiver $\bar{Q}(\Delta) = \big(Q(\Delta), I(\Delta)\big)$ is* gentle, *that is,*

(1) *every vertex of $Q(\Delta)$ is the source of at most two arrows and the target of at most two arrows;*

(2) *for every arrow α of $Q(\Delta)$, there is at most one arrow β whose source is the target of α and such that $\beta\alpha \in I$;*

(3) *for every arrow α of $Q(\Delta)$, there is at most one arrow β' whose source is the target of α and such that $\beta'\alpha \notin I$;*

(4) *for every arrow α of $Q(\Delta)$, there is at most one arrow γ whose target is the source of α and such that $\alpha\gamma \in I$;*

(5) *for every arrow α of $Q(\Delta)$, there is at most one arrow γ' whose target is the source of α and such that $\alpha\gamma' \notin I$.*

In other words, the configuration around each vertex v of $\bar{Q}(\Delta)$ is a full subquiver with relations of

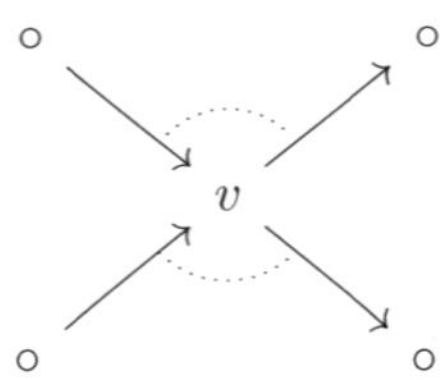

where dotted lines represent relations, and where some of the vertices ∘ may be identified.

The strength of the geometric model for the study of gentle algebras comes from the following result.

THEOREM 2.4 ([**OPS18, BCS18, PPP18**]). *Any basic gentle algebra (possibly of infinite-dimension) is isomorphic to an algebra of the form $A(\Delta)$, where Δ is an admissible ∘ -dissection of a surface with boundary and marked points (S, M, P). Moreover, the surface with dissection associated to a given gentle algebra is unique up to orientation-preserving homeomorphism.*

3. Objects in the derived category as graded curves on the surface

Let (S, M, P) be a surface with boundary and marked points, and let Δ be an admissible ∘-dissection. We will be interested in the bounded derived category $\mathcal{D}^b(\operatorname{mod} A(\Delta))$ of the category of finitely generated right $A(\Delta)$-modules. We will restrict ourselves to the full subcategory $K^b(\operatorname{proj} A(\Delta))$ of $\mathcal{D}^b(\operatorname{mod} A(\Delta))$ consisting of bounded complexes of projective modules. Using curves on the surface (S, M, P), we can describe all indecomposable objects of this category.

Call a *closed curve* on S a smooth map from the circle $\mathbb{S}^1$ to the interior of S.

DEFINITION 3.1. Let (S, M, P) and Δ be as before, and let Δ^* be the dual admissible •-dissection. A *graded curve* is a pair (γ, f), where γ is a ∘-arc or a closed curve (assumed to intersect the arcs of Δ^* minimally and transversally, which is possible up to homotopy) and f is a map assigning an integer to each intersection point of γ with the arcs of Δ^* , in such a way that the following local condition is satisfied. If p and q are two such successive intersection points, let $\gamma_{p,q}$ be the oriented segment of γ joining p to q. This segment lies in a connected component C of $S \setminus \Delta^*$; let $\bar{C}$ be the closure of this component. Then

$$f(q) = \begin{cases} f(p)+1 & \text{if the } \circ \text{ in } \bar{C} \text{ lies to the left of } \gamma_{p,q}; \\ f(p)-1 & \text{otherwise.} \end{cases}$$

If (γ, f) is a graded curve, define $(\gamma, f)[1] := (\gamma, f-1)$.

Finally, for any graded curve (γ, f) with intersection points p and q as above, note that each angle of C corresponds to an arrow of $Q(\Delta)$. Denote by $\sigma_{p,q}$ the path (or formal inverse of path) from p to q in $Q(\Delta)$ defined by composing these arrows as in the picture:

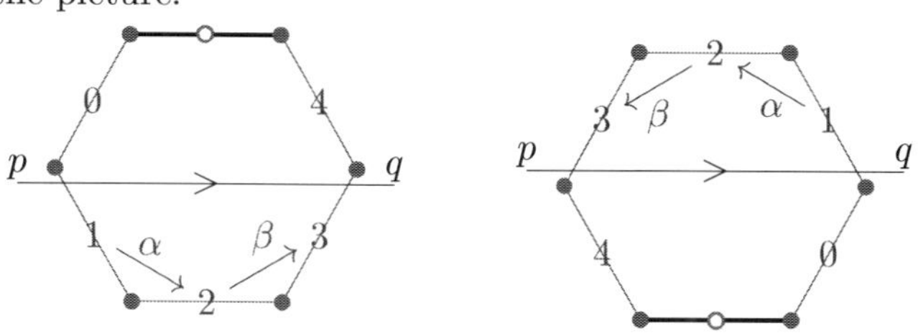

In the left picture, $f(q) = f(p)+1$ and $\sigma_{p,q} = \beta\alpha$, while in the right picture, $f(q) = f(p)-1$ and $\sigma_{p,q} = (\beta\alpha)^{-1}$.

The graded curves on S allow for the description of certain objects in $K^b(\operatorname{proj} A(\Delta))$.

DEFINITION 3.2. Let (S, M, P) and Δ be as before.

(1) Let (γ, f) be a graded $\circ$-arc. Let $p_1, \dots, p_r$ be the points of intersections of γ with the $\bullet$-arcs of Δ^*, in that order, and let $i_1, \dots, i_r$ be the corresponding vertices of $Q(\Delta)$. Then the *string object* $P^\bullet_{(\gamma,f)}$ is defined to be the $\mathbb{Z}$-graded $A(\Delta)$-module

$$P^\bullet_{(\gamma,f)} := \bigoplus_{j=1}^{r} P_{i_j}[f(p_j)]$$

with differential $d: P^\bullet_{(\gamma,f,M)} \to P^\bullet_{(\gamma,f,M)}[1]$ defined by the $(r\times r)$-matrix $(d_{x,y})$, where, for all $x \in \{1, \dots, r-1\}$,
- if $f(p_{x+1}) = f(p_x) + 1$, then $d_{x+1,x} = \sigma_{p_x,p_{x+1}}[f(p_x)]$;
- if $f(p_{x+1}) = f(p_x) - 1$, then $d_{x,x+1} = \sigma^{-1}_{p_x,p_{x+1}}[f(p_{x+1})]$;
- all other values of $d_{x,y}$ are 0.

(2) Let (γ, f) be a graded closed curve. Let $p_1, \dots, p_r$ be the points of intersections of γ with the $\bullet$-arcs of Δ^*, in that (cyclic) order, so that $p_1 = p_r$ are viewed as the starting/ending point of γ. Let $i_1, \dots, i_r$ be the corresponding vertices of $Q(\Delta)$. Define $\sigma_{p_x,p_{x+1}}$ as above, but this time with x taken modulo r. Finally, let V be an m -dimensional indecomposable representation of $k[t, t^{-1}]$, with matrix representation M.

Then the *band object* $P^\bullet_{(\gamma,f,M)}$ is defined to be the $\mathbb{Z}$-graded $A(\Delta)$-module

$$P^\bullet_{(\gamma,f,M)} := \bigoplus_{j=1}^{r-1} P^{\oplus m}_{i_j}[f(p_j)]$$

with differential $d : P^\bullet_{(\gamma,f,M)} \to P^\bullet_{(\gamma,f,M)}[1]$ defined by the $(r-1\times r-1)$-matrix $(d_{x,y})$, where, for all $x \in \{1, \dots, r-2\}$,
- if $f(p_{x+1}) = f(p_x) + 1$, then $d_{x+1,x} = \sigma_{p_x,p_{x+1}} \cdot Id_{P^{\oplus m}_{i_x}}[f(p_x)]$;
- if $f(p_{x+1}) = f(p_x) - 1$, then $d_{x,x+1} = \sigma^{-1}_{p_x,p_{x+1}} Id_{P^{\oplus m}_{i_{x+1}}}[f(p_{x+1})]$;
- set all other values of $d_{x,y}$ to 0,
- finally, for $x = r-1$:
  - if $f(p_1) = f(p_{r-1}) + 1$, then add $\sigma_{p_{r-1},p_1} \cdot M[f(p_{r-1})]$ to $d_{1,r-1}$;
  - if $f(p_1) = f(p_{r-1}) - 1$, then add $\sigma^{-1}_{p_{r-1},p_1} \cdot M[f(p_1)]$ to $d_{r-1,1}$.

The definitions of string and band objects look technical, but they are easy to use in practice.

EXAMPLE 3.3. As all objects appearing in the remainder of this note will be string objects, we only give an example of this type of object. Recall the dual dissection of the first surface of Example 1.2. Consider the curve γ depicted below.

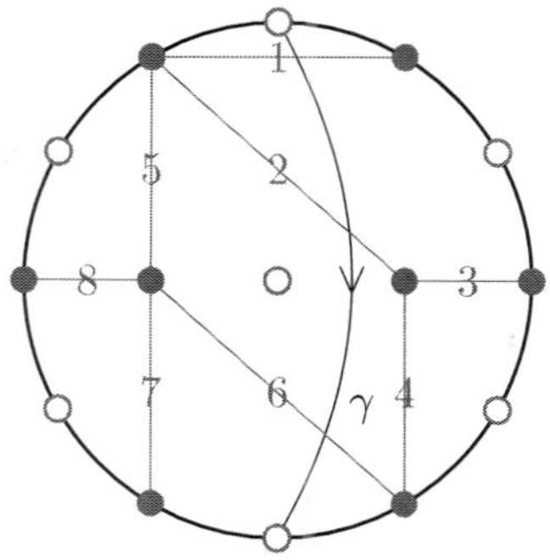

Define a grading f on γ by setting its intersection with the $\bullet$-arc 1 to be of degree 0. Then

$$P^{\bullet}_{(\gamma,f)} = \ldots \to 0 \to P_1 \oplus P_6 \xrightarrow{(\alpha,\delta\beta)} P_2 \to 0 \to \ldots,$$

where $\alpha : 1 \to 2$, $\beta : 6 \to 4$ and $\delta : 4 \to 2$ are arrows in $Q(\Delta)$.

THEOREM 3.4 ([**BM03, BD05, OPS18**]). *Assume that $P_\circ = \emptyset$. Then the string and band objects form a complete set of representatives for the isomorphism classes of indecomposable objects in $K^b(\operatorname{proj} A(\Delta))$.*

Moroever, a basis for the morphism spaces between objects in $K^b(\operatorname{proj} A(\Delta))$ can be completely described in terms of intersections. For the purposes of this note, it will be sufficient to restrict to morphisms between string objects.

THEOREM 3.5 ([**ALP16, OPS18**]). *Assume that $P_\circ = \emptyset$. Let $P^{\bullet}_{(\gamma,f)}$ and $P^{\bullet}_{(\delta,g)}$ be two string objects. Assume that γ and δ intersect each other and the arcs of Δ^* minimally and transversally. Then a basis of the morphism space $\operatorname{Hom}_{K^b}(P^{\bullet}_{(\gamma,f)}, P^{\bullet}_{(\delta,g)})$ is in bijection with the number of intersection points of γ and δ (possibly at their endpoints) satisfying the following local condition:*

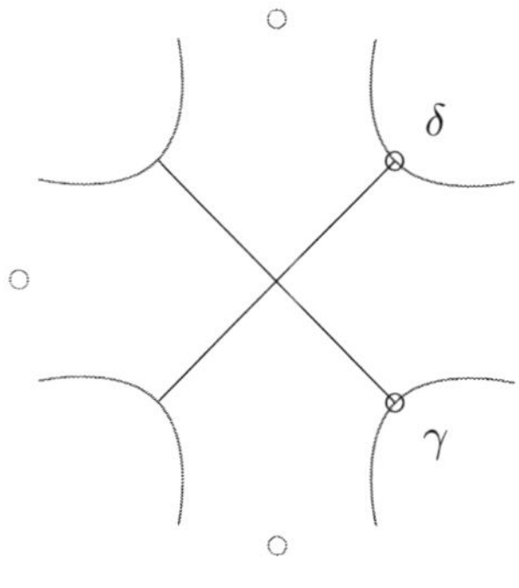

where the four rounded curves are (not necessarily pairwise distinct) arcs of Δ^*, the three $\circ$ are the allowed positions of the $\circ$-marked point in the polygon in which γ and δ intersect, and the two empty circles represent points p and q such that $f(p) = g(q)$.

4. Silting objects as dissections

Tilting and silting objects are central in the theory of derived equivalences.

DEFINITION 4.1. Let A be any finite-dimensional k algebra. An object X of $\mathcal{D}^b(\operatorname{mod} A)$ is *silting* if

(1) X is isomorphic to an object in the subcategory $K^b(\operatorname{proj} A)$;
(2) for any integer $i > 0$, we have that $\operatorname{Hom}_{\mathcal{D}^b}(X, X[i]) = 0$; and
(3) $K^b(\operatorname{proj} A)$ is the smallest full triangulated subcategory of $\mathcal{D}^b(\operatorname{mod} A)$ containing X and stable under taking direct summands.

An object satisfying conditions (1) and (2) is called *presilting*. A *tilting* object is a silting object X such that

(3') for any integer $i \neq 0$, we have that $\operatorname{Hom}_{\mathcal{D}^b}(X, X[i]) = 0$.

We now turn to the main result of this note:

THEOREM 4.2 ([**APS19**]). *Let (S, M, P) be a surface with marked points such that $P_\circ = \emptyset$. Let X be a silting object of $\mathcal{D}^b(\operatorname{mod} A(\Delta))$. Then there exist graded $\circ$-arcs $(\gamma_1, f_1), \dots, (\gamma_n, f_n)$ such that $X = \bigoplus_{i=1}^n P^\bullet_{(\gamma_i, f_i)}$ and $\Delta = \{\gamma_1, \dots, \gamma_n\}$ is an admissible $\circ$ -dissection of (S, M, P).*

Theorem 4.2 was independetly proved in [**Opp19**] for tilting objects, using *geometrically generating* collections of arcs instead of $\circ$-dissections.

In the proof of Theorem 4.2, the following result of independent interest is needed.

PROPOSITION 4.3 ([**APS19**]). *Let A be a basic gentle algebra, and let X be a basic presilting object of $\mathcal{D}^b(\operatorname{mod} A)$ whose number of indecomposable direct summands is the same as the number of indecomposable direct summands of A. Then X is silting.*

It is an open problem to determine whether the statement of Proposition 4.3 holds for any finite-dimensional **k**-algebra.

Theorem 4.2 also allows to reobtain the following known results on gentle algebras.

COROLLARY 4.4 ([**SZ03**]). *The class of gentle algebras is closed under derived equivalence.*

PROOF. Let A be gentle, and B be derived equivalent to A. By Rickard's Theorem [**Ric89**], there exists a tilting object T in $\mathcal{D}^b(\operatorname{mod} A)$ whose endomorphism algebra is isomorphic to B. By Theorem 4.2, the object T is isomorphic to an object $\bigoplus_{i=1}^n P^\bullet_{(\gamma_i, f_i)}$, where $\Delta = \{\gamma_1, \dots, \gamma_n\}$ is an admissible $\circ$-dissection of the surface associated with A. Using [**OPS18**], one sees that the endomorphism algebra of $\bigoplus_{i=1}^n P^\bullet_{(\gamma_i, f_i)}$ is gentle. □

We end this note by mentioning that, using a restriction of Theorem 4.2 to the class of tilting objects, we can refine the derived invariant of [**AAG08**] into a complete derived invariant for gentle algebras. This is done in [**APS19**, Theorem 4.1] and [**Opp19**, Theorem B]. Using the work of [**LP18**], one can express this a numerical derived invariant [**APS19**, Theorem 5.4].

5. Bibliographical notes

The definition of a gentle algebra was first given in [**AS87**], where it is proved that iterated tilted algebras of type $\widetilde{A}_n$ are gentle (see also [**AH81, AH82**] for iterated tilted algebras of type A_n).

The representation theory of gentle algebras has been extensively studied. The class of gentle algebras is a subclass of that of *special biserial algebras*, whose indecomposable modules were classified in [**WW85**] using the method of "functorial filtration" introduced in [**GP68**]. Gentle algebras are also special cases of *string algebras*, whose indecomposable finite-dimensional modules, irreducible maps and almost-split sequences were classified in [**BR87**]. All morphisms between finite-dimensional modules over a gentle algebras are known [**CB89, Kra91**].

Central to this note are the bounded derived categories of gentle algebras. They have also been studied in detail: their indecomposable objects were described in [**BM03, BD05**], their morphisms in [**ALP16**] and their cones in [**CPS16**]. Gentle algebras display several nice homological properties: they are Gorenstein [**GR05**], their class is closed under derived equivalence [**SZ03**], their categories of singularities are well-understood [**Kal15**], and they are Koszul [**BH08**]. A numerical derived invariant for gentle algebras was discovered in [**AAG08**], see also [**BH07**]. Their τ-tilting theory is also well-understood [**PPP17, BDM+17, Pla18, Mou19**].

The idea of a geometric model for gentle algebras takes its roots in the theory of cluster algebras from triangulated surfaces [**FG09, FST08**]. The first model for modules over gentle algebras from triangulations of surfaces appeared in [**ABCJP10**], and was generalized for any module over any gentle algebra in [**BCS18**] using partial triangulations. A model for the derived categories of gentle algebras was introduced in [**OPS18**], and is related to Fukaya categories of surfaces [**HKK17**]. This link has recently been exploited in [**LP18, APS19, Opp19**] to further the understanding of the derived categories of gentle algebras, and to refine the numerical derived invariant of [**AAG08**] into a complete derived invariant.

Acknowledgments

This note reflects a lecture given at the ARTA 7 held in México from September 24 to 28, 2018. The lecture was a report on joint works with Claire Amiot, Sebastian Opper and Sibylle Schroll [**OPS18, APS19**]. I would like to thank the organizers of the ARTA 7, namely Christof Geiss, Daniel Labardini-Fragoso, Octavio Mendoza and Corina Saénz, for inviting me to give this talk and for their kind hospitality.

References

[AAG08] Diana Avella-Alaminos and Christof Geiss, *Combinatorial derived invariants for gentle algebras*, J. Pure Appl. Algebra **212** (2008), no. 1, 228–243, DOI 10.1016/j.jpaa.2007.05.014. MR2355048

[ABCJP10] Ibrahim Assem, Thomas Brüstle, Gabrielle Charbonneau-Jodoin, and Pierre-Guy Plamondon, *Gentle algebras arising from surface triangulations*, Algebra Number Theory **4** (2010), no. 2, 201–229, DOI 10.2140/ant.2010.4.201. MR2592019

[AH81] Ibrahim Assem and Dieter Happel, *Generalized tilted algebras of type A_n*, Comm. Algebra **9** (1981), no. 20, 2101–2125, DOI 10.1080/00927878108822697. MR640613

[AH82] Ibrahim Assem and Dieter Happel. Erratum: "Generalized tilted algebras of type A_n" [Comm. Algebra **9** (1981), no. 20, 2101–2125]. *Comm. Algebra*, 10(13):1475, 1982.

[ALP16] Kristin Krogh Arnesen, Rosanna Laking, and David Pauksztello, *Morphisms between indecomposable complexes in the bounded derived category of a gentle algebra*, J. Algebra **467** (2016), 1–46, DOI 10.1016/j.jalgebra.2016.07.019. MR3545953

[APS19] Claire Amiot, Pierre-Guy Plamondon, and Sibylle Schroll, *A complete derived invariant for gentle algebras via winding numbers and Arf invariants*, `arXiv:1904.02555` [math.RT], 2019.

[AS87] Ibrahim Assem and Andrzej Skowroński, *Iterated tilted algebras of type* $\tilde{\mathbf{A}}_n$, Math. Z. **195** (1987), no. 2, 269–290, DOI 10.1007/BF01166463. MR892057

[BCS18] Karin Baur and Raquel Coelho Simões, *A geometric model for the module category of a gentle algebra*, `arXiv:1803.05802` [math.RT], to appear in IMRN, 2018.

[BD05] Igor Burban and Yuriy Drozd, *On derived categories of certain associative algebras*, Representations of algebras and related topics, Fields Inst. Commun., vol. 45, Amer. Math. Soc., Providence, RI, 2005, pp. 109–128. MR2146243

[BDM+17] Thomas Brüstle, Guillaume Douville, Kaveh Mousavand, Hugh Thomas, and Emine Yıldırım, *On the combinatorics of gentle algebras*, Preprint, `arXiv:1707.07665`, 2017.

[BH07] Christine Bessenrodt and Thorsten Holm, *q-Cartan matrices and combinatorial invariants of derived categories for skewed-gentle algebras*, Pacific J. Math. **229** (2007), no. 1, 25–47, DOI 10.2140/pjm.2007.229.25. MR2276501

[BH08] Christine Bessenrodt and Thorsten Holm, *Weighted locally gentle quivers and Cartan matrices*, J. Pure Appl. Algebra **212** (2008), no. 1, 204–221, DOI 10.1016/j.jpaa.2007.05.004. MR2355046

[BM03] Viktor Bekkert and Héctor A. Merklen, *Indecomposables in derived categories of gentle algebras*, Algebr. Represent. Theory **6** (2003), no. 3, 285–302, DOI 10.1023/A:1025142023594. MR2000963

[BR87] M. C. R. Butler and Claus Michael Ringel, *Auslander-Reiten sequences with few middle terms and applications to string algebras*, Comm. Algebra **15** (1987), no. 1-2, 145–179, DOI 10.1080/00927878708823416. MR876976

[CB89] W. W. Crawley-Boevey, *Maps between representations of zero-relation algebras*, J. Algebra **126** (1989), no. 2, 259–263, DOI 10.1016/0021-8693(89)90304-9. MR1024991

[CPS16] İlke Çanakçı, David Pauksztello, and Sibylle Schroll, *Mapping cones in the bounded derived category of a gentle algebra*, J. Algebra **530** (2019), 163–194, DOI 10.1016/j.jalgebra.2019.04.005. MR3940954

[FG09] Vladimir V. Fock and Alexander B. Goncharov, *Cluster ensembles, quantization and the dilogarithm* (English, with English and French summaries), Ann. Sci. Éc. Norm. Supér. (4) **42** (2009), no. 6, 865–930, DOI 10.24033/asens.2112. MR2567745

[FST08] Sergey Fomin, Michael Shapiro, and Dylan Thurston, *Cluster algebras and triangulated surfaces. I. Cluster complexes*, Acta Math. **201** (2008), no. 1, 83–146, DOI 10.1007/s11511-008-0030-7. MR2448067

[GP68] I. M. Gel′fand and V. A. Ponomarev, *Indecomposable representations of the Lorentz group* (Russian), Uspehi Mat. Nauk **23** (1968), no. 2 (140), 3–60. MR0229751

[GR05] Ch. Geiß and I. Reiten, *Gentle algebras are Gorenstein*, Representations of algebras and related topics, Fields Inst. Commun., vol. 45, Amer. Math. Soc., Providence, RI, 2005, pp. 129–133. MR2146244

[HKK17] F. Haiden, L. Katzarkov, and M. Kontsevich, *Flat surfaces and stability structures*, Publ. Math. Inst. Hautes Études Sci. **126** (2017), 247–318, DOI 10.1007/s10240-017-0095-y. MR3735868

[Kal15] Martin Kalck, *Singularity categories of gentle algebras*, Bull. Lond. Math. Soc. **47** (2015), no. 1, 65–74, DOI 10.1112/blms/bdu093. MR3312965

[Kra91] Henning Krause, *Maps between tree and band modules*, J. Algebra **137** (1991), no. 1, 186–194, DOI 10.1016/0021-8693(91)90088-P. MR1090218

[LP18] Yankı Lekili and Alexander Polishchuk, *Derived equivalences of gentle algebras via Fukaya categories*, Math. Ann. **376** (2020), no. 1-2, 187–225, DOI 10.1007/s00208-019-01894-5. MR4055159

[Mou19] Kaveh Mousavand, *τ-tilting finiteness of biserial algebras*, `arXiv:1904.11514` [math.RT], 2019.

[Opp19] Sebastian Opper, *On auto-equivalences and complete derived invariants of gentle algebras*, `arXiv:1904.04859` [math.RT], 2019.

[OPS18] Sebastian Opper, Pierre-Guy Plamondon, and Sibylle Schroll, *A geometric model for the derived category of gentle algebras*, Preprint, arXiv:1801.09659, 2018.

[Pla18] Pierre-Guy Plamondon, *τ-tilting finite gentle algebras are representation-finite*, Pacific J. Math. **302** (2019), no. 2, 709–716, DOI 10.2140/pjm.2019.302.709. MR4036747

[PPP17] Yann Palu, Vincent Pilaud, and Pierre-Guy Plamondon. Non-kissing complexes and tau-tilting for gentle algebras. to appear in *Memoirs of the AMS*, arXiv:1707.07574v3, [math.CO].

[PPP18] Yann Palu, Vincent Pilaud, and Pierre-Guy Plamondon, *Non-kissing and non-crossing complexes for locally gentle algebras*, J. Comb. Algebra **3** (2019), no. 4, 401–438, DOI 10.4171/jca/35. MR4046035

[Ric89] Jeremy Rickard, *Morita theory for derived categories*, J. London Math. Soc. (2) **39** (1989), no. 3, 436–456, DOI 10.1112/jlms/s2-39.3.436. MR1002456

[SZ03] Jan Schröer and Alexander Zimmermann, *Stable endomorphism algebras of modules over special biserial algebras*, Math. Z. **244** (2003), no. 3, 515–530, DOI 10.1007/s00209-003-0492-4. MR1992022

[WW85] Burkhard Wald and Josef Waschbüsch, *Tame biserial algebras*, J. Algebra **95** (1985), no. 2, 480–500, DOI 10.1016/0021-8693(85)90119-X. MR801283

Laboratoire de Mathématiques de Versailles, UVSQ, CNRS, Université Paris-Saclay, 78035, Versailles, France.

Email address: pierre-guy.plamondon@uvsq.fr

Contemporary Mathematics
Volume **761**, 2021
https://doi.org/10.1090/conm/761/15319

Socle deformations of selfinjective orbit algebras of tilted type

Andrzej Skowroński and Kunio Yamagata

Dedicated to José Antonio de la Peña on the occasion of his 60th birthday

ABSTRACT. We survey recent development of the study of finite-dimensional selfinjective algebras over a field which are socle equivalent to selfinjective orbit algebras of tilted type. Main aim is to present a characterization of these selfinjective algebras from [**53**] and, as an application, give a new characterization of selfinjective Nakayama algebras socle equivalent to selfinjective orbit algebras of tilted type.

1. Introduction

Throughout this article, by an algebra we mean a basic, connected, finite-dimensional associative K-algebra with identity over a fixed field K. For an algebra A, we denote by $\operatorname{mod} A$ the category of finite-dimensional right A-modules and $\operatorname{ind} A$ the full subcategory of $\operatorname{mod} A$ of indecomposable modules. We denote by Γ_A the Auslander-Reiten quiver of A, and by τ_A and τ_A^{-1} the Auslander-Reiten translations $D\operatorname{Tr}$ and $\operatorname{Tr} D$, respectively, where D is the standard duality $\operatorname{Hom}_K(-, K)$ on $\operatorname{mod} A$ and Tr is the transpose.

An algebra A is called *selfinjective* if A is injective in $\operatorname{mod} A$, or equivalently the projective modules in $\operatorname{mod} A$ are injective. Two selfinjective algebras A and A' are said to be *socle equivalent* if the quotient algebras $A/\operatorname{soc}(A)$ and $A'/\operatorname{soc}(A')$ are isomorphic. In this case, we also call A a *socle deformation* of A'. Moreover, two selfinjective algebras A and A' are called *stably equivalent* if their stable categories $\underline{\operatorname{mod}} A$ and $\underline{\operatorname{mod}} A'$ are equivalent as K-categories.

In the representation theory of selfinjective algebras an important role is played by the selfinjective algebras A which admit Galois covering of the form $\widehat{B} \to \widehat{B}/G = A$, where $\widehat{B}$ is the repetitive category of an algebra B and G is an admissible group of automorphisms of $\widehat{B}$. In this theory, the selfinjective orbit algebras $\widehat{B}/G$ given by algebras B of finite global dimension and infinite cyclic groups G are of special interest. Namely, frequently interesting selfinjective algebras are Morita equivalent

2010 *Mathematics Subject Classification.* Primary 16D50, 16G70; Secondary 16G10, 16E40.

The authors were supported by the research grant DEC-2011/02/A/ST1/00216 of the National Science Center Poland.

The second named author was also supported by JSPS KAKENHI Grant Numbers 16K05091, 19K03417.

The first author passed away on October 22, 2020.

to socle deformations of such orbit algebras, and we may reduce their representation theory to that for the corresponding algebras of finite global dimension. For example, it is the case for selfinjective algebras of polynomial growth over an algebraically closed field K (see Section 2 for details). This is also the case for the tame principal blocks of the enveloping algebras of restricted Lie algebras [**19**], or more general infinitesimal groups [**20**], over algebraically closed fields of odd characteristic. We also note that the prominent class of special biserial algebras over an algebraically closed field is formed by the orbit algebras of repetitive categories (see [**15**], [**18**], [**33**]).

For an arbitrary field K, even the structure of selfinjective algebras of finite representation type over K is far from being understood (see Section 6 for new results toward solution of this problem). In this article, we survey old and new results concerning selfinjective algebras A over a field which are socle equivalent to orbit algebras $\widehat{B}/G$ for an algebra B and G the infinite cyclic group generated by the composition $\varphi\nu_{\widehat{B}}$ of the Nakayama automorphism $\nu_{\widehat{B}}$ and a positive automorphism φ of $\widehat{B}$. In particular, we present, in Sections 3 and 4, criteria for a selfinjective algebra A to have this property. Sections 5 and 6 are devoted to presentations of interesting selfinjective algebras as socle deformations of orbit algebras of repetitive categories of tilted algebras. In the final Section 7 we show that, for the class of selfinjective algebras of tilted type, discussed in this article, stable equivalence and socle equivalence coincide.

For basic background on the representation theory discussed in this article we refer to the books [**3**], [**50**], [**52**], and the survey articles [**41**], [**49**].

2. Repetitive algebras and orbit algebras

Let B be an algebra and $1_B = e_1 + \cdots + e_n$ a decomposition of the identity of B into a sum of pairwise orthogonal primitive idempotents $e_1, \ldots, e_n$ of B. The algebra is regarded as a K-category, denoted by B again, whose objects are $e_1, \ldots, e_n$ and the morphism space $B(e_i, e_j) = \operatorname{Hom}_B(e_i, e_j)$ is $\operatorname{Hom}_B(e_iB, e_jB)$ for $i, j \in \{1, \ldots, n\}$. We then associate to B a selfinjective locally bounded K-category $\widehat{B}$, called the *repetitive category* of B (see [**24**]). The objects of $\widehat{B}$ are $e_{m,i}$ for $m \in \mathbb{Z}$, $i \in \{1, \ldots, n\}$, and the morphism spaces are defined as follows

$$\widehat{B}(e_{m,i}, e_{r,j}) = \begin{cases} e_jBe_i, & r = m, \\ D(e_iBe_j), & r = m+1, \\ 0, & \text{otherwise.} \end{cases}$$

Observe that $e_jBe_i = \operatorname{Hom}_B(e_iB, e_jB)$, $D(e_iBe_j) = e_jD(B)e_i$ and

$$\bigoplus_{(m,i)\in\mathbb{Z}\times\{1,\ldots,n\}} \widehat{B}(e_{m,i}, e_{r,j}) = e_jB \oplus D(Be_j),$$

for any $r \in \mathbb{Z}$ and $j \in \{1, \ldots, n\}$. The composition of morphisms in $\widehat{B}$ is naturally defined by the multiplication in the algebra B and the B-bimodule structure of $D(B)$.

An automorphism φ of the K-category $\widehat{B}$ is said to be *positive* or *rigid*, respectively, if for each $(m, i) \in \mathbb{Z} \times \{1, \ldots, n\}$,

$$\varphi(e_{m,i}) = e_{p,j} \text{ for some } p \geq m \text{ and } j \in \{1, \ldots, n\},$$

or

$$\varphi(e_{m,i}) = e_{m,j} \text{ for some } j \in \{1,\dots,n\}.$$

A *strictly positive* automorphism of $\widehat{B}$ is a positive but not rigid automorphism of $\widehat{B}$. In particular, the *Nakayama automorphism* of $\widehat{B}$, denoted by $\nu_{\widehat{B}}$, is the strictly positive automorphism defined by

$$\nu_{\widehat{B}}(e_{m,i}) = e_{m+1,i} \quad \text{for all} \quad (m,i) \in \mathbb{Z} \times \{1,\dots,n\}.$$

A group G of automorphisms of $\widehat{B}$ is said to be *admissible* if G acts freely on the set of objects of $\widehat{B}$ and has finitely many orbits. Following P. Gabriel [**21**], we may consider the orbit category $\widehat{B}/G$ of $\widehat{B}$ with respect to G, whose objects are the G-orbits of objects in $\widehat{B}$. The morphism space $(\widehat{B}/G)(a,b)$ for objects a,b in $\widehat{B}/G$ is the subspace of $\prod_{(x,y)\in a\times b} \widehat{B}(x,y)$ consisting of those elements $(f_{y,x})$ satisfying

$$gf_{y,x} = f_{gy,gx} \text{ for all } g \in G, (x,y) \in a \times b.$$

By the definition $\widehat{B}/G$ has finitely many objects and the morphism spaces are finite-dimensional. It therefore associates naturally the finite-dimensional selfinjective K-algebra $\bigoplus(\widehat{B}/G)$ which is the direct sum of all morphism spaces in $\widehat{B}/G$, called the *orbit algebra* of $\widehat{B}$ with respect to G (see[**24**]). A typical example is the trivial extension algebra $B \ltimes D(B)$ of an algebra B, which is isomorphic to the orbit algebra $\widehat{B}/(\nu_{\widehat{B}})$. More generally, the *r-fold trivial extension algebra* $T(B)^{(r)}$ of B is defined as the orbit algebra $\widehat{B}/(\nu^r_{\widehat{B}})$, for $r \geq 1$. Note that $T(B)^{(1)} = T(B)$.

A selfinjective algebra A is said to be *of tilted type* if A is isomorphic to an orbit algebra $\widehat{B}/G$ for a tilted algebra B and an admissible automorphism group G of $\widehat{B}$. An important remark is that in this case any admissible automorphism group of $\widehat{B}$ is an infinite cyclic group generated by a strictly positive automorphism of $\widehat{B}$ (see [**49**, Theorem 7.1]). In the paper, by a *tilted algebra* we mean the endomorphism algebra $B = \operatorname{End}_H(T)$ of a tilting module T in the module category $\operatorname{mod} H$ of a basic, connected, hereditary algebra H over a field K.

The classification of all representation-finite selfinjective algebras over an algebraically closed field was given in the early 1980's by C. Riedtmann (see [**13**], [**35**], [**36**], [**37**]) via a combinatorial classification of the Auslander-Reiten quivers of these algebras, based on the fundamental paper [**35**] linking them to the Dynkin quivers. Equivalently, the Riedtmann's classification can be presented in terms of the orbit algebras as follows (see [**41**, Section 3]).

THEOREM 2.1. *Let A be a non-simple, basic, connected, selfinjective algebra over an algebraically closed field K. Then A is of finite representation type if and only if A is socle equivalent to an orbit algebra $\widehat{B}/(\varphi)$, where B is a tilted algebra of Dynkin type and φ is a strictly positive automorphism of $\widehat{B}$.*

In the theorem, we may replace "socle equivalence" by "isomorphism" if $\operatorname{char}(K) \neq 2$, but it is not the case in general (see [**37**] and [**41**, Section 3]).

We would like to stress that a crucial role in proving the above interpretation of the Riedtmann's classification theorem is played by the Galois covering techniques introduced by P. Gabriel in [**21**] and the description of the module categories of repetitive categories of tilted algebras of Dynkin type given by D. Hughes and J. Waschbüsch in [**24**]. This was the starting point for the study of representation-infinite selfinjective algebras of polynomial growth in [**38**], where it was shown that all these algebras, having simply connected Galois coverings, are related to tilted algebras of Euclidean type and tubular algebras. Here, the new results on Galois coverings of representation-infinite algebras proved by P. Dowbor and A. Skowroński in [**14**], [**15**] are heavily applied.

The classification of arbitrary representation-infinite domestic selfinjective algebras was completed in the series of papers [**10**], [**11**], [**12**], [**27**] (see [**41**, Section 4]). In particular, we have the following theorem.

THEOREM 2.2. *Let A be a basic, connected, selfinjective algebra over an algebraically closed field K. Then A is representation-infinite domestic if and only if A is socle equivalent to an orbit algebra $\widehat{B}/(\varphi)$, where B is a tilted algebra of Euclidean type and φ is a strictly positive automorphism of $\widehat{B}$.*

We would like to stress that, for any algebraically closed field K, there are representation-infinite domestic selfinjective algebras which are not orbit algebras of repetitive categories of tilted algebras of Euclidean type (see [**12**] for description of these algebras). We also point out that the structure of the module categories $\operatorname{mod} A$ of the representation-infinite domestic selfinjective algebras A follows from Theorem 2.2 and the results proved in [**2**] and [**38**].

For an algebra A, the *infinite radical* $\operatorname{rad}_A^{\infty}$ of $\operatorname{mod} A$ is the intersection of all powers rad_A^i, $i \geq 1$, of the radical rad_A of $\operatorname{mod} A$. We note that, by a classical result of M. Auslander, $\operatorname{rad}_A^{\infty} = 0$ if and only if A is representation-finite. The following theorem is a consequence of Theorem 2.2 and the main result proved by O. Kerner and A. Skowroński in [**26**, Theorem].

THEOREM 2.3. *Let A be a basic, connected, representation-infinite selfinjective algebra over an algebraically closed field K. Then the infinite radical $\operatorname{rad}_A^{\infty}$ of $\operatorname{mod} A$ is nilpotent if and only if A is socle equivalent to an orbit algebra $\widehat{B}/(\varphi)$, where B is a tilted algebra of Euclidean type and φ is a strictly positive automorphism of $\widehat{B}$.*

The classification of arbitrary non-domestic selfinjective algebras of polynomial growth was completed in the series of papers [**6**], [**7**], [**8**], [**28**] (see [**41**, Section 5] for details). In particular, we have the following theorem.

THEOREM 2.4. *Let A be a basic, connected, selfinjective algebra over an algebraically closed field K. Then A is non-domestic of polynomial growth if and only if A is socle equivalent to an orbit algebra $\widehat{B}/(\varphi)$, where B is a tubular algebra and φ is a strictly positive automorphism of $\widehat{B}$.*

In the theorem, we may replace "socle equivalence" by "isomorphism" if $\operatorname{char}(K)$ is different from 2 and 3 (see [**8**] for details). We also mention that the structure of the module categories $\operatorname{mod} A$ of non-domestic selfinjective algebras A of polynomial growth follows from Theorem 2.4 and the results proved in [**31**] and [**38**].

Recall that an algebra A is called *periodic* if it is periodic with respect to the action of the syzygy operator Ω_{A^e} in the module category $\operatorname{mod} A^e$ over the enveloping algebra $A^e = A^{\mathrm{op}} \otimes_K A$. It is known that if A is periodic then A is selfinjective and all nonprojective indecomposable modules in $\operatorname{mod} A$ are periodic with respect to the action of the syzygy operator Ω_A in $\operatorname{mod} A$ (see [**50**, Proposition IV.11.18, Theorem IV.11.19]).

It is known that every non-simple, basic, connected, representation-finite selfinjective algebra A is periodic, by a theorem proved by A. Dugas in [**16**]. The following theorem is a consequence of Theorem 2.4 and the main result proved by J. Białkowski, K. Erdmann and A. Skowroński in [**5**, Theorem 1.1].

THEOREM 2.5. *Let A be a basic, connected, representation-infinite selfinjective algebra of polynomial growth over an algebraically closed field K. Then A is periodic if and only if A is socle equivalent to an orbit algebra $\widehat{B}/(\varphi)$, where B is a tubular algebra and φ is a strictly positive automorphism of $\widehat{B}$.*

We also note that the algebras B occurring in the above theorems are algebras of global dimension 1 or 2.

Recently, Theorems 2.1, 2.2 and 2.4 have been applied by S. Ariki, R. Kase, K. Miyamoto and K. Wada to provide in [**1**] a complete classification of selfinjective cellular algebras of polynomial growth, over algebraically closed fields K with $\mathrm{char}(K) \neq 2$.

We refer also to [**17**] (respectively, [**29**]) for the representation theory and homological properties of the orbit algebras $\widehat{B}/G$ of tilted algebras B of wild type (respectively, quasi-tilted algebras B of wild canonical type). These algebras B are also of global dimension 1 or 2.

In general, the following problem arises naturally.

PROBLEM. *Describe the basic, connected, selfinjective algebras over an arbitrary field K which are socle equivalent to an orbit algebra $\widehat{B}/G$, where B is an algebra of finite global dimension and G is an infinite cyclic group generated by a strictly positive automorphism of $\widehat{B}$.*

This seems to be a very hard problem in general. In the article, we present recent results concerning the above problem in the following case (coming naturally in several considerations):

- *B is a tilted algebra and G is generated by $\varphi\nu_{\widehat{B}}$ for a positive automorphism φ of $\widehat{B}$.*

To illustrate the situation, let us consider the following simple example.

EXAMPLE 2.6. Let Q_n be the cyclic quiver

$$1 \xrightarrow{\alpha_1} 2 \xrightarrow{\alpha_2} \cdots \xrightarrow{\alpha_{n-1}} n \xrightarrow{\alpha_n} 1, \quad n > 1,$$

and $A_n = KQ_n/N$, where KQ_n is the path algebra and N is the ideal generated by the compositions of all consecutive three arrows. Let $A = A_n$ for simplicity. Then A is a selfinjective Nakayama algebra with $\mathrm{rad}^3(A) = 0$. Take the right module $M = M_1 \oplus M_2$ where $M_1 = e_1A/\mathrm{rad}^2(e_1A)$ and $M_2 = e_2A/\mathrm{rad}\, e_2A$, and the right annihilator $I = r_A(M)$ of M in A. Then

$$I = \mathrm{rad}^2(e_1A) \oplus \mathrm{rad}(e_2A) \oplus e_3A \oplus \cdots \oplus e_nA.$$

The factor algebra $B = A/I$ is isomorphic to the path algebra of the quiver Δ of the form

$$1 \xrightarrow{\alpha_1} 2$$

of Dynkin type $\mathbb{A}_2$. It is easily seen that $A \cong \widehat{B}/(\varphi\nu_{\widehat{B}})$ for a positive automorphism φ of $\widehat{B}$. For example, in the case $n = 5$, we have $\varphi = \sigma\nu_{\widehat{B}}$ where σ is an automorphism satisfying $\sigma^2 = \nu_{\widehat{B}}$.

Thus the selfinjective algebra A_n for arbitrary n is always recovered from the path algebra $B = K\Delta$ and two indecomposable modules M_1, M_2 lying on a slice of Γ_{A_n}.

Before ending this section we would like to give a remark about difference between algebra isomorphism and socle equivalence.

EXAMPLE 2.7. Let F be a finite field extension of a field K such that the Hochschild extension group $H^2(F,F)$ is non-zero, where F is regarded as a K-algebra (see for example [**42**, Example 6.2]). Let Q be an arbitrary finite quiver without oriented cycles and without multiple arrows, and $H = FQ$ the path algebra of Q over F. We consider H as a K-algebra. We note that H being hereditary as an F-algebra is also hereditary as a K-algebra, because the right ideals of the K-algebra H coincide with the right ideals of the F-algebra H. Then it is proved in [**42**, Section 6] and [**45**, Proposition 5] that there exists a Hochschild extension algebra A of the K-algebra H by $D(H) = \operatorname{Hom}_K(H,K)$ such that A is socle equivalent to the trivial extension algebra $H \ltimes D(H)$, but not isomorphic to an orbit algebra of the form $\widehat{B}/(\varphi\nu_{\widehat{B}})$ where B is a K-algebra and φ is a positive automorphism of $\widehat{B}$. Here, note that $H \ltimes D(H) \cong \widehat{H}/(\nu_{\widehat{H}})$ as algebras.

3. Criterion theorems

We recall ring theoretical criterion theorems for a selfinjective algebra A to be socle equivalent to an orbit algebra of tilted type. Throughout this section, A denotes a selfinjective algebra and $\{e_1,\dots,e_r\}$ is a set of pairwise orthogonal primitive idempotents of A with $1_A = e_1 + \dots + e_r$.

For an algebra Λ and a subset X of a right or left Λ-module M, we denote by

$$r_\Lambda(X) = \{\lambda \in \Lambda \mid X\lambda = 0\}, \quad l_\Lambda(X) = \{\lambda \in \Lambda \mid \lambda X = 0\}$$

the right or left annihilator of X in Λ, respectively.

Let I be an ideal of A and $B = A/I$ the factor algebra. We can take an idempotent e of A such that $e + I$ is the identity of B, $e = e_1 + \dots + e_n$ $(n \le r)$ and $\{e_1,\dots,e_n\}$ is the set of all idempotents in $\{e_i \mid 1 \le i \le r\}$ not contained in I. Thus $1_A - e \in I$, and $e_1 + I, \dots, e_n + I$ are pairwise orthogonal primitive idempotents of B and $1_B = \bar{e}_1 + \dots + \bar{e}_n$, where $\bar{e}_i = e_i + I \in B$ for $i \in \{1,\dots,n\}$. By Krull-Schmidt theorem, such an idempotent e is uniquely determined up to inner automorphism of A, and is called a *residual identity* of $B = A/I$.

By Nakayama's theorem [**50**, Theorem IV.6.10] the annihilator operations l_A and r_A induce a Galois correspondence between the lattice $\mathcal{R}_A$ of right ideals and the lattice $\mathcal{L}_A$ of left ideals of A: $l_A : \mathcal{R}_A \to \mathcal{L}_A$, $r_A : \mathcal{L}_A \to \mathcal{R}_A$, each of which is the inverse to the other. An important consequence of this Galois correspondence is the following statement on the residual identity (see [**42**, Proposition 2.3] and [**47**, Lemma 5.1]).

LEMMA 3.1. *Let A be a selfinjective algebra and I an ideal of A. Then an idempotent e of A is a residual identity of $B = A/I$ if $l_A(I) = eI$ or $r_A(I) = Ie$. Moreover, in this case,* $\operatorname{soc} A \subseteq I$ *and $l_{eAe}(I) = eIe = r_{eAe}(I)$.*

Observe that the canonical correspondence $B \to eAe/eIe$, $a + I \mapsto eae + eIe$, is an algebra isomorphism.

In the following first criterion theorem, the implication (ii) $\Rightarrow$ (i) was proved in [**44**] and the converse in [**47**].

THEOREM 3.2. *For a basic, connected, selfinjective algebra A over a field K, the following statements are equivalent:*

(i) *A is isomorphic to an orbit algebra $\widehat{B}/(\varphi\nu_{\widehat{B}})$ for an algebra B and a positive automorphism φ of $\widehat{B}$.*

(ii) *There is an ideal of I of A and an idempotent e of A such that*
(a) $r_A(I) = eI$,
(b) *The canonical algebra homomorphism*

$$\rho : eAe \to eAe/eIe, \ eae \mapsto eae + eIe,$$

is a retraction, i.e,, there is an algebra homomorphism $\eta : eAe/eIe \to eAe$ with $\rho\eta = id_{eAe/eIe}$.

It should be noticed that the statement (ii)(b) holds always in case K is algebraically closed. Under the statement (ii), B in (i) may be chosen as A/I, and the idempotent e in (ii) is determined by I as a residual identity of A/I, see Lemma 3.1.

It was observed that in many important situations we may replace a selfinjective algebra A by its socle deformation satisfying the statement (ii)(b). In order to explain this we need the notion of deforming ideal from [**42**].

DEFINITION 3.3. Let A be a selfinjective algebra, I an ideal of A, and e a residual identity of A/I. Then I is said to be a *deforming ideal* of A if the following conditions are satisfied:

(D1) $l_{eAe}(I) = eIe = r_{eAe}(I)$,
(D2) the valued quiver $Q_{A/I}$ of A/I is acyclic.

The following statement is an immediate consequence of Lemma 3.1 and Definition 3.3, and is important for further considerations.

PROPOSITION 3.4. *Let A be a selfinjective algebra, I an ideal of A and $B = A/I$. Assume that*
(i) $r_A(I) = eI$ *for an idempotent e of A,*
(ii) *the valued quiver Q_B of B is acyclic.*
Then I is a deforming ideal of A and e is a residual identity of B.

Now assume that I is a deforming ideal of a selfinjective algebra A. Then the canonical correspondence $eAe/eIe \to A/I$, $eae + eIe \mapsto eae + I$, is an algebra isomorphism, which makes I as an eAe/eIe-bimodule and allows us to define a new algebra, denoted by $A[I]$, as follows. Let $A[I]$ be the direct sum of K-vector spaces $(eAe/eIe) \oplus I$ and define the multiplication in $A[I]$ by the rule

$$(b, x) \cdot (c, y) = (bc, by + xc + xy)$$

for $b, c \in eAe/eIe$ and $x, y \in I$. Then $A[I]$ is actually a K-algebra with the identity $(e + eIe, 1_A - e)$ and the ideal $\{(0, x) \mid x \in I\}$. By identifying $x \in I$ with $(0, x) \in A[I]$, we may regard I as an ideal of $A[I]$, so that $A[I]/I$ denotes the factor algebra of $A[I]$ by I with the residual identity $e = (e + eIe, 0)$. Thus, by identifying $e \in A$ with $(e, 0) \in A[I]$, we have

$$A[I]/I = eAe/eIe \cong A/I, \quad eA[I]e = (eAe/eIe) \oplus eIe.$$

Moreover, the canonical algebra epimorphism $eA[I]e \to eA[I]e/eIe$ is a retraction, which is obvious from the definition but the main reason why the algebra $A[I]$ has been introduced. In fact, $A[I]$ keeps other important properties of A as shown in the next theorem established in [**42**, Theorem 4.1], [**43**, Theorem 3] and [**48**, Lemma 3.1].

THEOREM 3.5. *Let A be a selfinjective algebra over a field K and I a deforming ideal of A. Then the following statements hold.*

(i) *$A[I]$ is a selfinjective algebra with the same Nakayama permutation as A and I is a deforming ideal of $A[I]$.*
(ii) *A and $A[I]$ are socle equivalent.*
(iii) *A and $A[I]$ are stably equivalent.*
(iv) *$A[I]$ is a symmetric algebra if A is a symmetric algebra.*

We note here that a socle equivalence does not imply a stably equivalence in general, as pointed out by J. Rickard (see [**32**], [**50**]). Moreover, a selfinjective algebra A with a deforming ideal I is not always isomorphic to $A[I]$ (see [**44**, Example 4.2]), but this is the case when K is an algebraically closed field ([**42**, Theorem 3.2]).

The following theorem proved in [**44**, Theorem 4.1] shows the importance of the algebras $A[I]$.

THEOREM 3.6. *Let A be a selfinjective algebra, I an ideal of A, $B = A/I$ and e an idempotent of A. Assume that $r_A(I) = eI$ and Q_B is acyclic. Then the following statements hold:*

(i) *$A[I]$ is isomorphic to an orbit algebra $\widehat{B}/(\varphi\nu_{\widehat{B}})$ for some positive automorphism φ of $\widehat{B}$.*
(ii) *A is a socle deformation of $\widehat{B}/(\varphi\nu_{\widehat{B}})$ for some positive automorphism φ of $\widehat{B}$.*

PROOF. First note that, by Proposition 3.4, the ideal I is a deforming ideal of A, so $A[I]$ is well defined.

(ii) follows from (i) and Theorem 3.5 (ii). To show (i), we apply Theorem 3.2 to $A[I]$ and its ideal I. In fact, it is seen that $r_{A[I]}(I) = eI$ in $A[I]$, and the canonical algebra homomorphism $eA[I]e \to eA[I]e/eIe$ is a retraction. As a result, the conditions (a) and (b) in (ii) of Theorem 3.2 are satisfied, which ensures the existence of an algebra isomorphism $A[I] \to \widehat{B}/(\varphi\nu_{\widehat{B}})$ for a positive automorphism φ of $\widehat{B}$. □

In the theorem, the algebra A is not necessarily isomorphic to an orbit algebra $\widehat{B}/(\varphi\nu_{\widehat{B}})$, where B is an algebra and φ is a positive automorphism of $\widehat{B}$ (see [**45**, Proposition 4]).

The following result proved in [**46**, Proposition 3.2] describes a situation when the algebras A and $A[I]$ are isomorphic.

THEOREM 3.7. *Let A be a selfinjective algebra over a field K, having a deforming ideal I, $B = A/I$, e be a residual identity of B, and ν the Nakayama permutation of A. Assume that $IeI = 0$ and $e_i \neq e_{\nu(i)}$, for any primitive summand e_i of e. Then the algebras A and $A[I]$ are isomorphic. In particular, A is isomorphic to an orbit algebra $\widehat{B}/(\varphi\nu_{\widehat{B}})$ for some positive automorphism φ of $\widehat{B}$.*

4. Hereditary stable slice

In this section we explain a new characterization from [**53**] of the socle deformations of selfinjective orbit algebras $\widehat{B}/G$ of tilted type where G is an admissible group generated by $\varphi\nu_{\widehat{B}}$ for a positive automorphism φ of $\widehat{B}$.

Let A be a selfinjective algebra over a field K and Γ_A^s the stable Auslander-Reiten quiver of A, obtained from Γ_A by removing the projective modules and the arrows attached to them. By $\operatorname{ind}\mathcal{P}(A)$ we understand the family of indecomposable projective modules in $\operatorname{mod} A$.

Following [**51**], a full valued subquiver Δ of Γ_A is said to be a *stable slice* if the following conditions are satisfied:

(1) Δ is connected, acyclic, and without projective modules.

(2) For any valued arrow $V \xrightarrow{(a,a')} U$ in Γ_A with U in Δ and V non-projective, V belongs to Δ or to $\tau_A\Delta$.

(3) For any valued arrow $U \xrightarrow{(b,b')} V$ in Γ_A with U in Δ and V non-projective, V belongs to Δ or to $\tau_A^{-1}\Delta$.

A stable slice Δ of Γ_A is said to be *right regular* (respectively, *left regular*) if Δ does not contain the radical $\operatorname{rad} P$ (respectively, the socle factor $P/\operatorname{soc} P$) of any P from $\operatorname{ind}\mathcal{P}(A)$. A stable slice Δ is said to be *almost right regular* if for any P from $\operatorname{ind}\mathcal{P}(A)$ with $\operatorname{rad} P$ lying on Δ, $\operatorname{rad} P$ is a sink of Δ, and *almost left regular* if for any $P \in \operatorname{ind}\mathcal{P}(A)$ with $P/\operatorname{soc} P$ lying on Δ, $P/\operatorname{soc} P$ is a source of Δ. Moreover, for a finite stable slice Δ, we denote by $M(\Delta)$ the direct sum of all modules lying on Δ and $H(\Delta) = \operatorname{End}_A(M(\Delta))$ the endomorphism algebra of $M(\Delta)$. Then a finite stable slice Δ of Γ_A is said to be *hereditary* if the endomorphism algebra $H(\Delta) = \operatorname{End}_A(M(\Delta))$ of $M(\Delta)$ is a hereditary algebra and its valued quiver $Q_{H(\Delta)}$ is the opposite quiver Δ^{op} of Δ.

The following theorem is proved in [**53**, Theorem 1.1] and extends results established in [**44**] and [**51**] to a general case, as explained in the next section.

THEOREM 4.1. *Let A be a selfinjective algebra over a field K. The following statements are equivalent:*

(i) *Γ_A admits a hereditary almost right regular stable slice.*

(ii) *Γ_A admits a hereditary almost left regular stable slice.*

(iii) *A is socle equivalent to an orbit algebra $\widehat{B}/(\varphi\nu_{\widehat{B}})$, where B is a tilted algebra and φ is a positive automorphism of $\widehat{B}$.*

It should be stressed that, if K is algebraically closed, we may replace in (iii) "socle equivalent" by "isomorphic", but the replacement is not possible in general.

IDEA OF THE PROOF. The equivalence of (i) and (ii) is easy, because a selfinjective algebra A satisfies (i) if and only if A^{op} satisfies (ii). Similarly, A satisfies (iii) if and only if A^{op} satisfies (iii) in view of the canonical algebra isomorphism

$$(\widehat{B}/(\varphi\nu_{\widehat{B}}))^{\mathrm{op}} \cong \widehat{B^{\mathrm{op}}}/(\psi\nu_{\widehat{B^{\mathrm{op}}}})$$

for a positive automorphism ψ of $\widehat{B^{\mathrm{op}}}$.

Now we explain how to define the algebra B in (iii) under the assumption (i) and conversely how to find an almost right regular stable slice being hereditary in (i) under the condition (iii).

Assume that a selfinjective algebra A has a hereditary and almost right regular stable slice Δ in the Auslander-Reiten quiver Γ_A. Take the direct sum M of all indecomposable modules in $\operatorname{mod} A$ lying on Δ, and let I be the right annihilator $r_A(M) = \{a \in A \mid Ma = 0\}$ of M, $B = A/I$ the factor algebra of A, and $H = \operatorname{End}_A(M)$. By the definition of hereditary stable slice, H is a hereditary algebra and the valued quiver Q_H of H is the opposite quiver Δ^{op} of Δ. Then it is shown in [**53**, Section 3] that B is a desired tilted algebra and I a deforming ideal of A satisfying $r_A(I) = eI$, so that Theorem 3.6 implies the assertion (iii).

Conversely, assume that a selfinjective algebra A satisfies the assertion (iii). We shall show how we find a hereditary and almost right regular stable slice Δ in Γ_A. Let A be socle equivalent to $\widehat{B}/(\varphi\nu_{\widehat{B}})$ as in (iii), and $\Lambda = \widehat{B}/(\varphi\nu_{\widehat{B}})$.

We observe that, if Λ has a stable slice Δ in Γ_Λ being hereditary and almost right regular, then A has a hereditary almost right regular stable slice, which corresponds to Δ under the given isomorphism $A/\operatorname{soc}(A) \to \Lambda/\operatorname{soc}(\Lambda)$, so A may be identified with Λ, that is, $A = \widehat{B}/(\varphi\nu_{\widehat{B}})$. We consider two cases:

(1) Assume A is of infinite representation type. Since B is not of finite representation type, by [**24**], [**23**], the tilted algebra B is not of Dynkin type. It follows from general theory developed in [**2**], [**17**], [**44**] and [**45**] that Γ_A admits an acyclic component $\mathcal{C}$ containing a right stable full translation subquiver $\mathcal{D}$ which is closed under successors in $\mathcal{C}$ and generalized standard (see below for definition). Then any stable slice Δ in $\mathcal{D}$ is a *hereditary* and *right regular* stable slice.

(2) Assume A is of finite representation type.

(a) First consider the case when A is a selfinjective Nakayama algebra. Then, for any indecomposable projective module P, the module $P/\operatorname{soc} P$ is always the radical of a projective module. For an arbitrarily chosen indecomposable projective module P, consider the sectional path

$$\Delta: \ \operatorname{soc} P = X_1 \to X_2 \to \cdots \to X_{n-1} \to X_n = \operatorname{rad} P$$

of canonical irreducible monomorphisms. Then Δ is a *hereditary* and *almost right regular* stable slice Γ_A.

(b) Next assume that A is of finite representation type but not a Nakayama algebra. Observe that then there exists an indecomposable projective module P such that $P/\operatorname{soc} P$ is not the radical of a projective module, which allows us to take the full subquiver Δ_P of Γ_A whose vertices are $\tau_A^{-1}(P/\operatorname{soc} P)$ and the indecomposable modules X such that there is a non-trivial sectional path in Γ_A^s from $P/\operatorname{soc} P$ to X. It is shown in [**53**, Proposition 4.4] that $\Delta = \tau_A(\Delta_P)$ is a *hereditary* and *right regular* stable slice [**25**, Theorem 3.1]. □

We recall some definitions from [**51**]. Let A be a selfinjective K-algebra. A stable slice Δ of Γ_A is said to be *semi-regular* if Δ is left or right regular, and *regular* if Δ is left and right regular. Moreover, a stable slice Δ is *double τ_A-rigid* if $\operatorname{Hom}_A(X, \tau_A Y) = 0$ and $\operatorname{Hom}_A(\tau_A^{-1} X, Y) = 0$ for all indecomposable modules X and Y from Δ. Note that a double τ_A-rigid stable slice Δ is finite ([**40**], [**3**, Lemma VIII.5.3]) and a full valued subquiver of a connected component $\mathcal{C}$ of Γ_A^s intersecting every τ_A-orbit in $\mathcal{C}$ exactly once. The following theorem proved in [**51**, Theorem 1] is a special case of Theorem 4.1.

THEOREM 4.2. *Let A be a basic, connected, selfinjective algebra over a field K. The following statements are equivalent:*

(i) *Γ_A admits a semi-regular double τ_A-rigid stable slice.*
(ii) *A has one of the following forms:*
  (a) *$A \cong \widehat{B}/(\varphi\nu_{\widehat{B}})$ as algebras, where $B = \operatorname{End}_H(T)$ for a hereditary algebra H and a tilting module T in $\operatorname{mod} H$ either without nonzero*

projective direct summand or without nonzero injective direct summand, and φ is a strictly positive automorphism of $\widehat{B}$.

(b) *A is socle equivalent to $\widehat{B}/(\varphi\nu_{\widehat{B}})$, where $B = \operatorname{End}_H(T)$ for a hereditary algebra H and a tilting module T in $\operatorname{mod} H$ without nonzero projective or injective direct summands, and φ is a rigid automorphism of $\widehat{B}$.*

Moreover, if K is an algebraically closed field, we may replace in (ii)(b) *"socle equivalent" by "isomorphic".*

We note that in general we cannot replace in (ii)(b) "socle equivalent" by "isomorphic", see comments below Theorem 5.2.

5. Selfinjective orbit algebras of infinite representation type

The first study of the socle deformations of a selfinjective orbit algebra over an arbitrary field K was given in [**42**], [**44**], where mainly representation-infinite selfinjective algebras are considered. In this section, we present two theorems on representation-infinite socle deformations of a selfinjective orbit algebra of tilted type.

Following [**39**], a full translation subquiver Σ of the Auslander-Reiten quiver Γ_Λ of an algebra Λ is said to be *generalized standard* if $\bigcap_{m>0} \operatorname{rad}^m \Sigma = 0$. The following theorem is from [**42**] and [**44**].

THEOREM 5.1. *Let A be a basic, connected, selfinjective algebra over a field K. The following statements are equivalent:*

(i) *Γ_A admits an acyclic generalized standard right stable full translation subquiver Σ which is closed under successors in Γ_A.*

(ii) *Γ_A admits an acyclic generalized standard left stable full translation subquiver Ω which is closed under predecessors in Γ_A.*

(iii) *A is socle equivalent to an orbit algebra $\widehat{B}/(\varphi\nu_{\widehat{B}})$, where B is a tilted algebra not of Dynkin type and φ is a positive automorphism of $\widehat{B}$.*

Moreover, if K is an algebraically closed field, we may replace in (iii) *"socle equivalent" by "isomorphic".*

In order to apply Theorem 4.1, let Σ be the generalized standard translation subquiver given in the statement (i), and let I be the annihilator $\operatorname{ann}_A \Sigma$ of Σ in A and $B = A/I$. Then it is shown in [**42**] that I is a deforming ideal of A such that

$$r_A(I) = eI \quad \text{and} \quad l_A(I) = Ie,$$

for a residual identity e of B. It should be noted that $\operatorname{ann}_A(\Sigma)$ is the same as the annihilator in A of any section in Σ.

The following theorem from [**51**, Theorem 2] shows another characterization of the socle deformations of a representation-infinite orbit algebra $\widehat{B}/(\varphi\nu_{\widehat{B}})$ of tilted type, and is a consequence of Theorem 4.2.

THEOREM 5.2. *Let A be a basic, connected, selfinjective algebra of infinite representation type over a field K. The following statements are equivalent:*

(i) *Γ_A admits a regular double τ_A-rigid stable slice.*

(ii) *A is socle equivalent to an orbit algebra $\widehat{B}/(\varphi\nu_{\widehat{B}})$, where B is a tilted algebra not of Dynkin type and φ is a positive automorphism of $\widehat{B}$.*

Moreover, if K is an algebraically closed field, we may replace in (ii) *"socle equivalent" by "isomorphic".*

In Example 2.7, if we take Q a quiver of non-Dynkin type and $T = H_H$, then $B = \mathrm{End}_H(H) = H$ is a tilted algebra not of Dynkin type and, as explained in Example 2.7, A is socle equivalent to $\widehat{H}/(\nu_{\widehat{H}})$ but not isomorphic to any orbit algebra of the form $\widehat{B}/(\varphi\nu_{\widehat{B}})$. Moreover, in this case, obviously, A is of infinite representation type, as H is. Thus, in the above theorems, "socle equivalent" cannot be replaced by "isomorphic" in general.

6. Selfinjective orbit algebras of finite representation type

In this section, we present recent results concerning the structure of selfinjective algebras of finite representation type, which are socle equivalent to selfinjective orbit algebras of tilted algebras of Dynkin type.

We first recall the following old result proved by C. Riedtmann [**35**] and G. Todorov [**54**] describing the structure of the stable Auslander-Reiten quiver of a selfinjective algebra of finite representation type (see [**50**, Theorem IV.15.6] for a proof).

THEOREM 6.1. *Let A be a non-simple, basic, connected, selfinjective algebra of finite representation type over a field K. Then the stable Auslander-Reiten quiver Γ_A^s of A is isomorphic to the orbit valued translation quiver $\mathbb{Z}\Delta/G$, where Δ is a Dynkin quiver of type $\mathbb{A}_n (n \geq 1)$, $\mathbb{B}_n (n \geq 2)$, $\mathbb{C}_n (n \geq 3)$, $\mathbb{D}_n (n \geq 4)$, $\mathbb{E}_6$, $\mathbb{E}_7$, $\mathbb{E}_8$, $\mathbb{F}_4$ and $\mathbb{G}_2$, and G is an admissible infinite cyclic group of automorphisms of $\mathbb{Z}\Delta$.*

We note that if $A = \widehat{B}/(\varphi)$ is an orbit algebra with B a tilted algebra of Dynkin type Δ over a field K and φ is a strictly positive automorphism of $\widehat{B}$ then $\Gamma_A^s \cong \mathbb{Z}\Delta/G$, where G is the infinite cyclic group of automorphisms of $\mathbb{Z}\Delta$ induced by φ. It would be interesting to know when a selfinjective algebra A' over an arbitrary field K is socle equivalent to a selfinjective orbit algebra $A = \widehat{B}/(\varphi)$ of Dynkin type. It follows from the classification result of C. Riedtmann, presented in Section 2, that all representation-finite selfinjective algebras over an algebraically closed field have this property. But for an arbitrary field K, this seems to be a hard problem.

We consider first selfinjective Nakayama algebras. A module X in a module category $\mathrm{mod}\, A$ is said to be *composition free* if all simple composition factors of X occur with multiplicity one. The following result is a consequence of Theorem 4.1.

THEOREM 6.2. *Let A be a non-simple, basic, connected, selfinjective Nakayama algebra over a field K. Then the following statements are equivalent:*

(i) *Any indecomposable projective module P in* $\mathrm{mod}\, A$ *has composition free radical* $\mathrm{rad}\, P$.
(ii) *Any indecomposable projective module P in* $\mathrm{mod}\, A$ *has composition free socle factor* $P/\mathrm{soc}\, P$.
(iii) *A is socle equivalent to an orbit algebra $\widehat{H}/(\varphi\nu_{\widehat{H}})$, where H is a hereditary Nakayama algebra and φ is a positive automorphism of $\widehat{H}$.*
(iv) *A is socle equivalent to an orbit algebra $\widehat{B}/(\varphi\nu_{\widehat{B}})$, where B is an algebra and φ is a positive automorphism of $\widehat{B}$.*

PROOF. The equivalence (i) $\Leftrightarrow$ (ii) and the implication (iii) $\Rightarrow$ (iv) are trivial.

Assume that (i) holds. Let P be an indecomposable projective module in $\operatorname{mod} A$. Since A is Nakayama, we have in Γ_A the sectional path

$$\Delta : \quad \operatorname{soc} P = X_1 \to X_2 \to \cdots \to X_{n-1} \to X_n = \operatorname{rad} P,$$

given by the irreducible inclusion monomorphisms, which is an almost right regular stable slice of Γ_A. Moreover, the slice Δ is hereditary, because $\operatorname{rad} P$ is composition free. Therefore, applying arguments from the proof of Theorem 4.1, we conclude that A is socle equivalent to an orbit algebra $\widehat{H}/(\varphi\nu_{\widehat{H}})$ for a hereditary Nakayama algebra H with the Gabriel quiver $Q_H = \Delta = \Delta^{\mathrm{op}}$ and a positive automorphism φ of $\widehat{H}$. Hence (iii) is satisfied.

Assume that (iii) holds. Let H be a hereditary Nakayama algebra and φ is a positive automorphism of $\widehat{H}$ such that the orbit algebra $\Lambda = \widehat{H}/(\varphi\nu_{\widehat{H}})$ is socle equivalent to A. Without loss of generality, we may assume that the quotient algebras $\Lambda/\operatorname{soc}(\Lambda)$ and $A/\operatorname{soc}(A)$ are equal. Let $1_H = e_1 + \cdots + e_n$ be a decomposition of the identity of H into a sum of pairwise orthogonal primitive idempotents. Then it follows from the definition of $\widehat{H}$ that every indecomposable projective $\widehat{H}$-module $e_{m,i}\widehat{H}$, $(m,i) \in \mathbb{Z} \times \{1,\dots,n\}$, is uniserial. Moreover, for any (m,i) and (r,j) in $\mathbb{Z} \times \{1,\dots,n\}$ with $m \neq r$, the socle factors $e_{m,i}\widehat{H}/\operatorname{soc}(e_{m,i}\widehat{H})$ and $e_{r,j}\widehat{H}/\operatorname{soc}(e_{r,j}\widehat{H})$ have disjoint simple composition factors. Then it follows that any indecomposable projective module P in $\operatorname{mod} \Lambda$ has composition free socle factor $P/\operatorname{soc} P$. Since the socle factors of indecomposable projective modules in $\operatorname{mod} \Lambda$ and $\operatorname{mod} A$ are exactly the indecomposable projective modules in $\operatorname{mod} \Lambda/\operatorname{soc}(\Lambda) = \operatorname{mod} A/\operatorname{soc}(A)$, we conclude that A satisfies (ii).

Assume now that (iv) holds, that is, A is socle equivalent to an orbit algebra $\Lambda = \widehat{B}/(\varphi\nu_{\widehat{B}})$, where B is an algebra and φ is a positive automorphism of $\widehat{B}$. Clearly, such an algebra B is basic and connected. We claim that B is a hereditary Nakayama algebra, and hence A satisfies (iii). Since A is a Nakayama algebra and $A/\operatorname{soc}(A) \cong \Lambda/\operatorname{soc}(\Lambda)$, we conclude that Λ is a non-simple selfinjective Nakayama algebra. We also note that B is a quotient algebra of Λ, because Λ is an orbit algebra of $\widehat{B}$ with respect to $(\varphi\nu_{\widehat{B}})$ with φ positive. Therefore, we obtain that B is a Nakayama algebra. Hence, we have only to show that any non-simple projective module in $\operatorname{mod} B$ has projective socle. Let $1_B = e_1 + \cdots + e_n$ be a decomposition of the identity of B into a sum of pairwise orthogonal primitive idempotents. Remember that $e_{m,i}$, $(m,i) \in \mathbb{Z} \times \{1,\dots,n\}$, are pairwise orthogonal primitive idempotents of $\widehat{B}$ and $e_{m,i}\widehat{B} = e_iB \oplus D(Be_i)$. For convenience, we write B_m for $1_m\widehat{B}1_m$ where $1_m = e_{m,1} + \cdots + e_{m,n}$, and $\mathbb{D}(\widehat{B}) = \bigoplus_{m\in\mathbb{Z}} D(B_m)$. Note that B_m is a copy of B, and observe that

$$\operatorname{rad} \widehat{B} = \Big(\bigoplus_{m\in\mathbb{Z}} \operatorname{rad} B_m\Big) \oplus \mathbb{D}(\widehat{B}).$$

Since $\Lambda = \widehat{B}/(\varphi\nu_{\widehat{B}})$ is a Nakayama algebra, each $e_{m,i}\widehat{B}$ and so e_iB is uniserial for all $m \in \mathbb{Z}$, $i \in \{1,\dots,n\}$. Now let P be a non-simple indecomposable projective module in $\operatorname{mod} B$, and assume $P = e_1B$, without loss of generality. Let

$$\operatorname{soc} P \cong \operatorname{top}(e_tB)$$

for some $t \in \{1, \dots, n\}$. Take the $\widehat{B}$-submodule $M = \operatorname{soc}(e_1 B) \oplus D(Be_1)$ of $e_{0,1}\widehat{B} = e_1 B \oplus D(Be_1)$. Then $M(\operatorname{rad}\widehat{B}) = D(Be_1)$, because $M/D(Be_1)$ is a simple $\widehat{B}$-module and M is uniserial. Let $f : e_{0,t}\widehat{B} \to e_{0,1}\widehat{B}$ be a $\widehat{B}$-homomorphism with $f(e_{0,t}\widehat{B}) = M$. It is shown that

$$M \cdot \mathbb{D}(\widehat{B}) = D(Be_1)$$

taking into account the uniseriality of $e_{0,t}\widehat{B}$. On the other hand, $f(e_{0,t}\widehat{B} \cdot \operatorname{rad}\widehat{B}) = M(\operatorname{rad}\widehat{B})$ and $f(e_{0,t}\widehat{B} \cdot \mathbb{D}(\widehat{B})) = M \cdot \mathbb{D}(\widehat{B})$. Consequently, we have $f(e_{0,t}\widehat{B} \cdot \operatorname{rad}\widehat{B}) = f(e_{0,t}\widehat{B} \cdot \mathbb{D}(\widehat{B}))$, which implies

$$e_{0,t}\widehat{B} \cdot \operatorname{rad}\widehat{B} = e_{0,t}\widehat{B} \cdot \mathbb{D}(\widehat{B}) + \operatorname{Ker} f = e_{0,t}\widehat{B} \cdot \mathbb{D}(\widehat{B}).$$

Indeed, since $f(e_{0,t}\widehat{B} \cdot \mathbb{D}(\widehat{B})) = D(Be_1) \neq 0$ and $e_{0,t}\widehat{B}$ is uniserial, $\operatorname{Ker} f$ is contained in $e_{0,t}\widehat{B} \cdot \mathbb{D}(\widehat{B})$. As a result, we conclude that

$$\operatorname{top}(e_t B) = e_{0,t}\widehat{B}/(e_{0,t}\widehat{B} \cdot \operatorname{rad}\widehat{B}) = e_{0,t}\widehat{B}/(e_{0,t}\widehat{B} \cdot \mathbb{D}(\widehat{B})) = e_t B,$$

and hence $\operatorname{soc} P$ is projective in $\operatorname{mod} B$, as desired. □

As in the last paragraph of the previous section, if in Example 2.7, we take Q a quiver of type $\mathbb{A}_n$, $n > 1$, with linear orientation of arrows, then H is a hereditary Nakayama algebra and hence it follows from [**55**, Corollary 2.6] (or [**52**, Theorem 7.12]) that the Hochschild extension algebra A of H by $D(H)$ is again a Nakayama algebra. As explained in Example 2.7, however, A is socle equivalent to $\widehat{H}/(\nu_{\widehat{H}})$ but not isomorphic to any orbit algebra of the form $\widehat{B}/(\varphi\nu_{\widehat{B}})$. Thus, in the above theorem, "socle equivalent" cannot be replaced by "isomorphic" in general.

We obtain also the following consequence of Theorems 3.7 and 6.2.

COROLLARY 6.3. *Let A be a non-simple, basic, connected, selfinjective Nakayama algebra over a field K. Then the following statements are equivalent:*

(i) *Any indecomposable projective module P in $\operatorname{mod} A$ is composition free.*
(ii) *A is isomorphic to an orbit algebra $\widehat{H}/(\varphi\nu_{\widehat{H}})$, where H is a hereditary Nakayama algebra and φ is a strictly positive automorphism of $\widehat{H}$.*
(iii) *A is isomorphic to an orbit algebra $\widehat{B}/(\varphi\nu_{\widehat{B}})$, where B is an algebra and φ is a strictly positive automorphism of $\widehat{B}$.*

Following [**34**] a *short cycle* in a module category $\operatorname{mod} A$ is a sequence $X \to Y \to X$ of non-isomorphisms between two indecomposable modules X and Y in $\operatorname{mod} A$. It was shown in [**34**, Corollary 2.2] that if an indecomposable module M does not lie on a short cycle in $\operatorname{mod} A$ then M is uniquely determined (up to isomorphism) by its composition factors, that is, by its image $[M]$ in the Grothendieck group $K_0(A)$. Moreover, it is known that if $\operatorname{mod} A$ has no short cycles, then A is of finite representation type [**22**]. We have also the following fact proved in [**25**, Lemma 3.2].

LEMMA 6.4. *Let A be a selfinjective algebra which does not admit a short cycle in $\operatorname{mod} A$. Then all stable slices in Γ_A are double τ_A-rigid.*

We also recall the following fact proved in [**25**, Theorem 3.1].

THEOREM 6.5. *Let A be a non-simple, connected, selfinjective algebra of finite representation type over a field K. The following statements are equivalent:*

(i) Γ_A *admits a semi-regular stable slice.*
(ii) A *is not a Nakayama algebra.*

The following theorem was proved by A. Jaworska-Pastuszak and A. Skowroński in [**25**].

THEOREM 6.6. *Let A be a non-simple, basic, connected, selfinjective algebra over a field K. The following statements are equivalent:*

(i) $\operatorname{mod} A$ *has no short cycles.*
(ii) A *is isomorphic to an orbit algebra $\widehat{B}/(\psi\nu_{\widehat{B}}^2)$, where B is a tilted algebra of Dynkin type and ψ is a strictly positive automorphism of $\widehat{B}$.*

It was shown in [**25**, Theorem 3.7] that the equivalence of (i) and (ii) holds if $A = \widehat{B}/G$ for a tilted algebra B of Dynkin type and an admissible group G of automorphisms of $\widehat{B}$, applying the fact that the push-down functor $F_\lambda : \operatorname{mod} \widehat{B} \to \operatorname{mod} A$, induced by the canonical Galois covering functor $F : \widehat{B} \to \widehat{B}/G = A$, is a Galois covering of module categories. We explain now why (i) implies (ii). Therefore, assume that $\operatorname{mod} A$ has no short cycles. By the above comments, it is enough to show that A is isomorphic to an orbit algebra $\widehat{B}/(\varphi\nu_{\widehat{B}})$ for a tilted algebra B of Dynkin type and a strictly positive automorphism φ of $\widehat{B}$. If A is not a Nakayama algebra, this follows from Theorems 4.2 and 6.5, and Lemma 6.4. Assume A is a Nakayama algebra. Since every indecomposable projective module P in $\operatorname{mod} A$ does not lie on a short cycle, we easily conclude that P is composition free, and then A has a required form $\widehat{B}/(\varphi\nu_{\widehat{B}})$ by Theorem 6.2.

In [**9**] M. Błaszkiewicz and A. Skowroński investigated the structure of selfinjective algebras of finite representation type whose module category admits maximal almost split sequences. For an algebra A and an almost split sequence

$$0 \longrightarrow \tau_A X \longrightarrow Y \longrightarrow X \longrightarrow 0$$

in $\operatorname{mod} A$, we may consider the numerical invariant $\alpha(X)$ of X being the number of summands in a decomposition $Y = Y_1 \oplus \cdots \oplus Y_r$ of Y into a direct sum of indecomposable modules. Then $\alpha(X)$ measures the complexity of homomorphisms in $\operatorname{mod} A$ with domain $\tau_A X$ and codomain X. It has been proved by R. Bautista and S. Brenner in [**4**] (see also [**30**] for an alternative proof) that if A is of finite representation type and X is an indecomposable nonprojective module in $\operatorname{mod} A$, then $\alpha(X) \leq 4$, and if $\alpha(X) = 4$, then the middle term Y of an almost split sequence in $\operatorname{mod} A$ with the right term X admits an indecomposable projective-injective direct summand. Recall that (by general theory) if P is an indecomposable projective-injective module in $\operatorname{mod} A$, then there is in $\operatorname{mod} A$ an almost split sequence of the form

$$0 \longrightarrow \operatorname{rad} P \longrightarrow (\operatorname{rad} P/\operatorname{soc} P) \oplus P \longrightarrow P/\operatorname{soc} P \longrightarrow 0.$$

If A is of finite representation type, then such an almost split sequence with $\alpha(P/\operatorname{soc} P) = 4$ is called a *maximal almost split sequence* in $\operatorname{mod} A$.

Let A be a selfinjective algebra of finite representation type, and assume that $\operatorname{mod} A$ admits an indecomposable projective-injective module P with $\alpha(P/\operatorname{soc} P) = 4$. We denote by Δ_P the full valued subquiver of Γ_A given by the module $\tau_A^{-1}(P/\operatorname{soc} P)$ and all indecomposable modules in $\operatorname{mod} A$ such that there is a non-trivial sectional path in Γ_A from $P/\operatorname{soc} P$ to X. It was shown in [**9**, Theorem 5.2] that Δ_P is a Dynkin quiver such that $\Gamma_A^s = \mathbb{Z}\Delta_P/G$ for an admissible automorphism group G

of the translation quiver $\mathbb{Z}\Delta_P$. We denote by M_P the direct sum of all modules lying on Δ_P. Then the main theorem in [**9**, Theorems 1 and 2] can be formulated as follows.

THEOREM 6.7. *Let A be a basic, connected, selfinjective algebra over a field K. Then the following statements are equivalent:*

(i) *A is of finite representation type and admits an indecomposable projective module P with $\alpha(P/\operatorname{soc} P) = 4$ and $\operatorname{Hom}_A(M_P, \tau_A M_P) = 0$.*

(ii) *A is socle equivalent to an orbit algebra $\widehat{B}/(\varphi\nu_{\widehat{B}}^m)$ for some positive integer m, a rigid automorphism φ of $\widehat{B}$, and a tilted algebra B of Dynkin type having an indecomposable projective module Q whose top is injective and the radical is a direct sum of three indecomposable projective modules.*

We note that, if K is algebraically closed, then the condition $\operatorname{Hom}_A(M_P, \tau_A M_P) = 0$ in (i) is superfluous, and we may replace in (ii) "socle equivalent" by "isomorphic". We refer to [**9**, Section 3] for a complete description of the tilted algebras B of Dynkin type occurring in (ii).

7. Stable equivalences of selfinjective orbit algebras

We end this article with the following combination of [**43**, Theorem 1] and [**48**, Theorem 1], and its consequence.

THEOREM 7.1. *Let A be a non-simple, basic, connected, selfinjective algebra over a field K. Then the following statements are equivalent:*

(i) *A is stably equivalent to an orbit algebra $\widehat{B}/(\varphi\nu_{\widehat{B}})$, where B is a tilted algebra and φ is a positive automorphism of $\widehat{B}$.*

(ii) *A is socle equivalent to an orbit algebra $\widehat{B}/(\varphi\nu_{\widehat{B}})$, where B is a tilted algebra and φ is a positive automorphism of $\widehat{B}$.*

Moreover, if K is an algebraically closed field, we may replace in (ii) *"socle equivalent" by "isomorphic".*

COROLLARY 7.2. *Let K be an algebraically closed field. Then the class of orbit algebras $\widehat{B}/(\varphi\nu_{\widehat{B}})$, where B is a tilted algebra over K and φ is a positive automorphism of $\widehat{B}$, is closed under stable equivalences.*

References

[1] S. Ariki, R. Kase, K. Miyamoto, and K. Wada, *Self-injective cellular algebras whose representation type are tame of polynomial growth*, Algebr. Represent. Theory **23** (2020), no. 3, 833–871, DOI 10.1007/s10468-019-09872-w. MR4109140

[2] Ibrahim Assem, Jerzy Nehring, and Andrzej Skowroński, *Domestic trivial extensions of simply connected algebras*, Tsukuba J. Math. **13** (1989), no. 1, 31–72, DOI 10.21099/tkbjm/1496161006. MR1003591

[3] Ibrahim Assem, Daniel Simson, and Andrzej Skowroński, *Elements of the representation theory of associative algebras. Vol. 1*, London Mathematical Society Student Texts, vol. 65, Cambridge University Press, Cambridge, 2006. Techniques of representation theory. MR2197389

[4] Raymundo Bautista and Sheila Brenner, *On the number of terms in the middle of an almost split sequence*, Representations of algebras (Puebla, 1980), Lecture Notes in Math., vol. 903, Springer, Berlin-New York, 1981, pp. 1–8. MR654699

[5] Jerzy Białkowski, Karin Erdmann, and Andrzej Skowroński, *Periodicity of self-injective algebras of polynomial growth*, J. Algebra **443** (2015), 200–269, DOI 10.1016/j.jalgebra.2015.05.033. MR3400401

[6] Jerzy Białkowski and Andrzej Skowroński, *Selfinjective algebras of tubular type*, Colloq. Math. **94** (2002), no. 2, 175–194, DOI 10.4064/cm94-2-2. MR1967373

[7] Jerzy Białkowski and Andrzej Skowroński, *On tame weakly symmetric algebras having only periodic modules*, Arch. Math. (Basel) **81** (2003), no. 2, 142–154, DOI 10.1007/s00013-003-0816-y. MR2009556

[8] Jerzy Białkowski and Andrzej Skowroński, *Socle deformations of selfinjective algebras of tubular type*, J. Math. Soc. Japan **56** (2004), no. 3, 687–716, DOI 10.2969/jmsj/1191334081. MR2071668

[9] Marta Błaszkiewicz and Andrzej Skowroński, *On selfinjective algebras of finite representation type with maximal almost split sequences*, J. Algebra **422** (2015), 450–486, DOI 10.1016/j.jalgebra.2014.09.018. MR3272087

[10] Rafał Bocian and Andrzej Skowroński, *Symmetric special biserial algebras of Euclidean type*, Colloq. Math. **96** (2003), no. 1, 121–148, DOI 10.4064/cm96-1-11. MR2013714

[11] Rafał Bocian and Andrzej Skowroński, *Weakly symmetric algebras of Euclidean type*, J. Reine Angew. Math. **580** (2005), 157–199, DOI 10.1515/crll.2005.2005.580.157. MR2130590

[12] Rafał Bocian and Andrzej Skowroński, *Socle deformations of selfinjective algebras of Euclidean type*, Comm. Algebra **34** (2006), no. 12, 4235–4257, DOI 10.1080/00927870600877902. MR2273702

[13] Otto Bretscher, Christian Läser, and Christine Riedtmann, *Self-injective and simply connected algebras*, Manuscripta Math. **36** (1981/82), no. 3, 253–307, DOI 10.1007/BF01322494. MR641977

[14] Piotr Dowbor and Andrzej Skowroński, *On Galois coverings of tame algebras*, Arch. Math. (Basel) **44** (1985), no. 6, 522–529, DOI 10.1007/BF01193992. MR797444

[15] Piotr Dowbor and Andrzej Skowroński, *Galois coverings of representation-infinite algebras*, Comment. Math. Helv. **62** (1987), no. 2, 311–337, DOI 10.1007/BF02564450. MR896100

[16] Alex S. Dugas, *Periodic resolutions and self-injective algebras of finite type*, J. Pure Appl. Algebra **214** (2010), no. 6, 990–1000, DOI 10.1016/j.jpaa.2009.09.012. MR2580674

[17] K. Erdmann, O. Kerner, and A. Skowroński, *Self-injective algebras of wild tilted type*, J. Pure Appl. Algebra **149** (2000), no. 2, 127–176, DOI 10.1016/S0022-4049(00)00035-9. MR1757729

[18] Karin Erdmann and Andrzej Skowroński, *On Auslander-Reiten components of blocks and self-injective biserial algebras*, Trans. Amer. Math. Soc. **330** (1992), no. 1, 165–189, DOI 10.2307/2154159. MR1144759

[19] Rolf Farnsteiner and Andrzej Skowroński, *Classification of restricted Lie algebras with tame principal block*, J. Reine Angew. Math. **546** (2002), 1–45, DOI 10.1515/crll.2002.043. MR1900992

[20] Rolf Farnsteiner and Andrzej Skowroński, *The tame infinitesimal groups of odd characteristic*, Adv. Math. **205** (2006), no. 1, 229–274, DOI 10.1016/j.aim.2005.07.008. MR2254312

[21] P. Gabriel, *The universal cover of a representation-finite algebra*, Representations of algebras (Puebla, 1980), Lecture Notes in Math., vol. 903, Springer, Berlin-New York, 1981, pp. 68–105. MR654725

[22] Dieter Happel and Shi Ping Liu, *Module categories without short cycles are of finite type*, Proc. Amer. Math. Soc. **120** (1994), no. 2, 371–375, DOI 10.2307/2159871. MR1164144

[23] Mitsuo Hoshino, *Trivial extensions of tilted algebras*, Comm. Algebra **10** (1982), no. 18, 1965–1999, DOI 10.1080/00927878208822817. MR674704

[24] David Hughes and Josef Waschbüsch, *Trivial extensions of tilted algebras*, Proc. London Math. Soc. (3) **46** (1983), no. 2, 347–364, DOI 10.1112/plms/s3-46.2.347. MR693045

[25] Alicja Jaworska-Pastuszak and Andrzej Skowroński, *Selfinjective algebras without short cycles of indecomposable modules*, J. Pure Appl. Algebra **222** (2018), no. 11, 3432–3447, DOI 10.1016/j.jpaa.2017.12.016. MR3806734

[26] Otto Kerner and Andrzej Skowroński, *On module categories with nilpotent infinite radical*, Compositio Math. **77** (1991), no. 3, 313–333. MR1092772

[27] Helmut Lenzing and Andrzej Skowroński, *On selfinjective algebras of Euclidean type*, Colloq. Math. **79** (1999), no. 1, 71–76, DOI 10.4064/cm-79-1-71-76. MR1671803

[28] Helmut Lenzing and Andrzej Skowroński, *Roots of Nakayama and Auslander-Reiten translations*, Colloq. Math. **86** (2000), no. 2, 209–230, DOI 10.4064/cm-86-2-209-230. MR1808677

[29] Helmut Lenzing and Andrzej Skowroński, *Selfinjective algebras of wild canonical type*, Colloq. Math. **96** (2003), no. 2, 245–275, DOI 10.4064/cm96-2-9. MR2010359

[30] Shi Ping Liu, *Almost split sequences for nonregular modules*, Fund. Math. **143** (1993), no. 2, 183–190, DOI 10.4064/fm-143-2-183-190. MR1240634

[31] Jerzy Nehring and Andrzej Skowroński, *Polynomial growth trivial extensions of simply connected algebras*, Fund. Math. **132** (1989), no. 2, 117–134, DOI 10.4064/fm-132-2-117-134. MR1002626

[32] Yosuke Ohnuki, *Stable equivalence induced by a socle equivalence*, Osaka J. Math. **39** (2002), no. 1, 259–266. MR1883923

[33] Zygmunt Pogorzały and Andrzej Skowroński, *Self-injective biserial standard algebras*, J. Algebra **138** (1991), no. 2, 491–504, DOI 10.1016/0021-8693(91)90183-9. MR1102821

[34] I. Reiten, A. Skowroński, and S. O. Smalø, *Short chains and short cycles of modules*, Proc. Amer. Math. Soc. **117** (1993), no. 2, 343–354, DOI 10.2307/2159166. MR1136238

[35] C. Riedtmann, *Algebren, Darstellungsköcher, Überlagerungen und zurück* (German), Comment. Math. Helv. **55** (1980), no. 2, 199–224, DOI 10.1007/BF02566682. MR576602

[36] Christine Riedtmann, *Representation-finite self-injective algebras of class A_n*, Representation theory, II (Proc. Second Internat. Conf., Carleton Univ., Ottawa, Ont., 1979), Lecture Notes in Math., vol. 832, Springer, Berlin, 1980, pp. 449–520. MR607169

[37] Christine Riedtmann, *Representation-finite self-injective algebras of class D_n*, Compositio Math. **49** (1983), no. 2, 231–282. MR704393

[38] Andrzej Skowroński, *Selfinjective algebras of polynomial growth*, Math. Ann. **285** (1989), no. 2, 177–199, DOI 10.1007/BF01443513. MR1016089

[39] Andrzej Skowroński, *Generalized standard Auslander-Reiten components*, J. Math. Soc. Japan **46** (1994), no. 3, 517–543, DOI 10.2969/jmsj/04630517. MR1276836

[40] Andrzej Skowroński, *Regular Auslander-Reiten components containing directing modules*, Proc. Amer. Math. Soc. **120** (1994), no. 1, 19–26, DOI 10.2307/2160162. MR1156473

[41] Andrzej Skowroński, *Selfinjective algebras: finite and tame type*, Trends in representation theory of algebras and related topics, Contemp. Math., vol. 406, Amer. Math. Soc., Providence, RI, 2006, pp. 169–238, DOI 10.1090/conm/406/07658. MR2258046

[42] Andrzej Skowroński and Kunio Yamagata, *Socle deformations of self-injective algebras*, Proc. London Math. Soc. (3) **72** (1996), no. 3, 545–566, DOI 10.1112/plms/s3-72.3.545. MR1376768

[43] Andrzej Skowroński and Kunio Yamagata, *Stable equivalence of selfinjective algebras of tilted type*, Arch. Math. (Basel) **70** (1998), no. 5, 341–350, DOI 10.1007/s000130050205. MR1612578

[44] Andrzej Skowroński and Kunio Yamagata, *Galois coverings of selfinjective algebras by repetitive algebras*, Trans. Amer. Math. Soc. **351** (1999), no. 2, 715–734, DOI 10.1090/S0002-9947-99-02362-4. MR1615962

[45] Andrzej Skowroński and Kunio Yamagata, *On selfinjective Artin algebras having nonperiodic generalized standard Auslander-Reiten components*, Colloq. Math. **96** (2003), no. 2, 235–244, DOI 10.4064/cm96-2-8. MR2010358

[46] Andrzej Skowroński and Kunio Yamagata, *On invariability of self-injective algebras of tilted type under stable equivalences*, Proc. Amer. Math. Soc. **132** (2004), no. 3, 659–667, DOI 10.1090/S0002-9939-03-07134-X. MR2019940

[47] Andrzej Skowroński and Kunio Yamagata, *Positive Galois coverings of self-injective algebras*, Adv. Math. **194** (2005), no. 2, 398–436, DOI 10.1016/j.aim.2004.07.001. MR2139919

[48] Andrzej Skowroński and Kunio Yamagata, *Stable equivalence of selfinjective Artin algebras of Dynkin type*, Algebr. Represent. Theory **9** (2006), no. 1, 33–45, DOI 10.1007/s10468-005-9005-y. MR2233115

[49] Andrzej Skowroński and Kunio Yamagata, *Selfinjective algebras of quasitilted type*, Trends in representation theory of algebras and related topics, EMS Ser. Congr. Rep., Eur. Math. Soc., Zürich, 2008, pp. 639–708, DOI 10.4171/062-1/15. MR2484737

[50] Andrzej Skowroński and Kunio Yamagata, *Frobenius algebras. I*, EMS Textbooks in Mathematics, European Mathematical Society (EMS), Zürich, 2011. Basic representation theory. MR2894798

[51] Andrzej Skowroński and Kunio Yamagata, *On selfinjective algebras of tilted type*, Colloq. Math. **141** (2015), no. 1, 89–117, DOI 10.4064/cm141-1-9. MR3373822

[52] Andrzej Skowroński and Kunio Yamagata, *Frobenius algebras. II*, EMS Textbooks in Mathematics, European Mathematical Society (EMS), Zürich, 2017. Tilted and Hochschild extension algebras. MR3585514

[53] Andrzej Skowroński and Kunio Yamagata, *Selfinjective algebras with hereditary stable slice*, J. Algebra **530** (2019), 146–162, DOI 10.1016/j.jalgebra.2019.04.010. MR3940953

[54] Gordana Todorov, *Almost split sequences for* TrD*-periodic modules*, Representation theory, II (Proc. Second Internat. Conf., Carleton Univ., Ottawa, Ont., 1979), Lecture Notes in Math., vol. 832, Springer, Berlin, 1980, pp. 600–631. MR607174

[55] Kunio Yamagata, *On extensions over Artinian rings with self-dualities*, Tsukuba J. Math. **4** (1980), no. 1, 67–75, DOI 10.21099/tkbjm/1496158792. MR597683

FACULTY OF MATHEMATICS AND COMPUTER SCIENCE, NICOLAUS COPERNICUS UNIVERSITY, CHOPINA 12/18, 87-100 TORUŃ, POLAND

Email address: `skowron@mat.uni.torun.pl`

INSTITUTE OF ENGINEERING, TOKYO UNIVERSITY OF AGRICULTURE AND TECHNOLOGY, NAKA-CHO 2-24-16, KOGANEI, TOKYO 184-8588, JAPAN

Email address: `yamagata@cc.tuat.ac.jp`

Published Titles in This Subseries

761 **Ibrahim Assem, Christof Geiß, and Sonia Trepode, Editors,** Advances in Representation Theory of Algebras, 2021

750 **A. Bourhim, J. Mashreghi, L. Oubbi, and Z. Abdelali, Editors,** Linear and Multilinear Algebra and Function Spaces, 2020

743 **H. Garth Dales, Dmitry Khavinson, and Javad Mashreghi, Editors,** Complex Analysis and Spectral Theory, 2020

680 **Sergei Gukov, Mikhail Khovanov, and Johannes Walcher, Editors,** Physics and Mathematics of Link Homology, 2016

655 **A. C. Cojocaru, C. David, and F. Pappalardi, Editors,** SCHOLAR—a Scientific Celebration Highlighting Open Lines of Arithmetic Research, 2015

654 **Carlo Gasbarri, Steven Lu, Mike Roth, and Yuri Tschinkel, Editors,** Rational Points, Rational Curves, and Entire Holomorphic Curves on Projective Varieties, 2015

638 **Javad Mashreghi, Emmanuel Fricain, and William Ross, Editors,** Invariant Subspaces of the Shift Operator, 2015

630 **Pierre Albin, Dmitry Jakobson, and Frédéric Rochon, Editors,** Geometric and Spectral Analysis, 2014

622 **S. Ejaz Ahmed, Editor,** Perspectives on Big Data Analysis, 2014

606 **Chantal David, Matilde Lalín, and Michelle Manes, Editors,** Women in Numbers 2, 2013

605 **Omid Amini, Matthew Baker, and Xander Faber, Editors,** Tropical and Non-Archimedean Geometry, 2013